Edited by Peter Jeschke, Wolfgang Krämer, Ulrich Schirmer, and Matthias Witschel

Modern Methods in Crop Protection Research

Related Titles

Krämer, W., Schirmer, U., Jeschke, P., Witschel, M. (eds.)

Modern Crop Protection Compounds

2012

ISBN: 978-3-527-32965-6

Walters, D.

Plant Defense

Warding off attack by pathogens, herbivores and parasitic plants

2010

ISBN: 978-1-4051-7589-0

Filho, V. C.

Plant Bioactives and Drug Discovery

Principles, Practice, and Perspectives

2012

ISBN: 978-0-470-58226-8

Tadros, T. F. (ed.)

Colloids in Agrochemicals

Colloids and Interface Science

Volume 5 of the Colloids and Interface Science Series

2009

ISBN: 978-3-527-31465-2

Edited by Peter Jeschke, Wolfgang Krämer, Ulrich Schirmer, and Matthias Witschel

Modern Methods in Crop Protection Research

WILEY-
VCH

WILEY-VCH Verlag GmbH & Co. KGaA

The Editors

Dr. Peter Jeschke
Bayer CropScience AG
BCS AG-R&D-CPR-PC-PCC-Chemistry 2
Bldg. 6510
Alfred-Nobel-Str. 50
40789 Monheim
Germany

Dr. Wolfgang Krämer
Rosenkranz 25
51399 Burscheid
Germany

Dr. Ulrich Schirmer
Berghalde 79
69126 Heidelberg
Germany

Dr. Matthias Witschel
BASF SE
GVA/HC-B009
67056 Ludwigshafen
Germany

Library of Congress Card No.: applied for

British Library Cataloguing-in-Publication Data
A catalogue record for this book is available from the British Library.

Bibliographic information published by the Deutsche Nationalbibliothek
The Deutsche Nationalbibliothek lists this publication in the Deutsche Nationalbibliografie; detailed bibliographic data are available on the Internet at <http://dnb.d-nb.de>.

Composition Laserwords Private Ltd., Chennai, India
Printing and Binding Markono Print Media Pte Ltd, Singapore
Cover Design Formgeber, Eppelheim

Print ISBN: 978-3-527-33175-8
ePDF ISBN: 978-3-527-65593-9
ePub ISBN: 978-3-527-65592-2
mobi ISBN: 978-3-527-65591-5
oBook ISBN: 978-3-527-65590-8

Printed in Singapore
Printed on acid-free paper

Contents

Preface

Today, modern agriculture is facing an enormous challenge – namely, that it ensure that sufficient high-quality food is available to meet the needs of a continuously growing population.

In 2011, the world's population exceeded seven billion people, and a prognosis by the United Nations has suggested that by the year 2050 – assuming moderate birth rates – this will increase to as many as 9.1 billion.

Beyond that, losses of agriculturally usable land, climate change, and changes in the eating habits of the peoples of newly industrialized countries will require major improvements to be made in agricultural productivity. In addition to the increasing demand for food in general, people are today requesting a greater protein intake, especially in countries undergoing transition, and this in turn will lead to a higher consumption of the cereals required as feed used for meat production. Coincidentally, these changing food demands are meeting new requests for bioenergy to be produced via agriculture. Climatic changes that influence the distribution of weeds, pests, and diseases, and their prospective consequences for agriculture, represent a further challenge for crop protection. Change in seed breeding and genetically modified (GM) crops demonstrate progressive solutions for better supplies of food by employing technological innovations from both biochemistry and biotechnology. Nevertheless, the traditional research and development of crop protection compounds remains the most effective method for combating losses in agricultural yields. Currently, such losses are in the range of 14% due to competition by weeds, 13% due to damage by fungal pathogens, and 15% by insect damage.

Another very important reason for employing crop protection compounds is to improve the quality of food. For example, mycotoxins produced by species of *Fusarium* (a fungus that causes damage to the ears of wheat) lead to increasing problems in food production. In addition, changes in rainfall, temperature, and relative humidity can each favor the growth of fungi that produce mycotoxins, so that crops such as groundnuts, wheat, maize, rice, and coffee may become unsuitable for consumption by both humans and animals. Thus, the need for effective research into new crop protection compounds can be fulfilled only by introducing new scientific approaches within the methodology of seeking new active ingredients, by improving the identification process of new targets, by studying aspects of bioavailability, and by improving the tools applied to risk assessment studies of toxicological and ecotoxicological aspects, utilizing new technologies.

This book, which is based partly on *Part IV: New Research Methods* of the First Edition of the textbook *Modern Crop Protection Compounds* (Wiley-VCH, 2007), provides details of the progress that has been made during the past few years towards new methods in modern crop protection research. This includes progress not only in chemical synthesis but also in physico-chemical research, the use of biological research progress and the knowledge and application of genetics and proteomics, and the use of mathematical methods in the design and risk assessment of new active ingredients. Consequently, this book will reflect the exclusively broad field of research in the areas of chemistry, biology, biochemistry, formulation research, toxicology, and ecotoxicology that have been used to identify and develop new chemical tools, such that "green" technology can enjoy further success.

The book, which provides a broad overview of a range of current methods used in modern crop protection research, is divided into four Parts that incorporate 15 chapters, each written by renowned experts at the R&D divisions of major agrochemical companies.

Part I presents methods for the design and optimization of new active ingredients. By using modern research techniques and serendipitous, highly specific biological screening systems, significant progress has been achieved during the past 25 years in computational methods for lead identification and optimization, based on molecular structure information and/or quantum chemistry. Additionally, *in-silico* toxicology approaches to estimate specific risk profiles of agrochemicals will have an emerging impact in the future. In the search for a so-called "optimal product" in modern crop protection in terms of efficacy, environmental safety, user friendliness, and economic viability, the halogen substitution of active ingredients is increasingly recognized as a very important tool.

In Part II are described new methods for identifying the modes of action of active ingredients. Reverse-genetic approaches such as RNA interference (RNAi) offer useful tools to elucidate modes of action, to identify novel targets for exploitation, or to help create new generations of crop protection technologies. For several years, the rapid identification of herbicidal modes of action has been possible via gene expression profiling, using DNA chips. An elucidation of the target sites of neuromuscular insecticides at an early stage in their discovery and development can play an important role in the prioritization of selected candidates. However, despite great technological progress having been made, the targeted discovery of novel fungicides remains an immense challenge because of the restrictions that have been posed on new active ingredients by the obligatory physico-chemical properties permitting a sufficient bioavailability that will, in turn, guarantee fungicidal activity.

In Part III, new methods are examined to improve the bioavailability of the active ingredients. According to novel trends in application technologies, an innovative formulation comprises a mixture of various molecules, besides the active ingredient. In this context, the influence of polymorphism and the organic solid state on the quality and efficiency of agrochemicals plays an important role. Molecular descriptors, as defined by Abraham, can be used to set up linear free energy relationships (LFERs) of relevance to agrochemical research and environmental fate.

Finally, modern methods for risk assessment are addressed in Part IV. Today, many tools are available that can be used to assimilate the knowledge required to evaluate human health and environmental safety, such as exposure modeling, *in vitro* models to evaluate phenotypic and gene expression changes, computational toxicology, bioinformatics, and systems biology. Despite its complexity and a lack of experience of its use, environmental effect modeling has a great potential for regulatory risk assessments with modern crop protection products, although at present its use is not yet fully accepted. In Chapter 14, entitled *Safety Evaluation of New Pesticide Active Ingredients: Enquiry-Led Approach to Data Generation*, attention is focused heavily on advances in molecular biology and biotechnology, and how these may be used in conjunction with computational toxicology and bioinformatics to make toxicity testing more relevant to low-level human exposures, to reduce the need for *in-vivo* testing in animal models, and to make the whole process of hazard data generation quicker and less expensive. In parallel, an evaluation of the endocrine disruption definition and screening aspects in light of the European Crop Protection Law has led to a proposal for decision criteria for endocrine-disrupting compound (EDC) regulatory agencies. This aspect is discussed, taking into consideration the scientific needs of the near future.

We hope that this book will prove to be an invaluable source of information for all of those people working in crop protection science – whether as governmental authorities, as researchers in agrochemical companies, scientists at universities, conservationists, and/or managers in organizations and companies involved with making improvements in agricultural production – to help nourish a continuously growing world population, and to advance the production of bioenergy.

Note

Within this book the authors have named the products/compounds preferably by their common names. Although, occasionally, registered trademarks are cited, their use is not free for everyone. In view of the number of trademarks involved, it was not possible to indicate each particular case in each table and contribution. We accept no liability for this.

May 2012

Peter Jeschke
Wolfgang Krämer
Ulrich Schirmer
Matthias Witschel

List of Contributors

Doerthe Ahlbory-Dieker
Metanomics GmbH
Tegeler Weg 33
10589 Berlin
Germany

Roland Buesen
BASF SE
Experimental Toxicology
and Ecology
Carl-Bosch-Str. 38
67056 Ludwigshafen
Germany

Marco Busch
Bayer CropScience SA
Site La Dargoire
14-20 rue Pierre Baizet
B.P. 9163
69263 Lyon Cedex 09
France

Eric D. Clarke
Syngenta Ltd
Jealott's Hill International
Research Centre
Bracknell
Berkshire RG42 6EY
UK

Daniel Cordova
DuPont Crop Protection
Chemical Genomics Group
Stine Haskell Research Center
1090 Elkton Rd.
Newark, DE 19714
USA

Mark Drewes
Bayer CropScience
BCS-R & D Bldg 6250
Alfred-Nobel-Str. 50
40789 Monheim am Rhein
Germany

Peter Eckes
Bayer CropScience AG
Biology Weed Control
Industriepark Hoechst
Bldg. H872N
65926 Frankfurt
Germany

Bernd Essigmann
Bayer CropScience SA
Site La Dargoire
14-20 rue Pierre Baizet
B.P. 9163
69263 Lyon Cedex 09
France

Eric Fabian
BASF SE
Experimental Toxicology
and Ecology
Carl-Bosch-Str. 38
67056 Ludwigshafen
Germany

Burkhard Flick
BASF SE
Experimental Toxicology
and Ecology
Carl-Bosch-Str. 38
67056 Ludwigshafen
Germany

Michael Herold
Metanomics GmbH
Tegeler Weg 33
10589 Berlin
Germany

Peter Jeschke
Bayer CropScience
BCS AG-R&D-CPR-PC-PCC
Chemistry 2
Bldg. 6510
Alfred-Nobel-Str. 50
40789 Monheim am Rhein
Germany

Hennicke G. Kamp
BASF SE
Experimental Toxicology and
Ecology
Carl-Bosch-Str. 38
67056 Ludwigshafen
Germany

Susanne N. Kolle
BASF SE
Experimental Toxicology
and Ecology
Carl-Bosch-Str. 38
67056 Ludwigshafen
Germany

Gerhard Krennrich
BASF SE
Experimental Toxicology
and Ecology
Carl-Bosch-Str. 38
67056 Ludwigshafen
Germany

Edgar Leibold
BASF SE
Experimental Toxicology
and Ecology
Carl-Bosch-Str. 38
67056 Ludwigshafen
Germany

Ralf Looser
Metanomics GmbH
Tegeler Weg 33
10589 Berlin
Germany

Laura J. Mallon
Syngenta Ltd
Jealott's Hill International
Research Centre
Bracknell
Berkshire RG42 6EY
UK

Werner Mellert
BASF SE
Experimental Toxicology
and Ecology
Carl-Bosch-Str. 38
67056 Ludwigshafen
Germany

Britta Olenik
Bayer HealthCare AG
Friedrich-Ebert-Str. 333
42069 Wuppertal
Germany

Eric Paget
Bayer CropScience SA
Site La Dargoire
14-20 rue Pierre Baizet
B.P. 9163
69263 Lyon Cedex 09
France

Paul Parsons
Syngenta Limited
Toxicology and Health Science
Jealott's Hill International
Research Centre
Bracknell
Berkshire RG42 6EY
UK

Rolf Pontzen
Bayer CropScience AG
Formulation Technology
Alfred-Nobel-Straße 50
40789 Monheim am Rhein
Germany

Alexandre Prokoudine
Metanomics GmbH
Tegeler Weg 33
10589 Berlin
Germany

Tzutzuy Ramirez
BASF SE
Experimental Toxicology
and Ecology
Carl-Bosch-Str. 38
67056 Ludwigshafen
Germany

Bennard van Ravenzwaay
BASF SE
Experimental Toxicology
and Ecology
Carl-Bosch-Str. 38
67056 Ludwigshafen
Germany

Michael Schindler
Bayer CropScience AG
Research Building 6500
Alfred-Nobel-Str. 50
40789 Monheim
Germany

Klaus-Jürgen Schleifer
BASF SE
Computational Chemistry
and Biology
Carl-Bosch-Str. 38
67056 Ludwigshafen
Germany

Frédéric Schmitt
Bayer CropScience SA
Site La Dargoire
14-20 rue Pierre Baizet
B.P. 9163
69263 Lyon Cedex 09
France

Walter Schmitt
Bayer CropScience AG
Environmental Safety
Alfred-Nobel-Str. 50
40789 Monheim am Rhein
Germany

Peter Schreier
Bayer CropScience AG
BCS-R-DCM Bldg. 6240
Alfred-Nobel-Str. 50
40789 Monheim am Rhein
Germany

Thomas C. Sparks
Dow AgroSciences
9330 Zionsville Road
Indianapolis, IN 46268
USA

Volker Strauss
BASF SE
Experimental Toxicology
and Ecology
Carl-Bosch-Str. 38
67056 Ludwigshafen
Germany

Gerhard Thielking
Bayer CropScience AG
Alfred-Nobel-Str. 50
42117 Wuppertal
Germany

Klaus Tietjen
Bayer CropScience AG
BCS-R-DCM Bldg. 6240
Alfred-Nobel-Str. 50
40789 Monheim am Rhein
Germany

Arnoldus W.P. Vermeer
Bayer CropScience AG
Formulation Technology
Alfred-Nobel-Straße 50
40789 Monheim am Rhein
Germany

Tilmann Walk
Metanomics GmbH
Tegeler Weg 33
10589 Berlin
Germany

Jan Wiemer
Metanomics GmbH
Tegeler Weg 33
10589 Berlin
Germany

Part I
Methods for the Design and Optimization of New Active Ingredients

Modern Methods in Crop Protection Research, First Edition.
Edited by Peter Jeschke, Wolfgang Krämer, Ulrich Schirmer, and Matthias Witschel.

1
High-Throughput Screening in Agrochemical Research

Mark Drewes, Klaus Tietjen, and Thomas C. Sparks

1.1
Introduction

Efficient and economical agriculture is essential for sustainable food production fulfilling the demands for high-quality nutrition of the continuously growing population of the world. To ensure adequate food production, it is necessary to control weeds, fungal pathogens, and insects, each of which poses a threat of yield-losses of about 13–15% before harvest (Figure 1.1). Although a broad range of herbicides, fungicides and insecticides already exists, shifts in target organisms and populations and increasing requirements necessitate a steady innovation of crop-protection compounds.

Weeds, fungal pathogens and insects belong to evolutionary distinct organism groups (Figure 1.2), which makes it virtually impossible to have a single crop-protection compound capable of addressing all pest control problems. On closer examination, even the grouping of pests simply as insects, fungi and weeds is, in many cases, still an insufficient depiction. Although the term "insecticide" is often used for any chemical used to combat insects, spider mites or nematodes, the differences between these organisms are so significant that it is more precise to speak of insecticides, acaricides, and nematocides. Among plant pathogenic fungi, the evolutionary range is even much broader and oomycetes are not fungi at all, although oomyceticides commonly are also commonly referred to as *"fungicides"*. Hence, the agrochemical screening of fungicides and insecticides requires a substantial range of diverse species. The situation for herbicide screening is, in some ways, the reverse, but is no easier. Indeed, the close genetic similarity between crop and weed plants generates challenges with regards to the specificity of herbicidal compounds, in differentiating between crop and weed plants. This also results in a need to use a range of different crop and weed plants in screening programs.

In light of the above circumstances, agrochemical screening has employed, in both laboratory and glass-house trials, a wide spectrum of model and pest species. The recent developments described in this chapter, however, have allowed an even higher throughput not only in glass-house tests on whole organisms, but

Modern Methods in Crop Protection Research, First Edition.
Edited by Peter Jeschke, Wolfgang Krämer, Ulrich Schirmer, and Matthias Witschel.

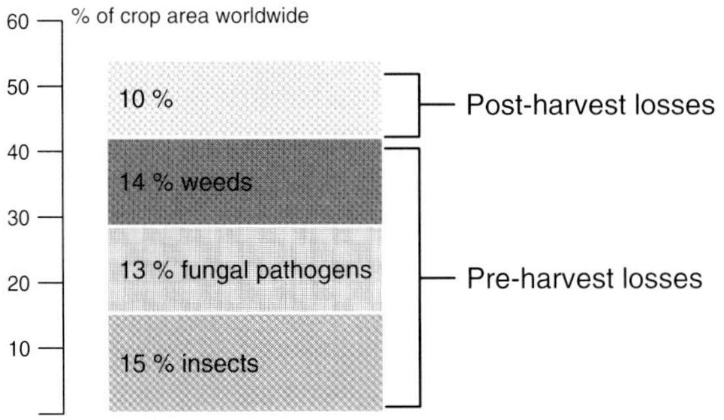

Figure 1.1 Losses of potential agricultural harvest of major crops due to different pests, diseases, and weeds [1, 2]. Non-treated, approximately 50% of the harvest would be lost.

also the exploitation of biochemical (*in vitro*) target tests. Not surprisingly, the implementation of molecular screening techniques and the "omics" technologies – functional genomics, transcriptomics and proteomics, etc. – into agrochemical research has been a major challenge due to the high diversity of the target organisms [5].

Molecular agrochemical research with biochemical high-throughput target screening commenced with several model species, each of which was chosen mainly because of their easy genetic accessibility or specific academic interests. These first favorite model organisms of geneticists and molecular biologists were largely distinct from the most important pest species in agriculture, however. Nonetheless, recent progress in genome sequencing has led to a steadily growing knowledge about agronomically relevant organisms (Figure 1.3 and Table 1.1).

The situation is relatively simple for weeds, as all plants are closely related (Figure 1.2). The first model plant to be sequenced, *Arabidopsis thaliana*, is genetically not very distinct from many dicotyledonous weeds, and the monocotyledonous crops are closely related to the monocotyledonous weeds which, in turn– starting several thousand years ago – formed the foundation for today's cereals species. The first sequenced insect genome of *Drosophila melanogaster*, a dipteran insect, was exploited extensively in both genetic and molecular biological research. To better reflect relevant pest organisms such as lepidopteran pests or aphids, species such as *Heliothis virescens* (tobacco budworm) and *Myzus persicae* (green peach aphid) have been investigated by the agrochemical industry, while *Bombyx mori*, *Acyrthosiphon pisum* and *Tribolium castaneum* have been sequenced in public projects (Table 1.1). Baker's yeast, *Saccharomyces cerevisiae*, has long been the most commonly used model fungus, while the ascomycete *Magnaporthe grisea* and the ustilaginomycete *Ustilago maydis* have been the first sequenced relevant plant pathogens. It is certain that, within the next few years, even the broad evolutionary range of the many different plant pathogenic fungi and oomycetes (see Figure 1.2) will be included in genome projects.

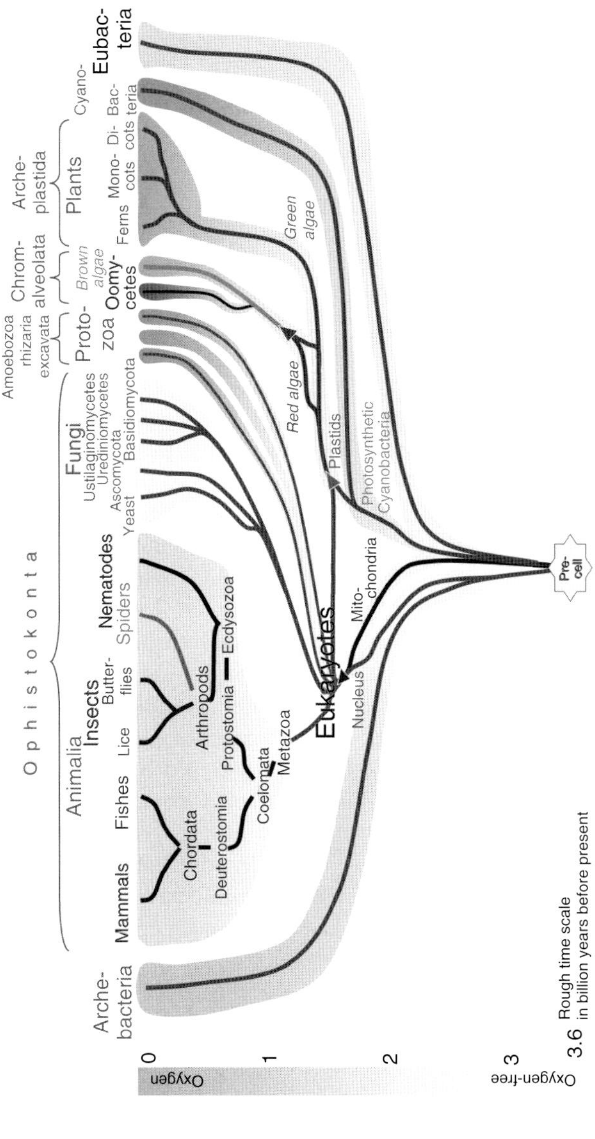

Figure 1.2 Modern evolutionary tree of life. The view is based on Refs [3, 4]; for a more detailed view of fungi, see Ref. [5].

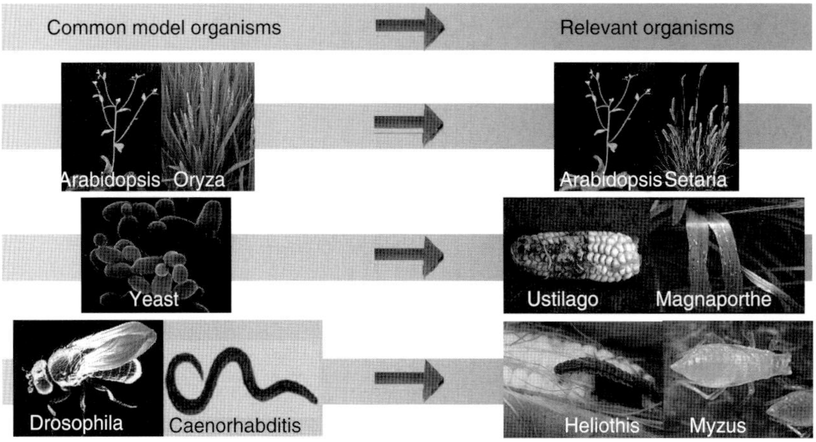

Figure 1.3 Model organisms in molecular biology and agronomically relevant target species.

1.2
Target-Based High-Throughput Screening

1.2.1
Targets

The progress of molecular biology of agronomically relevant organisms has enabled the introduction of target-based biochemical (*in vitro*) high-throughput screening (HTS), which has significantly changed the approach to the screening for agrochemicals during the past 15 years. Target-based HTS is a technology utilized in the agrochemical industry to deliver new actives with defined modes of action (MoA) [6].

Most major research-based agrochemical companies have established biochemical HTS, often conducted in cooperation with companies having special expertise in specific fields of biotechnology. The first wave of genomics – which included genome-wide knock-out programs of model organisms – indicated that about one-quarter of all genes are essential; that is, they were lethal by knock-out [6–8]. The resulting high number of potential novel targets for agrochemicals must be further investigated to clarify the genes' functions (reverse genetics) and to better understand their role in the organism's life cycle. Although the technology of genome-wide knock-out itself was highly efficient and well established, it transpired that even the knock-out of some known relevant targets were not lethal, either because of genetic or functional redundancy, counter-regulation, or because a knock-out does not perfectly mimic an agonistic drug effect on, for example, ion channels. Consequently, knock-out data are today reviewed critically with respect to as many aspects as possible of the physiological roles of potential targets and, as a result, they are taken as just one argument for a gene to be regarded as

Table 1.1 Agronomically relevant organisms with completed or ongoing genome sequencing projects.

Organisms		
Plants	**Fungi and oomycetes**	**Insects and nematodes**
Dicotyledonous plants	**Ascomycetes**	**Diptera**
Arabidopsis thaliana[a]	*Saccharomyces cerevisiae*[a]	*Drosophila melanogaster*[a]
Brassica oleracea	*Alternaria brassicicola*	*Musca domestica*
Glycine max	*Aspergillus oryzae*[a]	**Aphids**
Lotus corniculatus	*Botryotinia fuckeliana*	*Acyrthosiphon pisum*
Solanum tuberosum[a]	*Gibberella zea*	**Lepidoptera**
Monocotyledonous plants	*Magnaporthe grisea*[a]	*Bombyx mori*[a]
Oryza sativa[a]	*Mycosphaerella graminicola*	**Coleoptera**
Sorghum bicolor	*Neurospora crassa*	*Tribolium castaneum*
Triticum aestivum	*Podospora anserina*[a]	**Hymenoptera**
Zea mays	*Sclerotinia sclerotiorum*	*Aphis melifera*
Brachypodium distachyon	**Ustilaginomycetae**	**Nematodes**
Setaria italica	*Ustilago maydis*[a]	*Caenorhabditis elegans*[a]
Hordeum vulgare	**Uredinomycetae**	*Meloidogyne incognita*
	Puccinia graminis	
	Phakopsora pachyrhizi	
Oomycetes	**Basidiomycetes**	
Hyaloperonospora arabidopsis	*Phanerochaete chrysosporium*	
Phytophtora infestans[a]	*Laccaria bicolor*	
Pythium ultimum	**Zygomycota**	
	Rhizopus oryzae	

[a]Completed or close to completion, otherwise: in progress.

an interesting potential target. It must also be considered that clarification of the genes' functions is a challenging and resource-consuming task, and that attention is perhaps more often focused on targets with a sound characterization of their physiological role.

The best proof for an interesting agrochemical target is the "chemical validation" by biologically active compounds. This is true for all the established targets. However, new chemical hits acting on such targets must have an advantage over the already known compounds. This may be a chemical novelty, a novel binding site, an increased performance, or providing a means to overcome resistance. From the standpoint of innovation and the chance to open new areas, novel targets are of particular interest, especially when active compounds are already known, such as a natural product (e.g., the ryanodine receptor for insecticides). Most interesting are novel and proprietary targets which arise from genetics programs or from MoA discovery. MoA elucidation for biological hits has, therefore, become much more important.

Modern analytical methods such as high-performance liquid chromatography/mass spectrometry (HPLC/MS), electrophysiology, imaging, and others build a gateway to today's novel target discovery. The benefit of electrophysiology for clarifying neurophysiological effects is obvious. Cellular imaging techniques complement electrophysiology and are, furthermore, a general approach for MoA studies. For metabolic targets, such as those of sterol biosynthesis, direct target identification may be possible by metabolite analysis [9, 10]. For such compounds gene expression profiling has also proved to be a valuable tool for the MoA classification [11, 12]. When used as fingerprint methods, metabolite profiling and gene expression profiling allows a rapid and reliable detection method for known MoA, and a clear identification and classification of unknown modes of action. Yet, despite the extensive progress in technology, MoA elucidation of novel targets is still – and will be for the near future – a highly demanding challenge. Only the combination of all available methodologies, with emphasis placed on traditional careful physiological and biochemical examinations, will reveal a clearly identified novel molecular target [13].

During the past decades, the identification of resistance mutations to pesticides has provided one of the most clear-cut approaches to target clarification. Although the technological progress has considerably fostered throughput in screening for mutations with a certain phenotype – so-called "forward genetics" [14] – it yet does not seem to be a reliable source of novel targets.

Once a target has been envisaged, further criteria for a "good" target are applied. Clearly, the most important criterion is the druggability of a target, which means accessibility by agro-like chemicals (see below) [15]. It is no coincidence, that the best druggable targets have preexisting binding niches, favoring ligands that comply with certain physico-chemical properties. Furthermore, the target should be relevant during the damaging life phase of a pest, and the destructive effect on a weed or pest under practical conditions should occur within a short period of time after treatment.

Having cleared all of these hurdles, an interesting target must be assayable in order to be exploited, which in turn makes assay technology capabilities a crucial asset. Overall, the number of promising targets remaining is at least two orders of magnitude lower than the number of potential targets found by gene knock-out [6]. Yet, even after having made such great efforts it still difficult to predict whether or not a new active ingredient will be identified, and whether a novel target finally will be competitive in the market.

Often, pharmaceutical research is systematically concentrated towards particular target classes, an example being protein kinases in cancer research [16]. Thereby, know-how can be accumulated and specialized technology can be concentrated for a higher productivity [17]. A successful target triggers the attention to the next similar targets, leading to a considerable understanding of, for example, the human kinome [18, 19]. A similar approach in agrochemical research is of limited value, as there are no such privileged target classes (Figure 1.4). In fact, the common denominator of the diverse agrochemical targets often is the destructive character of the physiological consequences of interference with the

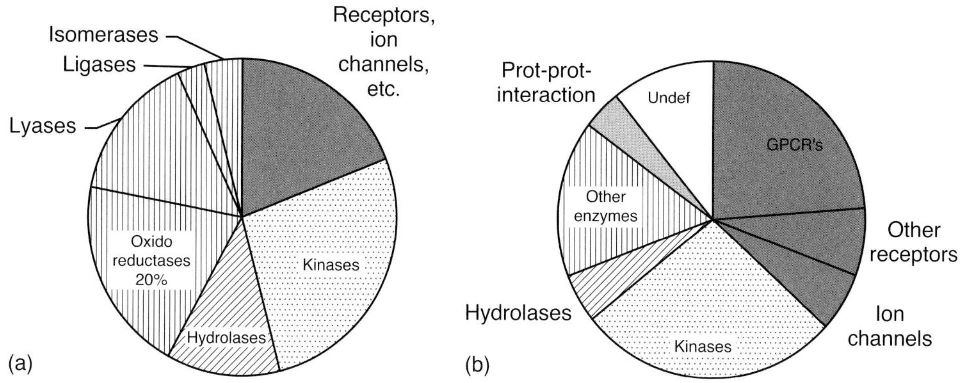

Figure 1.4 Classification by function of (a) agrochemical and (b) pharmaceutical targets (b) [23] for HTS.

target's function, sometimes even being a "side-effect," such as the generation of reactive oxygen species (ROS) [6]. Nevertheless, there are exceptions – one of which is the class of protein kinases – which have been identified as a promising target class for fungicides [20–22].

1.2.2
High-Throughput Screening Techniques

In pharmaceutical research, HTS [24] has proven to be a major source of new lead structures [23], thereby motivating agrochemical research to incorporate – at least in part – this approach into the pesticide discovery process. At Bayer CropScience for example, the first HTS systems were set up during the late 1990s, after which the screening capacity expanded rapidly to more than 100 000 data points per day on a state-of-the-art technology platform. This included fully automated 384-well screening systems, a sophisticated plate replication and storage concept, a streamlined assay validation, and a quality control workflow. An expansion of the compound collection with the help of combinatorial chemistry and major investments in the development of a suitable data management and analysis system was also initiated.

The concept allows the screening of large numbers of compounds as well as large numbers of newly identified targets, thus yielding a corresponding number of hits. The simultaneously developed quality control techniques were able to separate valid hits from false-positives and/or uninteresting compounds due to various reasons (e.g., unspecific binding). Interestingly, several target assays deliver considerable numbers of *in vivo* active compounds, while for other *in vitro* HTS assays the often remarkable target inhibition was not transferred into a corresponding *in vivo* activity. In some cases, this can be attributed to an insufficient target lethality of more speculative targets or "Agrokinetic" factors for *in vivo* species. As discussed earlier, the value of a thorough validation of (i) targets, (ii) assays, and (iii) chemical hits becomes evident.

The extended target validation led to increased numbers of target screens with *in vivo* active compounds. Hence, even more time could be spent on the hit validation, namely the introduction of control tests to eliminate, for example, readout interfering compounds (i.e., hits that were found only due to their optical properties or chemical interference with assay components).

The process of continuous improvement has to date shifted to an extended characterization of hits with respect to reactivity, binding modes [25] (competitive/non-competitive, reversible/irreversible, and so on [26]), speed of action and erratic inhibition due to "promiscuous" behavior of the compound class among others [27]. At the same time – if feasible – the hits or hit classes are submitted to orthogonal assays such as electrophysiology in case of neuronal targets, that help to further classify and validate the hits independent of the readout.

During the late 1990s, Bayer CropScience followed the trend introduced by pharmaceutical companies of conducting genomic projects in collaboration with a biotech-partner. Unfortunately, however, although this genomic approach provided more than 100 new screening assays, it did not deliver the desired output.

Hence, about five years ago the target-based screening approach was redirected, with the new direction subsequently leading to the following favorable changes:

- The screening of known MoA with validated inhibitors.
- A more stringent validation process together with indication biochemistry to ensure better starting points for chemistry.
- A cleansing of the screening library to increase the sample quality as well as the structural diversity of the collection.
- The screening of new, validated modes of action to help innovative areas such as plant stress or malaria.

Of great interest also was the observation that the relative percentage of enzyme assays compared to cell-based assays has increased (Figure 1.5) over the past 10 years. This finding reflects not only technological progress that has been made, but also the increasing back-concentration on ion channel targets for insecticides, which are especially highly validated targets.

At the same time, the chemical libraries at Bayer CropScience became more diversity-oriented, with major efforts being made to further increase the quality of the compound collections (see below), with especially careful quality checks of the hit compounds.

All of these measures together have greatly increased the proportion of true hits, so that finally the chemistry capacities are concentrated on fewer, albeit well-characterized, hit classes with a clearly increased likelihood of a successful hit-to-lead optimization.

The huge amount of data and information generated during the various phases of HTS and subsequent validation processes has triggered the development of sophisticated data analysis tools [28] that help biologists and chemists to select and

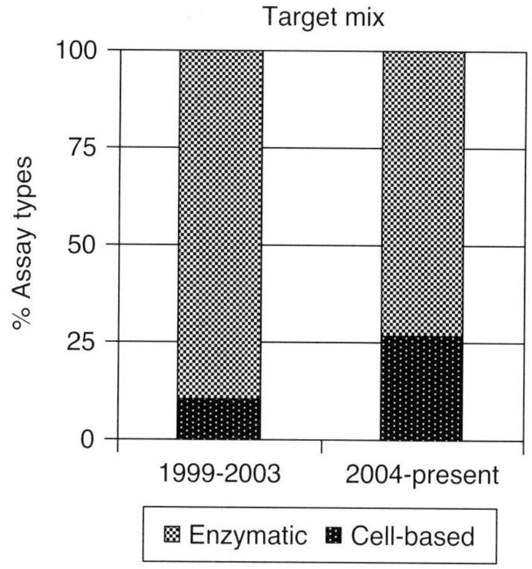

Figure 1.5 Proportions of assay types screened at Bayer Crop Science HTS facility. Between 1999 and 2003 (age of genomic targets) these included 10% cell-based assays (including cell-growth assays) and 90% enzyme assays. From 2004 to the present day (age of chemically validated targets), they included 26% cell-based assays (exclusively ion channels) and 74% enzyme assays.

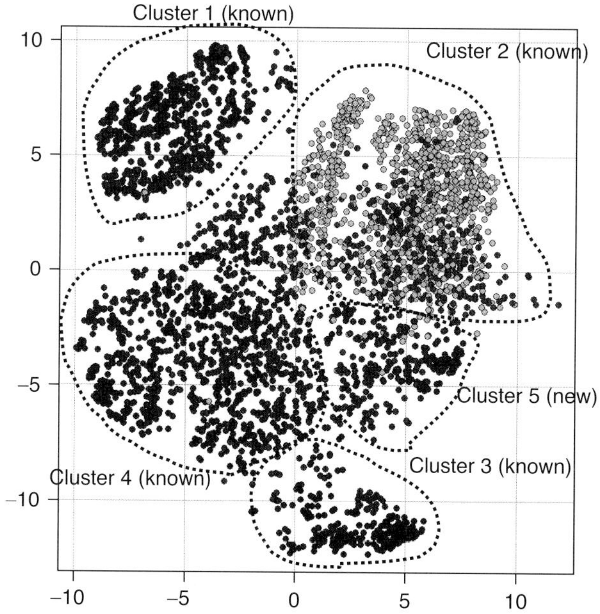

Figure 1.6 Example of the visualization of the chemical space of hits and similar inactive structures from a target assay. Light gray = inactive; gray to black = decreasing IC_{50}.

prioritize the most promising hits or hit classes (cluster of similar compounds) (Figure 1.6).

Biochemical *in vitro* screening may deliver compounds that, despite a clear target activity, are unable to exhibit *in vivo* activity due to, for example, unfavorable physico-chemical properties (lacking bioavailability), rapid metabolism, insufficient stability, or a poor distribution in the target organism. Nevertheless, these chemical classes are still of interest to chemists because such properties reflect the characteristics of the compounds that may be overcome by chemical optimization. As a consequence, "agrokinetics" has led to the identification of pure *in vitro* hits as such, and also helped to elucidate the reasons for failure in the *in vivo* test, thus guiding the *in vitro* to *in vivo* transfer of hit classes. Such observations underline the fact that the *in vitro* and *in vivo* screening processes can be complementary, and together can be used to broadly characterize the activity of test compounds within the early discovery process.

Currently, two trends can be observed among the high-throughput community: (i) miniaturization into the nanoliter dispensing regime; and (ii) new high-content screening (HCS) techniques. The small-volume screening (either on 1536-well plates or the recently introduced low-volume 384-well plates) clearly is also of interest for agrochemical research, since the enzymes and substrates of new target proteins are often difficult and costly to produce in larger quantities. Due to the above-mentioned screening strategy this process is not so much driven by the need to further increase the capacity, but rather by cost efficiency, the standard reaction volume having decreased from more than 50 μl to 5–10 μl (Figure 1.7). Moreover, further reductions are possible with new pipetting equipment having now reached a robust quality with inaccuracies of below 5% in the 1 μl range.

Other very important aspects of ion channel screening are the recently developed automated and medium-throughput patch clamping systems that perfectly meet the increased demand for in-depth hit characterization. Yet, the future role of HCS – fully automated confocal life cell microscopy imaging systems – is less clear than in pharmaceutical research, where it has become *the* validation and screening method development of the past few years [29]. Nonetheless, the applicability of HCS to agrochemical research will need to be evaluated in the future.

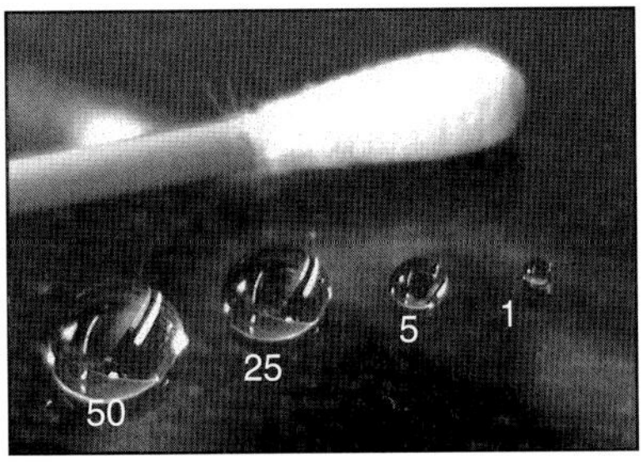

Figure 1.7 Size comparison of water drops between 50 and 1 μl as compared to a cosmetic tip.

1.3
Other Screening Approaches

1.3.1
High-Throughput Virtual Screening

During the past two decades, computational chemistry has become a key partner in drug discovery. Indeed, one of its main contributions to high-throughput methods is that of virtual screening [24, 30], a computational method that can be applied to large sets of compounds with the goal of evaluating or filtering those compounds against certain criteria, prior to or *in lieu* of *in vivo* or *in vitro testing*. In this regard, some methods consider target structure information while others are based solely on ligand similarity to complex model systems. Additionally, when three-dimensional information is incorporated into an analysis, the calculation becomes more demanding, especially if a flexible target protein is considered. Although massive screening with fully flexible models is not yet feasible, the so-called flexible *docking* of huge (both real and virtual) compound collections into a rigid binding pocket has today become routine [31]. The most obvious advantage of the latter method over the relatively fast similarity searches is that any compound which has binding site complementarity will be identified, and that no similarity to a known ligand is needed. This stands in contrast to similarity-based screening, where completely new scaffolds are rarely found.

In order to have a reasonable hit enrichment when using docking methods, computational chemistry must start with high-resolution protein structures; if possible, more than one ligand co-crystal would be used to construct the binding domain. In addition, some programs are capable of handling a certain degree of target flexibility through *ensemble formations* of binding domains from various experimental structures [32]. Whilst the quality of the results will obviously improve, a greater computational effort will be required as a consequence.

Virtual-target-based screening can be applied in many ways, the most obvious being the screening of huge libraries in order to prioritize the synthesis, acquisition and/or biochemical screening, or to select reactants for combinatorial libraries that show highest hit likeliness. These applications do yield target-focused libraries, and can be extended to families of targets, such as kinases or G-protein-coupled receptors (GPCRs).

1.4
In Vivo High-Throughput Screening

Since the very beginning of the search for new agrochemicals, *in vivo* screening has been the primary basis for agrochemical research, leading to the identification and characterization of new active chemistries and their subsequent optimization. In 1956, 1800 compounds needed to be evaluated for every one that became a product, a number that had risen to 10 000 by 1972 [33]. By 1995, the number has risen

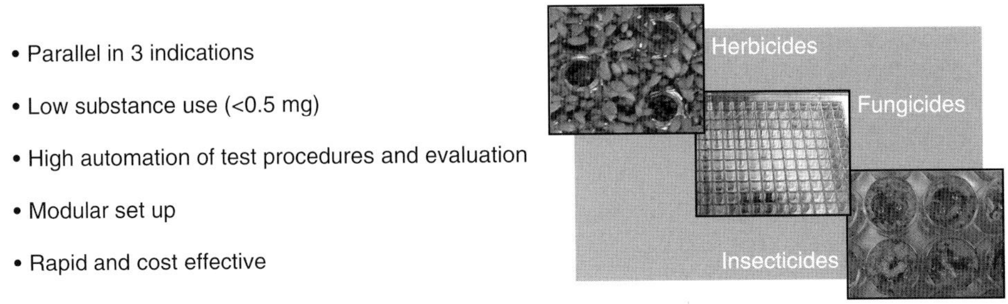

• Parallel in 3 indications
• Low substance use (<0.5 mg)
• High automation of test procedures and evaluation
• Modular set up
• Rapid and cost effective
Herbicides
Fungicides
Insecticides

Figure 1.8 The advantages of high-throughput screening.

to more than 50 000, and today is about 140 000 compounds tested per product discovered [34]. In part, this rise is due to the increasing demands with regard to the need for increased biological activity, improved mammalian and environmental safety, as well as a variety of economic considerations. Beginning in the mid-1990s, most of the major agrochemical companies established *in vivo* HTS systems [15, 35–38], an interest which coincided with the development and expanding use of combinatorial libraries. In the HTS systems, the numbers of compounds screened each year are reported as ranging between 100 000 and 500 000, with most programs utilizing less than 0.5 mg of substance to produce relevant answers for a targeted set of plants, insects and fungi, using either 96-well or 384-well microtiter plates (MTPs) (Figure 1.8). Such HTS systems can produce a large number of hits, all of which are dependent on the screening dose, pass criteria, and the number and type of test species used. The quality of the hits from the HTS can be improved through the addition of extra dose rates and replicates [15], which can in turn improve the quality of the hits delivered to relevant follow-up screens.

HTS programs are based on automation, miniaturization, and often also the use of model organisms or systems which are easy to handle and adaptable to the MTP format. In pesticide discovery programs, model systems using *Aedes aegypti*, *D. melanogaster*, *A. thaliana*, *Caenorhabditis elegans* [39] or cell-growth-based fungicide assays can be successful in identifying a large number of hits. These model systems, using species that can be highly sensitive, are primarily intended to identify biological activity. However, in follow-up tests with agriculturally relevant species the number of interesting compounds often decreases dramatically due to a weak translation between the model organisms and the real pest species. As such, HTS systems with model organisms can potentially miss relevant hits (Figure 1.9).

As a consequence of this less-than-ideal translation, there has been an evolution among *in vivo* HTS systems to incorporate more relevant target organisms [40], particularly for insecticides and fungicides. For example, 96-well MTP assays involving pest lepidopteran larvae are widely used [15, 41, 42], while leaf-disc assays have been developed [6, 33, 43] that have been adapted by many companies for sap-feeding insects such as aphids.

HTS systems for fungicides utilize cell growth tests, but also cover only a part of the relevant target organisms; all obligate pathogens such as mildews or rusts cannot be tested. Additionally, such cell tests do not test the relevant phases of the

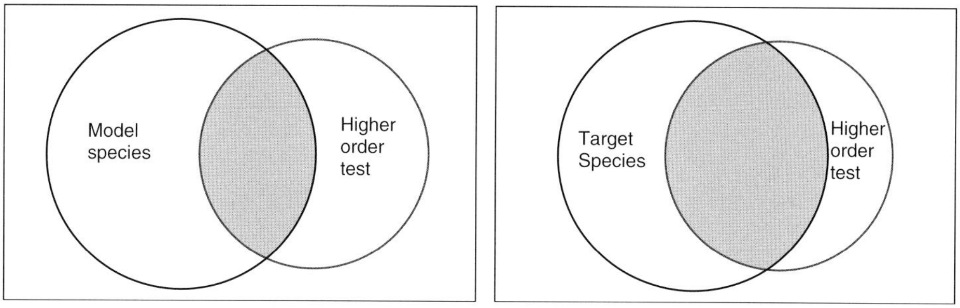

Figure 1.9 The overlap of mutual active chemical hits found in model species tests versus target species tests.

development of fungal pathogens on living plant tissues. However, this gap can be closed by using leaf discs [44, 45] or whole plants with relevant fungal species.

The development and further improvement of the more relevant HTS assays using target pest species for insecticides and fungicides is an on-going challenge. In many cases, these assays can be significantly more complex, and the time and effort required to run target organism assays can be greater than was required for previous model systems. As such, the number of species screened in an *in vivo* HTS has often been reduced to just a few, with one or two model species as general indicators of biological activity, plus perhaps a couple of specific pests that represent major product areas. For example, in the case of insecticides many discovery programs focus on one or two lepidopteran species that serve as indicators for a broad range of chewing pests, and an aphid species that is an indicator for a broad range of sap-feeding insect pests. While these two product areas do not denote the total insecticide market, they do capture the largest segments. Thus, the use of these more complex HTS systems requires a balance relative to throughput and dedicated resources for an *in vivo* HTS program. The net result is that better-characterized compounds with a more relevant biological profile are derived from HTS programs that focus on representative pest insects.

1.4.1 Compound Sourcing and *In-Silico* Screening

In order to achieve the ambitious goals of HTS, a large number of compounds are needed to satisfy the capacities of the tests. Consequently, many of the major chemical companies – both pharmaceutical and agrochemical – began to buy large numbers of "off-the-shelf" compounds [46] from so-called "bulkers" on a worldwide basis. Further, the boom triggered by combinatorial chemistry also helped to satisfy the need for large numbers of new substances, and this in turn led to the founding of several new companies that synthesized such materials (e.g., ArQule, BioFocus, or ChemBridge) to meet the demand. The compounds initially purchased were predominantly driven by availability and convenience. However, in spite of the increased throughput of compound screening, the number of new

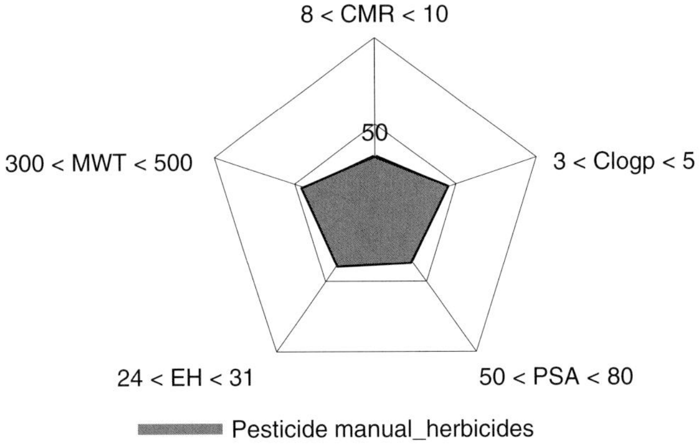

Figure 1.10 Percentages of herbicides in the *Pesticide Manual* [50] within constraint range. CMR, molar refractivity; EH, equivalent hydrocarbons; PSA, polar surface area.

biologically active classes of herbicides, fungicides and insecticides did not increase correspondingly. It was quickly recognized that for both pharmaceutical and agricultural compounds, certain constraints were needed on the types of compound acquired to obtain an effective level of relevant biological activity (Figure 1.10). Subsequently, in pharmaceutical research two general approaches emerged to resolve these problems, namely fragment-based screening and diversity-oriented synthesis [47, 48]. Agrochemistry commonly favors diversity to be early, in accord with the constraints posed on compounds. These constraints, along with (substructural) fingerprints as descriptors [49] for molecular similarity, have been applied to select chemical collections for agrochemical discovery.

A further refinement of the agro-like constraints [51, 52], assisted by *in-silico* screening, has further improved the diversity [53] of the collections. Importantly, with these and other *in-silico* approaches to refining and targeting the types and numbers of desired molecules [54], the requirement for screening vast numbers of compounds has been potentially reduced. Thus, improvements in the quality and relevance of the inputs to an HTS program should increase the number of potentially interesting compounds that emerge from that program.

In the area of combinatorial chemistry a significant realignment has occurred, with the starting points used for the libraries having changed from "blue sky" chemistries to more relevant scaffolds with a biological background [6, 55, 56]. Such considerations entail more intricate synthetic routes, which in turn can lead to a reduction in the size of the libraries. However, various studies have indicated that with a correct design, very large libraries are unnecessary for the adequate sampling of a desired chemical space, and that smaller libraries can be just as effective [55, 57]. With these considerations, the probability of obtaining better-quality hits is improved, thereby providing a better path forward in the early phases of lead finding. In the future, it is likely that a combination of agro-likeness tools and

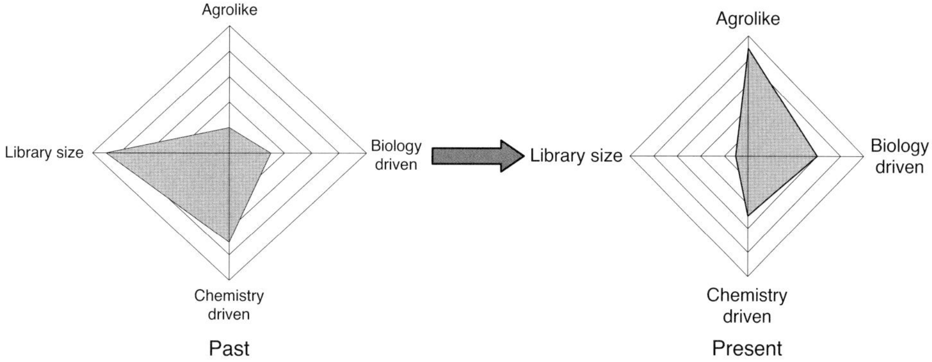

Figure 1.11 Higher input of agro-likeness and biological input in combinatorial chemistry scaffolds.

carefully chosen biological scaffolds will be among the approaches giving rise to new leads and, ultimately, to products for the agrochemical industry (Figure 1.11).

1.5 Conclusions

During the past 15 years, HTS has been adopted by the agrochemical industry as an essential component of the early discovery phase, in part to address the increasingly challenging requirements in the development of new pesticides and the declining success rates in the identification and development of new products. In contrast to the pharmaceutical industry, which extensively employs *in vitro* target-based HTS in its discovery programs, the agrochemical industry has the added advantage of being able to capitalize on *in vivo* HTS using, in part, the pest species of interest. The *in vivo* HTS programs have been developed using the experience of classical and well-established biological screening. In agrochemical research, the broad diversity of the target organisms presents a specific and complex challenge which must be carefully considered and addressed for each screening program. Fed by high-throughput chemistry, functional genomic projects and significant progress in robotic screening systems, procedures have been successfully established that allow agrochemical companies to test large numbers of compounds very efficiently and with a broad set of test organisms, including newly identified and well-established targets.

As an effective pesticide discovery program is continuously evolving, it is essential to continuously evaluate and incorporate the experiences concerning the advantages and limitations of new and established technologies and approaches. With modern agrochemical research platforms undergoing continuous and dynamic changes, adjustments to such platforms must be aimed at integrating the most promising parts of the many approaches that are currently available.

With the continued implementation of new technologies into the standard screening and testing workflows for both early and late research phases, a broad knowledge has been gained which by far exceeds the specific HTS approach alone. Moreover, such knowledge is being translated to overall improvements in agrochemical research. Finally, it is to be expected that, as a result of these new technologies, innovative products will emerge to meet the needs of modern agriculture.

Acknowledgments

The authors wish to thank R. Klein (BCS), M. Adamczewski (BCS), and H.-J. Dietrich (BCS) for providing us with insight for this chapter.

References

1. Oerke, E.C., Dehne, H.C., Schönbeck, F., and Weber, A. (1994) *Crop Production and Crop Protection*, Elsevier, Amsterdam.
2. Yudelman, M., Ratta, A., and Nygaard, D. (1998) *Pest Management and Food Production. Looking into the Future*, International Food Policy Research Institute, Washington, DC.
3. Adl, S.M., Simpson, A.G.B., Farmer, M.A., Andersen, R.A., Anderson, O.R., Barta, J.R., Bowser, S.S., Brugerolle, G., Fensome, R.A., Fredericq, S., Sergei, T.Y.J., Karpov, K.P., Krug, J., Lane, C.E., Lewis, L.A., Lodge, J., Lynn, D.H., Mann, D.O., McCourt, R.M., Mendoza, L., Moestrup, O., Mozley, S.E., Standridge, N.T.A., Shearer, C.A., Smirnov, A.V., Spiegel, F.W., and Taylor, M.F.J.R. (2005) *J. Eukaryot. Microbiol.*, **52**, 399–451.
4. Embley, T.M. and Martin, W. (2006) *Nature*, **440**, 623–630.
5. Schindler, M., Sawada, H., and Tietjen, K. (2007) in *Modern Crop Protection Compounds* (eds W. Krämer and U. Schirmer), Wiley VCH Verlag GmbH, Weinheim, pp. 683–707.
6. Tietjen, K., Drewes, M., and Stenzel, K. (2005) *Comb. Chem. High Throughput Screening*, **8**, 589–594.
7. Berg, D., Tietjen, K., Wollweber, D., and Hain, R. (1999) *Brighton Crop Prot. Conf. Weeds*, **2**, 491–500.
8. Lein, W., Börnke, F., Reindl, A., Ehrhardt, T., Stitt, M., and Sonnewald, U. (2004) *Curr. Opin. Plant Biol.*, **7**, 219–225.
9. Hole, S.J., Howe, P.W.A., Stanley, P.D., and Hadfield, S.T. (2000) *J. Biomol. Screening*, **5**, 335–342.
10. Ott, K.H., Aranibar, N., Singh, B., and Stockton, G.W. (2003) *Phytochemistry*, **62**, 971–985.
11. Kunze, C.L., Meissner, R., Drewes, M., and Tietjen, K. (2003) *Pest Manag. Sci.*, **59**, 847.
12. Eckes, P., van Almsick, C., and Weidler, M. (2004) *Pflanzenschutz-Nachr. Bayer (Bayer CropScience AG)*, **57**, 62–77.
13. Grossmann, K. (2005) *Pest Manag. Sci.*, **61**, 423–431.
14. Beffa, R. (2004) *Pflanzenschutz-Nachr. Bayer (German edition)*, **57**, 46–61.
15. Smith, S.C., Delaney, S.J., Robinson, M.P., and Rice, M.J. (2005) *Comb. Chem. High Throughput Screening*, **8**, 577–587.
16. Cohen, P. (2002) *Nat. Rev. Drug Discov.*, **1**, 309–315.
17. Fischer, P.M. (2004) *Curr. Med. Chem.*, **11**, 1563–1583.
18. Griffin, J.D. (2005) *Nat. Biotechnol.*, **23**, 308–309.
19. Downey, W. (2009) Hit the Target, World Pharmaceutical Frontiers *http://www.worldpharmaceuticals.net/editorials/015_march09/WPF015_hitthetarget.pdf* (accessed 000000).

20. Irmler, S., Rogniaux, H., Hess, D., and Pillonel, C. (2006) *Pestic. Biochem. Physiol.*, **84**, 25–37.
21. Pillonel, C. (2005) *Pest Manag. Sci.*, **61**, 1069–1076.
22. Tueckmantel, S., Greul, J., Janning, P., Brockmeyer, A., Gruetter, C., Simard, J., Gutbrod, O., Beck, M., Tietjen, T., Rauh, D., and Schreier, P. (2011) *ACS Chem. Biol.*, **6**, 926–933.
23. Fox, S. (2005) *High Throughput Screening 2005: New Users, More CellBased Assays, and a Host of new Tools*, HighTech Business Decisions, Moraga, CA, October 2005.
24. Devlin, J. (1997) *High Throughput Screening: The Discovery of Bioactive Substances*, Marcel Dekker, New York.
25. Swinney, D.C. (2004) *Nat. Rev.*, **3**, 801–808.
26. Galasinski, S. (2004) (Amphora Discovery), Comprehensive analysis of inhibitor behavior for lead identification and optimization. Presented at MipTec, 3–6 May 2004, Basel.
27. McGovern, S.L., Helfand, B.T., Feng, B., and Shoichet, B.K. (2003) *J. Med. Chem.*, **46**, 4265–4272.
28. Modelling: Tripos: *http://www.tripos.com/, Accelrys: http://www.accelrys.com/, Visualisation and Datamanagement: Spotfire: http://www.spotfire.com/, Genedata: http://www.genedata.com/, IDBS: http://www.idbs.com/, MDL http://www.mdl.com/.*
29. Soderholm, J., Uehara-Bingen, M., Weis, K., and Heald, R. (2006) *Nat. Chem. Biol.*, **2**, 55–58.
30. Böhm, H.J. and Schneider, G. (2000) Virtual screening for bioactive molecules, *Methods and Principles in Medicinal Chemistry*, vol. 10, Wiley-VCH Verlag GmbH, Weinheim.
31. Autodock: *http://www.scripps.edu/mb/olson/doc/autodock/, Gold:http://www.ccdc.cam.ac.uk/products/life_sciences/gold/, FlexX:http://www.biosolveit.de/FlexX/, DOCK: http://mdi.ucsf.edu/DOCK_availability.html.*
32. Claußen, H., Buning, C., Rarey, M., and Lengauer, T. (2001) *Mol. Biol.*, **308**, 377–395.
33. Metcalf, R.L. (1980) *Annu. Rev. Entomol.*, **25**, 219–256.
34. CropLife (2011). Available at: *http://www.croplifeamerica.org/crop-protection/pesticide-facts.*
35. Hermann, D., Hillesheim, E., Gees, R., and Steinrücken, H. (2000) 52. Deutsche Pflanzenschutztagung, 9–12 October 2000, Freising-Weihenstephan, p. 124.
36. Stuebler, H. (2000) 52. Deutsche Pflanzenschutztagung, 9–12 October 2000, Freising-Weihenstephan, p. 123.
37. Steinrucken, H.C. and Hermann, D. (2000) *Chem. Ind.*, **7**, 246–249.
38. Short, P. (2005) *Chem. Eng. News*, **83**, 19–22.
39. MacRae, C.A. and Petersen, R.T. (2003) *Chem. Biol.*, **10**, 901–908.
40. Ridley, S.M., Elliott, A.C., Young, M., and Youle, D. (1998) *Pestic. Sci.*, **54**, 327–337.
41. Jansson, R.K., Halliday, W.R., and Argentine, J.A. (1997) *J. Econ. Entomol.*, **90**, 1500–1507.
42. Choung, W., Lorsbach, B.A., Sparks, T.C., Ruiz, J.M., and Kurth, M.J. (2008) *Synlett*, **19**, 3036.
43. Smith, S. (2003) *Pestic. Outlook*, **14**, 21–25.
44. Eckes, P. (2005) BASF press interview August 2005.
45. Mueller, U. (2002) *Pure Appl. Chem.*, **74**, 2241–2246.
46. *http://www.timtec.net/timtec/articles/0821-2000-HTS.htm.*
47. Galloway, W., Isidro-Llobet, A., and Spring, D. (2010) *Nat. Commun.*, **1**, 80.
48. Hajduk, P., Galloway, W., and Spring, D. (2011) *Nature*, **470** (7332), 42–43.
49. Todeschini, R. and Consonni, V. (2000) *Handbook of Molecular Descriptors*, Wiley-VCH Verlag GmbH, Weinheim.
50. Tomlin, C.D.S. (2000) *Pesticide Manual*, 12th edn, British Crop Protection Council, Kent, UK.
51. Tice, C.M. (2001) *Pest Manag. Sci.*, **57**, 3–16.
52. Schleifer, K.J. (2007) in *Pesticide Chemistry* (eds H. Ohkawa, H. Miyagawa, and P.W. Lee), Wiley-VCH Verlag GmbH, Weinheim, Germany, pp. 77–100.
53. Poetter, T. and Matter, H. (1998) *J. Med. Chem.*, **41**, 478.

54. Boehm, M. (2011) in *Virtual Screening, Principles, Challenges and Practical Guidelines* (ed. C. Sotriffer), Wiley-VCH Verlag GmbH, Weinheim, Germany, pp. 3–33.

55. Ruiz, J.M. and Lorsbach, B.A. (2005) in *New Discoveries in Agrochemicals*, vol. 892 (eds J.M. Clark and H. Ohkawa), ACS, Washington, DC, pp. 99–108.

56. Lindell, S.D., Pattenden, L.C., and Shannon, J. (2009) *Bioorg. Med. Chem.*, **17**, 4035–4046.

57. Wright, T., Gillet, V.J., Green, D.V.S., and Pickett, S.D. (2003) *J. Chem. Inf. Comput. Sci*, **43**, 381–390.

2
Computational Approaches in Agricultural Research

Klaus-Jürgen Schleifer

2.1
Introduction

As the stronger guidelines of registration authorities in terms of risk assessment will, in time, squeeze many currently available products from the market, there exists a great opportunity for agrochemical companies to substitute the upcoming gaps with innovative novel compounds. However, in order to fulfill the specific requirements for future registration, new strategies in the R&D process must be implemented, taking into account not only the classical lead identification and optimization process, with a special focus on biological activity, but also the risk assessment of compounds at the very early stages of the process. Concomitant with a permanent cost pressure, efficient strategies must consider inexpensive computational approaches instead of additional extensive laboratory-based experiments to support this enormous effort.

In this chapter, a general overview will be provided of the current computational techniques used for lead identification and lead optimization, based on molecular structure information. Additionally, the first *in silico* toxicology approaches for the estimation of specific risk profiles will be discussed that will undoubtedly have an emerging impact in future.

2.2
Research Strategies

Two general screening strategies are followed to identify potential lead structures:

- In the first strategy, chemicals are directly tested on harmful organisms (e.g., weeds), and relevant phenotype modifications are rated (e.g., bleaching). This *in vivo* approach indicates biological effects without any knowledge of the addressed mode of action (MoA). Optimization strategies must consider that several modes of action may be involved, and that during synthetic optimization

Modern Methods in Crop Protection Research, First Edition.
Edited by Peter Jeschke, Wolfgang Krämer, Ulrich Schirmer, and Matthias Witschel.

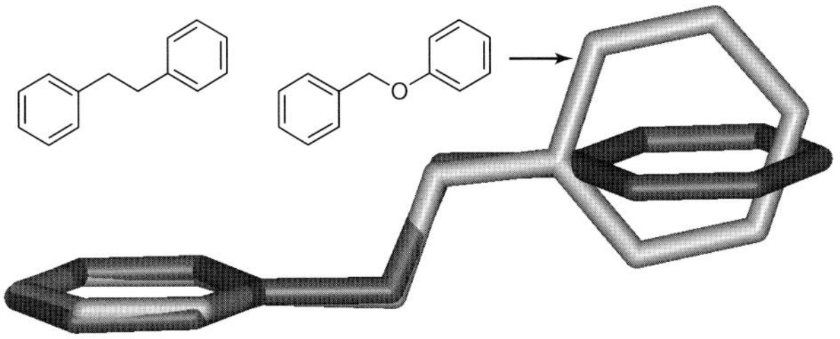

Figure 2.1 Chemical structures and superimposed X-ray coordinates of 1,2-diphenylethane (dark, CSD-code DIBENZ04) and benzyloxybenzene (bright, CSD-code MUYDOZ), indicating the different orientation of one phenyl ring induced by the substitution of methylene with an ether function.

the original MoA might be changed. In addition, all observed effects reflect a combination of the target-activity and bioavailability of the compounds.
- A second strategy – the so-called *mechanism-based* approach – allows specific target activity optimization. A fundamental condition for this procedure is the availability of a molecular target protein and a suitable biochemical assay to study the protein's function in the presence of screening compounds. In this case, the main challenge is the transfer of activity from the biochemical assay to the biological system.

This clearly reflects that – independent of the screening strategy – hits rarely fulfill all necessary criteria for a new lead structure. Therefore, medicinal chemists have to analyze the screening results (usually structural formulas with corresponding biological or biochemical data) in order to derive a first structure–activity relationship (SAR) hypothesis.

Occasionally, two-dimensional (2-D) analyses are not sufficient to clarify the *real* situation, which is in Nature three-dimensional (3-D). Consequently, minor chemical variations may completely change the geometry of a molecule (Figure 2.1), while even diverse substances (from a 2-D view) may bind to a common binding site (e.g., acetylcholinesterase inhibitors).

Nowadays, molecular modeling packages are applied to calculate the relevant conformations of a molecule via an energy function (i.e., force fields [1]) that is adjusted to experimentally derived reference geometries (mostly X-ray structures). *Van der Waals* and Coulomb terms define steric and electrostatic features, and each mismatch to reference values is penalized.

2.3
Ligand-Based Approaches

In order to identify molecular features crucial for biological activity, all compounds of a common hit cluster must be superimposed to yield a pharmacophore model. Since this is done in 3-D space, relevant conformers of each ligand and critical molecular functions must be determined. X-ray crystal structures of the ligands

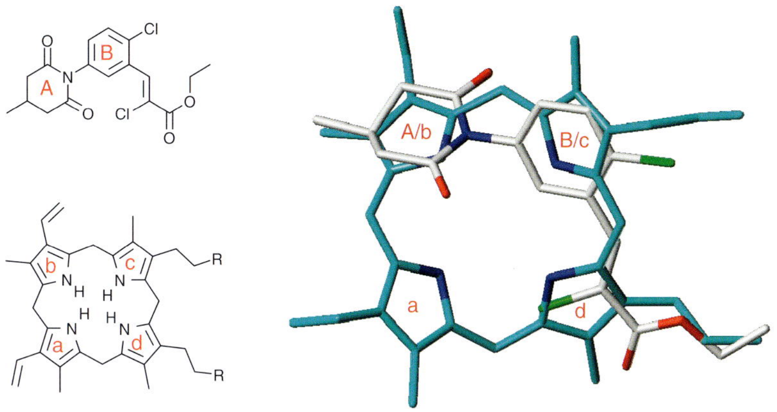

Figure 2.2 Superposition of a Protox inhibitor from pyridinedione-type on a calculated protoporphyrinogen-like template (cyan). For reasons of clarity, corresponding ring systems are indicated and hydrogen atoms are omitted. Atoms are color-coded as: carbon, gray; nitrogen, blue; oxygen, red; chlorine, green.

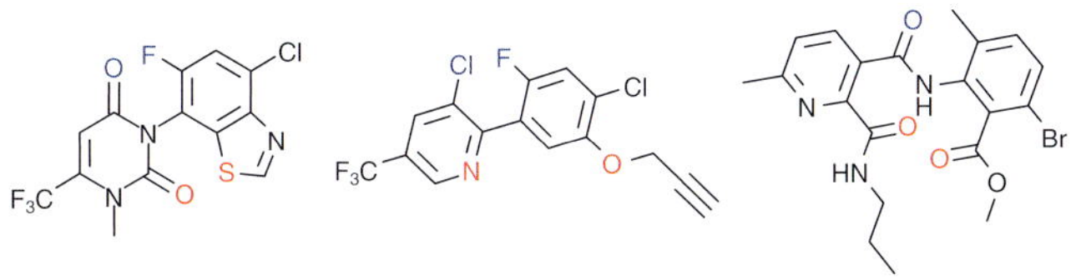

Figure 2.3 Common interaction pattern of potent Protox inhibitors from uracil- (left) and pyridine-type. Each molecule comprises two ring systems and electron-rich functions on both sides of the linked rings (blue- and red-colored).

(or of congeners) can be helpful to solve the conformational problem, since they indicate at least one potential minimum conformation. Even more helpful can be the 3-D structure of the physiological endogenous substrate or a postulated transition state of an enzyme reaction (Figure 2.2).

Sometimes, however, there are no experimental data available at all, and in this situation a theoretical exploration of relevant conformers must be performed, taking into consideration all rotational degrees of freedom (e.g., systematic conformational search). The derived conformations are evaluated with respect to their potential energy. Corresponding to Boltzmann's equation, low energy values indicate greater chances to resemble reality. Very often, several distinct conformers are assessed as being energetically similar, and in this case the most rigid highly active ligand will serve as a template molecule to superimpose all other minimized ligands (i.e., an active analog approach).

The identification of crucial functions – which should be present (at least in part) in all active ligands – takes place via an SAR analysis of all compounds of the cluster. Hypotheses derived from SAR (Figure 2.3) may be experimentally validated by testing compounds with an absent or optimized substitution pattern.

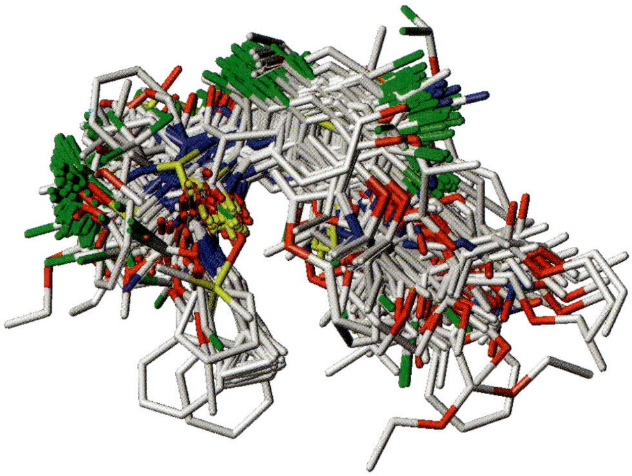

Figure 2.4 Pharmacophore model of 318 Protox inhibitors. Atoms are color-coded as: carbon, gray; nitrogen, blue; oxygen, red; sulfur, yellow; chlorine, green.

To superimpose all ligands in an appropriate manner, essential groups (e.g., carbonyl groups, aromatic rings, etc.) of energetically favorable conformers are chosen as fit points. The yielded pharmacophore model characterizes the common bioactive conformations because similar functional groups (e.g., hydrogen bond acceptors) of all molecules are pointing to the same 3-D space (Figure 2.4). The lack of one or several of these functions is usually associated with a drop in activity.

Pharmacophore models may be used to derive ideas for the substitution of one group (e.g., hydroxyl) against another chemical group with similar features (e.g., an amine group as a hydrogen-bond donor and acceptor). This is a helpful indication that facilitates planned synthesis strategies or a guided compound purchase. Modeling tools such as CoMFA (comparative molecular field analysis) [2], CoMSIA (comparative molecular similarity indices analysis) [3], or PrGen [4] even allow an estimation of the effects on a quantitative level. These so-called 3-D QSAR (three-dimensional quantitative structure–activity relationship) studies require the pharmacophore model to determine significantly different interaction patterns that are directly associated with experimental data (e.g., activity). The statistical machinery behind is mainly based on principal component analysis (PCA) and partial least squares (PLS) regression. The PCA transforms a number of (possibly) correlated variables into a (smaller) number of uncorrelated variables called *principal components*. PLS regression is probably the least restrictive of the various multivariate extensions of the multiple linear regression models. In its simplest form, a linear model specifies the (linear) relationship between a dependent (response) variable Y, and a set of predictor variables, the X's, so that

$$Y = b_0 + b_1X_1 + b_2X_2 + \ldots + b_pX_p$$

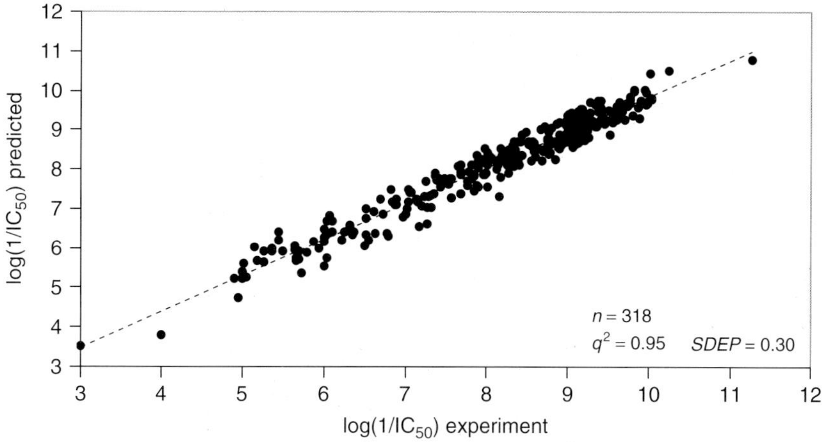

Figure 2.5 The correlation of experimental and predicted IC_{50} values yielded by a "leave-one-out" cross-validation ($q^2 = 0.95$) for the pharmacophore model shown in Figure 2.4.

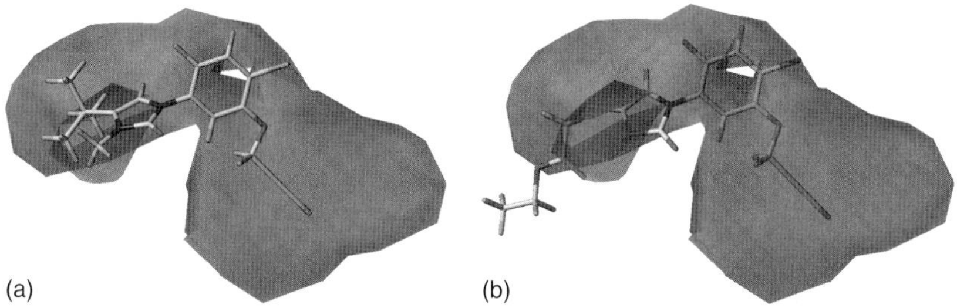

Figure 2.6 Contour map derived by a 3-D QSAR study. The clouds indicate the favorable space to be occupied by potent Protox inhibitors. While the highly active imidazolinone derivative (a) fits almost perfectly, the ethylcarboxylate residue of the weaker ligand protrudes from the preferred region (b).

In this equation b_0 is the regression coefficient for the intercept and the b_i values are the regression coefficients (for variables 1 through p) computed from the data.

The correlation of experimental and calculated activities assesses the quality of 3-D QSAR models. The squared correlation coefficient (r^2) yielded by this statistics is a measure of the goodness of fit. The robustness of the model is tested via cross-validation techniques (leave-x%-out), indicating the goodness of prediction (q^2). Models with q^2 values >0.4–0.5 are considered to yield reasonable predictions for hypothetical or as-yet untested molecules that are structurally comparable to those compounds used to establish the model (Figure 2.5).

Both, CoMFA and CoMSIA not only derive a mathematical equation but also generate contour maps (e.g., steric or electrostatic fields) that should or should not be occupied by new compounds with optimized characteristics (Figure 2.6).

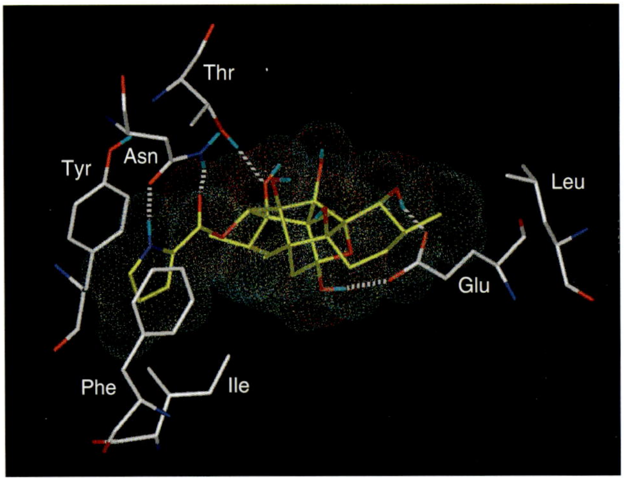

Figure 2.7 Pseudoreceptor model for insecticidal ryanodine derivates constructed with the program PrGen [4]. The binding site model is composed of seven amino acid residues and contains the structure of ryanodine [5]. Hydrogen bond interactions are indicated by dashed lines.

PrGen [4] creates a pseudoreceptor model around the pharmacophore representing an image of the hypothetical binding site (Figure 2.7). Ligand–pseudoreceptor site interactions, solvation and entropic energy terms are calculated to correlate the experimental and computed free binding energies. The binding site construction may take into account experimentally determined amino acid residues of the real binding site, or just those residues with complementary features to the ligands.

New hypothetical compounds may be introduced in the validated pseudoreceptor model to estimate free binding energies, and thus, to prioritize the laboratory capacities.

A common drawback of ligand-based approaches is the fact that data derived from screening hits may only be interpolated to somehow similar compounds. If any structural information is not present in the training set compounds, then a transfer to totally new structures is generally not possible [6].

2.4 Structure-Based Approaches

New scaffolds for active ingredients are classically obtained by an experimental random screening. Essential for this high-throughput experiment is a multitude of compounds that must be either purchased or synthesized and handled. For capacity reasons, it is desirable not to test all available compounds, but only those with a high chance of success. One helpful strategy to focus a compound library to a particular target is based on the molecular structure of this protein.

At present, highly sophisticated analytical methods such as X-ray crystallography, nuclear magnetic resonance (NMR) or cryoelectron microscopy are applied to solve the 3-D structures of enzymes, ion channels, G-protein-coupled receptors (GPCRs), and other proteins. A collection of more than 70 000 protein coordinates is freely available at the Protein Data Base (PDB) [7], and in some cases, even ligand–protein co-crystal structures are resolved. Coordinates derived from co-crystals unambiguously localize the binding site and provide an insight into the binding mode of a bound ligand; this in turn allows the computational chemists to characterize specific interaction patterns as being crucial for tight binding.

Equipped with this information, the binding site may be used much like a lock in order to identify the best fitting key, either by virtually screening diverse compound libraries (i.e., lead identification) or by increasing the specific fit of weak binders (lead optimization). The automation of this so-called (protein) *structure-based approach* [8, 9] is typically divided into a docking and a scoring step [10]. While docking yields the pose(s) of a ligand in the complex, scoring is necessary to discriminate between good and bad binders by calculating the free energies of binding for each generated conformer of a ligand.

In this context, it is common to differentiate between *empirical* and *knowledge-based* scoring functions [11]. The term "*empirical scoring function*" stresses that these quality functions approximate the free energy of binding, $\Delta G_{\text{binding}}$, as a sum of weighted interactions that are described by simple geometric functions, f_i, of the ligand and receptor coordinates r (Equation 2.1). Most empirical scoring functions are calibrated with a set of experimental binding affinities obtained from protein–ligand complexes; that is, the weights (coefficients) ΔG_i are determined by regression techniques in a supervised fashion. Such functions usually consider individual contributions from hydrogen bonds, ionic interactions, hydrophobic interactions, and binding entropy. As with many empirical approaches, the difficulty with empirical scoring arises from inconsistent calibration data.

$$\Delta G_{\text{binding}} = \Sigma \Delta G_i, f_i\,(r_{\text{ligand}}, r_{\text{receptor}}) \tag{2.1}$$

Knowledge-based scoring functions have their foundation in the inverse formulation of the Boltzmann law, computing an energy function that is also referred to as a "potential of mean force" (PMF). The inverse Boltzmann technique can be applied to derive sets of atom-pair potentials (energy functions) favoring preferred contacts and penalizing repulsive interactions. The various approaches differ in the sets of protein–ligand complexes used to obtain these potentials, the form of the energy function, the definition of protein and ligand atom types, the definition of reference states, distance cut-offs, and several additional parameters [12].

An extension of docking procedures is *de novo* design [13] with BASF's archetype LUDI [14]. Here, molecular fragments are composed inside a given binding pocket in order to design a perfectly matched new molecule.

Both attempts rely on an accurate binding site characterization, an appropriate ligand/binding site complex generation, and a reliable estimation of the free binding energies. The principle of a docking and scoring procedure is illustrated in Figure 2.8.

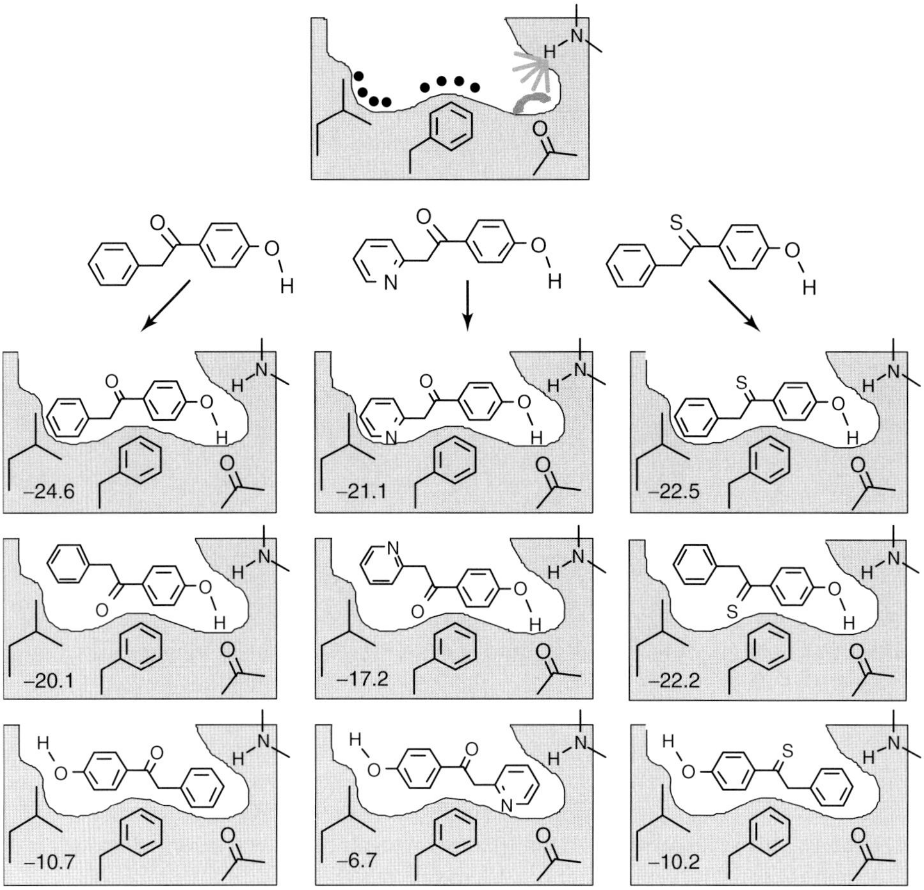

Figure 2.8 Protocol of a classical docking and scoring procedure. The binding site cavity is characterized via, for example, hydrophobic (filled circles), hydrogen-bond donor (lines), and hydrogen-bond acceptor properties (circle segment). Each compound of a database (or real library) is flexibly docked into the binding site and the free binding energy ($kJ\,mol^{-1}$) for each of the derived poses (indicated in the lower left corner of each pose) is estimated by a mathematical scoring function.

In order to demonstrate a docking application, the crystal structure of mitochondrial protoporphyrinogen IX oxidase (Protox) from common tobacco complexed with an acidic phenylpyrazole inhibitor (INH) and a non-covalently bound flavin adenine dinucleotide (FAD) cofactor was chosen (PDB ID code 1SEZ [15]). A salt bridge primarily fixes the inhibitor from the carboxylate group to a highly conserved arginine (Arg98) at the entrance of the binding niche. Further stabilization is due to hydrophobic contacts to Leu356, Leu372, and Phe392 in the core region.

In a first step, INH was extracted from the binding site cavity and a commercial docking program (FlexX [16]) was applied to determine whether the original binding

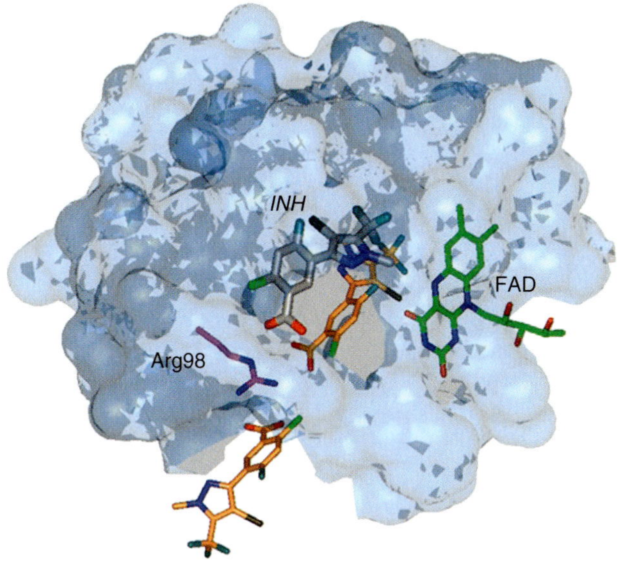

Figure 2.9 X-ray crystallographically determined binding site of Protox [15], including the cocrystallized inhibitor (INH; for structural formula, see Figure 2.10) and a part of the cofactor FAD. Highlighted is Arg98 at the entrance of the binding site cavity, interacting with INH, and almost all solutions of the FlexX approach via electrostatic and hydrogen bond interactions. Two docking poses representing a cluster of yielded solutions are indicated, one at the outside and one inside the binding site cavity (orange-colored carbon atoms).

pose of the X-ray structure could be re-found. For this calculation, only a volume with a radius of 10 Å around the binding site was considered, not the complete protein.

The program detects 98 favorable docking solutions within an energy range of 10.0 kJ mol^{-1}($\Delta\Delta G$). Except for two poses, all solutions strictly interact with their acidic function to the basic Arg98; although only 20 of them are actually located in the binding niche. The energetically most favorable proposals fix the guanidinium group of Arg98 from the solvent side (Figure 2.9), while a further 13 solutions block the gorge to the binding site.

In order to rationalize the docking process and to circumvent non-realistic solutions (i.e., outside the known binding region), two pharmacophore-type constraints may be set. First, an interacting group in the receptor site may be specified (i.e., interaction constraint). During the simulation each docking solution is checked to see whether there is any contact between the ligand and this particular hot spot; if not, the solution is discarded. The second type is a spatial constraint, where a spherical volume is defined in the active site and a specified atom or group of atoms from the ligand must lie within this sphere in the docking solution. FlexX-Pharm [17] offers both constraint types that even may be combined.

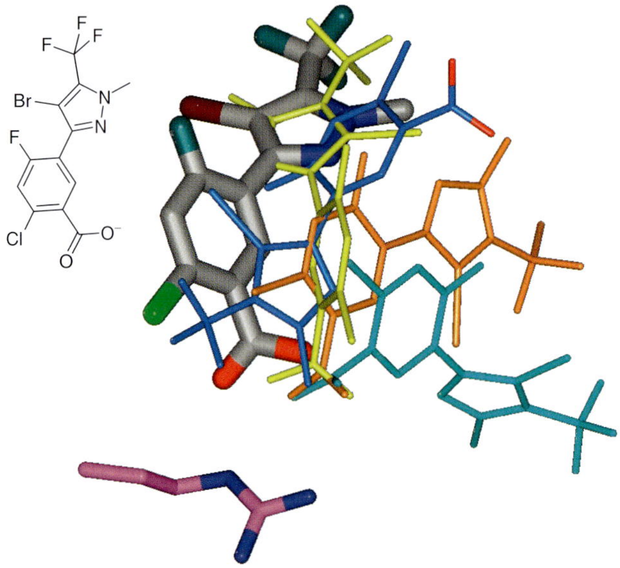

Figure 2.10 Structural formula of the cocrystallized inhibitor (INH) and comparison of the poses derived from FlexX docking (single-colored) and crystallization experiment (thick). Indicated is the crucial Arg98 that stabilizes all poses with the exception of the blue-colored solution, which interacts with the acid group (red-colored oxygen atoms) to the opposite side (i.e., Asn67).

Taking into consideration only the 20 accurate docking solutions, it must be stated that the original pose of INH is not perfectly found. Although most acidic groups interact with Arg98, the binding mode is different compared to the experimentally solved X-ray structure. Only the more hydrophobic pyrazole ring matches (in some cases) its reference counterpart. Furthermore, two docking solutions are totally different, their acid function interacting with the terminal amide group and the backbone NH of Asn67, which is opposite to Arg98 (Figure 2.10).

It is worth mentioning here that the docking procedure used does not take into account the flexibility of the binding site residues; rather, only the ligand is considered flexible in an energetically restricted range. However, some programs allow a concerted consideration of flexibility for ligand and binding site residues to simulate induced-fit docking (e.g., GLIDE, FlexE).

In contrast to the charged INH inhibitor used for the co-crystallization experiment, all of BASF's in-house compounds presented in the above-mentioned 3-D QSAR study are uncharged. Therefore, a second docking study with a neutral uracil derivative UBTZ (Figure 2.11) should clarify how ligands without acid function bind to this target site.

By applying the default parameters, FlexX produced 122 solutions predominantly located in the binding cavity. The two energetically most favorable solutions are compared with the original pose of INH (Figure 2.11). It is interesting to note, that each pose of UBTZ has a direct contact to Arg98. In one pose, a carbonyl oxygen

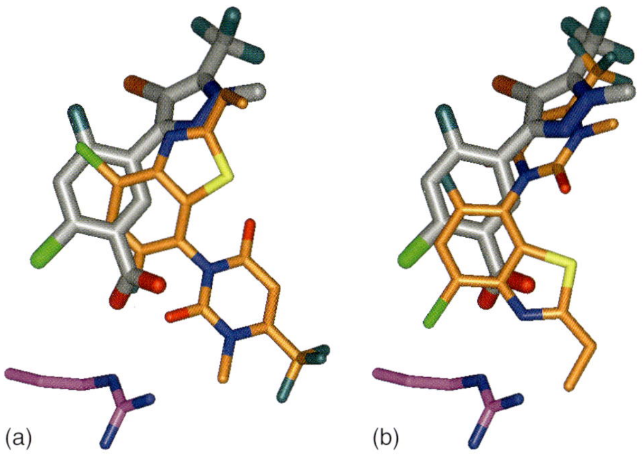

Figure 2.11 Comparison of two docking solutions for BASF's uracil benzothiazole derivative (UBTZ) with the bound inhibitor (INH). UBTZ interacts with Arg98 over the carbonyl oxygen of uracil and a fluorine of the benzothiazole ring (a) or the nitrogen atom of the benzothiazole ring (b).

of the uracil and a fluorine of the benzothiazole ring are involved; alternatively, the nitrogen atom of the benzothiazole ring is directed to the positively charged Arg98, and this docking solution shows a better total overlap with INH. Interestingly, although chemically diverse, both types of inhibitor (INH and UBTZ) obviously mimic similar binding properties necessary for complex formation.

In a next step, attempts were made to dock the physiological substrate, protoporphyrinogen IX, to the INH and Triton-X100-cleaned enzyme. Since this failed, the maximum overlap volume and the clash factors were modified in such a way that the narrow binding pocket was apparently relaxed and, subsequently, even the cofactor FAD was (non-physiologically) removed. Only the product of the enzyme reaction, protoporphyrin IX – which sterically is less demanding than the substrate – was inserted into the FAD-free binding site cavity (Figure 2.12). The yielded solution interacts loosely with one propionate group to Arg98, while the second acid group protrudes into the solvent region. Although the cofactor was not present during the calculation, the final pose indicates that carbon atom C20 of protoporphyrin IX is in close proximity to the electron-accepting nitrogen atom N5 of the flavin ring. This is in general agreement with the results obtained by Koch *et al.* [15] and Layer [18], which propose the initial hydride transfer at C20, followed by hydrogen rearrangements in the whole ring system by enamine–imine tautomerizations.

One possible explanation for the failed docking of the tetrapyrrole derivatives under physiological conditions (i.e., in the presence of FAD) might be the reference topology of the binding site. During the cocrystallization experiment, the ligand and the binding pocket adapt to each other perfectly to form a ligand-specific complex structure. Via this induced fit, the narrow cleft is unable to incorporate much larger ligands, and therefore only the flattened protoporphyrin IX could be

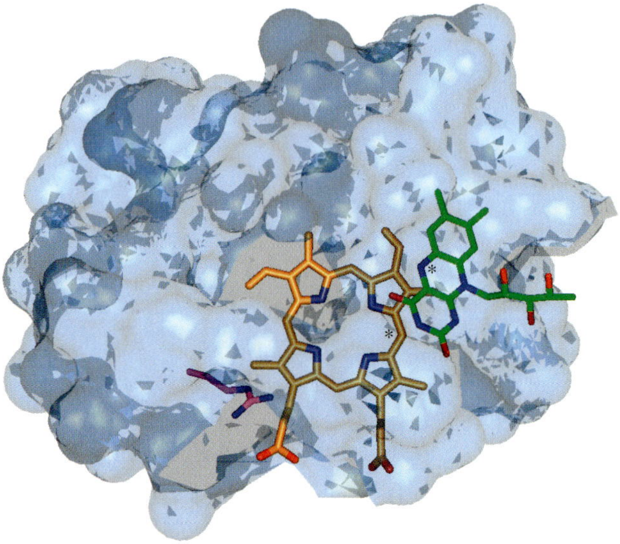

Figure 2.12 Docking solution for protoporphyrin IX in the Protox binding site. One propionic acid is close to Arg98, but does not form an explicit hydrogen bond. The asterisks indicate the proposed reaction centers C20 of protoporphyrin IX and N5 of FAD (see text for details).

introduced, though not in the intuitively expected manner (i.e., completely buried in the binding site cavity with a tight contact to Arg98). In order to circumvent such ligand-specific binding topology for a general docking approach, the ligand-free protein (apoprotein or holoprotein) may be relaxed applying a molecular dynamics (MD) simulation, which is a computer simulation of physical movements of atoms and molecules. For a given protein, the atoms are allowed to interact for a period of time, thus providing a view of the motion of the atoms. In the most common version, the trajectories of the atoms are determined by numerically solving the Newton's equations of motion for a system of interacting particles, where the forces between the particles and potential energy are defined by molecular mechanics force fields [19]. As a result of this simulation, several diverse protein conformers may be chosen as reference input structures for a docking approach.

Besides the binding site flexibility, water molecules may also play a crucial role in protein–ligand binding. Taking the above-mentioned Protox as an example, there are two relevant X-ray structures available in the Protein Databank. The first structure, as used for the modeling study, is from tobacco and has a low resolution of only 2.9 Å; within its binding pocket only one water molecule can be seen that does not show any interaction to the cocrystallized ligand or the cofactor. Consequently, this water molecule was not considered in the docking approach. In contrast to that, the human Protox X-ray structure (as published seven years later; pdb-code 3NKS) has a high resolution of 1.9 Å, where the binding pocket encloses the ligand, the cofactor and a cluster of about 10 water molecules, thus forming a hydrogen bond network to the ligand, the cofactor, and among each other [20].

In this case, explicit water molecules would have to be taken into account as fixed anchor points to elucidate reasonable binding poses for the ligands in docking experiments.

The relevance of such particular water molecules for lead optimization was described in a recent example for the herbicidal target protein IspD [21]. In this study, a water molecule forming hydrogen bonds to the ligand and a residue of the binding pocket was replaced by an additional substituent of a new ligand to form almost the same interaction pattern as the water molecule. The introduction of a nitrile group, as a hydrogen bond acceptor, yielded a fourfold increase in binding affinity. Unexpectedly, a carbonic acid function at the same position, also mimicking an H-bond acceptor, decreased the binding affinity of the ligand by a factor of almost 2000.

These examples indicate typical challenges of structure-based approaches starting from the need for a highly resolved target structure, a multitude of yielded docking poses, and problematic estimations of free binding energies. An enormous advantage of this technique is its unbiased use, with results obtained for a particular target site providing information for new chemical structures without any prior expert knowledge or selection. From a technological viewpoint, there is ongoing improvement to achieve more realistic docking solutions (e.g., interaction and spatial constraints or a post-processing step). Additionally, the quality of the energy estimation may be increased by tailor-made scoring functions, although this requires extensive experimental data (e.g., cocrystal data and IC_{50} values) for a particular family of targets (e.g., kinases) in order to perform the calibration.

In summarizing this topic, it is clear that structure-based methods are extremely helpful for creating ideas for new scaffolds and further optimization strategies.

2.5 Estimation of Adverse Effects

Even a compound with the highest efficacy can fail because of undesired adverse effects. The new Plant Protection Products Regulation (Regulation EC No. 1107/2009) clearly indicates a strict guideline for relevant toxicological endpoints. Furthermore, there are defined cut-off (restriction) criteria for hazardous properties which will result in a substance being banned even if it can be applied in a safe manner. This comprises compounds proven to be carcinogenic, mutagenic, toxic for reproduction (CMR), endocrine disruptors, and persistent (bio-accumulative). Most of this information can only be produced by conducting costly higher-tier studies; moreover, especially in the early research stages such experiments would be prohibitive and in contradiction to the 3R (Replacement, Refinement, and Reduction) philosophy to replace animal studies. Therefore, cell-based indication studies and/or even cheaper *in silico* techniques will need to be developed to provide first insights into the risk potential of novel compounds.

2.6
In-Silico Toxicology

In-silico toxicology indicates a variety of computational techniques which relate the structure of a chemical to its toxicity or fate with the advantages of cost-effectiveness, speed compared to traditional testing, and reduction in animal use. Based on experimental data, two general techniques are followed for toxicity predictions: (i) rule-based expert systems that rely on a set of chemical structure alerts; and (ii) correlative SAR methods based on statistical analysis.

2.7
Programs and Databases

Several software tools are available that allow the statistical estimation of some critical endpoints. The commercial computer program DEREK (Deductive Estimation of Risk from Exixting Knowledge; Lhasa Inc.) is designed to assist chemists and toxicologists in predicting likely areas of possible toxicological risk for new compounds, based on an analysis of their chemical structures. DEREK indicates whether a specific toxic response may occur – it does not provide a quantitative estimate of the prediction. The program accepts as input a "target" (the molecule to be analyzed) drawn in the language of structural formulae that is common to all organic chemists. DEREK scans a "rule base" of substructures which are known to have adverse toxicological properties, looking for matches to substructures in the target molecule. "Hits" in the rule base are shown to the user on a graphical display and summarized in tabular form for hardcopy output.

TOPKAT (toxicity prediction by komputer-assisted technology) quantifies electronic, bulk, and shape attributes of a structure in terms of electrotopological state (E-states) values of all possible two-atom fragments, atomic size-adjusted E-states computed from rescaled count of valence electrons, molecular weight, topological shape indices, and symmetry indices. The methodology is an extension of classic QSARs. Leadscope tools are used to analyze the datasets of chemical structures and related biological or toxicity data. Structures and/or data can be loaded from an SD file and data can be loaded from a text file.

Leadscope provides a number of ways to group a set of compounds, including the chemical feature hierarchy (27 000 named substructures), recursive partitioning/simulated annealing (a method for identifying active classes characterized by combinations of structural features), structure-based clustering, and dynamically generated significant scaffolds or substructures. Based on the presence or absence of the substructures, models are generated with training sets of compounds associated with wanted (activity) or unwanted (toxicological endpoint) biological data. Test compounds with unknown biological activity are then classified according to relevant substructures (i.e., descriptors) of the validated models, and the activity is estimated. In addition, Leadscope automatically calculates the following properties for all imported compounds: aLog*P*; polar surface area; the number of hydrogen bond donors; the number of hydrogen bond acceptors; the number of rotatable bonds; molecular weight; number of atoms; and Lipinski score.

A further example of a knowledge-based system is CASE (computer automated structure evaluation) and its successor, MCASE/MC4PC (formerly MultiCASE), designed for the specific purpose of organizing biological/toxicological data obtained from the evaluation of diverse chemicals. These programs can automatically identify molecular substructures that have a high probability of being relevant or responsible for the observed biological activity of a learning set comprised of a mix of active and inactive molecules of diverse composition. New, untested molecules can then be submitted to the program, and an expert prediction of the potential activity of the new molecule is obtained. A further program is Case Ultra, which was especially developed with an objective to meet the current and most updated regulatory needs for the safety evaluation of compounds.

Like DEREK, OncoLogic [22], HazardExpert [23], and ToxTree [24] are further knowledge-based expert systems for the prediction of toxicological endpoints.

Independent of the particular program, the crucial basis for good models and predictions is the quality of the data. In the best-case scenario, all data are obtained by in-house experiments for the toxicological endpoints of interest; however, in many cases there will not be enough experiments to develop a general model, but only a tailor-made scaffold-related model for a certain endpoint. Therefore, external data might be of value to provide a better coverage of chemical space. Publicly available data sources exist for endpoints such as human health cancer and mutagenicity (ISSCAN, CPDB, and OASIS Genetox), skin sensitization (local lymph node assay, guinea pig maximization test, and ECETOC skin sensitization), mammalian single/repeated dose toxicity studies (Japan EXCHEM, RepDose), eye irritation (ECETOC), skin irritation (OECD toolbox), and skin penetration (EDETOX). The OECD (Q)SAR toolbox, which is a software application intended to be used by governments, chemical industry, and other stakeholders in filling gaps in (eco)toxicity data needed for assessing the hazards of chemicals, allows the further estimation of bioaccumulation, acute aquatic toxicity, and estrogen receptor binding.

Parallel to these freely available databases and tools there are also commercially compiled databases of toxicological information. Typically addressed endpoints are carcinogenicity, genetic toxicity, chronic and subchronic toxicity, acute toxicity as well as reproductive and developmental toxicity, mutagenicity, skin/eye irritation, and hepatotoxicity. Providers are companies such as Leadscope (Leadscope database), Lhasa Ltd. (VITIC Nexus), TerraBase Inc. (TerraTox), or MDL (RTECS). In order to combine a number of databases and resources together, several tools have become available. For example, TOXNET, the OECD's eChemPortal and the US Environmental Protection Agency (EPA) ACToR (Aggregated Computational Toxicology Resource) resources are most likely to be useful for obtaining information about single compounds. ACToR is a freely available collection of databases from more than 200 sources. The data include chemical structure, physico-chemical properties, *in vitro* assays, and *in vivo* toxicology data for industrial chemicals with mostly high and medium production volumes, pesticides, and potential groundwater or drinking water contaminants. Based on the analysis of Judson *et al.* [25], and using approximately 10 000 substances (industrial chemicals and pesticide ingredients), it was shown that acute hazard data are available for 59% of the surveyed chemicals, testing information for carcinogenicity for 26%, developmental toxicity for 29%, and reproductive toxicity for only 11%. In order to fill the toxicity data gaps, EPA has designed the ToxCast screening and prioritization program. ToxCast is profiling over 300 well-characterized chemicals (mostly pesticides) in over 400 HTS endpoints. These endpoints include

- Biochemical assays of protein function
- Cell-based transcriptional reporter assays
- Multi-cell interaction assays
- Transcriptomics on primary cell cultures, and
- Developmental assays in zebrafish embryos.

Almost all the compounds have been tested in traditional toxicology tests, including developmental toxicity, multigeneration studies, subchronic, and chronic rodent bioassays. These data, collected in the Toxicity Reference Database (ToxRef DB; *http://www.epa.gov/ncct/toxrefdb/*) will be used to build computational models to forecast the potential human toxicity of chemicals with the aim of leading a more efficient use of animal testing.

2.7.1
In-Silico Toxicology Models

SAR and QSAR for toxicological endpoints were applied only sporadically to drug discovery during the 1960s and 1970s. This was due primarily to the combination of a lack of detailed understanding of most mechanisms of toxicology, a lack of systematically generated datasets around specific toxicities, and a lack of general directives from regulators for standard tests that should be performed prior to drug testing in humans. With the advent of the *Salmonella* reverse mutation assay (Ames test) during the early 1970s, however, this picture began to change. In the Ames test [26], frameshift mutations or base-pair substitutions may be detected by the exposure of histidine-dependent strains of *Salmonella typhimurium* to a test compound. When these strains are exposed to a mutagen, reverse mutations that restore the functional capability of the bacteria to synthesize histidine enable bacterial colony growth on a medium that is deficient in histidine ("revertants"). In some cases, there is a need to activate the compounds via a mammalian metabolizing system, which contains liver microsomes (with S9 mix). Ames-positive compounds significantly induce revertant colony growth at least in one out of usually five strains, either in the presence or absence of the S9 mix. A compound is judged Ames-negative if it does not induce significant colony growth in any reported strain. The adoption of the assay to assess the mutagenic potential of chemicals provided a relatively consistent data source previously unknown to toxicology. In 2009, Hansen *et al.* [27] collected 6512 non-confidential compounds (3503 Ames-positive and 3009 Ames-negative),

Table 2.1 Comparison of all classifiers for mutagenicity, as applied from Hansen *et al.* [27].

	50% false positives	43% false positives	36% false positives	Model
Sensitivity	0.93 ± 0.01	0.91 ± 0.01	0.88 ± 0.01	SVM
	0.89 ± 0.01	0.86 ± 0.01	0.83 ± 0.02	GP
	0.90 ± 0.02	0.87 ± 0.02	0.84 ± 0.03	Random Forest
	0.86 ± 0.02	0.86 ± 0.02	0.81 ± 0.02	k-Nearest Neighbor
	–	–	0.84 ± 0.02	Pipeline Pilot
	0.73 ± 0.01	–	–	DEREK
	–	0.78 ± 0.01	–	MultiCASE

True-positive predictions of mutagens (i.e., sensitivity) relative to false-positive rates (50, 43, and 36%) for trained support vector machines (SVM), Gaussian process classification (GP), Random Forest, and k-Nearest Neighbor in comparison to commercial programs.

together with their biological activity, to form a new benchmark data set for the *in-silico* prediction of Ames mutagenicity. The data set contains 1414 compounds from the World Drug Index (i.e., drugs) and has a mean molecular weight (MW) of 248 ± 134 (median MW 229). With these data, the commercial software programs DEREK, MultiCASE, and Pipeline Pilot (Accelrys) as well as four non-commercial machine learning implementations (i.e., support vector machines (SVM), Gaussian process classification (GP), Random Forest, and k-nearest neighbor) were evaluated. Molecular descriptors were extracted from DragonX version 1.2 [28]. The final statistical results indicate the general applicability of all programs (Table 2.1).

All commercial programs have fixed sensitivity levels, whereas the sensitivity of all parametric classifiers (e.g., SVM) can be calculated to arbitrary levels of specificity. A sensitivity value of 0.93 for the SVM model at a 50% false positive rate indicates that, for a given test set of 200 compounds with 100 mutagens and 100 non-mutagens, 93 mutagens will be classified as true positives together with 50 false positives. A shift to the 36% false positives level yields 88 true positives, together with only 36 false positives. Of the commercial programs, Pipeline Pilot performs best; one reason for this is the explicit training with the given validation data set, whereas DEREK and MultiCASE are based on a fixed set of mainly 2-D descriptors (MultiCASE) or a static system of rules derived from a largely unknown data set and expert knowledge (DEREK). However, the latter two programs provide structure–activity and/or mechanistic information essential for structure optimization and regulatory acceptance. The parametric classifiers outperform the commercial programs and may be applied for larger screening campaigns. On the other hand, these classifiers do not provide hints as to why a certain compound was predicted to be a mutagen, and thus optimization guidance is not given.

In a second case study [29], aquatic toxicity was predicted for a data set of 983 unique compounds tested in the same laboratory against *Tetrahymena pyriformis* [30]. For model validation, 644 compounds were chosen randomly and the remaining 339 compounds were used as a first external test set (external test I).

Table 2.2 Statistical results for selected QSAR models predicting aquatic toxicity [29].

		Modeling set (n = 644)			Validation set I (n = 339)			Validation set II (n = 110)		
Model	Group	Q_{abs}^2	MAE	Cov. (%)	R_{abs}^2	MAE	Cov. (%)	R_{abs}^2	MAE	Cov. (%)
kNN-Dragon	UNC	0.92	0.22	100	0.85	0.27	80.2	0.72	0.33	52.7
kNN-Dragon	UNC	0.92	0.22	100	0.84	0.29	100	0.59	0.43	100
ISIDA-SVM	ULP	0.95	0.15	100	0.76	0.32	100	0.38	0.50	100
ASNN	VCCLAB	0.83	0.31	83.9	0.87	0.28	87.4	0.75	0.32	71.7
ASNN	VCCLAB	0.83	0.31	83.9	0.85	0.30	100	0.66	0.38	100
ConsMod I		0.92	0.23	100	0.85	0.29	100	0.67	0.39	100

kNN-Dragon: k-nearest neighbor with descriptor set from Dragon; ISIDA-SVM: Support Vector Machine applied at molecular fragments calculated with the ISIDA program [31]; ASNN: Associative neuronal network; UNC: University of North Carolina; ULP: University of Louis Pasteur; VCCLAB: Virtual Computational Chemistry Laboratory; Q_{abs}^2: linear regression coefficient for external validation set.

In addition, a second test set (external test II) was used with compounds recently reported by the same laboratory. Six independent academic groups developed 15 different types of QSAR models with a particular focus on the predictive power for the external test sets. Each group relied on its own QSAR modeling approaches to generate toxicity models, using the same data sets. For all models the applicability domain was calculated, thus yielding a measure whether compounds of the test sets may be predicted, or not. In case that totally novel compounds in the test set have no equivalents in the training set, prediction is restricted and this unique compound will not be considered. The use of applicability domains will lead to lower coverage rates (i.e., <100%), but typically to better predictions of the test sets.

The internal prediction accuracy for the modeling set ranged from 0.76 to 0.95 as measured by the leave-one-out cross-validation correlation coefficient (Q_{abs}^2).The prediction accuracy for the external validation sets I and II ranged from 0.71 to 0.87 (linear regression coefficient R_{absI}^2) and from 0.38 to 0.83 (R_{absII}^2), respectively. Finally, several consensus models were developed by averaging the predicted aquatic toxicity of all 15 models. The results of several individual models and the best consensus model with respect to coverage and predicting power are exemplified in Table 2.2.

The consensus model (ConsMod I) has a coverage of 100% (i.e., all compounds of the training and test sets were considered), and shows the most robust prediction behavior compared to each other method applied in this study.

The third use case describes a QSAR model for the prediction of carcinogenicity [32]. A carcinogen is a type of mutagen that specifically leads to cancer. The authors trained counter propagation artificial neuronal networks (CP ANN) with eight MDL descriptor (model A) and 12 Dragon descriptors (model B) for a dataset of 805 non-congeneric chemicals extracted from a Carcinogenic Potency Database (CPDBAS) [33]. The main advantage of neuronal network modeling is that complex, nonlinear relationships can be modeled without assumptions about the form of the

Table 2.3 Statistical performance of both carcinogenicity models classified by ANN [32].

	Model A (8 MDL descriptors)		Model B (12 Dragon descriptors)	
Internal validation	Training ($CV_{l20\%o}$)	Test 1/2	Training ($CV_{l20\%o}$)	Test 1/2
Accuracy (%)	91 (66)	73/61.4	89 (62)	69/60.0
Sensitivity (%)	96	75/64.0	90	75/61.8
Specificity (%)	86	69/58.9	87	61/58.4

Accuracy: total number of non-carcinogens and carcinogens correctly predicted among all compounds; Sensitivity: correctly classified carcinogens among all carcinogens; Specificity: correctly classified non-carcinogens among all non-carcinogens; Training was performed with 644 compounds; test sets 1 and 2 were composed of 161 and 738 compounds, respectively. $CV_{l20\%o}$: cross-validation of the training set compounds leaving several times 20% of the compounds out for (internal) prediction.

model and they can cope with noisy data. The dataset was divided into a training set of 644 compounds and a test set of 161 compounds (test 1). With the training set compounds, both models were fed to identify the best architecture for optimum predictions. Based on this evaluation step, prediction for the test set compounds was performed. In a following step, a second external test set of 738 compounds (test 2) with known activity was investigated in order to yield a more realistic assessment of the robustness of the models.

The trained models are able to classify 91% (model A) and 89% (model B) of all carcinogens and non-carcinogens (i.e., accuracy) of the training set compounds (Table 2.3). Model A outperforms model B with respect to the classification of carcinogens – that is, it is more sensitive by 96 versus 90%, whereas in terms of classifying non-carcinogens (i.e., specificity) model B is slightly better (87% versus 86%). Accuracy for the prediction of test set 1 (73% versus 69%) and test set 2 (61.4% versus 60.0%) is generally lower; however, it is still in a comparable range relative to the leave-20%-out cross-validation results ($CV_{l20\%o}$) of the training set compounds (66% versus 62%). Obviously, test set 2 (738 external compounds) seems to be more diverse compared to the original training set data, yielding a significant reduction in accuracy, sensitivity, and specificity. Nevertheless, the models are useful for setting priorities among chemicals for further testing, and are publicly accessible at the CAESAR (Computer Assisted Evaluation of industrial chemical Substances According to Regulation) internet site (*http://www.caesar-project.eu*).

2.8 Conclusion

Ligand-based and structure-based approaches are valuable tools for the identification and optimization of lead compounds. However, whilst each strategy will need special prerequisites, it will also have strengths and weaknesses. In some cases, the strengths of both methods may be combined for a joint approach, which is referred to as a *structure-based pharmacophore alignment*, where the receptor site serves as a complement to build the pharmacophore model, and sophisticated statistical methods from 3-D QSAR (PCA and PLS) are applied to the prediction of activity [34, 35].

The high efficacy of a candidate is necessary, but not sufficient to obtain a registration. Only "clean" compounds that fulfill all criteria of the new Plant Protection Products Regulation (Regulation EC No. 1107/2009) will have access to the market in future. This means that, even at an early research phase, the critical toxicological endpoints of both hits and leads should be indicated in order to initiate toxicological indicator studies and/or new synthesis strategies to circumvent this risk.

For both the optimization of activity and the prediction of toxicological alerts, computer-aided strategies are (partly) available and implemented in modern research processes. However, especially for the prediction of critical endpoints such as developmental and reproduction toxicity, as well as endocrine disruption, there is a need not only for a larger quantity of data but also for better data in order to allow the design of models for *in-silico* estimation.

References

1. For references, see: *http://vesta.chem.umn.edu/classes/ch8021s06/FF_refs.htm.*
2. Cramer, R.D. III, Patterson, D.E., and Bunce, J.D. (1988) *J. Am. Chem. Soc.*, **110**, 5959–5967.
3. Klebe, G., Abraham, U., and Mietzner, T. (1994) *J. Med. Chem.*, **37**, 4130–4146.
4. Zbinden, P., Dobler, M., Folkers, G., and Vedani, A. (1998) *Quant. Struct.-Act. Relat.*, **17**, 122–130.
5. Schleifer, K.J. (2000) *J. Comput.-Aided Mol. Des.*, **14**, 467–475.
6. Bordás, B., Komives, T., and Lopata, A. (2003) *Pest Manag. Sci*, **59**, 393–400.
7. Protein Data Base. Available at : *http://www.rcsb.org.*
8. Waszkowycz, B. (2002) *Curr. Opin. Drug Discov. Dev.*, **3**, 407–413.
9. Taylor, R.D., Jewsbury, P.J., and Essex, J.W. (2002) *J. Comput.-Aided Mol. Des.*, **3**, 151–166.
10. Kitchen, D.B., Decornez, H., Furr, J.R., and Bajorath, J. (2004) *Nat. Rev. Drug Discov.*, **11**, 935–949.
11. Gohlke, H. and Klebe, G. (2002) *Angew. Chem. Int. Ed.*, **41**, 2644–2676.
12. Gohlke, H., Hendlich, M., and Klebe, G. (2000) *J. Mol. Biol.*, **295**, 337–356.
13. Schneider, G. and Fechner, U. (2005) *Nat. Rev. Drug Discov.*, **4**, 649–663.
14. Böhm, H.J. (1992) *J. Comput.-Aided Mol. Des.*, **6**, 61–78.
15. Koch, M., Breithaup, C., Kiefersauer, R., Freigang, J., Huber, R., and Messerschmidt, A. (2004) *EMBO J.*, **23**, 1720–1728.
16. Rarey, M., Kramer, B., and Lengauer, T. (1995) *Proc. Int. Conf. Intell. Syst. Mol. Biol.*, **3**, 300–308.
17. Hindle, S.A., Rarey, M., Buning, C., and Lengauer, T. (2002) *J. Comput.-Aided Mol. Des.*, **16**, 129–149.
18. Layer, G., Jahn, D., and Jahn, M. (2011) Heme Biosynthesis: Protoporphyrin IX Oxidase (PPO), in *Handbook of Porphyrin Science with Application to Chemistry, Physics, Materials Science, Engineering, Biology and Medicine* (eds K.M. Kadish, K.M. Smith, R. Guilard), Vol. 15, World Scientific Publishing Co. Pte. Ltd, Singapore, pp. 197–200.
19. Hansson, T., Oostenbrink, C., and van Gunsteren, W. (2002) *Curr. Opin. Struct. Biol.*, **2**, 190–196.
20. Qin, X., Tan, Y., Wang, L., Wang, B., Wen, X., Yang, G., Xi, Z., and Shen, Y. (2011) *FASEB J.*, **25**, 653–664.
21. Witschel, M.C., Höffken, H.W., Seet, M., Parra, L., Mietzner, T., Thater, F., Niggeweg, R., Röhl, F., Illarionov, B., Rohdich, F., Kaiser, J., Fischer, M., Bacher, A., and Diederich, F. (2011) *Angew. Chem.*, **123**, 8077–8081.

22. Woo, Y.T. and Lai, D.Y. (2005) in *Predictive Toxicology* (ed. C. Helma), CRC Press, Boca Raton, FL, pp. 385–413.
23. Lewis, D.F.V., Bird, M.G., and Jacobs, M.N. (2002) *Hum. Exp. Toxicol.*, **21**, 115–122.
24. Patlewicz, G., Jeliazkova, N., Safford, R.J., Worth, A.P., and Aleksiev, B. (2008) *SAR QSAR Environ. Res.*, **19**, 495–524.
25. Judson, R., Richard, A., Dix, D.J., Houck, K., Martin, M., Kavlock, R., Dellarco, V., Henry, T., Holderman, T., Sayre, P., Tan, S., Carpenter, T., and Smith, E. (2009) *Environ. Health Perspect.*, **117**, 685–695.
26. Ames, B.N., Lee, F.D., and Durston, W.E. (1973) *Proc. Natl Acad. Sci. USA*, **70**, 782–786.
27. Hansen, K., Mika, S., Schroeter, T., Sutter, A., ter Laak, A., Steger-Hartmann, T., Heinrich, N., and Müller, K.R. (2009) *J. Chem. Inf. Model.*, **49**, 2077–2081.
28. Todeschini, R. and Consonni, V. (2002) *Handbook of Molecular Descriptors*, 1st edn, Wiley-VCH Verlag GmbH, Weinheim.
29. Zhu, H., Tropsha, A., Fourches, D., Varnek, A., Papa, E., Gramatica, P., Oberg, T., Dao, P., Cherkasov, A., and Tetko, I.V. (2008) *J. Chem. Inf. Model.*, **48**, 766–784.
30. Schultz, T.W. and Netzeva, T.I. (2004) in *Modeling Environmental Fate and Toxicity*, Chapter 12, vol. 4 (eds M.T. Cronin and D.J. Livingstone), CRC Press, Boca Raton, FL, pp. 265–284.
31. Varnek, A., Fourches, D., Solov'ev, V.P., Baulin, V.E., Turanov, A.N., Karandashev, V.K., Fara, D., and Katritzky, A.R. (2004) *J. Chem. Inf. Comput. Sci.*, **44**, 1365–1382.
32. Fjodorova, N., Vracko, M., Novic, M., Roncaglioni, A., and Benfenati, E. (2010) *Chem. Cent. J.*, **4** (Suppl. 1), S3.
33. Fitzpatrick, R.B. (2008) *Med. Ref. Serv. Q*, **27**, 303–311.
34. Christmann-Franck, S., Bertrand, H.O., and Goupil-Lamy, A., der Garabedian, P.A., Mauffret, O., Hoffmann, R., and Fermandjian, S. (2004) *J. Med. Chem.*, **47**, 6840–6853.
35. Schlegel, B., Stark, H., Sippl, W., and Höltje, H.D. (2005) *Inflamm. Res.*, **54** (Suppl. 1), 50–51.

3
Quantum Chemical Methods in the Design of Agrochemicals

Michael Schindler

3.1
Introduction

One of the central paradigms in rational design states that there exists a relationship between the physico-chemical properties of a molecule and its *in vivo* activity. While this relation is obscured by a plethora of processes taking place during an active ingredient's route from its point of administration to its molecular target in insects, weeds, or fungi, a detailed analysis of the various steps reveals that the paradigm holds for each of these steps separately. It must be admitted, however, that the present understanding of the complex dynamical networks in biological systems is still far from complete. Hence, it is only our limited knowledge that prevents the design of insecticides, herbicides, or fungicides having the necessary properties for optimum performance in the field.

To identify quantitative structure–activity relationships (QSARs) and to optimize compounds according to these rules are the main tasks of research in the design of modern agrochemicals. The complexity of the structure–activity problem is reduced, however, if a restriction is placed on the use only of *in vitro* data when measuring biological activity; for example, of enzyme assays where transport processes and metabolism are avoided.

Basic to the efficient use of SARs is the question of whether it is possible to make predictions; that is, to provide the physico-chemical properties – and hence the activities – of virtual compounds in order to prioritize the synthesis of new substances.

This is where quantum chemistry comes into play. While there exist many fast and reliable prediction systems, starting from simple increment systems to sophisticated expert systems – the rules of which are based on available experimental databases – these tools reach their limits when the targeted molecule or property cannot be estimated by interpolation in the property space spanned by the experimental data. *Extrapolation* is the domain of *ab initio* quantum chemistry.

In this chapter, the status of computational quantum chemistry in the field of rational design of agrochemicals will be reviewed. In the past, several attempts have been made to cover the diversity of computational chemistry in general [1],

Modern Methods in Crop Protection Research, First Edition.
Edited by Peter Jeschke, Wolfgang Krämer, Ulrich Schirmer, and Matthias Witschel.

and in order to obtain a continuously up-to-date overview of the field, the reader is referred to the excellent series *Reviews in Computational Chemistry* [2], which started in 1990. Many aspects of quantum chemical approaches to biochemical topics are addressed in Ref. [3]. In addition, introductory chapters on quantum chemical methods will form part of any modern textbook on molecular modeling (e.g., Ref. [4]).

Here, emphasis will be placed on ligand-based approaches; that is, the calculation of molecular properties and their use in establishing QSARs or solving chemical problems. The text will be restricted to computational techniques that can routinely be applied to rather large sets of molecules. Consequently, the exciting fields of *ab initio* molecular dynamics [5, 6] and quantum mechanical/molecular mechanics (QM/MM) approaches [7], which combine the QM description of the most interesting part of a drug–enzyme system with MM for the rest, will not be discussed. The computational investigation of chemical reaction mechanisms [8] or catalytic events in, for example, cytochrome P_{450} is also beyond the scope of this review.

3.2
Computational Quantum Chemistry: Basics, Challenges, and New Developments

The quality of quantum chemical computations depends mainly on two factors: the method chosen, and the so-called "basis set" used. In order to understand the hierarchy of methods, a brief outline will be provided of the underlying assumptions and approximations which lead from the Schrödinger equation to quantum chemical program packages.

Within the Born–Oppenheimer approximation, which permits the separation of nuclear and electronic motions of a molecule, quantum chemistry provides approximate solutions of the time-independent electronic Schrödinger equation

$$\hat{H}\Psi = E\Psi \tag{3.1}$$

where $\hat{H}$ is the electronic Hamilton operator acting on the molecular wavefunction Ψ resulting in the molecular energy E as the eigenvalue of Ψ. For a molecule with N electrons and M nuclei, $\hat{H}$ consists of the kinetic and potential energy operators $\hat{T}$ and $\hat{V}$ of the form

$$\hat{H} = \hat{T} + \hat{V} = -\frac{1}{2}\sum_{i=1}^{N}\nabla_i^2 - \sum_{i=1}^{N}\sum_{A=1}^{M}\frac{Z_A}{r_{iA}} + \sum_{i<j=1}^{N}\frac{1}{r_{ij}} \tag{3.2}$$

in atomic units, that is, $m_e = \hbar = |e| = 1$. The Schrödinger equation is an inhomogeneous second-order differential equation depending on the positions (and spins) of all electrons, the exact solution of which is possible only in very rare cases.

Imposing the Pauli principle and approximating the multi-electron wavefunction Ψ by an antisymmetrical product (Slater-determinant) of one-electron molecular orbitals (MOs) ψ which are represented by linear combinations of atomic orbitals (AOs) ϕ, each of which is constructed from linear combinations of so-called basis functions χ, leads to the simplified algebraic form of the Schrödinger equation, the Roothan–Hall or Hartree–Fock equation, which describes the motion of one electron in the mean field of the remaining electrons of the molecule. This pseudo-eigenvalue equation has to be solved iteratively until self-consistency is reached – that is, until the electron density does not change during the iterations. The time-consuming step in the solution of the Hartree-Fock self consistent field (HF-SCF) equations is the calculation of the electron–electron interaction; that is, the two-electron, four-center classical Coulomb- and non-classical Exchange-Integrals over the basic functions which scales formally as n^4, where n is the number of basic functions.

The natural choice for these basis functions are the exact solutions of one-electron Schrödinger equations of hydrogen-like atoms, the Slater-type orbitals (STOs). They are qualitatively correct, having a cusp at the positions of the nuclei ($r = 0$) and decreasing exponentially for large distances. However, the calculation of their electron–electron integrals is extremely cumbersome, and this is the reason why almost all quantum chemistry codes [9–13] use Gaussian-type orbitals (GTOs) as basic functions. Their advantage of the fast calculation of 2e-integrals outweighs by far the wrong behavior at $r = 0$ and the much too steeply decreasing density at large r, which needs to be compensated by a much larger number of GTOs than STOs. The slow convergence of the GTOs toward complete basis sets led to the development of explicitly correlated basis functions r_{12}[14, 15] and f_{12} [16], which are becoming increasingly efficient in conjunction with high-end *ab initio* methods.

To reduce the computational burden, semi-empirical methods such as MNDO, AM1, or PM3 [17] approximate the core electrons by empirical functions and concentrate only on the valence electrons which are formally described by a minimal basis set of STOs. Additional simplifications make the solution of the eigenvalue problem, scaling as n^3, the time-consuming step in semi-empirical calculations.

The HF-SCF equations represent the simplest form of the *ab initio* level of theory. They are the starting point for improvements of the wavefunction in order to describe the correlated movement of the electrons. In the main, three different routes lead finally to the nonrelativistic limit:

- Configuration interaction (CI) variationally determines the expansion coefficients of a linear expansion in a basis of electron configurations.
- Many-body perturbation theory (MBPT) handles the electron correlation as a perturbation of the HF-wavefunction.
- The coupled cluster (CC) approach generates the correlated wavefunction via an exponential substitution operator acting on the SCF-wavefunction.

For the relative merits of each approach, see Refs [18, 19]. Common to all approaches is the tremendously increasing demand on CPU-time and disc-space.

In fact, the "gold standard" of non-relativistic quantum chemical methods is the CC singles, doubles, and approximate triples ansatz CCSD(T), which reaches chemical accuracy. However, as it scales as n^7 it can be used only for benchmark calculations on rather small systems. In addition, as a rule of thumb, the better the computational method, the larger the basis set must be.

Routinely affordable is second-order Møller-Plesset perturbation theory (MP2), which formally scales as n^5, though the "workhorses" of computational chemistry are methods based on density functional theory (DFT).

Based on the Hohenberg–Kohn theorems [20], which state that the energy and all molecular properties are uniquely defined by the ground state charge density $\rho(r)$, DFT concentrates on $\rho(r)$ as the central molecular entity. Given the external potential $V_{Ne}(r)$ of the nuclei, the energy can be written as a functional of $\rho(r)$:

$$E_0[\rho_0] = F_{HK}[\rho_0] + \int \rho_0(r) V_{Ne} dr \tag{3.3}$$

with the Hohenberg–Kohn functional $F_{HK}[\rho_0]$ comprising the kinetic energy and the electron–electron interaction, the bottleneck being that the functional form of both terms is unknown. By introducing the kinetic energy $T_S[\rho_r]$ of a non-interacting reference state, Kohn and Sham [21] divided the functional $F_{HK}[\rho_0]$ into three parts:

$$F_{HK}[\rho_r] = T_S[\rho_r] + J[\rho_r] + E_{XC}[\rho_r] \tag{3.4}$$

Only $T_S[\rho_r]$ and the Coulomb term $J[\rho_r]$ are known. The art in DFT consists of finding the optimal form of the exchange-correlation term $E_{XC}[\rho_r]$ which contains the kinetic energy correction, the non-classical electrostatic, the correlation, and the self-interaction terms of the potential energy; hence, everything which is unknown.

Formally analogous to the HF SCF ansatz, starting from a non-interacting reference system and introducing Kohn–Sham orbitals, leads to a set of effective one-electron equations, the Kohn–Sham equations, which must be solved iteratively until self-consistency is reached.

The original restriction of the validity of DFT to ground states only could be removed by the introduction time-dependent density functional theory (TDDFT) [22]. This opens the way to excited states and the calculation of spectroscopic data [23], again at relatively low computational costs as compared to *ab initio* methods such as CC2 [24] or even CASPT2 [25], the multireference second-order perturbation theory based on a complete active space self-consistent field (CASSCF) wavefunction.

Even from the very cursory discussion given above, it becomes clear that it is not at all easy to improve the DFT ansatz systematically, and the choice of the "optimum" functional seems to be empirical. Although the different approaches to find the correct form of $E_{XC}[\rho_r]$ will not be discussed at this point, mention must be made that many of them lead to results comparable to MP2, albeit with an effort which is lower by two orders of magnitude.

Another important aspect concerns the incorporation of bulk effects by so-called "continuum models" (for an introduction, see Ref. [26]). Gas-phase chemistry differs from "real-life" chemistry in solution since, in most cases, the inclusion of explicit solvent molecules in the calculation is prohibitive due to the unfortunate scaling of the algorithms. Instead, solvent molecules are represented by a continuum characterized by its bulk dielectric constant. In polarized continuum models (PCMs) [27], reaction fields are induced by charges at the surface of the solvated molecule. With the advent of this and similar approaches such as the conductor-like screening models (COSMOs) [28] and their extension to real solvents (COSMO-RSs) [29], an important step towards a realistic description of molecules in solution has been achieved, without the need for vast amounts of computational resources.

3.3 Minimum Energy Structures and Potential Energy Surfaces

The three-dimensional (3-D) structure of a molecule is its most important property. Besides its equilibrium geometry, the conformational space that is accessible at room temperature is often required to understand its biological activity. This translates to the task of computing the potential energy hypersurface (PES) of a molecule with sufficient accuracy.

In general, force fields do a good job in scanning PESs in organic chemistry. Especially in the design of agrochemicals, however, in addition to carbon, hydrogen and oxygen, nitrogen, sulfur, and the halogens play a prominent role, and several crucial motifs are not well parameterized in force fields. This may lead to unreliable rotation barriers and even to incorrect assignments of minimum energy conformations.

This is demonstrated for the thioether bonds of a fungicidal dithiazole-dioxide [30] (Figure 3.1).

The torsional energy surface $E(\phi, \psi)$ of the two thioether bonds, where ϕ and ψ denote the rotation of the heterocycle and the phenyl ring, respectively, is shown for different levels of theory in Figure 3.2: the Merck force field MMFF94 [31–34], the semi-empirical MNDO ansatz [17, 35], MP2, and DFT-d [36, 37], an approach combining DFT with an empirical dispersion term.

Both, the force field and the semi-empirical PES are not only quantitatively but also qualitatively different from the *ab initio* MP2 and the DFT-d surfaces. According to MMFF94, both rings are rather free to rotate, MNDO finds an almost free rotation of the heterocycle and barriers for the phenyl rotation, while MP2 and DFT-d locate different minima and pronounced barriers. If an ether had been investigated rather than a thioether, only the MNDO-surface would have been qualitatively wrong.

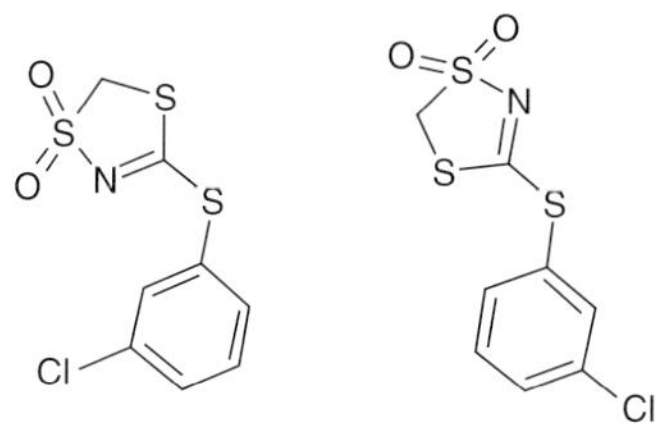

Figure 3.1 Thioether bonds in dithiazole-dioxide as templates for torsional energy surfaces.

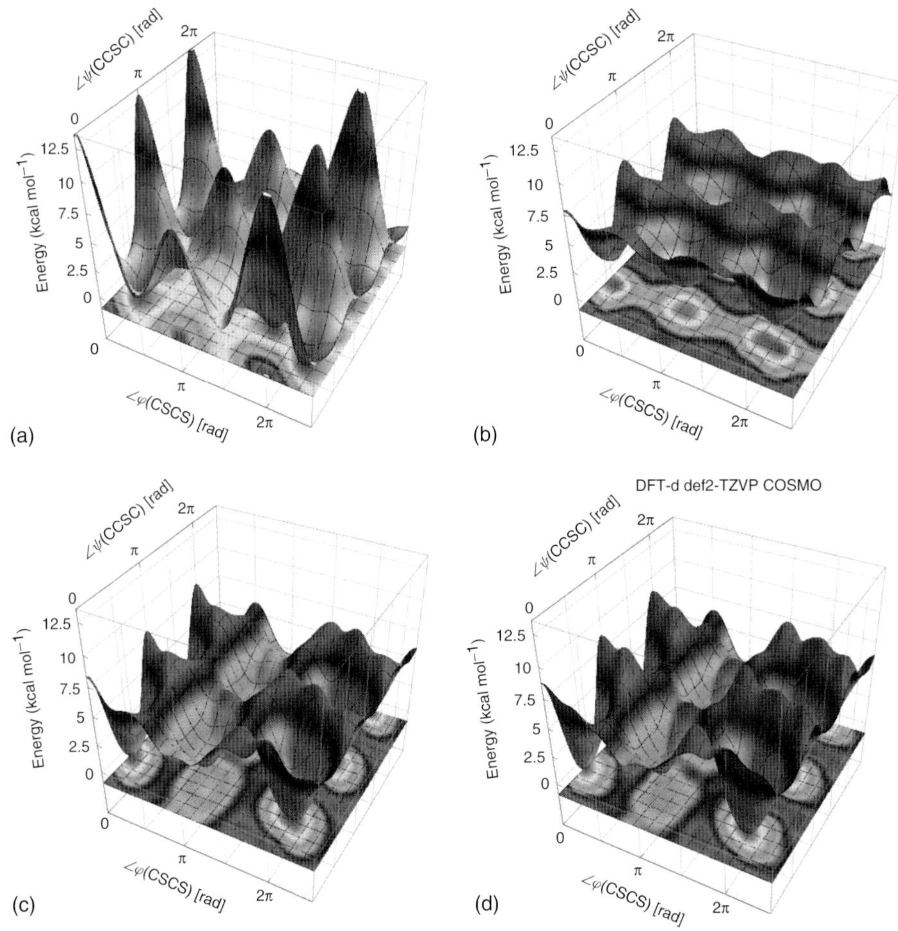

Figure 3.2 Torsional PES of thioether bonds, obtained at different levels of theory. (a) MMFF94S forcefield; (b) MNDO; (c) MP2//def2-TZVPP; (d) DFT-d/COSMO//def2-TZVP.

Whilst rotations about single bonds pose no problem to DFT, rotations about double bonds are beyond the scope of simple quantum chemical methods; rather, this is the domain of high-level methods such as CCSD(T).

The question of possible transition between *E*- and *Z*-isomers is just such a type of problem, and common to the neonicotinoid class of insecticides [38]. When taking clothianidin as an example, the objective was to identify the minimum energy path between the experimentally known *E*-conformer and a proposed *Z*-conformation (see Figure 3.3) by scanning its PES.

As data for the barrier of rotation of the formal C=N double bond in imines or oximes were missing, this was initially estimated to be of the order of an amide bond, ~20 kcal mol^{-1}, as compared to ~65 kcal mol^{-1} of an isolated C=C bond. In order

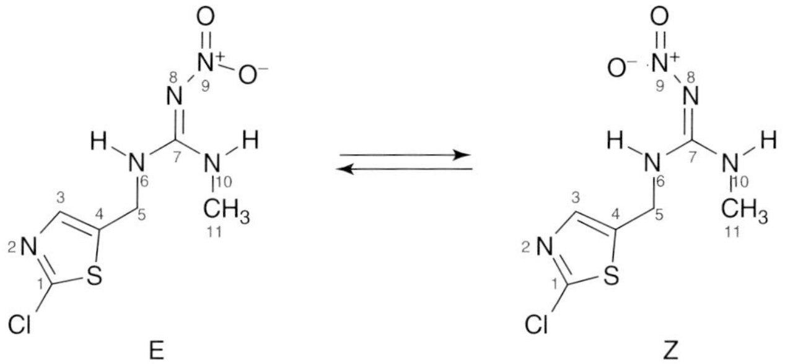

Figure 3.3 The *E*- and proposed *Z*-isomers of clothianidin.

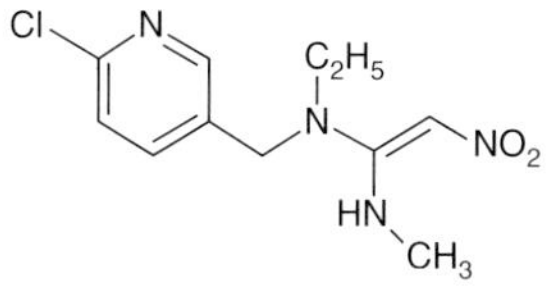

Figure 3.4 Structure of nitenpyram.

to check whether methods such as MP2 or DFT could be applied, CCSD(T) reference calculations were made for the model system nitroguanidine at MP2/def2-TZVPP optimized geometries, with planar and perpendicular NO_2-groups representing the ground-states and transition-states (TSTs).

Single-point CCSD(T)/def2-QZVPP and CCSD(T)/aug-cc-pVXZ (X = T,Q) calculations revealed energy differences of $16.7 \pm 0.1{-}0.2\ \text{kcal mol}^{-1}$. This indicates that scanning the PES of clothianidin using DFT would lead to reliable results. In fact, in the *N*-nitro-guanidine moiety the electrons are delocalized and the formal C=N double bond is actually only a resonance-stabilized single bond. Hence, a discrimination between single and double bonds is not possible here. However, substituting the "doubly bound" N by CH, as seen in the nitromethylene insecticide nitenpyram [38] (Figure 3.4), reveals much larger energy differences ($\sim$30 kcal mol^{-1}) which, depending on the method used, vary by 1.0–2.0 kcal mol^{-1}, indicating that in this case computations at the DFT- or MP2-levels would be borderline.

Hence, a scan of the PES of clothianidin in the gas phase and in water was performed by optimizing the molecule at all $18^3 = 5832$ combinations of the three fixed torsional angles $\alpha(\text{RNHC=NNO}_2)$, $\beta(\text{RNH-C(NNO}_2))$, and $\gamma(\text{RNHC-NHCH}_3)$ from 0° to 340° in steps of 20°.

Out of several possible sequences of rotations leading from *E*- to *Z*-clothianidin with barrier heights of similar magnitude, the minimum energy path in water, which is slightly preferred to other possibilities, starts with a rotation of the CH_2-thiazolyl moiety as a first step, followed by rotation of the C=N "double bond," and finally by the rotation of N-CH_3 as a third step. This is different from the gas phase, where the order of the steps is reversed, as the C=N rotation is easier (16 versus 21 kcal mol^{-1}) with the N-CH_3-group already rotated.

A graphical illustration of the four-dimensional dataset $E(\alpha, \beta, \gamma)$ is given in Figure 3.5. The first three dimensions are the three torsion angles (in radians, i.e.,

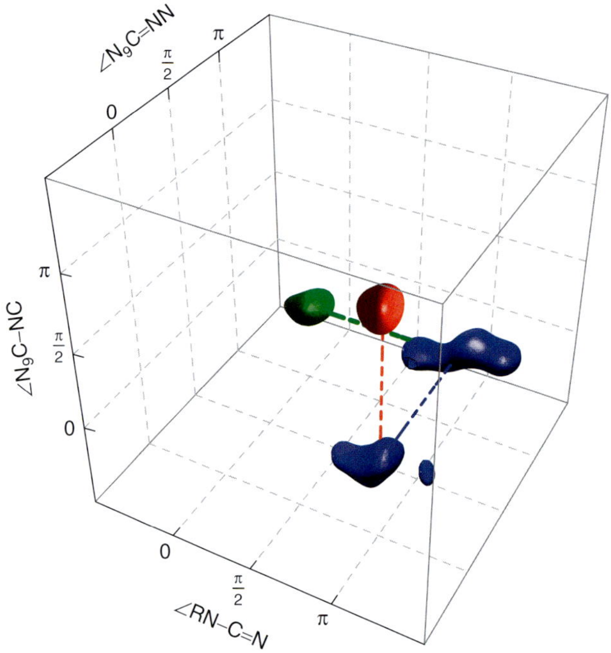

Figure 3.5 Four-dimensional torsional PES E(α, β, γ) of the clothianidin *E*- to *Z*-transition. *E*-region: green, *Z*-region: red, local minima: blue (DFT-d/COSMO).

$0^\circ = 0$, $180^\circ = \pi$, and $360^\circ = 2\pi$), while the fourth dimension is the color of the iso-energy levels. They are chosen at 1 kcal mol^{-1} (green), 3 kcal mol^{-1} (red), and 6 kcal mol^{-1} (blue), with additional local minima being omitted for clarity. Hence, it is possible to locate regions on the torsion-space PES relating to *E*-clothianidin (green), *Z*-clothianidin (red), and intermediate energy minima (blue).

TSTs between the minima, confirmed by calculating the eigenvalues of the Hessian, are found on the dashed lines connecting them. Reported energies of all stationary points are corrected for zero-point-vibrations.

Figure 3.6 shows a simplified energy profile of the conformational changes along the *E*/*Z*-reaction coordinate. The barrier heights are roughly inversely proportional to the mean bond lengths: 1.356 Å (RHN-C), 1.364 Å (C=N), and 1.347 Å (C-$NHCH_3$).

In conclusion, the *Z*-isomer is ~2 kcal mol^{-1} above the absolute minimum-energy structure, and separated by a barrier of approximately 15 kcal mol^{-1} from the *E*-isomer. Whilst it should be possible to isolate the *Z*-isomer at very low temperatures, it should not be found under physiological conditions.

Of course, these types of investigation are not restricted to molecules in their ground state. Photochemical stability depends on the accessibility of low-lying excited states and the energetics and topology of states involved in the possible deactivation or degradation processes. TDDFT permits the structure elucidation of, for example, the first excited singlet S_1 and triplet T_0 states, whose energies

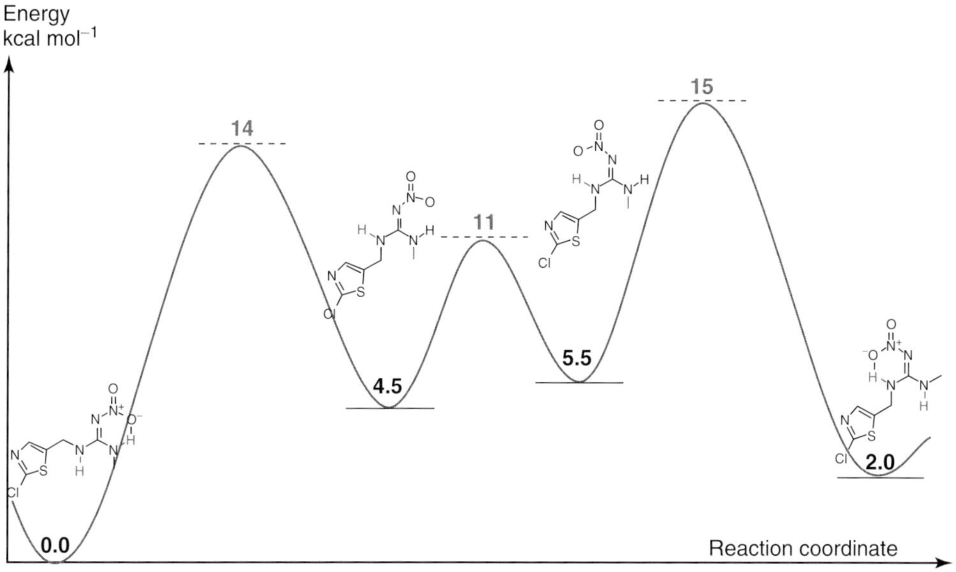

Figure 3.6 Simplified reaction path for the *E/Z* transition of clothianidin.

and structures often give hints to answer these questions. An example of this is deltamethrin, a potent pyrethroid insecticide, where TDDFT calculations can explain the observed photochemical interconversion of the eight isomers. The calculations correctly predict, for example, the observed [39] ester cleavage upon radiative excitation to S_1, followed by opening of the cyclopropyl ring, a 90° rotation of the Br-C-Br plane in T_0 or a loss of one bromine S_1 (Figure 3.7). On the other hand, however, it is clear that the photophysical properties of an active ingredient alone are less relevant, as significant changes in its behavior can be observed in formulations and inside biological membranes.

Another word of caution is on order: although, during the past 20 years, TDDFT has become a popular tool in computational photochemistry [40], there remain sufficient examples where there is a need to resort to (approximate) CC methods such as CCSD, CC2, or ADC(2) [41, 42], or even to (perturbative) multi-reference approaches, for example, CASPT2, to obtain reliable excited state energies. Regrettably, with the possible exception of resolution of identity (RI)-ADC(2), these approaches are far outside the routine applications in "real-life chemistry".

3.4
Physico-Chemical Properties

The intrinsic biological activity of a molecule is a necessary, but not a sufficient, condition for that molecule to become a potent agrochemical. Many factors – some of which are summarized under the acronym ADMET (adsorption, distribution,

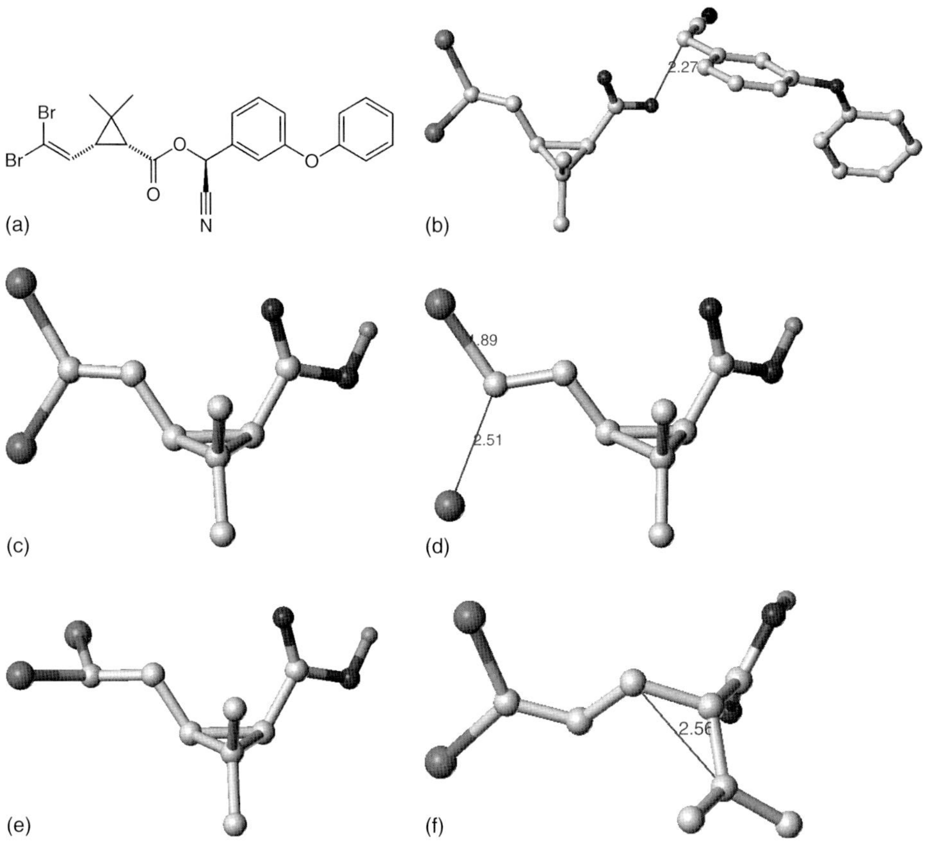

Figure 3.7 Computational photochemistry of deltamethrin-(1*R*3*R*αS): Ester cleavage upon excitation to S_1, followed by loss of Br in S_1 or 90° C=C rotation or cyclopropane ring opening in T_0 of fragment. (a) Deltamethrin structure; (b) S_1: ester cleavage; (c) Fragment S_0; (d) S_1: loss of Br; (e) T_0: C=C rotation; (f) T_0: cyclopropyl ring opening.

metabolism, excretion, toxicity) – can influence a molecule's fate before it reaches its point of action. Hence, it is of utmost importance to have reliable estimates of the underlying physico-chemical properties at hand. In addition to being valuable molecular descriptors as such, calculated properties are useful as quantum chemical descriptors in QSARs [43–45].

At this point, some examples are mentioned where the calculated properties provide the information that drives synthesis decisions. For example, the famous Kleier diagram [46] connects systemicity with dissociation state and lipophilicity, the preferred regions for nucleophilic or electrophilic attacks are identified by Fukui functions, the chemical reactivity in terms of hardness and softness is provided by DFT calculations, while hints towards photochemical stability can be taken from calculated ultraviolet spectra, and the absolute configuration of chiral molecules can be assigned by their optical rotatory dispersion (ORD) spectra.

The starting points for all calculations are reliable molecular geometries and energies. These are vital for property calculations, as all molecular properties are the responses of a molecule to external or internal perturbations such as electromagnetic fields or nuclear moments. Thus, they can be defined as derivatives of the molecular energy with respect to these perturbations.

Stationary points on the potential energy surface (minima, maxima, TSTs) are characterized by the eigenvalues of the Hessian; that is, the matrix of the second derivatives of the energy with respect to atomic positions. Infra-red spectra then visualize the eigenvectors of the Hessian.

Electric properties such as dipole- and higher multipole moments or polarizabilities depend on external electric fields $\vec{E} = -\vec{\nabla}\Phi$, while magnetic properties such as magnetic susceptibilities or nuclear magnetic shieldings depend either on external magnetic fields $\vec{B} = \vec{\nabla} \times \vec{A}$ or on fields arising from the atomic nuclear moments. The external fields appear in the Schrödinger equation by adding the corresponding potentials Φ or $\vec{A}$ to the Hamiltonian, as Φ and $\vec{A}$ have the proper dimension of an energy.

The photostability of molecules often depends on the accessibility and stability of their excited states. Although the validity of the Born–Oppenheimer approximation must be carefully checked when investigating questions of this type, often it is sufficient just to know the structure and energy of the lowest excited singlet and triplet states. Ultraviolet, circular dichroism (CD), and ORD spectra also require knowledge relating to the excited states of the molecules, which can be obtained with a modest computational effort as response properties from TDDFT instead of solving the Schrödinger equations for the excited states using coupled cluster CC2 [47] or more advanced methods.

Conceptual DFT [48] provides the link to well-known chemical concepts of reactivity descriptors such as softness and hardness, electronegativity, and electron affinity. While these properties indicate how reactive (parts of) the molecules are, Fukui functions describe where within the molecules potential changes occur; that is, they evaluate variations of the electron density with the number of electrons.

Last, but not least, bulk properties that are related to equilibrium thermodynamics such as dissociation constants (pK_a), lipophilicities ($\log P$), or Henry constants, which require Boltzmann averages over sets of conformations, are accessible if reliable conformational energies are available.

3.4.1
Electrostatic Potential, Fukui Functions, and Frontier Orbitals

The molecular electrostatic potential

$$\Phi(\mathbf{r}) = \sum_{A=1}^{M} \frac{Z_A}{|\mathbf{R}_A - \mathbf{r}|} - \int d\mathbf{r}' \frac{\rho(\mathbf{r})}{|\mathbf{r}' - \mathbf{r}|} \tag{3.5}$$

is the sum of a nuclear and an electronic contribution, has its origin in the unsymmetrical charge distribution within a molecule, and is responsible for

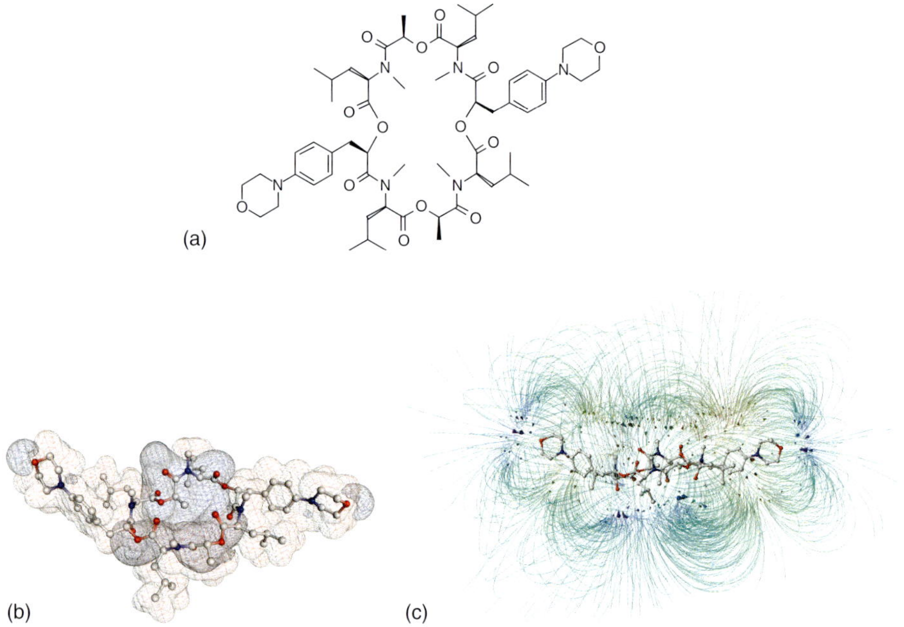

Figure 3.8 Emodepside: electrostatic potential shown in different representations. (a) Structure of emodepside; (b) ESP: attractive (blue) and repulsive (red) regions; (c) ESP: field line representation.

molecule–molecule recognition at large distances, while van der Waals interactions at short distances arise from overlapping electron densities.

The acaricidal and nematicidal compound emodepside [49], which is mainly used as a potent anthelminticum against gastrointestinal nematodes [50–53], may serve as an example: $\Phi(\mathbf{r})$ of emodepside is contoured at equivalent attractive and repulsive levels in Figure 3.8. It consists of a core which is attractive for positively charged probes, and a repulsive periphery with attractive spots at the two morpholine moieties.

A fundamental property is the molecular electron density ρ, which describes the shape and size of a molecule. Changes in the electron density within a molecule upon varying the number of electrons resulting from electrophilic or nucleophilic attacks, are reflected in the Fukui functions, $f^+(\mathbf{r})$ and $f^-(\mathbf{r})$ [54, 55], which are defined by

$$f^{\pm}(\mathbf{r}) = \left(\frac{\partial}{\partial N}\rho(\mathbf{r})\right)^{\pm}_{V(\mathbf{r})} \tag{3.6}$$

Practically, the Fukui function is calculated by finite differences:

$$\begin{aligned} f^+(\mathbf{r}) &\approx \rho(N+1,\mathbf{r}) - \rho(N,\mathbf{r}) \\ f^-(\mathbf{r}) &\approx \rho(N,\mathbf{r}) - \rho(N-1,\mathbf{r}) \end{aligned} \tag{3.7}$$

with all densities evaluated at a fixed external potential $V(\mathbf{r})$.

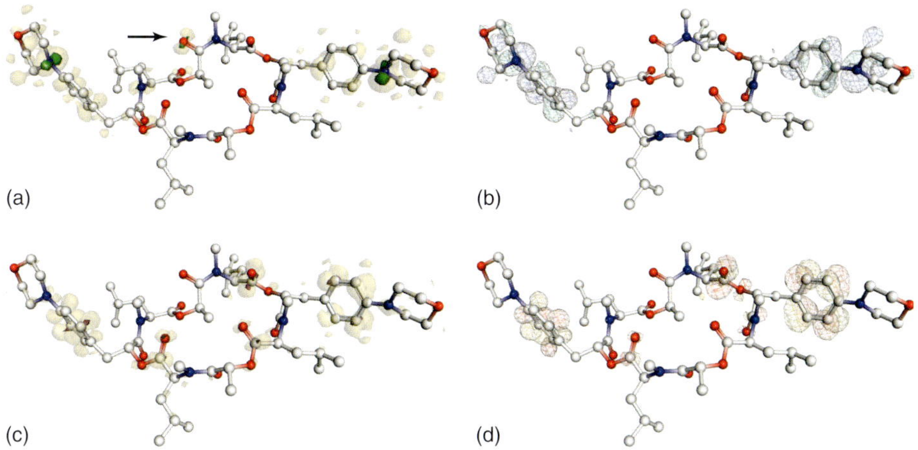

Figure 3.9 Emodepside: Fukui functions and frontier orbitals. (a) f_{elec}; (b) HOMO and HOMO-1; (c) f_{nuc}; (d) LUMO and LUMO + 1.

The maxima of $f^{+} = f_{nuc}$ and $f^{-} = f_{elec}$ correspond to regions in the molecule which are susceptible to attack by a nucleophile or an electrophile, respectively. In other words, f^{+} indicates where increasing ρ is energetically favorable, while f^{-} is maximal where lowering ρ is preferred. Frontier orbital densities $\rho^{HOMO-n}(\mathbf{r})$ or $\rho^{LUMO+m}(\mathbf{r})$ can be considered as frozen orbital approximations to $f^{\pm}(\mathbf{r})$.

In Figure 3.9 the Fukui functions f_{elec} and f_{nuc} are compared to the frontier orbitals. It is clear that the combination of HOMO and HOMO-1 in the morpholine parts is an approximation to f_{elec}, and that the combination of LUMO+1 and LUMO in the phenyl parts mimics f_{nuc} of emodepside [56, 57]. However, one important detail of the electrophilic Fukui function is missing in these approximations, namely the susceptibility of the carbonyl-oxygen of the *cis*-amide bond to electrophilic attacks. Indeed, it is precisely this oxygen which is selectively thionated in the synthesis of thioamide analogs [58].

3.4.2
Magnetic Properties

The electronic and nuclear magnetic moments of a molecule can interact with each other, and also with external magnetic fields. While the various spin-interactions lead to spin-spin couplings of nuclear magnetic resonance (NMR) and electron spin resonance (ESR) spectroscopy, magnetic susceptibilities χ, and chemical shieldings σ arise from interactions with external magnetic fields. Due to the problems of accurately describing the spin density at the positions of the nuclei, reliable spin coupling constants are extremely difficult to calculate, and will not be discussed at this point.

The NMR shieldings σ are defined as mixed derivatives of E with respect to the external magnetic field $\vec{B}$ and the atomic nuclear moments $\vec{\mu}_{Nk}$, and are calculated relative to the shieldings of the bare nuclei while experimental shifts δ are recorded relative to a reference. Conversion from the absolute to relative shift scales is achieved by subtracting the calculated shielding values from the absolute shieldings of reference molecules $\delta(k) = \sigma_{ref} - \sigma(k)$, calculated at the same level of theory [tetramethylsilane (TMS): ^{1}H, ^{13}C, ^{29}Si; H_2O: ^{17}O; NH_3: ^{14,15}N; FH: ^{19}F; H_2S: ^{33}S].

The calculation of magnetic properties is complicated by the so-called "gauge problem." The vector potential $\vec{A} = 1/2\vec{B} \times \vec{r}$ of the magnetic field is only determined up to the gradient of a scalar function. The requirement that calculated data must be independent of gauge transformations (in particular, of the choice of the gauge origin $\vec{r}$) is fulfilled only in the limit of complete basis sets; this means in practical terms that the calculated NMR data will depend on the choice of $\vec{A}$. To resolve this problem, distributed gauge origins have been introduced, the most popular being GIAOs (gauge including atomic orbitals) [59] or IGLOs (individual gauges for localized molecular orbitals) [60]. A comprehensive overview on the calculation of NMR and EPR parameters is given in [61].

In NMR spectroscopy, solvent effects play a role for all atoms at the molecular surfaces. In particular, proton NMR shifts are sensitive to solvents, and pronounced effects are observed for those protons that are capable of building hydrogen bonds.

As an example, consider the *E*/*Z*-conformations of clothianidin (c.f. Figure 3.3). Here, NMR-shielding calculations at the MP2-GIAO/def2-TZVP level of theory for the DFT-optimized *E*- and *Z*-geometries revealed that representing the solvent by a continuum model was insufficient. It transpired that the inclusion of one explicit solvent molecule (dimethylsulfoxide or acetonitrile), interacting with the hydrogen atom not involved in the internal H-bond, was essential. The addition of a second explicit solvent molecule had almost no effect, but led to tremendously increased CPU-times. The averaged shieldings of the methyl protons, which give separate signals in the calculation, are shown in Table 3.1, whereas the signals of the methylene protons (which also are equivalent on the NMR timescale) differ much more and are listed individually.

Initiated by the calculations, the *Z*-form of clothianidin was detected for the first time by state-of-the-art low-temperature experiments [62]. The predicted shift inversion of the NH protons resulting from internal H-bonds unambiguously differentiates between the *E*- and *Z*-isomers of clothianidin and was nicely confirmed. However, the low sensitivity of carbon and nitrogen did not permit the identification of corresponding data for these nuclei experimentally. Calculations which include an acetonitrile molecule are closest to the observed data; these describe the drastic changes in the NH proton shifts (experimental: 3 ppm, calculated: 4 ppm; see Table 3.1) between the *E*- and *Z*-forms. At room temperature only the *E*-isomer can be identified [63].

In general, high-quality calculations of magnetic properties are still too costly to be performed on a routine basis. However, for some challenging analytical questions, the combined efforts of theory and experiment can lead to convincing solutions of the problems.

Table 3.1 ^{1}H-NMR spectra of *E*- and *Z*-clothianidin, experimental [62, 63] and calculated data.

Atom	δ_H [62]	δ_H [63] (CD_3OD)	δ_H [63] ($CDCl_3$)	δ_{Hcalc} (gas)	δ_{Hcalc} (ACN)	δ_{Hcalc} (DMSO)
E-conformer						
H_3	7.50	7.54	7.46	7.71	7.62	7.68
H_5	4.51	4.61	4.67	4.20, 5.46	4.38, 5.60	4.38, 5.57
H_{11}	2.81	2.93	2.97	4.35	3.09	3.08
NH_6	6.47	–	5.1 to 4.9	4.35	6.56	7.96
NH_{10}	9.18	–	9.6 to 9.4	10.01	10.26	10.26
Z-conformer						
H_3	7.56	–	–	7.76	7.95	8.36
H_5	4.55	–	–	4.77, 4.53	4.53, 4.84	4.38, 5.18
H_{11}	2.73	–	–	2.99	2.91	2.91
NH_6	9.67	–	–	10.84	10.87	10.79
NH_{10}	5.97	–	–	3.97	5.94	7.96

3.4.3 pK_a Values

From the law of mass action, the equilibrium constant of a dissociation reaction can be written as

$$\mathrm{p}K_a^i = \frac{1}{2.303RT}(\Delta G_{neutral}^i - \Delta G_{ion}^i) \tag{3.8}$$

The calculation of pK_a values amounts to estimating the equilibrium constant for an ensemble of molecular conformations in a solvent via a Boltzmann averaging of their free energy contributions. This is by no means a trivial task [64], and a recent survey [65] reviewed the current methods for predicting pK_as of proteins and small molecules. Referring to *ab initio* methods, it is stated that continuum solvation model-based approaches reach levels of accuracy that are unavailable to empirical methods. An internal evaluation using inhouse data at Bayer CropScience corroborated this finding [66]. Boltzmann ensembles of all tautomers of all neutral and singly charged species are generated in a multistep procedure, starting with tautomer creation, followed by the Monte Carlo generation of conformers and DFT/COSMO-RS optimization of unique structures, leading to an estimate of ΔG which is used in the final QSAR formula for the pK_a-values [67]

$$\mathrm{p}K_a^i = c_0 + c_1(\Delta G_{neutral}^i - \Delta G_{ion}^i) \tag{3.9}$$

where $c_0 = -148.9$ and $c_1 = 0.5481$ for BP-TZVP-COSMO calculations. Equation (3.9) differs from the definition equation (3.8), indicating that the recipe for the calculation of pK_a values could be improved.

From Figure 3.10, which shows p$K_{a(acid)}$ (upper row) and p$K_{a(base)}$ (lower row), it becomes clear that prediction methods such as ACD [68] and EPIK [69], which

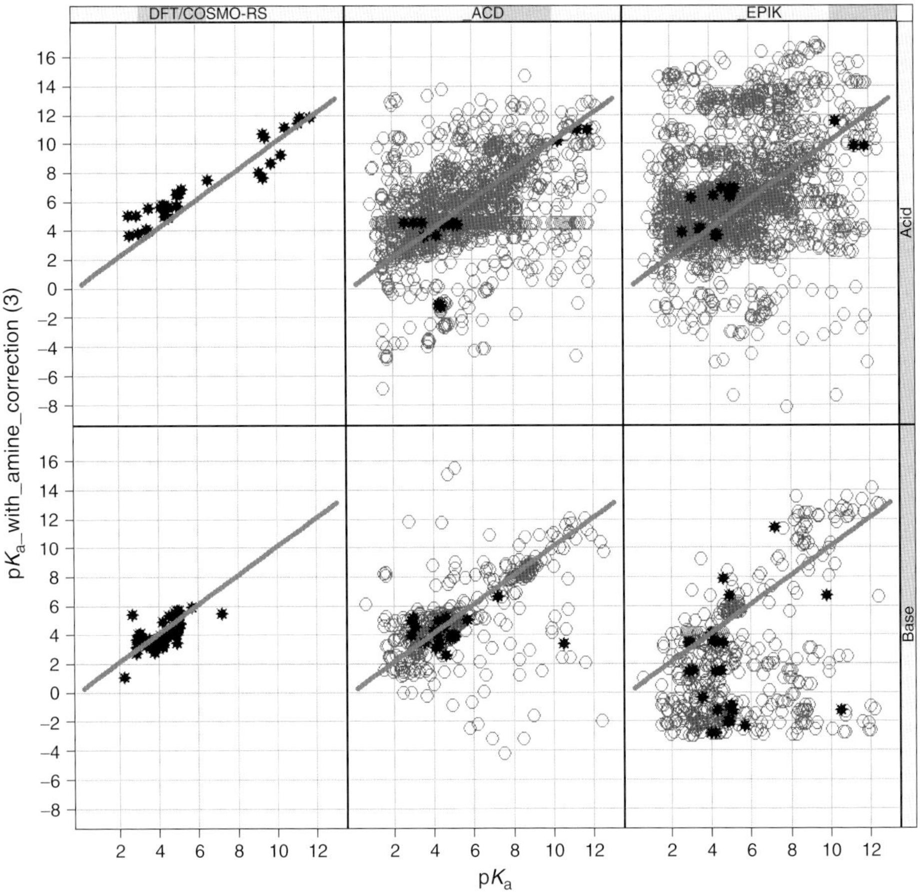

Figure 3.10 Comparison of pK_a-prediction methods: DFT/COSMO-RS, ACD, and EPIK.

are trained on external sets of data, have difficulties with proprietary compound classes, documented in particular by the suspicious ACD-bar at p$K_{a(acid)} = 4.3$. Although unbiased methods such as COSMO-RS perform quite well, there is still some room for improvement, especially for secondary and tertiary amines.

At this point, the performance of the COSMO-RS approach is documented for a set of sulfonylurea herbicides inhibiting acetolactate synthase (ALS) [70] (see Table 3.2).

Bearing in mind that the calculations provide the first dissociation step of some diprotic compounds, the agreement between calculation and measurements is satisfactory. The prediction of pK_as for several classes of amines using a PCM model for the solvent showed maximum errors of ± 1.5pK_a units [71]; predictions by *ab initio* G3(MP2) methods of gas- and solution-phase acidities [72] of strong Brønsted acids revealed a split picture for gas-phase acidities, an excellent agreement for highly acidic ($\geq$304 kcal mol^{-1}), and large errors for less

Table 3.2 Calculated (COSMO-RS) and observed (see Ref. [70]) pK_a-values of sulfonylureas.

Name	$pK_{a\,(acid)\,calc}$	$pK_{a\,obs}$	Comment
Azimsulfuron	4.1	3.6	–
Cyclosulfamuron	4.4	5.0	–
Ethoxysulfuron	2.8	5.3	–
Flucetosulfuron	3.5	3.5	–
Flupyrsulfuron-methyl	5.5	4.9	–
Foramsulfuron	5.6	4.6	–
Iodosulfuron-methyl	3.4	3.2	–
Mesosulfuron-methyl	4.2	4.4	–
Orthosulfamuron	4.3	−1.4;0.7;3.5;9.6;11.5	Degradation
Oxasulfuron	5.6	5.1	–
Trifloxysulfuron	4.7	4.8	–
Tritosulfuron	5.3	4.7	–

acidic ($\leq$302 kcal mol^{-1}) molecules, while the errors were typically within ± 2 pK_a units for aqueous solutions. Similar ranges of errors are obtained by alternatives to QM-continuum models such as density functional-based molecular dynamics (DFMD) calculations (c.f. Ref. [73] and references cited therein).

3.4.4
Solvation Free Energies

The prediction of solubilities is a challenging task, the best indication of which is an absence of methods that are able reliably to predict the water-solubility of drugs – with the possible exception of sets of homologous compounds. Crystallization effects bring an additional complexity to the thermodynamic cycles involved. Hence, the calculation of free energies of solvation ΔG_S, which possibly includes phase transitions between crystallinic-, liquid-, and gas-phases, is at least as demanding as the calculation of pK_a-values. Even if melting processes are excluded, the calculation of pure gas-to-liquid equilibria remains difficult enough. Henry's law connects the concentration c of a solute in a solvent with its partial pressure p above the solution, $p = k_H c$. Henry's constant $k_H(T)$ is hence a – temperature dependent – quantitative measure of a compound's volatility, and can be calculated within the COSMO-RS framework.

Experimental free energies of solvation are rather rare, despite their importance in calculating binding energies and the like. The results of a recent blind challenge, based on a list of 63 crop protection compounds, showed encouraging results however [74], when root mean square errors of 2.76 kcal mol^{-1} for the *ab initio* results [75] indicated that there is still room for improvement. Prominent examples of other "solvation" equilibria are estimates of the octanol–water partition coefficients (log *P*).

3.4.5
Absolute Configuration of Chiral Molecules

Chiroptical properties describe the molecular responses to electromagnetic fields. In the case of chiral molecules, they can be used to assign absolute configurations of isomers. Before the advent of TDDFT, their calculation had been restricted to rather small model systems; however, TDDFT subsequently established itself as a reliable tool for the computation of linear response properties also of larger molecules. The process permitted not only the calculation of excitations via absorption of radiation, the UV-spectra, but also the perturbations of the electric or magnetic dipole moments by time-dependent magnetic and electric fields (i.e., CD- and ORD-spectra). TDDFT can, therefore, be used almost routinely to assign the absolute configuration of chiral molecules in cases where X-ray investigations are either impossible or too costly. Examples of the performance of TDDFT with respect to chiroptical property calculations are provided in Refs [76, 77]. The latter review focuses in particular on post-2005 reports, and provides numerous examples of TDDFT calculations on drug-like molecules.

In order to obtain reliable assignments for non-rigid, larger drug-like molecules, it is advisable to start with a conformation analysis and to include contributions from low-energy conformations according to their Boltzmann weights. In some cases, solvent effects which might even change the protonation state of a molecule, may require the inclusion of explicit solvents, as has been shown for NMR calculations.

3.5
Quantitative Structure-Activity Relationships

Since the early days of Hammett [78], Hansch and Fujita [79], and Free and Wilson [80], thousands of QSARs have been reported in drug research. The majority of these have used linear free-energy relationships (LFERs) that relate the negative logarithm of a concentration, $log1/c$, which is linearly connected to the free energy of the reaction, ΔG, to physico-chemical parameters such as lipophilicity (log P) or electronic effects (Hammett's electronic parameter σ):

$$log1/c = k_1 logP + k_2 (logP)^2 + k_3\sigma + \ldots \quad (3.10)$$

The coefficients k_i are determined by linear multiple regression analysis. An advantage of this approach is the direct physical meaning of the descriptors, and hence the straightforward interpretation of the model terms. However, as many of the descriptors are intercorrelated, considerable ambiguity remains which should be eliminated by a principal component analysis (PCA).

The 3-D-structures of drugs are used in approaches such as comparative molecular field analysis (CoMFA) [81] or the comparative molecular similarity approach (CoMSIA) [82]. Interaction fields of molecules with probe atoms on regular grids form the basis for setting up QSARs [83], thereby eliminating the bottleneck of missing parameters, which is present in the Hansch and Free–Wilson approaches, especially for the prediction of new compounds. Whether implicitly or explicitly, it is assumed that the compounds under investigation bind to the same target and do so in at least similar ways, though the validity of such an assumption should be checked very carefully.

As compared to a 2-D-QSAR, some additional decisions must be taken prior to setting up a CoMFA, the most important being the choice of the "active conformation" of a molecule, a superposition rule by which the set of molecules will be aligned, the types of interaction fields, and the grid size of the computational box.

A typical CoMFA or CoMSIA equation constitutes a linear model in the field values at all grid points of the computational box with grid-spacings of 1–2 Å.

$$\log 1/c_i = \sum_k^{grid} a_k S_{ik} + \sum_k^{grid} b_k E_{ik} + \sum_k^{grid} c_k H_{ik} + \ldots \qquad (3.11)$$

Partial least squares (PLS; also projection to latent structures) analysis [84], or variants thereof [85], are used to solve these equations relating very few independent activity variables y to $O(10^4)$ descriptor variables x. The interpretation of the PLS-model is mainly achieved visually by the inspection of contour maps.

3.5.1
Property Fields, Wavelets, and Multi-Resolution Analysis

Instead of the rather arbitrary construction of the CoMFA and COMSIA interaction fields, QM-3-D approaches such as WAVE3D [86] use molecular property fields [54, 55, 87–89], thus removing any ambiguities concerning probe atoms or assumptions of locally isotropic fields. The inclusion of quadratic or higher-order terms of the property fields is simple, and the property fields are intuitively interpretable, which is important for their use as QSAR descriptors.

The most important fields are the molecular electron density ρ, the Fukui functions $f^{\pm}(\mathbf{r})$, and the electrostatic potential Φ. Of minor importance are the first and second derivatives of the electron density, its gradient $\nabla\rho(\mathbf{r})$, and its Laplacian $\nabla^2\rho(\mathbf{r})$.

Frontier orbital densities $\rho^{HOMO-n}(\mathbf{r})$ or $\rho^{LUMO+m}(\mathbf{r})$ as reactivity descriptors [90–92] have already been applied on a semi-empirical level in a herbicidal 3-D-QSAR study [93].

In CoMFA and COMSIA studies, low-resolution grids often lead to better models [94]. For property fields – and especially those involving spatial derivatives or derivatives with respect to electron number – finer sampling is essential, although a higher grid resolution leads to a cubic increase in the effort for the learning algorithm. It is for this is reason that these PLS models are generated in wavelet space [86]. Recently, a detailed analysis of the application of discrete wavelet transform/multi-resolution analysis (DWT/MRA) to conventional GRID interaction fields [95, 96] for 3-D-QSAR studies, has been described [97].

After applying DWT/MRA to a molecular field on a 3-D grid, its variance is focused on relatively few coefficients in the wavelet domain. Almost without information loss, it is possible to perform model building only on the relevant wavelet coefficients. For a typical training set, and an original grid size of $64^3 = 262144$ grid points, less than 1–5% of the original grid points are relevant. The standard NIPALS algorithm [98] benefits from the compression of the signal space, thus allowing the generation of many models for high-resolution grids without any compromise with respect to model quality.

The strength of CoMFA and other 3-D-QSAR methods, namely the ability to visualize and interpret the resulting models directly, is not lost when the model is built in a wavelet domain [99].

The application of DWTs is not restricted to PLS modeling. Rather, it significantly speeds up the calculation of quantum similarities such as the Carbo index, which is invariant to wavelet transform. The Carbo index C_{ab} for two fields $\sigma_a(\mathbf{r})$ and $\sigma_{ab}(\mathbf{r})$ ranges from 0 (no common features) to 1 (identical fields):

$$C_{ab} = \frac{\langle a|b\rangle}{\sqrt{\langle a|a\rangle\langle b|b\rangle}} \tag{3.12}$$

where $\langle a|b\rangle = \int d\mathbf{r}\sigma_a(\mathbf{r})\sigma_{ab}(\mathbf{r})$.

The sparsity of the wavelet representation makes the numerical evaluation of such integrals highly efficient, and allows the calculation of full similarity matrices for large training sets at high grid resolutions.

A QSAR model for N molecules is characterized by the number of latent variables (LVs), by the quality of fit, given by r^2 and the standard error s, and by the quality of prediction q^2, as defined by the predictive residual sum of squares (PRESS) and the sum of squared residuals (SS), and by the standard error of prediction s_{press}, obtained by the leave-one-out procedure. Occasionally, the cross-validated root mean square error (RMSE) for the test set is given, obtained from a leave-one-out or leave-several-out procedure with max(10, $N/10$) cross-validation groups.

$$r = \frac{\sum_i^N (x_i - \overline{x})(y_i - \overline{y})}{\sqrt{\sum_i^N (x_i - \overline{x})^2 \sum_i^N (y_i - \overline{y})^2}} \tag{3.13}$$

$$q^2 = 1 - \frac{\sum_i^N (y_{i_{pred}} - y_i)^2}{\sum_i^N (y_i - \overline{y})^2} \tag{3.14}$$

$$s_{press} = \sqrt{\sum_i^N (y_{i_{pred}} - y_i)^2/(N - LV - 1)} \tag{3.15}$$

$$RMSE = \sqrt{\sum_i^N (y_{i_{pred}} - y_i)^2/N} \tag{3.16}$$

As a rule of thumb, in small datasets of N molecules the number of LVs should not exceed $N/6$ to $N/5$.

In Ref. [86], the WAVE3D approach is applied to the well-known CoMFA-steroid dataset, which has become a benchmark set for new methods, and also to a more recent dataset from the crop science literature, to highlight different aspects of 3-D-QSARs.

In all WAVE3D examples the molecules have been optimized by RI-DFT/COSMO [28, 100], using the BP86 functional and def2-TZVP basis sets [101] as implemented in Turbomole V5.10 [10]. Property fields were calculated on rectangular grids of 64^3 grid points each, leading to grid spacings of 0.2–0.5 Å.

3.5.2
The CoMFA Steroid Dataset

CoMFA was introduced in an investigation of the binding affinities of a set of 31 steroids to human corticosteroid binding globulins (CBGs) and testosterone binding globulins (TBGs) [81], and has since become a reference standard for all new 3-D-QSAR methods. As steroids are rather rigid, alignment problems (which are critical to many other 3-D-models) play no role; however, a reinvestigation [102] emphasized that the correct choice of training and test sets would lead to a marked improvement in the quality of these models.

Two training sets of 21 molecules – those of Cramer and Kubinyi – were considered in the WAVE3D approach. The f^--based similarity matrix of the complete set is shown in Figure 3.11, where the color coding is such that the similarity between molecules increases from green via yellow to red. A comparison of the selected sets (left, Cramer; right, Kubinyi in Figure 3.11) reveals that more diverse selections are possible.

Cramer's training set gives an s_{press} of 0.66, Kubinyi's selection an s_{press} of 0.406, and WAVE3D models which use ρ and Φ only, have an s_{press} close to 0.3 25. The best WAVE3D model achieves an s_{press} of 0.195, but comprises different types of field

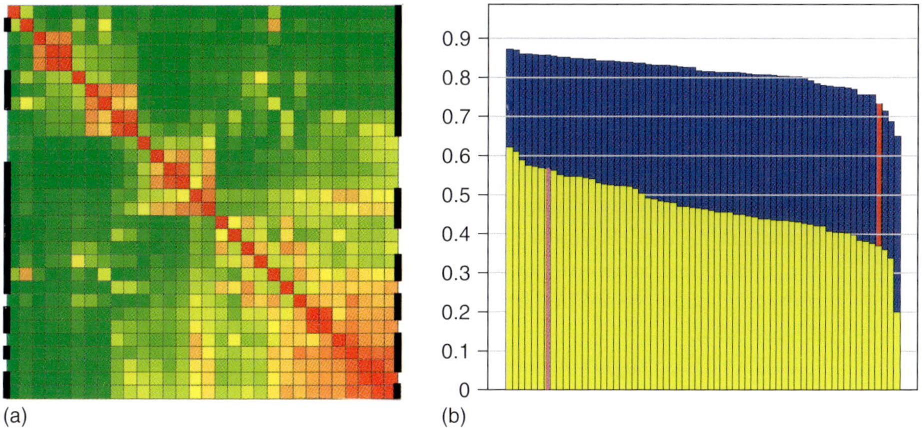

Figure 3.11 The steroid CoMFA dataset. Test set selection and model quality. (a) Training sets of Cramer (left bars) and Kubinyi (right bars), compared to the similarity matrix $C_{f_a^- f_b^-}$; (b) q^2 for 126 linear models, based on Cramer (yellow) and Kubinyi (blue) selections. Models using Φ and ρ only are marked in magenta and red.

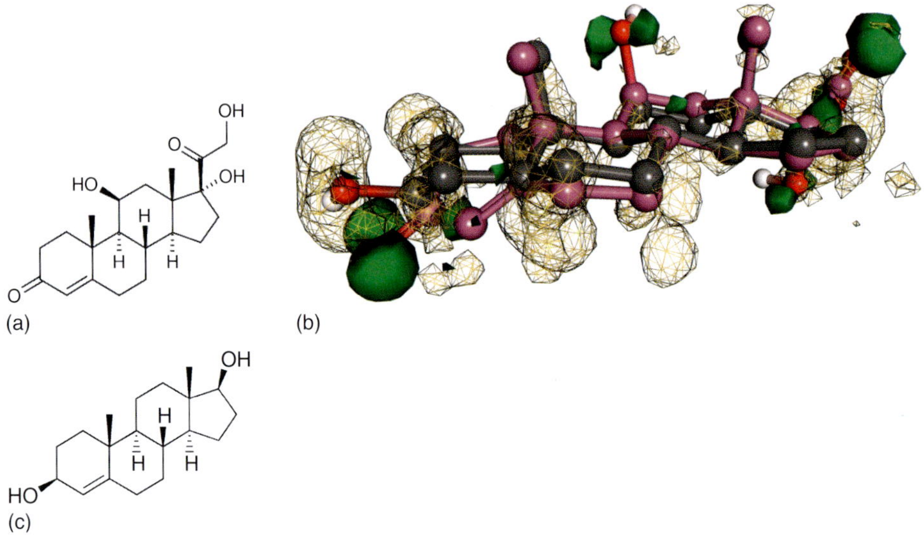

Figure 3.12 WAVE3D model of steroid CBG activity, based on f^-. Solid green surfaces and yellow meshes indicate favorable/unfavorable regions. Shown are cortisol (magenta) and androstenediol (black) as examples of active and inactive compounds.

not available to Refs [81] and [102]. Figure 3.12 shows the graphical analysis of a good model, based on f^- alone, with three LVs, $r^2 = 0.97$ and $q^2 = 0.87$. Favorable (solid) and unfavorable (grid) regions are mapped nicely to parts of cortisol (potent) and androstenediol (weak).

3.5.3
A Neonicotinoid Dataset

The second example of WAVE3D applications focuses on a set of 32 cyclic and acyclic chloronicotinyl insecticides (CNIs) [38] from a CoMFA study of the binding affinity to *Musca domestica* nicotinic actetylcholine receptors (*n*AChRs) [103]. The compounds were superimposed by a root mean square (RMS) fit of four key atoms to the reference conformation of imidacloprid, based on its X-ray structure. Three CoMFA models were derived, one each for the acyclic and cyclic compounds and a combined model (upper three lines in Table 3.3). Interestingly, one acyclic compound had to be omitted from the set to produce statistically significant results (Figure 3.13).

In WAVE3D, a slightly different superposition rule was used, such that different properties as similarity descriptor fields led to different groupings of the set (Figure 3.14). While all CNIs were similar with respect to their total electron density ρ, a much better discrimination could be found by considering their frontier orbitals only (ρ^{LUMO}) or the Fukui functions f_{elec}.

One CoMFA and $2^{10} - 1 = 1023$ WAVE3D models were built from combinations of ρ, $\nabla\rho$, ρ^{LUMO}, ρ^{LUMO+1}, Φ, their squares, and f_{elec} and f_{nuc}. None of the compounds

Table 3.3 Quality of 3-D-QSAR models for neonicotinoids.

Number of LVs	n	s	r^2	s_{press}	q^{2a}	RMSE	Method	Reference	Alignment
3	12	0.277	0.957	0.829	0.615	–	CoMFA 1.5 Å	[103]	4 pt model
3	19	0.451	0.901	0.857	0.643	–	CoMFA 1.5 Å	[103]	4 pt model
6	31	0.226	0.979	0.792	0.746	–	CoMFA 1.5 Å	[103]	4 pt model
3	32	0.550	0.863	0.968	0.574	–	CoMFA 1.0 Å	[86]	3 pt model
4	32	0.325	0.945	0.672	0.772	0.662	WAVE 3D	[86]	3 pt model
1	32	1.004	0.509	1.117	0.392	–	CoMFA 1.0 Å	[86]	Rotated
2	32	0.838	0.635	1.020	0.461	1.020	WAVE 3D	[86]	Rotated

[a]Cross-validation groups: CoMFA: n, WAVE3D: 10.

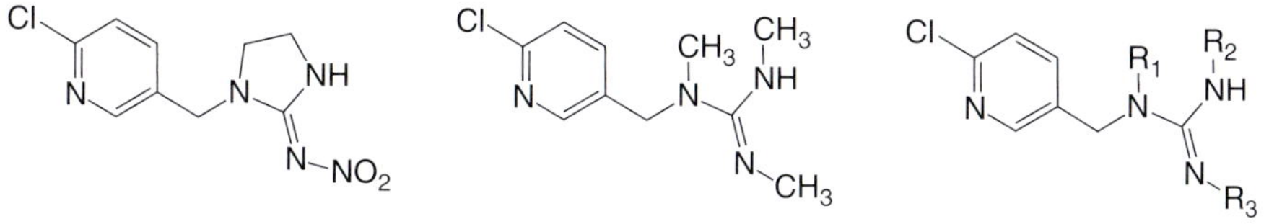

Figure 3.13 Cyclic and acyclic neonicotinic *n*AChR agonists (from Table 3.1 of Ref. [104]).

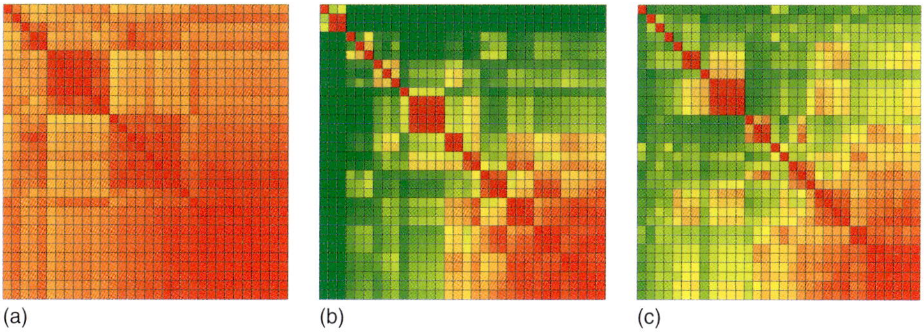

Figure 3.14 Similarity matrices of neonicotinoids with respect to different physico-chemical properties. (a) Total density ρ; (b) ρ^{LUMO}; (c) f_{elec}.

had to be excluded in any of the models, and the best models used three (CoMFA) or four (WAVE3D) LVs (instead of six, as noted in Ref. [103]), which in turn led to a considerable improvement in the stability of the QSARs. In total, 994 models had $q^2 \geq 0.65$, 448 had $q^2 \geq 0.70$, and 40 had $q^2 \geq 0.75$. Although CoMFA needed only three LVs, the statistical parameters of the best 40 WAVE3D models indicated their clearly superior performance (Table 3.3).

The optimum WAVE3D model (Figure 3.15) uses the electron density ρ and ρ^2, the Fukui function f_{elec}, and the LUMO density, while a four LV model with ρ

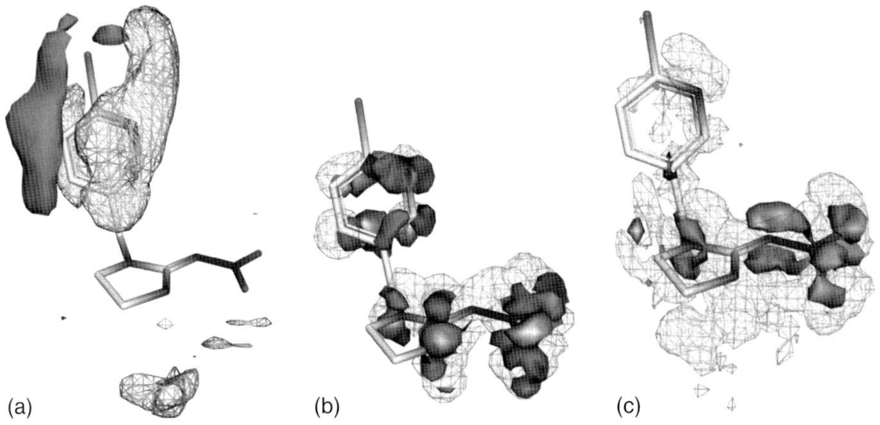

Figure 3.15 Contributions to a WAVE3D model for neonicotinoids: solid surfaces indicate favorable regions, and grids unfavorable regions. (a) Total density ρ; (b) ρ^{LUMO}; (c) f_{elec}.

and Φ, which is most easily compared with CoMFA fields, has a $q^2 = 0.585$ and an $s_{press} = 0.893$, where s_{press} is calculated using 10 cross-validation (CV) groups instead of 32 as in CoMFA.

The importance of choosing the "correct" superposition rules cannot be overestimated. In the example given here, a simple 180° rotation of the pyridyl rings in the set led to structures with energies that were almost indistinguishable from the first set of conformations. However, the resulting CoMFA and WAVE3D models (which are denoted as "rotated" in Table 3.3) were much worse than the original models.

In summary, the choice of alignment rule and the selection of a training set are both at least as important as the method used to generate the 3-D-QSAR model. Given these provisions, it is possible to generate better 3-D-QSAR models than previously thought possible, simply by using quantum chemistry-based molecular property fields instead of interaction fields. The computational disadvantages that result from a need to utilize much finer grids than in conventional approaches such as CoMFA, can be avoided by transforming to wavelet space. As a byproduct, large numbers of combinations of property fields in both linear and nonlinear models can be tested to identify the optimum model, without creating a heavy computational burden.

3.6 Outlook

The progress in computational hardware and methodological developments [104] that has been made within the past 20 years has extended the range of applications of *ab initio* and DFT methods to normal agrochemical-like molecules. The efficient electron density estimate and the (multipole accelerated) RI approach [105] have reduced demands for CPU time by up to two orders of magnitude as compared to conventional codes. The application of this approach to the explicitly correlated basis functions r_{12} or f_{12} has also reduced the computational burden by additional orders of magnitude, as significantly smaller f_{12} double-ζ basis sets are sufficient to achieve quadruple-ζ quality, and as the evaluation of the necessary integrals is only slightly more involved.

With regards to hardware, the past few years have witnessed the development of graphical processing units (GPUs) in quantum chemistry that offer the potential to accelerate computations by two orders of magnitude. In a few years time, this should permit CCSD(T) GPU calculations [106] for "real" agrochemicals to be performed routinely, assuming that theory is capable of providing efficient parallel algorithms [107] for this type of architecture.

Beyond handling larger sets of agrochemical-like molecules by quantum chemical methods (DFT and MP2), these developments will extend the applicability of computational quantum chemistry methods in two important directions:

- To investigate chemical reaction mechanisms by using high-level methods (CCSD(T) with f_{12} basis sets) [108]; that is, rather small systems at a level of theory reaching chemical accuracy.
- To tackle processes that proceed on time-scales and spatial dimensions which previously were beyond scope, including the solvation of drugs and their interaction with membranes; that is, extended systems at a medium level of theory.

Clearly, there is a long way to go before the quantum computational handling of "real life" chemistry in condensed media will be as reliable as it is today for gas-phase problems.

References

1. Schleyer, P.v.R., Allinger, N.L., Clark, T., Gasteiger, J., Kollman, P.A., Schaefer, H.F. III, and Schreiner, P.R. (eds) (1998) *Encyclopedia of Computational Chemistry*, vols **1–5**, John Wiley & Sons, Ltd, Chichester.
2. Lipkowitz, K.B. and Boyd, D.B. (1990–2011) *Reviews in Computational Chemistry*, vol. 1–27, Wiley-VCH Verlag GmbH.
3. Matta, C.F. (ed.) (2010) *Quantum Biochemistry*, Vols **1+2**, Wiley-VCH Verlag GmbH, Weinheim.
4. Leach, A.R. (2001) *Molecular Modelling, Principles and Applications*, 2nd edn, Pearson Education EMA.
5. Car, R. and Parrinello, M. (1985) *Phys. Rev. Lett.*, **55**, 2471–2474.
6. Marx, D. and Hutter, J. (2000) in *Ab initio Molecular Dynamics: Theory and Implementation*, NIC Series, Vol. 1 (ed. J. Grotendorst), Forschungszentrum, Jülich, pp. 301–449.
7. Warshel, A. and Levitt, M. (1976) *J. Mol. Biol.*, **103**, 227–249.
8. Chen, H., Lai, W., and Shaik, S. (2011) *J. Phys. Chem. B*, **115**, 1727–1742.
9. Frisch, M.J., Trucks, G.W., Schlegel, H.B., Scuseria, G.E., Robb, M.A., Cheeseman, J.R., Montgomery, J.A. Jr, Vreven, T., Kudin, K.N., Burant, J.C., Millam, J.M., Iyengar, S.S., Tomasi, J., Barone, V., Mennucci, B., Cossi, M., Scalmani, G., Rega, N., Petersson, G.A., Nakatsuji, H., Hada, M., Ehara, M., Toyota, K., Fukuda, R., Hasegawa, J., Ishida, M., Nakajima, T., Honda, Y., Kitao, O., Nakai, H., Klene, M., Li, X., Knox, J.E., Hratchian, H.P., Cross, J.B., Bakken, V., Adamo, C., Jaramillo, J.,

Gomperts, R., Stratmann, R.E., Yazyev, O., Austin, A.J., Cammi, R., Pomelli, C., Ochterski, J.W., Ayala, P.Y., Morokuma, K., Voth, G.A., Salvador, P., Dannenberg, J.J., Zakrzewski, V.G., Dapprich, S., Daniels, A.D., Strain, M.C., Farkas, O., Malick, D.K., Rabuck, A.D., Raghavachari, K., Foresman, J.B., Ortiz, J.V., Cui, Q., Baboul, A.G., Clifford, S., Cioslowski, J., Stefanov, B.B., Liu, G., Liashenko, A., Piskorz, P., Komaromi, I., Martin, R.L., Fox, D.J., Keith, T., Laham, M.A., Peng, C.Y., Nanayakkara, A., Challacombe, M., Gill, P.M.W., Johnson, B., Chen, W., Wong, M.W., Gonzalez, C., and Pople, J.A. (2004) *Gaussian 03, Revision C. 02*, Gaussian, Inc., Wallingford, CT.

10. Ahlrichs, R., Bär, M., Häser, M., Horn, H., and Kölmel, C. (1989) *Chem. Phys. Lett.*, **162**, 165, TURBOMOLE V6. 3, 2011. Available at: *http://www.turbomole.com*.
11. Neese, F. (2008) *ORCA, an Ab initio, DFT and Semiempirical SCF-MO Package*. Lehrstuhl für Theoretische Chemie, Bonn.
12. Helgaker, T., Jensen, H., Jørgensen, P., Olsen, J., Ruud, K., Ågren, H., Auer, A.A., Bak, K.L., Bakken, V., Christiansen, O., Coriani, S., Dahle, P., Dalskov, E.K., Enevoldsen, T., Fernandez, B., Hättig, C., Hald, K., Halkier, A., Heiberg, H., Hettema, H., Jonsson, D., Kirpekar, S., Kobayashi, R., Koch, H., Mikkelsen, K.V., Norman, P., Packer, M.J., Pedersen, T.B., Ruden, T.A., Sanchez, A., Saue, T., Sauer, S.P.A., Schimmelpfennig, B., Sylvester-Hvid, K.O., Taylor, P.R., and Vahtras, O. (2001) Dalton, a Molecular Electronic Structure Program, Release 1.2. Available at: *http://daltonprogram.org*.
13. Werner, H.-J., Knowles, P.J., Knizia, G., Manby, F.R., Schütz, M. *et al.* (2010) MOLPRO, Version 2010, 1, a Package of Ab initio Programs. Available at: *http://www.molpro.net*.
14. Kutzelnigg, W. and Klopper, W. (1991) *J. Chem. Phys.*, **94**, 1985–2001.
15. Klopper, W. (1991) *Chem. Phys. Lett.*, **186**, 583–585.
16. Ten-no, S. (2004) *Chem. Phys. Lett.*, **398**, 56–61.
17. Stewart, J.J.P. (1990) *J. Comput.-Aided Mol. Des.*, **4**, 1–105.
18. Bartlett, R.J. and Stanton, J.F. (1994) in *Reviews in Computational Chemistry*, Vol. 5 (eds K.B. Lipkowitz and D.B. Boyd), VCH Publishers, Inc., Weinheim, pp. 65–169.
19. Crawford, T. and Schaefer, H.F. III (2000) in *Reviews in Computational Chemistry*, Vol. 14 (eds K.B. Lipkowitz and D.B. Boyd), Wiley-VCH Verlag GmbH, Weinheim, pp. 33–136.
20. Hohenberg, P. and Kohn, W. (1964) *Phys. Rev. B*, **136**, 864–871.
21. Kohn, W. and Sham, L.J. (1965) *J. Chem. Phys.*, **140**, A1133–A1138.
22. Runge, E. and Gross, E.K.U. (1984) *Phys. Rev. Lett.*, **52**, 997–1000.
23. Grimme, S. (2004) in *Reviews in Computational Chemistry* (eds K.B. Lipkowitz, R. Larter, and T.R. Cundari), Wiley-VCH, Verlag GmbH, Weinheim, John Wiley & Sons, Inc., Hoboken, NJ, pp. 153–218.
24. Christiansen, O., Koch, H., and Jørgensen, P. (1995) *Chem. Phys. Lett.*, **243** (5-6), 409–418.
25. Andersson, K. and Roos, B.O. (1995) in *Modern Electronic Structure Theory* (ed. D.R. Jarkony), World Scientific, pp. 55–109.
26. Cramer, C.J. and Truhlar, D.G. (1995) in *Reviews in Computational Chemistry*, Vol. 6 (eds K.B. Lipkowitz and D.B. Boyd), VCH Publishers, Inc., pp. 1–72.
27. Miertus, S., Scricco, E., and Tomasi, J. (1981) *J. Chem. Phys.*, **55**, 117–129.
28. Klamt, A. and Schüürmann, G. (1993) *J. Chem. Soc., Perkin Trans. 2*, **5**, 799–805.
29. Klamt, A. (1995) *J. Phys. Chem.*, **99**, 2224–2235.
30. Uhr, H., Stenzel, K., Kugler, M., and Schrage, H. (1997) Preparation of 3-arylthio-1,4,2-dithiazole-1,1-dioxides as agrochemical microbicides. Bayer AG, Ger. Offen. DE Pat. 19545635.
31. Halgren, T.A. (1996) *J. Comput. Chem.*, **17**, 490–519.
32. Halgren, T.A. (1996) *J. Comput. Chem.*, **17**, 520–552.

33. Halgren, T.A. (1996) *J. Comput. Chem.*, **17**, 553–586.
34. Halgren, T.A. and Nachbar, R.B. (1996) *J. Comput. Chem.*, **17**, 587–615.
35. Dewar, M.J.S. and Thiel, W. (1977) *J. Am. Chem. Soc.*, **99**, 4899–4907.
36. Grimme, S. (2004) *J. Comput. Chem.*, **25**, 1463–1473.
37. Grimme, S., Antony, J., Ehrlich, S., and Krieg, H. (2010) *J. Chem. Phys.*, **132**, 154104 - 1–154104-19.
38. Jeschke, P., Nauen, R., Schindler, M., and Elbert, A. (2011) *J. Agric. Food Chem.*, **59**, 2897–2908.
39. Ruzo, L.O., Holmstead, R.L., and Casida, J.E. (1977) *J. Agric. Food Chem.*, **29**, 1385–1394.
40. Olivucci, M. (ed.) (2005) *Computational Photochemistry*, Theoretical and Computational Chemistry, Vol. 16, Elsevier, Amsterdam.
41. Schirmer, J. (1995) *J. Phys. B*, **26**, 2395–2416.
42. Trofimov, A.B. and Schirmer, J. (1995) *J. Phys. B*, **28**, 2299–2324.
43. Karelson, M., Lobanov, V.S., and Katritzky, A.R. (1996) *Chem. Rev.*, **96**, 1027–1043.
44. Karelson, M. (2004) in *Computational Medicinal Chemistry for Drug Discovery* (eds P. Bultinck, H. De Winter, W. Langenaeker, and J.P. Tollenaere), Marcel Dekker, Inc., New York, Basel, pp. 641–667.
45. Todeschini, R. and Consonni, V. (2009) *Molecular Descriptors for Chemoinformatics*, Vols. I and II, 2nd edn, Wiley-VCH Verlag GmbH, Weinheim.
46. Kleier, D.A. (1988) *Plant Physiol.*, **86**, 803–810.
47. Hättig, C. and Weigend, F. (2000) *J. Chem. Phys.*, **113**, 5154–5161.
48. Geerlings, P., DeProft, F., and Langenaeker, W. (2003) *Chem. Rev.*, **103**, 1793–1873.
49. Von Samson-Himmelstjerna, G., Nentwig, G., and Jeschke, P. (2001) Acaricides for dust mite control. Bayer AG, Ger. Offen. DE Pat. 19930076.
50. Dyker, H., Andersch, W., Erdelen, C., Laesel, P., and Nauen, R. (2001) Pesticidal cyclodepsipeptide PF Pat. 1022-221, Bayer AG, Germany, PCT Int. Appl. WO Pat. 2001045511 A1 20010628.
51. Nishiyama, H. *et al.* (1993) WO Pat. 93/19053, Fujisawa Pharm. Co., Ltd.
52. Harder, A., Dye-Holden, L., Walker, R., and Wunderlich, F. (2005) *Parasitol. Res.*, **97**, S1–S10.
53. Jeschke, P., Iinuma, K., Harder, A., Schindler, M., and Murakami, M. (2005) *Parasitol. Res.*, **97**, S11–S16.
54. Parr, R.G. and Yang, W. (1989) *Density Functional Theory of Atoms and Molecules*, Clarendon Press.
55. Parr, R.G. and Yang, W. (1984) *J. Am. Chem. Soc.*, **106**, 4049–4050.
56. Beck, M.E. (2006) Estimation of molecular reactivity by quantum chemical calculations and its application to biotic degradation. 11th IUPAC International Congress of Pesticide Chemistry, Kobe, Japan.
57. Beck, M.E. and Schindler, M. (2007) in *Pesticide Chemistry* (eds H. Ohkawa, H. Miyagawa, and P.W. Lee), Wiley-VCH Verlag GmbH, pp. 227–238.
58. Jeschke, P., Harder, A., Etzel, W., Gau, W., Thielking, G., Bonse, G., and Iinuma, K. (2001) *Pest Manag. Sci.*, **57**, 1000–1006.
59. Ditchfield, R. (1974) *Mol. Phys.*, **27**, 789–807.
60. Schindler, M. and Kutzelnigg, W. (1982) *J. Chem. Phys.*, **76**, 1919–1933. P
61. Kaupp M., Bühl M., and Malkin V.G. (eds) (2004) *Calculation of NMR and EPR Parameters*, Wiley-VCH, Weinheim, Germany.
62. Jank, M. (2011) NMR Analysis. Bayer CropScience Internal Report, 8 March 2011.
63. Jeschke, P., Uneme, H., Benet-Buchholz, J., Stölting, J., Sirges, W., Beck, M.E., and Etzel, W. (2003) *Planzenschutz-Nachr. Bayer*, **56**, 5–25.
64. Zhang, S., Baker, J., and Pulay, P. (2010) *J. Phys. Chem. A*, **114**, 425–431.
65. Lee, A.C. and Crippen, G.M. (2009) *J. Chem. Inform. Model.*, **49**, 2013–2033.
66. Beck, M. and Schindler, M. (2006) Benchmark of pK_a-Prediction Methods. Bayer CropScience Internal Report.

67. Beck, M.E. and Bürger, T. (2003) in *Conference Proceedings of Designing Drugs and Crop Protectants* (eds J. Dearden and H. Van de Waterbeemd), Blackwell Publications, pp. 446–450.
68. Advanced Chemistry Development (ACD) (2011) Scientific Software. Available at: *http://www.acdlabs.com.*
69. Schrödinger (2011) Scientific Software. Available at: *http://www.schrodinger.com.*
70. Ort, O. (2011) in *Modern Crop Protection Compounds*, 2nd edn (eds W. Krämer and U. Schirmer), Wiley-VCH Verlag GmbH, Weinheim, pp. 227–238.
71. Kallies, B. and Mitzner, R. (1997) *J. Phys. Chem. B*, **101**, 2959–2967.
72. Gutkowski, K.E. and Dixon, D. (2006) *J. Phys. Chem. A*, **110**, 12044–12054.
73. Mangold, M., Rolland, L., Constanzo, F., Sprik, M., Sulpizi, M., and Blumberger, J. (2011) *J. Chem. Theory Comput.*, **7**, 1951–1961.
74. Guthrie, J.P. (2009) *J. Phys. Chem. B*, **113**, 4501–4507.
75. Klamt, A., Eckert, F., and Diedenhofen, M. (2009) *J. Phys. Chem. B*, **113**, 4508–4510.
76. Crawford, T.D., Tam, M.C., and Abrams, M.L. (2007) *J. Phys. Chem. A*, **111**, 12057–12068.
77. Autschbach, J., Nitscg-Velasquez, L., and Rudolph, M. (2011) in *Electronic and Magnetic Properties of Chiral Molecules and Supramolecular Architectures* (eds R. Naaman, D.N. Beratan, and D.H. Waldeck), Springer, pp. 1–98.
78. Hammett, L.P. (1937) *J. Am. Chem. Soc.*, **59**, 96.
79. Hansch, C. and Fujita, T. (1964) *J. Am. Chem. Soc.*, **86**, 1616–1626.
80. Free, S.M. Jr and Wilson, J.W. (1964) *J. Med. Chem.*, **7**, 395–399.
81. Cramer, R.D., Patterson, D.E., and Bunce, J.D. (1988) *J. Am. Chem. Soc.*, **110**, 5959–5967.
82. Klebe, G., Abraham, U., and Miezner, T. (1994) *J. Med. Chem.*, **37**, 849–857.
83. Kubinyi, H. (1993) *3D QSAR in Drug Design. Theory, Methods, and Applications*, ESCOM Science, Leiden.
84. Wold, S., Johansson, E., and Cocchi, M. (1993) in *3D QSAR in Drug Design: Theory, Methods, and Applications* (ed. H. Kubinyi), ESCOM Science, Leiden, pp. 523–550.
85. Bush, B.L. and Nachbar, R.B. (1993) *J. Comput.-Aided Mol. Des.*, **7**, 587–619.
86. Beck, M.E. and Schindler, M. (2009) *Chem. Phys.*, **356**, 121–130.
87. Yang, W., Parr, R.G., and Pucci, R.J. (1984) *Phys. Rev.*, **81**, 2862–2863.
88. Ayers, P.W. and Parr, R.G. (2000) *J. Am. Chem. Soc.*, **122**, 2010–2018.
89. Ayers, P.W. and Levy, M. (2000) *Theor. Chem. Acc.*, **103**, 353–360.
90. Hoffmann, R. and Woodward, R.G. (1968) *Acc. Chem. Res.*, **1**, 17–22.
91. Woodward, R.G. and Hoffmann, R. (1969) *Angew. Chem. Int. Ed. Engl.*, **21**, 797–869.
92. Fukui, K. (1982) *Science*, **218**, 747–754.
93. Durst, G.L. (1998) *Quant. Struct.-Act. Relat.*, **17**, 419–426.
94. Cramer, R.D., DePriest, S.A., Patterson, D.E., and Hecht, P. (1993) in *3D-QSAR in Drug Design* (ed. H. Kubinyi), ESCOM Science Publication, pp. 443–485.
95. Goodford, P.J. (1985) *J. Med. Chem.*, **28**, 849–857.
96. Cruciani, G. (2006) *Molecular Interaction Fields: Applications in Drug Discovery and ADME Prediction*, Wiley-VCH Verlag GmbH, Weinheim.
97. Martin, R.L., Gardiner, E., Gillet, V.J., Mucoz-Muriedas, J., and Senger, S. (2010) *Mol. Inform.*, **29**, 603–620.
98. Massart, D.L., Vanderginste, B.G.M., Buydens, L.M.C., de Jong, S., Lewi, P.J., and Smeyers-Verbeke, J. (1887) *Handbook of Chemometrics and Qualimetrics, Parts A and B*, Data Handling in Science and Technology, Vols. 20A and 20B, Elsevier.
99. Vidakovic, B. (1999) *Statistical Modelling by Wavelets*, John Wiley & Sons, Ltd.
100. Eichkorn, K., Treutler, O., Öhm, H., Häser, M., and Ahlrichs, R. (1995) *Chem. Phys. Lett.*, **242**, 652.
101. Weigend, F. and Ahlrichs, R. (2005) *Phys. Chem. Chem. Phys.*, **7**, 59.
102. Kubinyi, H. (1998) in *Encyclopedia of Computational Chemistry*

(eds P.R. vanSchleyer, N.L. Allinger, T. Clark, J. Gasteiger, P.A. Kollman, H.F. Schaefer III, and P.R. Schreiner), John Wiley & Sons, Inc., New York, pp. 448–460.

103. Okazawa, A., Akamatsu, M., Nishiwaki, H., Nakagawa, Y., Miyagawa, H., Nishimura, K., and Ueno, T. (2000) *Pest Manag. Sci.*, **56**, 509–515.

104. Sherrill, C.D. (2010) *J. Chem. Phys.*, **132**, 110902.

105. Sierka, M., Hogekamp, A., and Ahlrichs, R. (2003) *J. Chem. Phys.*, **118**, 9136–9148.

106. Ma, W., Krishnamoorthy, S., and Kowalski, K. (2011) *J. Chem. Theory Comput.*, **7**, 1316–1335.

107. Janssen, C.L. and Nielsen, I.M.B. (2008) *Parallel Computing in Quantum Chemistry*, CRC Press, Boca Raton, FL.

108. Bachorz, R.A., Bischoff, F.A., Glöss, A., Hättig, C., Höfener, S., Klopper, W., and Tew, D.P. (2011) *J. Comput. Chem.*, **32**, 2492–2513.

4 The Unique Role of Halogen Substituents in the Design of Modern Crop Protection Compounds

Peter Jeschke

4.1 Introduction

The past 30 years have witnessed a period of significant expansion in the use of halogenated compounds in the field of modern agrochemical research and development [1–3]. Interestingly, there has been a significant rise in the number of commercial products containing "mixed" halogens, for example, one or more fluorine, chlorine, bromine, or iodine atoms, in addition to one or more further halogen atoms (Figure 4.1) [4]. An extrapolation of the current trend indicates that a definite growth is to be expected in fluorine-substituted commercial products throughout the twenty-first century.

A survey of the new active ingredients (the total number up to November 2010 was 139) used as modern crop protection compounds (insecticides/acaricides, fungicides, and herbicides), provisionally approved by ISO during the past 12 years (1998–2010), shows that around 79% of these are halogen-substituted (Br, I < Cl/F, Cl < F). During this time, approximately twofold more halogen-containing insecticides, acaricides and fungicides, as well as around one-and-a-half-fold more herbicides, were approved than non-halogenated active ingredients.

According to data from Phillips McDougall, in each area of crop protection (insecticides, fungicides, and herbicides), 12 halogen-containing products (about 60%) are among the 20 best-selling compounds. These are the insecticides imidacloprid, thiamethoxam, clothianidin, λ-cyhalothrin, fipronil, the fungicides pyraclostrobin, trifloxystrobin, chlorothalonil, tebuconazole, and prothioconazole, and the herbicides acetochlor and 2,4-D, each of which achieved sales in 2009 of between US$ 370 and 1.05 million (i.e., total sales of US$ 6430 million).

Today, around two-thirds of all known crop protection compounds contain halogen-substituted aryl and hetaryl moieties. The correct selection and

Modern Methods in Crop Protection Research, First Edition.
Edited by Peter Jeschke, Wolfgang Krämer, Ulrich Schirmer, and Matthias Witschel.

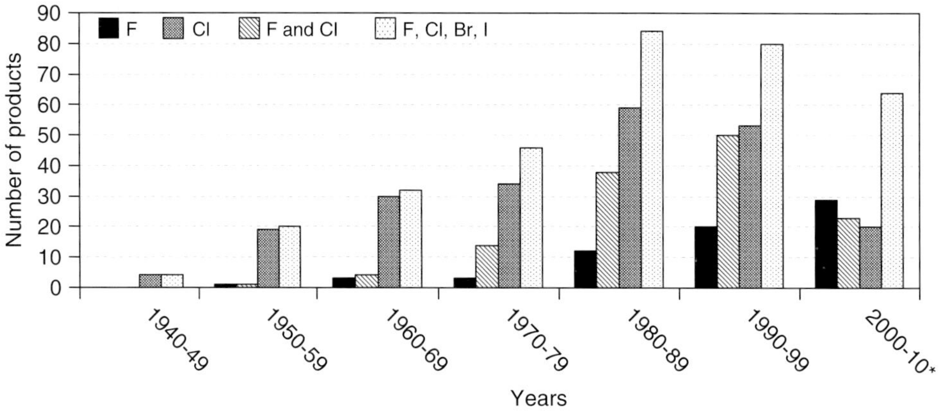

Figure 4.1 Launch of halogenated commercial products in the timeframe 1940–2010; *, inclusive compounds provisionally approved by ISO until November 2010 (*www.alanwood.net*).

modification of the appropriate substituents at the periphery of a molecule, and their substitution pattern, often play a decisive role in the achievement of an excellent biological activity [5]. Some halogenated pyridyl moieties are used broadly in various crop protection areas, because they can strengthen the biological activity, as shown for the 3-chloro-5-trifluoromethyl-2-pyridinyl moiety in fluazinam or fluopicolide (fungicides), in the benzoyl phenyl urea (BPU) fluazuron (acaricide) and in haloxyfop-P-methyl (herbicide), respectively [2].

Outstanding progress has been made in synthetic methods for particular halogen-substituted key intermediates that previously were prohibitively expensive (e.g., fluoroorganic chemistry – the development of novel reagents/methodologies usable on a kilogram scale: deoxyfluorination with SF_4), electrophilic fluorination using F^+ reagents including enantioselective variants [6–9], the modified Balz-Schiemann reaction/fluorodediazotization [9], C/F exchange reactions/aromatic trifluoromethylation with metal complexes [7, 9–11]/HF chemistry including Halex reaction [9], selective direct fluorination with F2 [11].

The combination with other core technologies (e.g., chlorination, catalytic hydrogenation, Cl/N- and Cl/O-exchange, Sandmeyer reactions, Suzuki cross-coupling, and others) allows the synthesis of a broad variety of new building blocks containing fluorine. These efforts also included the introduction of fluorinated aryl moieties, so-called "fluoroaromatics," such as F_2HCO- or F_3CO-aryl fragments and other moieties into modern agrochemicals [12, 13]. Today, the basic raw material for these products – trifluoromethoxybenzene – is produced on an industrial scale; consequently, several agrochemicals from different crop protection areas produced from trifluoromethoxybenzene and its derivatives are known, such as indoxacarb (**28**; insecticide), triflumuron (**35**; insect growth regulator; IGR), thifluzamide (**68**; fungicide), flurprimidol (**88**; plant growth regulator), and flucarbazone-sodium (**115**; herbicide).

4.2 The Halogen Substituent Effect

What, then, is the rationale behind using halogen atoms and/or halogen-containing substituents in the design of modern crop protection compounds? The influence of halogens on the efficacy of a biological active molecule can be exciting and remarkable [14]. This can be demonstrated with examples from Bayer CropScience, derived from the different areas of agrochemistry (Figure 4.2).

- Whereas, the unsubstituted triazole (**1**) shows only low fungicidal activity, the incorporation of chlorine into the *para*-position of the phenyl moiety leads to the highly active cereal triazole fungicide, triadimenol (**2**; 1980, Baytan®) [15].
- Starting with **3**, herbicidal activity is strongly increased by the introduction of chlorine into the *ortho*-position, giving the selective paddy amide herbicide fentrazamide (**4**; 2000, Lecs®) [16] with excellent crop compatibility, even on young seedlings.

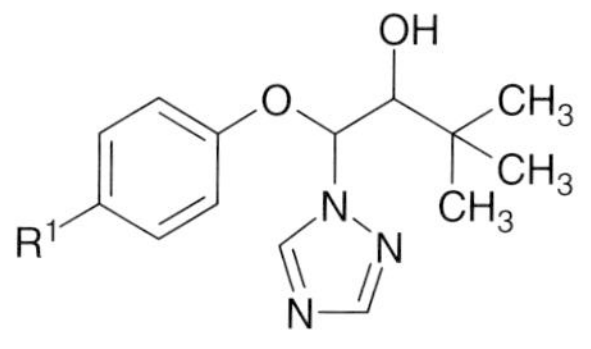

1 $R^1 = H$ << **2** $R^1 = Cl$

Fungicidal activity

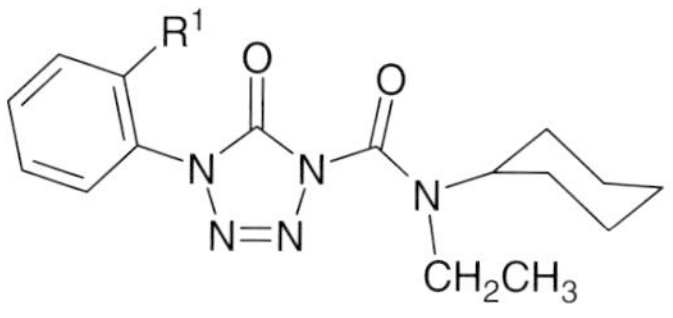

3 $R^1 = H$ << **4** $R^1 = Cl$

Herbicidal activity

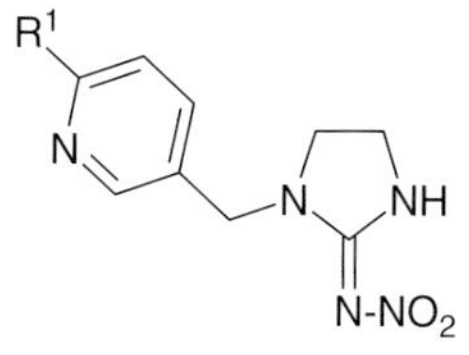

5 $R^1 = H$ << **6** $R^1 = Cl$

Insecticidal activity

Figure 4.2 Commercial products (**2**, **4**, and **6**) obtained by the incorporation of chlorine (R^1) into the aryl(hetaryl) moiety.

- One of the most important structural requirements for active nicotinic acetylcholine receptor (*n*AChR) effectors (neonicotinoids) such as imidacloprid (**6**; 1991, Gaucho®) [17] is the incorporation of chlorine into the 6-position of the pyridin-3-ylmethyl substituent of **5** [18].

The minimal variation of the chemical structure substitution with halogens such as chlorine into the aryl(hetaryl) moiety led to a commercial fungicide (**2**), herbicide (**4**), and insecticide (**6**).

The significant and increasingly important role of halogen atoms and/or halogen-containing substituents can be attributed to the well-known physicochemical effects arising from the introduction of fluorine, chlorine, bromine, or iodine, and/or halogenated substituents into biologically active molecules such as commercial products.

As expected for electronegative elements with accessible ion pairs, halogens can act as hydrogen-bond acceptors; however, during the 1950s it became clear that halogens could also form complexes with H-bond acceptors [19, 20], a behavior which has been outlined based on molecular electrostatic potential surfaces [21]. Fluorine behaves much like a ball of negative charge, and thus can act only as a H-bond acceptor, whereas the other halogens display a more positive region on the surface opposite to the direction of the C–halogen bond (Hal = Cl, Br, I), as well as an equatorial belt of negative potential [22]. As result, they can act as either H-bond donors or acceptors, depending on the angle of approach. The magnitude and area of the zone of positive potential increases with the size of the appropriate halogen atom (F < Cl < Br < I), which means that iodine in particular can make relatively strong interactions with H-bond donors. On the other hand, the so-called "fluorine-factor" (as described in the literature several years ago) stems from the unique combination of properties associated with the fluorine atom itself, namely its high electronegativity and moderately small size, its three tightly bound ion-pair electrons, and the excellent match between its 2s and 2p orbitals and those of carbon.

In the following sections, a few examples are provided to illustrate how halogen substitution is used successfully in contemporary agrochemistry.

4.2.1
The Steric Effect

The C–halogen bond lengths increase in the order C–F < C–Cl < C–Br < C–I (Table 4.1). With a van der Waals radius of 1.47 Å [23], covalently bound fluorine occupies a smaller volume than a methyl, amino, carbonyl, or C-O-group (1.52 Å), but a substantially larger volume than a hydrogen atom (1.20 Å). Nevertheless, the substitution of a hydrogen atom by a fluorine atom is described as one of the most commonly applied bioisosteric replacements [24, 25].

The fluorine atom was introduced [26], for example, as a chemical isostere of the essential tertiary hydroxy group (–OH versus –F, isoelectronic; bioisosterism) into the demethylation inhibitor (DMI) triazole fungicide flutriafol (**7**; 1984, Impact®, ICI/Zeneca, now Syngenta) [27] (Figure 4.3).

Table 4.1 Bond lengths, van der Waals radii, and total size of carbon halogen bonds.

Bond	Length (Å)	van der Waals radius (Å)	Total size (Å)
O-H	0.96	1.20	2.16
C-H	1.09	1.20	2.29
C=O	1.23	1.50	2.73
C-F	1.35	1.47	2.82
C-O-	1.43	1.52	2.95
C-Cl	1.77	1.80	3.57
C-Br	1.93	1.95	3.88
C-I	2.14	2.15	4.29

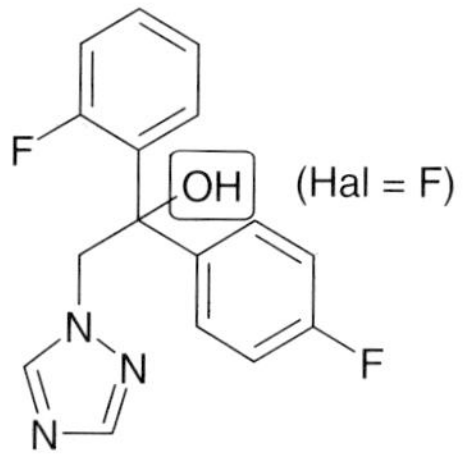

Figure 4.3 Flutriafol (**7**) – replacement of the tertiary hydroxyl group by fluorine.

Notably, fluorine may also exert a substantial effect on the conformation of a molecule [28]. On the other hand, an excellent match is found for the carbonyl group (Table 4.1) [29, 30]. The short C–F bond length is in the range of the C–O bond length, which suggests an isosteric behavior (mimic effect) of the hydroxy group in a bioactive compound with respect to steric requirements at receptor sites or enzyme substrate recognition [31]. Increasingly, these largely recognized aspects of fluorine substitution are used to enhance the binding affinity to the target protein [32].

In contrast, the so-called Bondi volumes ($cm^3\ mol^{-1}$) [23] for halogen atoms attached directly to phenyl rings are: 5.8 (F) < 12.0 (Cl) < 15.12 (Br) < 19.64 (I). This ranking of halogen atoms can be exemplified by the structure–activity relationship (SAR) of halogenated phthalic acid diamides (F < Cl < Br < I) in flubendiamide (**55**; 2006, Belt®, Hal = I; Nihon Nohyaku Co., Ltd./Bayer CropScience) [33], which activates selective ryanodine-sensitive intracellular Ca^{2+} release channels in insects [34]. The introduction of a bulky and moderate lipophilic halogen such as iodine into the 3-position of the phthalic acid aryl moiety led to a considerable increase in insecticidal activity (Figure 4.4).

A beneficial steric halogen effect of fluorine in both the 2-position and in the 6-position on the inhibition of insecticidal chitin synthase [35], and the difference in soil degradation half-life (DT_{50}) caused by the presence of these atoms, have been discussed for diflubenzuron (**34a**) [36] and its *N*-2,6-dichlorobenzoyl derivative (**34b**) (Figure 4.5) [37, 38]. *N*-2,6-Difluorobenzoyl-*N*′-phenyl ureas such as **34a** are stable at acidic pH values, but are hydrolyzed at pH 9–10 to give 2,6-difluorobenzoic acid and

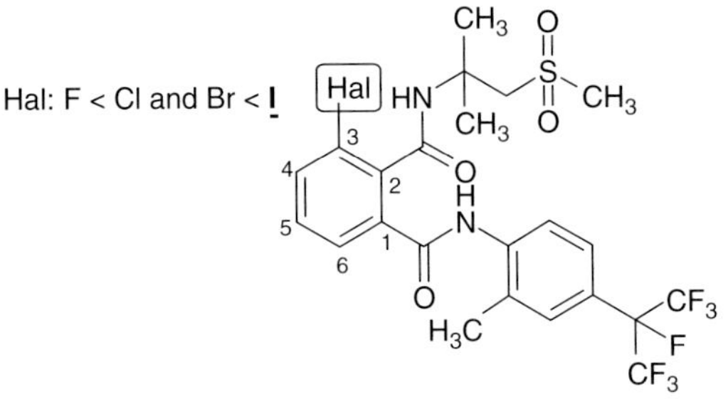

Figure 4.4 Structure–activity relationship (SAR) of halogen-containing phthalic acid diamides, flubendiamide (**55**; Hal = I).

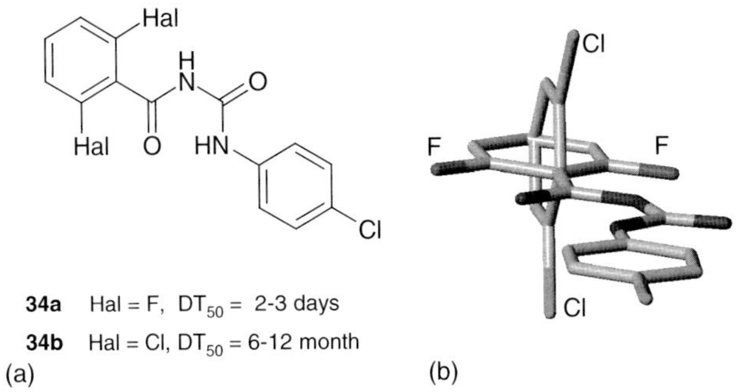

Figure 4.5 (a) Soil degradation half-life (DT_{50}) of BPUs (**34a**,**b**); (b) The effect of halogen substituents.

a *N*-4-chloro-phenylurea. In contrast to the conformation of the less-active analog **34b**, which degrades in between 6 and 12 months, the 2,6-difluorobenzoyl moiety in **34a** is in-plane with the whole urea structure (Figure 4.5). As a consequence, different metabolic pathways are observed for these two ureas.

The perfluoroalkyl group CF_3 has a relatively large van der Waals volume – larger than methyl, mono-fluoromethyl, and di-fluoromethyl, but between those of the *iso*-propyl and the *tert*-butyl groups [39]. The latter is comparable in size to the heptafluoro-*iso*-propyl group.

4.2.2
The Electronic Effect

4.2.2.1 Electronegativities of Halogens and Selected Elements/Groups on the Pauling Scale

The replacement of hydrogen by the most electronegative element fluorine (P ~ H = 2.1 < C ~ S ~ I = 2.5 < C_6F_5 ~ Br = 2.8 < N ~ Cl = 3.0 < CF_3 = 3.3 < O = 3.5 < F = 4.0) affords bonds that possess a high ionic character and are strongly polarized $^{\delta+}C{-}F^{\delta-}$ [40]; this alters the steric and electronic properties of the molecules, affecting their physico-chemistry, such as the basicity or acidity of neighboring functional groups, and also strengthens all neighboring bonds [41].

Furthermore, the fluorine substituent with three tightly bound non-bonding electron pairs is associated with a set of electronic effects that encompass both "push" effects, such as $+M$ or $+I\pi$ effects in aromatic systems and stabilization of α-carbocations ($=C^+{-}F \leftrightarrow =C{=}F^+$; relative stability: $^+CHF_2 > ^+CH_2F > ^+CF_3 > ^+CH_3$), and "pull" effects, such as destabilization of β-carbocations and possibly negative (or anionic) hyperconjugation. The influence of fluorine regarding stabilization of tetrahedral transition states (e.g., CF_3 group) and prevention of decomposition through proteolysis, by forming a vicinal positive charge, has also been described [42]. The CF_3 group has an electronegativity similar to that of oxygen [43] and a large hydrophobic parameter [44].

4.2.2.2 **Effect of Halogen Polarity of the C–Halogen Bond**

Halogens such as chlorine and fluorine, when connected to a carbon atom, withdraw electrons from other parts of the molecule and can create a large dipole moment (μ) of the C–halogen bond (C–I, $\mu = 1.29$ Debye (D) $<$ C–Br, $\mu = 1.48$ D $<$ C–F, $\mu = 1.51$ D $<$ C–Cl, $\mu = 1.56$ D), and overall reactivity and chemical inertness. The theoretical basis for using the dipole moment as a free energy related parameter in studying drug–receptor interaction and quantitative structure–activity relationship (QSAR) has been described for aromatic substituents of mono-substituted benzene derivatives [45].

4.2.2.3 **Effect of Halogens on pK_a Value**

Depending on the position of the fluorine substituent relative to the acidic or basic group in the molecule, a pK_a shift of several log units can be observed, which can again improve absorption properties [46]. For instance, the pK_as of acids and alcohols are considerably reduced by several units when they bear a CF_3 group and, as a consequence, the hydrogen-bonding ability of fluoralcohols is enhanced compared with that of their non-fluorinated counterparts. This may result in an enhanced intrinsic activity and, finally, may induce a reinforcement of the binding between an active ingredient and a biological target.

On the other hand, linear and cyclic amines become much less basic with both β- and γ-fluorine substitution (cf. X-CH_2-NH_3^+: X $= CF_3$ (p$K_a = 5.7$) $<$ X $= CHF_2$(p$K_a = 7.3$) $<$ X $= CH_2F$ (p$K_a = 9.0$) $<$ X $= CH_3$ (p$K_a = 10.7$). Whilst these inductive effects influence pK_a even with δ-substitution [47], quite often a change in pK_a can have a strong influence on different parameters in lead optimization, including physico-chemical properties (solubility, log *D*), binding affinities (potency, selectivity), and pharmacokinetic properties (adsorption, distribution, metabolism, etc.), and also safety issues. Halogen bonds in the active ingredients of modern agrochemicals clearly demonstrate the potential significance of this interaction in ligand binding and recognition [22].

Unlike the hydroxy group, organic fluorine is a very poor H-bond acceptor [48], and is not a H-bond donor at all. The replacement of a hydroxy group by a fluorine atom totally perturbs the interaction pattern; however, fluorine can participate in hydrogen-bonding interactions with H–C, even if the bonds to C–F are definitely much weaker than those observed to oxygen or nitrogen [49]. Such C–F H–C interactions have been proposed as a design principle for crystal engineering [50]. Nevertheless, controversy persists regarding the existence of hydrogen bonds between the C–F group and –OH or –NH donors [51].

4.2.2.4 **Improving Metabolic, Oxidative, and Thermal Stability with Halogens**

Extensive surveys of structures in the Cambridge Structural Database [52], coupled with *ab initio* calculations, have led to a characterization of the geometry of halogen bonds in small molecules, and have shown that the interaction is primarily electrostatic, with contributions from polarization, dispersion, and charge transfer.

The stabilizing potential of halogen bonds is estimated to range from about half to slightly greater than that of an average hydrogen bond in directing the self-assembly of organic crystals [53, 54]. In comparison with C–H (416 kJ mol^{-1}), C–C (348 kJ mol^{-1}), C–N (305 kJ mol^{-1}), and other C–halogen bonds (C–Cl, 338 kJ mol^{-1}; C–Br, 276 kJ mol^{-1}; C–I, 238 kJ mol^{-1}), the high C–F bond energy of 485 kJ mol^{-1} causes a significant influence on metabolic degradation, oxidative, and thermal stability [55]. In mono-halogenoalkanes, the C–F bond is about 100 kJ mol^{-1} stronger than the C–Cl bond, and the difference in heterolytic bond dissociation energies is even greater (~130 kJ mol^{-1}). The high C–F bond strength, in connection with the poor nucleofugality of F^-, make alkyl mono-fluorides poor substrates in typical S_N1 solvolysis or S_N2 displacement reactions (alkyl chlorides are 10^2- to 10^6-fold more reactive) [56].

Although fluorination has little effect on C–F bonds, it significantly strengthens C–H bonds; for example, the C–H bond in $(CF_3)_2C\underline{H}$ is estimated to be at least 60 kJ mol^{-1} stronger than the C–H bond in $(CH_3)_2C\underline{H}$, which makes it stronger than C–H in methane.

The oxidative metabolism of phenyl rings is a common problem, and fluorine substitution (usually at the 4-position) has become a widespread practice to increase stability in various substance classes. A plot of the Hammett σ coefficients against stability for various aromatic ring substituents shows that halogen atoms and halogen-containing substituents more strongly influence the relative stability towards oxidation, hydrolysis, and/or soil degradation than do any other residues. Electron-withdrawing, halogen-containing groups (e.g., CCl_3, CF_3, OCF_3, $OCHF_2$, $COCF_3$, or SO_2CF_3) can stabilize an aromatic ring system to oxidative (or electrophilic) attacks, though too the presence of many withdrawing groups may bring about a susceptibility to nucleophilic attack. On the other hand, halogens and halogen-containing groups such as CCl_3, CF_3, or OCF_3 are themselves very stable to attack, and consequently an increased degradation stability is observed for biologically active molecules or fragments containing this special group of substituents.

Metabolic stability is one of the key factors in determining the bioavailability of active ingredients. Rapid oxidative metabolism – for example, by the cytochrome

P_{450} enzymes – can often lead to a limited bioavailability however, though a frequently employed strategy to overcome this problem is to block the reactive site by the introduction of halogen atoms. The replacement of hydrogen atoms by fluorine atoms at an oxidizable site protects against hydroxylation processes mediated by the cytochrome P_{450} enzymes. The metabolic stability of the C–F bond can be exploited to produce a proinsecticide, for example, 29-fluorostigmasterol [57].

The different metabolic pathways of the triazolopyrimidine herbicide diclosulam (**120**; 1997, Spider®, Dow AgroSciences) are guided by the substituent at the 7-position on the triazolopyrimidine ring system [58]. Typically, the predominance of one pathway is very crop-specific (Scheme 4.1); for example, in cotton **120** is metabolized by a displacement of the 7-fluoro substituent on the triazolopyrimidine ring by a hydroxy group, forming **122**. The soybean selectivity of this compound is attributed to a facile conjugation with *homo*-glutathione (*homo*GSH), which displaces the 7-fluoro substituent (**123**). In maize and wheat, however, **120** is detoxified by hydroxylation at the 4'-position on the 2,6-dichloroaniline moiety (**124**), followed by subsequent glycosidation.

The special role of the substitution pattern has been demonstrated in the case of the selective pre- and early post-emergence oxyacetamide herbicide flufenacet (**8**; 1998, Axiom®, Bayer CropScience) [59], which is a selective inhibitor of cell growth and cell division (Figure 4.6). Whereas, the unsubstituted phenyl moiety provides a good herbicidal activity against *Echinochloa crus galli*, the selectivity achieved is

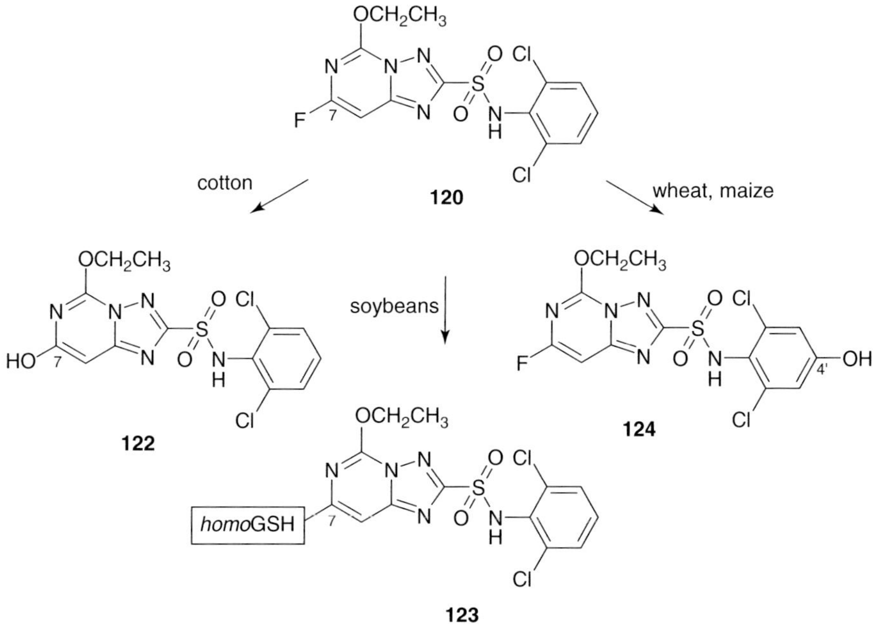

Scheme 4.1 Metabolism of diclosulam (**120**) in various crops.

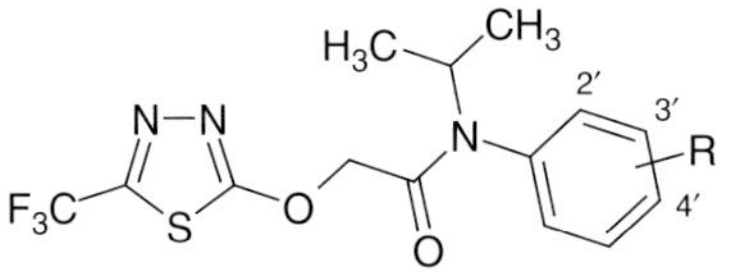

Figure 4.6 Structure–activity relationship (SAR) of oxyacetamides; flufenacet (**8**; R = 4′ – F).

insufficient for soybeans and maize. However, by incorporating halogens such as chlorine or fluorine, the selectivity of the oxyacetamides can be significantly increased, although with chlorine this improvement correlated with a reduced herbicidal activity. Subsequently, only the 4-fluorophenyl-containing compound **8** showed good herbicidal efficacy and selectivity against grasses.

Furthermore, the stronger C–F bonds, compared to other C–halogen bonds such as the C–Cl, serve as the actual thermodynamic driving force for "Halex reactions" towards the "fluoroaromatics" [60]. The Halex reaction is a nucleophilic aromatic substitution (S_NAr), in which chlorine atoms activated by an electron-withdrawing group are displaced by fluorine upon reaction with a metal fluoride under polar aprotic conditions [61].

4.2.3 Effect of Halogens on Physico-Chemical Properties

4.2.3.1 Effect of Halogens on Molecular Lipophilicity

Lipophilicity is a key parameter that governs the absorption and transport of active ingredients *in vivo* – and hence also their bioavailability. Notably, the presence of halogen substituents in biologically active molecules enhances their lipophilicities, whereby the substituents can influence pharmacokinetic behavior such as their uptake *in vivo*, by enhancing the passive diffusion of active ingredients across membranes and their transport *in vivo*. It seems that this effect is often relevant for fluorinated aryl(hetaryl) systems having interaction with π-electrons, however.

The incorporation of halogen atoms (especially chlorine or fluorine) can be important for "fine-tuning" the distribution of bioactive substances between aqueous and fatty media. In this context, the poor polarizability of fluorine-substituted groups plays a crucial role in phase behavior. For example, numerous insecticides which act on the central nervous system (CNS) contain a fluorophenyl moiety or one of the most lipophilic functional groups such as trifluoromethyl and X–CF_3 (X = O; S), which contribute to the insecticide's overall pharmacological activity by enhancing its CNS penetration [62]. The reason why the F_3CO-aryl moiety is so attractive in this respect is its ability to improve the membrane permeability of the compound in which it is embedded.

While halogen atoms such as chlorine, bromine and iodine (as well as CF_3 and OCF_3) substituents invariably boost lipophilicity, single fluorine atoms may alter this parameter in either direction (Table 4.2) [63]. This can be exemplified

Table 4.2 Lipophilicity increments π as assessed for *mono*-substituted benzenes H_5C_6-X.

Substituent	π	Substituent	π
X = OH	−0.68	X = Cl	+0.71
X = COOH	−0.32	X = Br	+0.86
X = OCH_3	−0.02	X = CF_3	+0.88
X = H	0.00	X = OCF_3	+1.04
X = F	+0.14	X = I	+1.12
X = SO_2CF_3	+0.55	X = SF_5	+1.23
X = CH_3	+0.56	X = SCF_3	+1.44

by the lipophilic 4-trifluoromethyl-phenyl moiety in pyrasulfotole (**9**, 2008, Precept® Bayer CropScience) [64], a new herbicidal 4-hydroxyphenylpyruvate dioxygenase (HPPD) inhibitor which shows a binding preference for the hydrophilic niche of the open conformation of the target enzyme (Figure 4.7) [2, 65, 66].

If the halogen occupies a vicinal or *homo*-vicinal position with respect to a hydroxy, alkoxy, or carbonyl oxygen atom, it enhances the solvation energy in water more than in organic solvents and hence lowers the lipophilicity. Conversely, a fluorine atom placed near a basic nitrogen center will diminish the donor capacity of the latter and, as a consequence, cause a strong log *D* (log *P*) increase.

The increased lipophilicity (π), and a superior metabolic stability compared to the methyl analog, often leads to an improved activity profile. The mono-fluorination

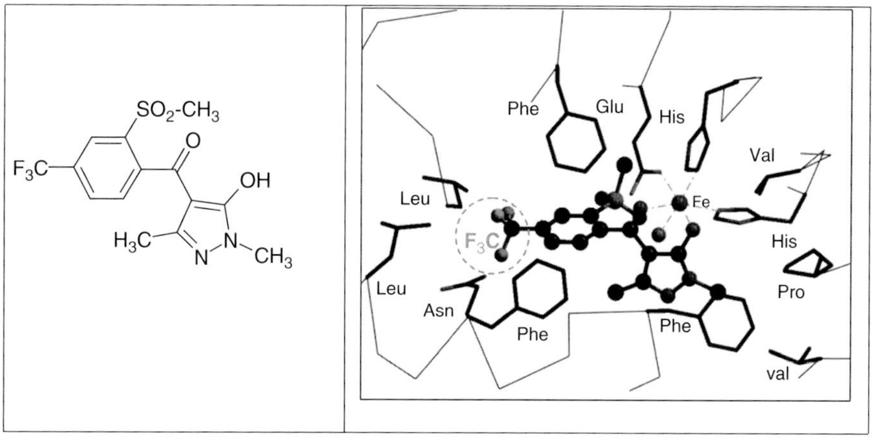

Figure 4.7 Pyrasulfotole (**9**; Precept®) and its molecular interaction with the herbicide target HPPD. Data taken from Freigang *et al.*, 2008 [66].

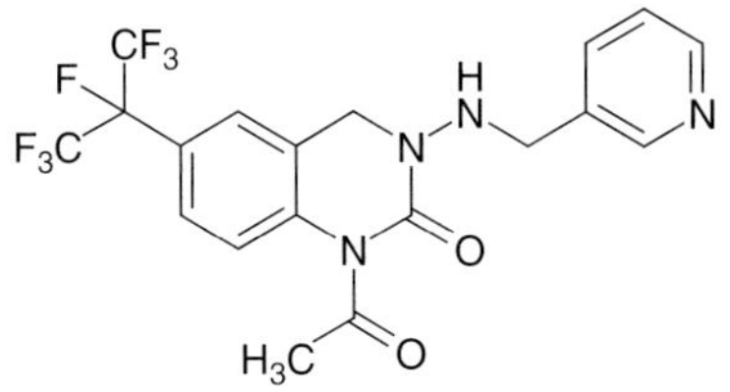

Figure 4.8 Pyrifluquinazon (**10**, Colt®).

and trifluorination of saturated aliphatic groups normally decrease lipophilicity, whereas higher fluoroalkyl groups (perfluoroalkyl groups) are introduced mainly to increase lipophilicity [67] and to decrease polarizability, as shown for the 4′-heptafluoro-*iso*-propyl-2-methyl-phenyl-amide fragment (4′-position: halogen < fluoroalkoxy <CF_3 < C_{2-4}-fluoroalkyl) in flubendiamide (**55**; 2006, Belt®, Nihon Nohyaku Co., Ltd./Bayer CropScience) (for more details, see Ref. [68]) and recently described for the insecticide pyrifluquinazon (**10**; 2010, Colt®, Nihon Nohyaku) [69] (Figure 4.8) and heptafluoro-*iso*-propyl-containing benzoylurea structures [70].

4.2.3.2 **Classification in the Disjoint Principle Space**

The systematic variation of substituents in a molecule has been the subject of various studies [71]. Besides synthetic feasibility and economic considerations, halogen atoms and halogen-containing substituents are chosen on the basis of properties such as polarity, size, and H-bonding capacity. The disjoint principle properties (DPPs), derived from a large set of property descriptors for substituents including halogen atoms and/or halogen-containing substituents, can be used to make rational and effective choices (e.g., from the following similarities: (i) F ≈ SH, C≡CH; (ii) Br, Cl, I ≈ CF_3, NCS; (iii) SO_2CF_3 ≈ SO_2CH_3, SO_2NH_2; (iv) OCF_3 ≈ $COOCH_3$, $NHCOCH_3$; and (v) SCF_3 ≈ O-phenyl, CO-phenyl). Several excellent examples are described in Section 4.6.2.1, concerning sulfonylurea and triazolone herbicides; these examples include the successful exchange of the following:

- The 3-ethylsulfonyl group in the 2-pyridyl ring of rimsulfuron (**104**) (R = $SO_2CH_2CH_3$) with the 3-trifluoromethyl group to give flazasulfuron (**106**) (R = CF_3) (see Figures 4.24 and 4.25 below);
- The 2-methoxycarbonyl group of propoxycarbazone-sodium (**114**) (R^1 = $COOCH_3$) with the 2-trifluoromethoxy group of flucarbazone-sodium (**115**) (R^1 = OCF_3) (see Figure 4.27 below).

4.2.4
Effect of Halogens on Shift of Biological Activity

In some cases, substitution with suitable halogen-containing residues can result in a so-called "*shift*" of biological activity, as exemplified by herbicidal *N*-aryl-pyrazole and fungicidal β-methoxyacrylate (strobilurine) chemistry (Figure 4.9) [2].

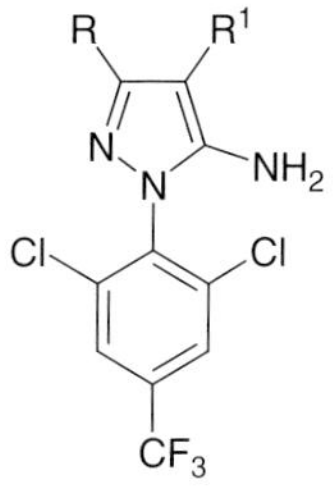

Shift from herbicidal activity

R = H, $R^1 = NO_2$

↓

to insecticidal activity

R = CN, $R^1 =$ **CF_3SO**

Fipronil (**29**; 1994, ... Regent®, Rhone-Poulenc)

MoA: GABA-Cl channel blocker

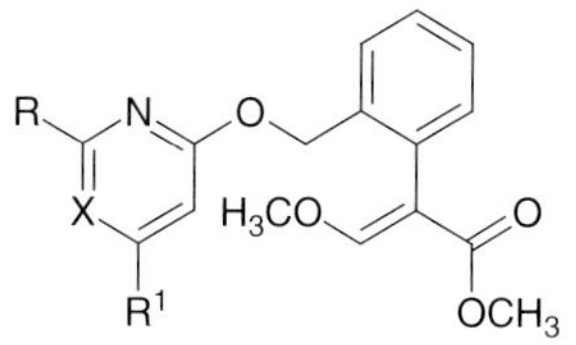

Shift from fungicidal activity

$R = CF_3$, R = H; X = CH (pyridine ring)

Picoxystrobin (**79**; 2001, ... Acanto®, Syngenta)

↓

to acaricidal activity

$R = (CH_3)_2CHO$, **$R = CF_3$**; X = N (pyrimidine ring)

Fluacrypyrim (**52**; 2002, ... Titaron®, Nippon Soda/BASF)

MoA: Q_o-site / complex III

Figure 4.9 Examples of herbicidal *N*-aryl-pyrazoles and fungicidal methoxyacrylates for halogen group-induced shift of biological activity (Jeschke, 2010) [2].

Optimization of the herbicidally active *N*-(2,6-dichlor-4-trifluoromethylphenyl)-4-nitropyrazole (JKU 0422, Bayer CropScience) [72] by replacement of NO_2 by the SCF_3 group leads to a highly active insecticidal derivative, which already shows structural similarity in the 4-position $R^1 = SCF_3$ versus $SOCF_3$) to the commercial insecticide fipronil (**29**; Regent®, Rhone Poulenc, later BASF) [73] (see Section 4.3.3).

On the other hand, the formal replacement of the 6-trifluoro-2-pyridinyl moiety in the arylalkyl ether side chain of the β-methoxyacrylate fungicide picoxystrobin (**79**; 2002, Acanto®; Syngenta) [74] by a 2-iso-propyloxy-4-trifluoro-6-pyrimidinyl moiety results in the acaricide fluacrypyrim (**52**; 2002, Titaron®, Nippon Soda) [75] (see Section 4.3.5.2). This compound is more lipophilic (difference log $P_{OW} \sim 0.9$), and has about a 10-fold lower water solubility than **79**.

Therefore, the strong influence of halogen atoms and/or halogen-containing substituents can lead to biological superiority of halogenated active ingredients over their non-halogenated analogs. Various selected commercial products testify to the successful utilization of halogens in the design of modern agrochemicals, in particular insecticides/acaricides, fungicides, plant growth regulators, and herbicides.

Generally, the corresponding biochemical targets or modes of action have been described. These are:

1) Voltage-gated sodium channel (vgSCh); γ-aminobutyric acid (GABA) receptor/chloride ionophore complex; chitin biosynthesis pathways; mitochondrial respiratory chain; and ryanodine receptor (RyR) for insecticides.
2) Sterol biosynthesis (sterol-C_{14}-demethylase); mitochondrial respiratory chain (complex I, II and the so-called Q_o-site of complex III) germination and hyphal growth; protein kinase (PK) for fungicides.
3) Gibberellin biosynthesis pathway for plant growth regulators.
4) Carotenoid biosynthesis – (phytoene desaturase; PDS), acetolactate synthase (ALS), and protoporphyrinogen IX oxidase (PPO) for herbicides.

4.3 Insecticides and Acaricides Containing Halogens

Insecticides that act on targets of the insect CNS (cf. vgSoCh, GABA receptor/chloride ionophore complex) are the most effective compounds in preventing crop damage. Generally, halogen substitution is a useful tool for improving the intrinsic properties of active ingredients, for example, by enhancing their CNS penetration. However, halogenated ligands are significant also in the case of the calcium channel (RyR), which appears to be involved in muscle contraction (see Section 4.3.6).

4.3.1 Voltage-Gated Sodium Channel (vgSCh) Modulators

The development of synthetic pyrethroids provides a significant historical illustration of agrochemicals regarding the introduction of halogen atoms (F, Cl, Br) and halogen-containing substituents such as CF_3, $OCHF_2$, $OCBrF_2$, $COO(CF_3)_2$, and halogenated aryl residues.

During the 1950s, shortening and simplifying the pentadienyl side chain of the natural insecticide pyrethrin I (**11**) [76] led to more stable synthetic pyrethroids, which block nerve signals by prolonging the opening of the vgSCh [77, 78]. With 43 members, the synthetic pyrethroid insecticides represent the largest single group in the chemical class. In addition to natural pyrethrins, pyrethroid esters of general structure types A and B, as well as non-esters of type C, are currently registered worldwide (Scheme 4.2; for simplification here, all of the chirality centers have been deleted) [79].

Today, about 40% of the pyrethroids contain no halogen substituents, while the remaining 60% are halogen substituted – mainly by "mixed" halogens such as fluorine/chlorine and chlorine or fluorine.

4.3.1.1 Pyrethroids of Type A

More than 20 years later, replacement of the cyclopentene alcohol group, the insertion of an α-cyano substituent R^1 at the phenoxybenzyl alcohol, and the

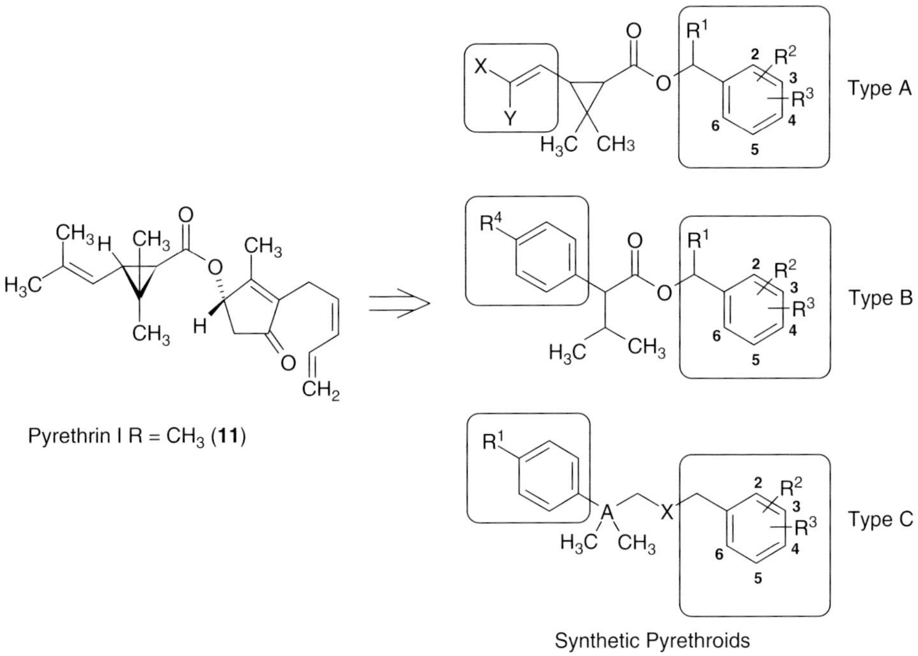

Scheme 4.2 Structural evolution of synthetic pyrethroids (types A–C) from pyrethrin I (**11**).

introduction of the dihalogenvinyl moiety resulted, for example, in permethrin (**12**) [80], deltamethrin (**13**) [81], and cypermethrin (**14**) [82] (Table 4.3).

In 1980, the first fluorine-containing pyrethroid, cyfluthrin (**15**) [83] was launched; this contained the 4-fluoro-3-phenoxybenzyl substituent, and was the remarkable result of a program directed at the synthesis of all possible isomers with fluorinated alcohol modifications [84]. Compared to cypermethrin (**14**), cyfluthrin (**15**) was shown to need less than one-third of the usage rate in order to control cotton pests [85]. The more active form of **15** for interaction at receptor sites involves a conformation in which the 3-phenoxy substituent (R^2) is twisted because of the fluorine effect in the 4-position (Figure 4.10).

In contrast, a different orientation of the 4-chloro-3-phenoxy-benzyl moiety and a lower insecticidal activity were observed. The commercialization of F_3C-containing pyrethroids began during the 1980s with λ-cyhalothrin (**16**) [86] from ICI/Zeneca (now Syngenta), which represents the optimum choice of fluorine-containing substituents for activity: ($F_3C(Cl)C{=}$) > ($F_3C(F)C{=}$) = ($Cl(Cl)C{=}$) = ($F(F)C{=}$) > ($F_3C(F_3C)C{=}$). The presence of the trifluoromethyl group in **16** also leads to insecticidal effects on phytophagous mites.

A comparison of the physical and chemical environment-related properties of structurally similar pyrethroids demonstrates the influence of both the fluorine atom at the phenyl moiety in the 4-position, and substitution on the vinyl side chain such as hydrogen, halogen, or the trifluoromethyl group [87].

Table 4.3 Halogen-substituted pyrethroids of type A (**12–20**) (see Scheme 4.2 for basic structure).

Compound no.	Common name, trade name	Manufacturer (year introduced)	X	Y	R[1]	R[2]	R[3]
12	Permethrin, Chinetrin®	ICI/Zeneca[a] (1977)	Cl	Cl	H	H	3-OPh
13	Deltamethrin, Decis®	Roussel Uclaf[b] (1977)	Br	Br	CN	H	3-OPh
14	Cypermethrin, Viper®	ICI/Zeneca[a] (1978)	Cl	Cl	CN	H	3-OPh
15	Cyfluthrin, Baythroid®	Roussel Uclaf[b] (1980)	Cl	Cl	CN	4-F	3-OPh
16	λ-Cyhalothrin, Banish®	ICI/Zeneca[a] (1984)	CF_3	Cl	CN	H	3-OPh
17	Bifenthrin, Brigade®	FMC Corp. (1986)	CF_3	Cl	H	2-CH_3	3-Ph
18	Tefluthrin, Attack®	ICI/Zeneca[a] (1988)	CF_3	Cl	H	2,3,5,6-F_4	4-CH_3
19	Acrinathrin, Ardent®	Roussel Uclaf[b] (1991)	H	CO-OCH$(CF_3)_2$	CN	H	3-OPh
20	Meperfluthrin[c]	Youth Chemical	Cl	Cl	H	2,3,5,6-F_4	4-CH_2OCH_3

[a] Now Syngenta.
[b] Now Bayer CropScience.
[c] Provisionally approved by ISO, development product.

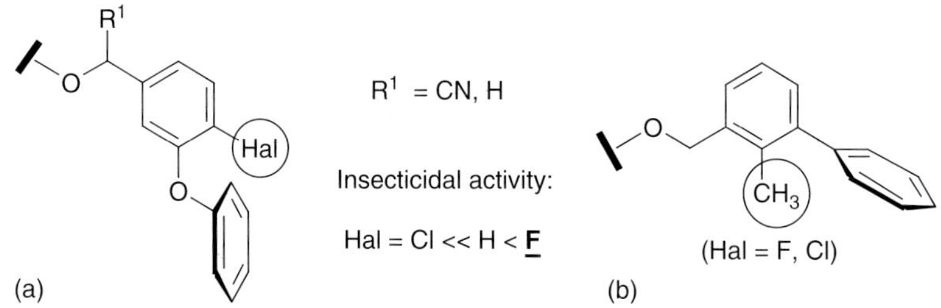

Figure 4.10 Pyrethroid alcohol modification: conformation of preferred substituent pattern, for example, in **15** (a) and **17** (b).

The biphenyl-type pyrethroid bifenthrin (**17**) [88] is available in a partially resolved (*Z*)-(1*RS*)-*cis*-isomer mixture [89], and has become one of the most important termiticides. In **17** the 3-phenyl substituent is twisted because of the effect of the methyl group in the *ortho*-position (F, Cl << CH_3; Figure 4.10). The extended exploitation of the acidic part of **16** resulted in a soil-applicable (*Z*)-(*1RS*)-*cis*-isomer of tefluthrin (**18**) [90], which is optimized in terms of stability, volatility, fast penetration, and water solubility.

Replacement of the dihalogenvinyl moiety with a hexafluoro-*iso*-propyl vinyl ester group led to acrinathrin (**19**) [91], which controls the larval and adult stages of a broad range of phytophagous mites, as well as various sucking insects such as aphids, thrips, and psyllids.

Finally, the development product meperfluthrin (**20**; provisionally approved by ISO) [92] is structurally similar to **18** and is useful for the control of mosquitoes, flies, and other household insect pests.

4.3.1.2 Pyrethroids of Type B

The racemate fenvalerate (**21**) [93], a non-systemic insecticide and acaricide with contact and stomach actions, shows efficacy against chewing, sucking, and boring insects such as lepidoptera (cotton: 30–150 g a.i.ha^{-1}), coleoptera (potatoes: 100–200 g a.i. ha^{-1}), and others (Table 4.4).

Today, so-called "chiral switches" [94], which exploit single enantiomers of already commercialized racemic mixtures, are an important feature of active ingredient development portfolios. Within this context, esfenfalerate (**23**) [95], the (2*S*,α*S*)-fenvalerate isomer was introduced by Sumitomo Chem. Co., Ltd into the market, and has an enhanced insecticidal activity against lepidopteran pests (cotton: 20–30 g a.i. ha^{-1}) [95]. Flucythrinate (**22**) [96], the difluoromethoxy derivative of **21** is a highly active insecticide (cotton: 30–75 g a.i. ha^{-1}) and shows broad-spectrum activity with excellent residual efficacy. The introduction of bromine into the *para* position of the phenoxy residue of **22** leads to the less-toxic flubrocythrinate (**24**) [acute oral LD_{50} for rats: >1000 mg kg^{-1} (**24**) versus 67–81 mg kg^{-1} (**22**)] [97], which shows additional activity against spider mites such as European red spider mite *Panonychus ulmi* (Koch), two-spotted spider mite, *Tetranychus telarius* L., hawthorn red spider mite, and *T. veinnensis* Zach. Flubrocythrinate is also active against eggs, larvae, and has a residual activity which lasts for more than three weeks.

Table 4.4 Halogen-substituted pyrethroids of type B (**21**–**24**) (see Scheme 4.2 for basic structure).

Compound no.	Common name, trade name	Manufacturer (year introduced)	R^1	R^2	R^3	R^4
21	Fenvalerate, Belmark®	Sumitomo (1976)	CN	H	3-Oph	Cl
22	Flucythrinate, Pay off®	ACC/BASF (1981)	CN	H	3-OPh	$OCHF_2$
23	Esvenvalerate, Samurai®	Sumitomo (1986)	CN	H	3-OPh	Cl
24	Flubrocythrinate, Lubrocythrinate®	Shanghai Zhongxi (1992)	CN	H	3-O(4-Br-Ph)	$OCHF_2$

Table 4.5 Halogen-substituted pyrethroids of type C (**25–27**) (see Scheme 4.2 for basic structure).

Compound no.	Common name, trade name	Manufacturer (year introduced)	X	A	R^1	R^2	R^3
25	Etofenprox, Fogger®	Mitsui (1986)	O	C	OCH_2CH_3	H	H
26	Eflusilanate, Silonen®	Hoechst[a] (1991)	CH_2	Si	OCH_2CH_3	4-F	3-OPh
27	Halfenprox, Prene EL®	Mitsui (1993)	O	C	OCF_2Br	H	3-OPh

[a] Now Bayer CropScience.

4.3.1.3 **Pyrethroids of Type C**

The achiral etofenprox (**25**; X = O, A = carbon; $R^1 = OCH_2CH_3$) containing no halogen atom is a non-ester displaying pyrethroid-like activity, and is highly advantageous with regards to the rice insecticide market (Table 4.5).

The incorporation of one fluorine and two bromine atoms in the ethoxy group R^1 leads to a shift in the spectrum of activity. The resultant halfenprox (**27**; $R^1 = OCBrF_2$) displays a good acaricidal activity and has a similar short environmental persistence in the soil ($DT_{50} = 10$ days versus ~6 days for **25**).

Finally, organosilicon pyrethroids such as eflusilanate (**26**; X = CH_2; A = silicon; $R^1 = OCH_2CH_3$) are obtained by replacing the quaternary carbon atom (A) with the appropriate isosteric silicon atom, and by replacing oxygen (X) with methylene. The latter has an extremely low fish and mammalian toxicity, combined with an insecticidal activity that is comparable to that of the parent compounds.

4.3.2 **Voltage-Gated Sodium Channel (vgSCh) Blockers**

Evolution of the insecticidal pyrazoline moiety (numerous halogenated pyrazolines have been described [98], which act by blocking the vgSCh of neurons [99]; no commercial example) has led to the discovery of the proinsecticide indoxacarb (**28**; 1998, Steward®, DuPont) [100]. The SAR for 1,3,4-oxadiazines against the fall armyworm, *Spodoptera frugiperda*, are outlined in Figure 4.11.

Generally, the existence of halogens and/or halogen-containing substituents R^1 in the 4- or 5-position of the annellated benzo ring A, such as chlorine, bromine or trifluoroethoxy, and trifluoromethyl, leads to derivatives with the highest activity. The angular R^2-group was either 4-fluorophenyl or methoxycarbonyl in the most active analogs. Preferred R^3 substituents, such as trifluoromethoxy or trifluoromethyl groups, were best at the *para*-position of the phenyl ring B.

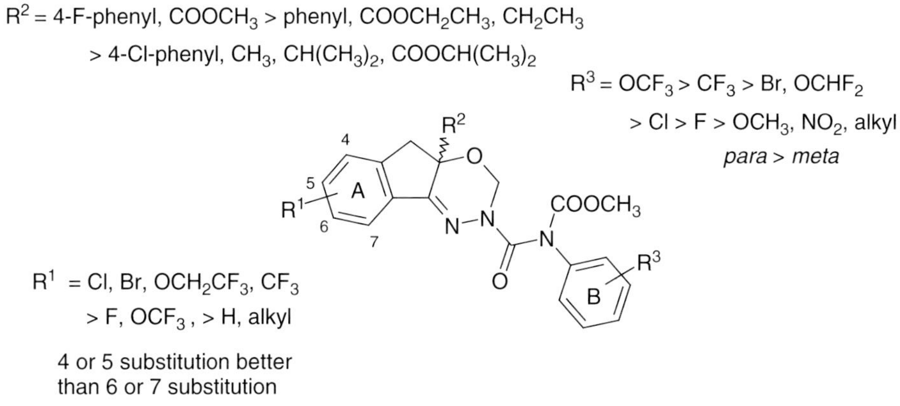

28 R^1= Cl, R^2= $COOCH_3$, R^3 = 4-OCF_3 (*S*) > (*R*) enantiomer

Figure 4.11 Insecticidal activity (SAR) for 1,3,4-oxadiazines against *Spodoptera frugiperda*.

4.3.3
Inhibitors of the γ-Aminobutyric Acid (GABA) Receptor/Chloride Ionophore Complex

The GABA receptor/chloride ionophore complex, which is located in the insect CNS and also in peripheral nerves, has been the focus of intense interest as both a target of insecticidal action and in its role in resistance [101].

One of the most important non-competitive GABA agonists in insects belongs to the pyrazole insecticide class represented by the trifluoromethyl sulfoxide-containing fipronil (**29**; $n = 1$, $R^1 = H$, $R^2 = CF_3$) (Figure 4.12, Table 4.6) [73]. This *N*-2,6-dichloro-4-trifluoromethyl-phenyl-substituted 5-aminopyrazole ($R^1 =$ H) either acts by interacting with an allosteric binding site or by irreversible binding [102], and has a wide margin of safety because it exhibits minimal activity at the corresponding mammalian channel [103]. Fipronil (**29**) is a broad-spectrum insecticide that is systemic in plants and highly active against lepidopterous larvae and numerous soil and foliar insects. It is also used as a household insecticide and for veterinary use [104].

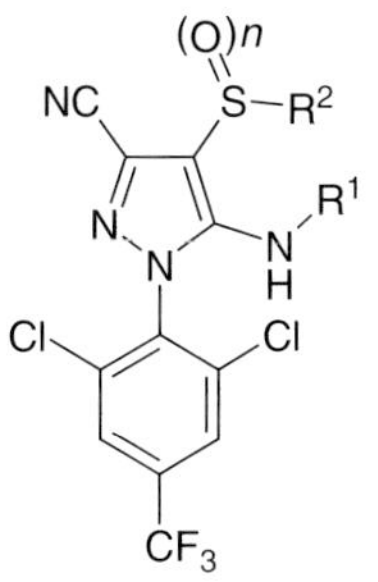

Figure 4.12 Pyrazole insecticide class.

Table 4.6 N-2,6-Dichloro-4-trifluoromethyl-phenyl-substituted 5-amino-pyrazoles (**29–33**) (see Figure 4.12 for basic structure).

Compound no.	Common name, trade name	Manufacturer (year introduced)	R^1	R^2	N	Use
29	Fipronil, Regent®	Rhone-Poulenc (1993)	H	CF_3	1	Foliar, soil, rice seedling box
30	Ethiprole, Curbix®	Bayer CropScience (2005)	H	CH_3	1	Foliar, seed treatment
31	Pyrafluprole[a]	Nihon Nohyaku	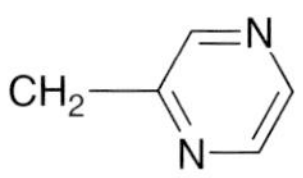	CH_2F	0	n.d.
32	Pyriprole[b]	Nihon Nohyaku	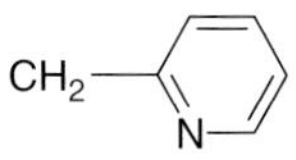	CHF_2	0	n.d.
33	Flufiprole[b]	Faming Zhuanli Shenqing Gongkai Shuomingshu	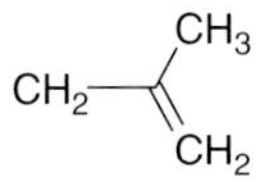	CF_3	1	n.d.

[a] Prac-tic®, developed by Novartis Animal Health Inc. as spot-on treatment for fleas and ticks by once-a-month administration.
[b] Provisionally approved by ISO, development product.
n.d. = not described.

The trifluoromethyl sulfoxide group at the 4-position ($n = 1$, $R^2 = CF_3$) of fipronil (**29**) can undergo cytochrome P_{450}-catalyzed oxidation in insects to give the corresponding trifluoromethyl sulfone metabolite ($n = 2$, $R^2 = CF_3$), which is slightly more toxic and two- to sixfold more active on the GABA receptor. Consequently, this conversion of the pro-insecticide **29** could confer negative cross-resistance in insect strains having elevated cytochrome P_{450} detoxification activity. The trifluoromethyl sulfoxide group is a remarkable trigger for insecticidal activity.

All new pyrazoles that are either commercialized or are in the developmental stage contain the essential *N*-dichloro-4-trifluoromethyl-phenyl fragment. They differ mainly regarding their functional group at the 3-position, such as ethiprole (**30**; n=1, R^1 = H, $R^2 = CH_3$) [105]; whereas mono- and di-fluoromethylthio-substituted pyrafluprole (**31**; $n = 0$, R^1 = H, $R^2 = CH_2F$) and pyriprole (**32**; $n = 0$, R^1 = H, $R^2 = CHF_2$) have an additional *N*-hetarylalkyl substituent at the 5-amino group (Table 4.6). The latter pyrazole is effective for the control of coleopteran, hemipteran pests, and also exhibits fungicidal activity against *Pyricularia oryzae*.

Finally, the so-called butene-fipronil flufiprole (**33**; $n = 1$, $R^1 = CH_2(=CH_2)CH_3$, $R^2 = CF_3$) [106] is a development product for use mainly against rice pests such as brown plant hopper or leafy beetle in rice and vegetables (Table 4.6).

4.3.4 Insect Growth Regulators (IGRs)

Over the past three decades, the BPUs have been developed and used as commercial IGRs which act by inhibiting chitin biosynthesis [107], thereby causing abnormal endocuticular deposition and abortive molting [108].

Until now, 14 BPUs have been commercialized or are in late-stage development as chitin synthesis inhibitors that contain both fluorine (2–9) and chlorine (1–3) atoms. Early studies of the SARs of BPUs reflected little scope for the variation of substituents at the *N*-benzoyl moiety, with only derivatives with at least one *ortho*-substituent retaining insecticidal activity. Such an *ortho*-substituent (R^1) can be methyl, OCF_3, or OC_2F_5 and lead to active derivatives. However, all commercialized products have *ortho*-halogen atoms, and the insecticidal or acaricidal activity generally follows in the order (Hal, R^1): 2,6-F_2 > 2-Cl, 6H > 2,6-Cl_2 > 2-F, 6H (Figure 4.13, Table 4.7).

Whilst the *N*′-arylamino moiety allows a broader variation, QSAR studies have shown that for optimum activity the *N*′-arylamino ring must be substituted by electron-withdrawing groups such as halogen, halogenoalkyl, α-fluoroalkoxy, or halogenated pyridin-2-yl. In this case, the *para*-position of the *N*′-arylamino moiety is preferable for high activity. Besides chlorine, the *N*-arylamine moiety of these ureas contains fluorine in most cases, sometimes together with various types of fluorinated substituent, such as $OCF_2CHFOCF_3$, OCF_2CHFCF_3, OCF_2CHF_2, F_3C giving a substitution pattern that often has extended the pesticidal spectrum to include mites and ticks.

Starting with diflubenzuron (**34a**; Hal, R^1 = F; Aryl = 4′-Cl-phenyl), the intense search for potent BPUs provided further compounds containing chlorine and/or fluorine, such as teflubenzuron (**36**) [109] or flucycloxuron (**37**) [110], which was the first BPU to shown an ability to control rust mites. Chlorfluazuron (**38**) [111] controls chewing insects on cotton and *Plutella* spp., thrips, and other on vegetables, and can also be used on fruit, potatoes, ornamentals, and tea (2.5 g a.i. ha^{-1}). Flufenoxuron (**39**) [112] controls the eggs, larvae, and nymphs of spider mites, as well as some insect pests, while bistrifluron (**44**) [113] has activity against whitefly and lepidopterous insects at 75–400 g a.i. ha^{-1}. Hexaflumuron (**40**; 4-OCF_2CF_3) [114], lufenuron (**41**; 4-OCF_2CHFCF_3) [115], novaluron (**42**; 4-$OCF_2CHFOCF_3$) [116], and noviflumuron (**43**; 4-OCF_2CHFCF_3) [117], each of which contain an α-fluoroalkoxy residue in the *para*-position, are insecticides which are especially active against

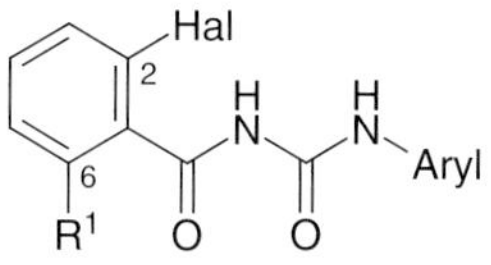

Figure 4.13 *N*-Benzoyl-*N*′-phenyl ureas (BPUs).

Table 4.7 Halogenated *N*-benzoyl-*N*′-phenyl ureas (BPUs) (**34a–44**) (see Figure 4.13 for basic structure).

Compound no.	Common name, trade name	Manufacturer (year introduced)	R[1]	Halogen	Aryl	Application rate cotton (g a.i. ha^{-1})
34a	Diflubenzuron, Dimilin®	Philips Duphar (1975)	F	F	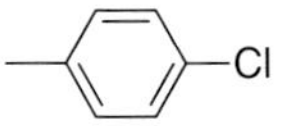	25–150
35	Triflumuron, Alsystin®	Bayer CropScience (1979)	H	Cl	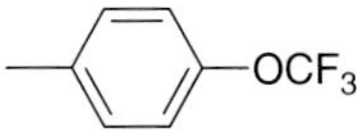	100–200
36	Teflubenzuron, Nomolt®	Celamerck (1986)	F	F	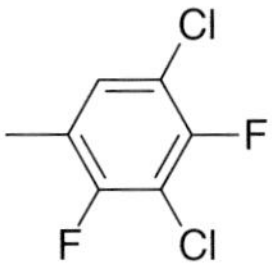	15–75
37	Flucycloxuron, Andalin®	Philips Duphar (1988)	F	F	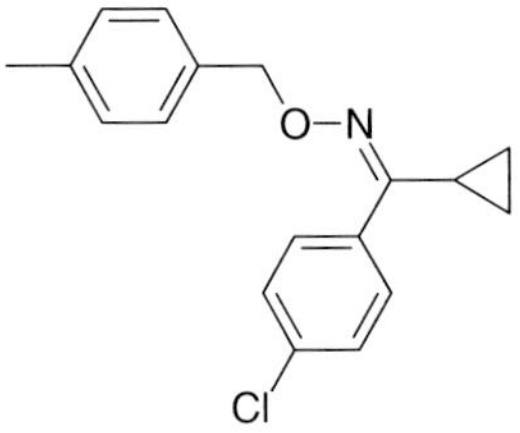	70–150
38	Chlorfluazuron, Aim®	Syngenta (1989)	F	F	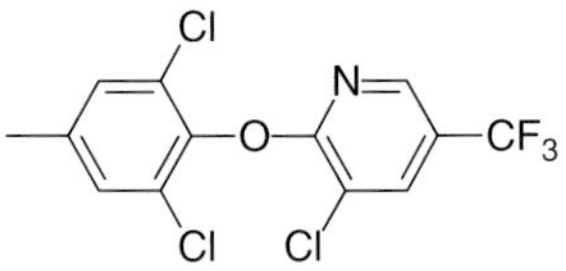	25–200
39	Flufenoxuron, Cascade®	Shell/BASF (1989)	F	F	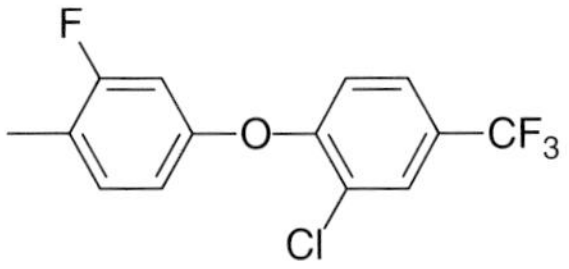	20–100
40	Hexaflumuron, Ridel®	Dow AgroSciences (1989)	F	F	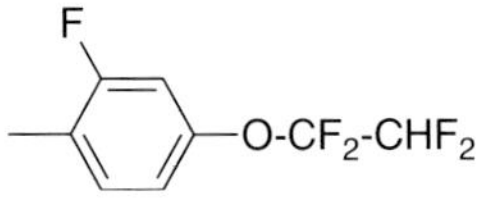	25–100
41	Lufenuron, Match®	Syngenta (1993)	F	F	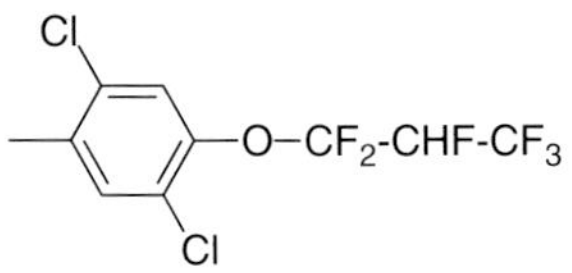	10–15

Table 4.7 (*continued.*)

Compound no.	**Common name, trade name**	**Manufacturer (year introduced)**	**R^1**	**Halogen**	**Aryl**	**Application rate cotton (g a.i. ha^{-1})**
42	Novaluron, Rimon®	Dow AgroSciences (1998)	F	F	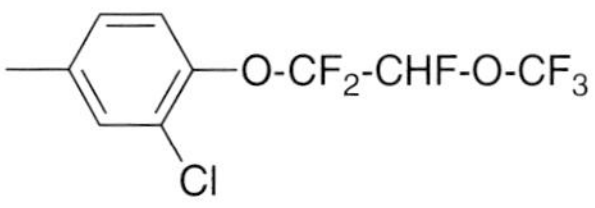	25–50
43	Noviflumuron, Recruit III®	Dow AgroSciences (2003)	F	F	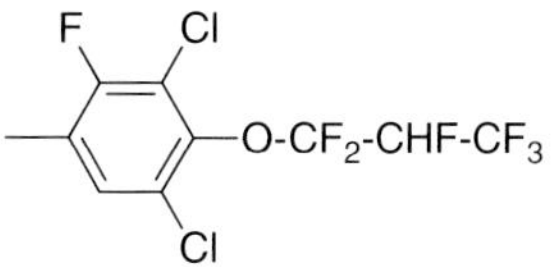	[a]
44	Bistrifluron Hanaro®	Dongbu Hannong Chemical	F	F	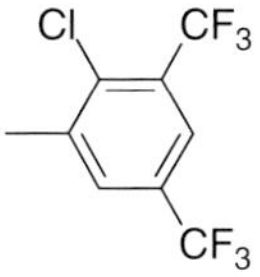	75–400

[a]Termiticide.

Hymenoptera such as ants, cockroaches, fleas, and termites (**43**) (Table 4.7). Notably, they are all more potent against various agricultural pests than diflubenzuron (**34a**) [118].

Among the BPUs currently available commercially, only the triflumuron (**35**; Hal = Cl; R^1 = H) [119] does not have the typical 2,6-difluoro substitution pattern (Hal, R^1 = F). During optimization of the *N′*-arylamino moiety, the pseudohalogenic trifluoromethoxy group in the 4-position was found to be beneficial for a broad insecticidal activity, combined with a strong feeding and contact action against chewing pests such as *S. frugiperda* and activity against coleopteran pests such as *Phaedon cochleariae* [120].

Because of their non-toxicity to vertebrates, the BPUs **35**, **40**, and **41** are also used in veterinary medicine (**35**; Staricide®, **41**; Program®) and at home (**35**; Baycidal®, **43**; Recruit III®) against animal and human health pests such as fleas, ticks, and cockroaches.

The oxazoline etoxazole (**45**; 1998, Baroque®, Yashima/Sumitomo) [121] is an acaricidal IGR, and also possess the 2,6-difluorophenyl moiety, similar to **46** and the BPU class (Figure 4.14).

The mode of action (MoA) of **45** appears to be an inhibition of the molting process during mite development, similar to that of the BPUs [122].

Closely related is the development product diflovidazin (**46**; 1996, Flumite®, Chinoin) [123], a 1,2,4,5-tetrazine acaricide that contains the 2,6-difluorophenyl

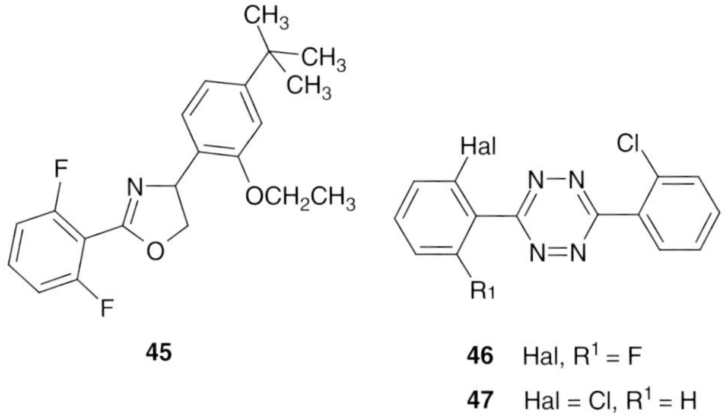

Figure 4.14 Etoxazole (**45**) and 1,2,4,5-tetrazine acaricides (**46** and **47**).

group as a common feature of BPUs (cf. acaricidally active **37** and **39**) and **46** (Figure 4.14). The MoA of **46** involves inhibition of mite development at both the egg and chrysalis stages, but the mechanism by which this occurs has not been clarified [124]. The mite growth inhibitor clofentezine (**47**; 1983, Apollo®, Schering) [125] was launched as the first compound of the 1,2,4,5-tetrazine type. The design of the 2,6-difluorophenyl group-containing heterocycles are described as mimics of the best conformation of the BPUs [126] based on a comparison of the X-ray structures of teflubenzuron (**36**) and (**46**).

4.3.5 Mitochondrial Respiratory Chain

4.3.5.1 Inhibitors of Mitochondrial Electron Transport at Complex I

During the past few years, mitochondrial respiration has been targeted by several new structurally diverse acaricides and insecticides [127]. Beside non-halogenated compounds such as the pyrazole fenpyroximate (1991, Danitron®, Nihon Nohyaku) [128] or the quinazoline fenazaquin (1993, Magister®, Dow AgroSciences) [129], *mono*-chlorinated heterocyclic inhibitors of the mitochondrial electron transport of complex I (NADH dehydrogenase) have been described as so-called mitochondrial electron transport inhibitor (METI) acaricides. The first example is pyridaben (**48**; 1991, Sunmite®, Nissan) [130] and the second, an acaricide from pyrazole chemistry, tebufenpyrad (**49**; 1993, Masai®, Mitsubishi) [131]. The pyrimidine amine insecticide pyrimidifen (**50**; 1995, Miteclan®, Sankyo/Ube Ind.) [132], is active not only against all stages of spider mites (as are the former agents) but also against the diamondback moth, *Plutella xylostella*. Activity against aphids and whitefly identified the racemic development product flufenerim (**51**; Flumfen®, Ube Industries) (Figure 4.15).

The 5-chloropyrimidine system of **51** contains a 6-α-fluoroethyl group (R^1 = F) as well as a novel 4-trifluoromethoxy-phenethylamino side chain (R^2), and is structurally closely related to the acaricide (**50**).

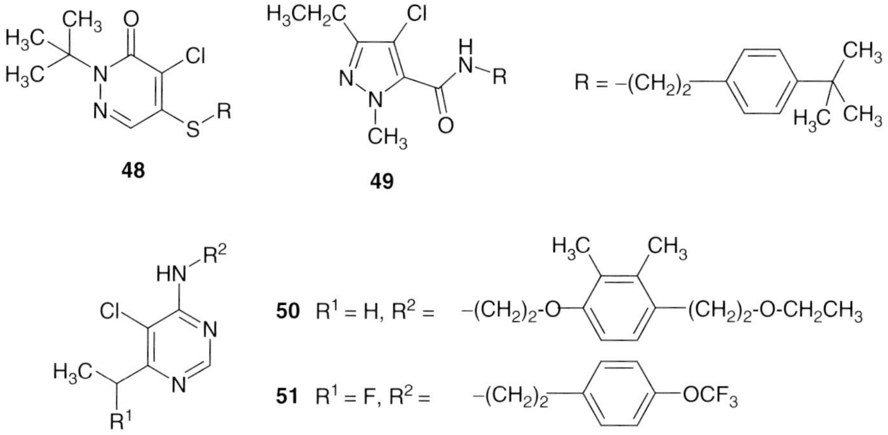

Figure 4.15 Inhibitors of mitochondrial electron transport at complex I (**48**–**51**).

4.3.5.2 **Inhibitors of Q_o Site of Cytochrome bc1 – Complex III**

Fluacrypyrim (**52**; 2002, Titaron®, Nippon Soda) (Figure 4.9) [75] is the first strobilurin analog to be marketed as an acaricide rather than a fungicide; notably, it inhibits mitochondrial electron transport at complex III of the respiratory chain (see Section 4.2.4). Moreover, it is active against all growth stages of spider mites, and shows an acaricidal contact and stomach action against *Panonychus ulmi* and *Tetranychus urticae* on citrus fruits and apples, as well as against spider mites on pears.

4.3.5.3 **Inhibitors of Mitochondrial Oxidative Phosphorylation**

Chlorfenapyr (**53**; 1995, Pirate®, ACC/BASF) [133], a potent uncoupler of mitochondrial oxidative phosphorylation [134], is based on a trifluoromethyl substituted pyrrole core (Scheme 4.3). It was modeled according to the fungicidal pyrrole natural product dioxapyrrolomycin [135], and contains three different halogen atoms and/or halogenated groups in the molecule (CF_3, 4-Cl-Ph, Br).

As the proinsecticide, chlofenapyr (**53**) [136, 137] is activated by an oxidative removal of the *N*-ethoxymethyl group, forming the *N*-dealkylated

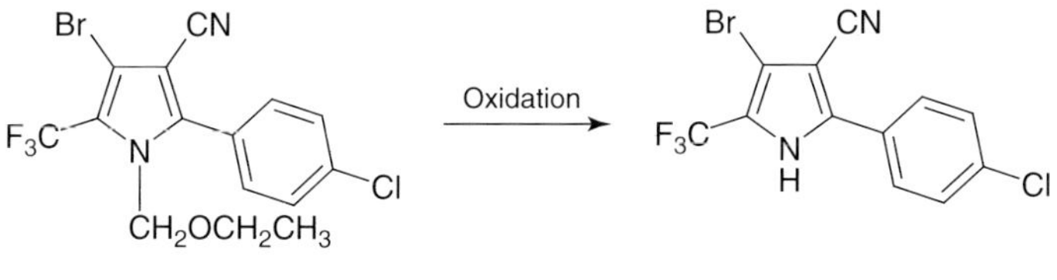

53 **54**

Scheme 4.3 Proacaricide chlorfenapyr (**53**) and its *N*-dealkylated metabolite (**54**).

metabolite **54**, which is a potent uncoupler of mitochondrial oxidative phosphorylation [138].

4.3.6 Ryanodine Receptor (RyR) Effectors

Flubendiamide (**55**; 2006, Belt®; Nihon Nohyaku Co., Ltd./Bayer CropScience) [139–141] (Figure 4.16) with a heptafluoro-*iso*-propyl moiety in the anilide part of the molecule, induces ryanodine-sensitive cytosolic Ca^{2+} transients that were independent of extracellular Ca^{2+} concentration in isolated neurons from the pest insect *Heliothis virescens*, as well as in transfected Chinese hamster ovary (CHO) cells expressing the RyR from *Drosophila melanogaster*. Binding studies on microsomal membranes from *H. virescens* flight muscle revealed that **55** interacts with a site distinct from the ryanodine binding site, and disrupted the Ca^{2+} regulation of ryanodine binding by an allosteric mechanism.

A second class of RyR effectors – the structurally different anthranilic amide derivative chlorantraniliprole (**56a**; 2007, Rynaxypyr®; DuPont) – was found also to be active against different species of lepidoptera, such as *P. xylostella*, *S. frugiperda*, and *H. virescens*. When the effect of different heterocyclic moieties and halogen atoms was investigated, radioligand-binding studies with **56a** and derivatives [142, 143] in *Periplaneta americana* skeletal muscle demonstrated a single saturable binding site, which was also distinct from that of ryanodine (Figure 4.16) [144].

In the search for analogs with enhanced systemic properties, the replacement of halogen by cyano in position R^3 was carried out. The resulting cyantraniliprole (**56b**; Cyazypyr™ provisionally approved by ISO; DuPont) [145] has an improved plant mobility and shows a broad activity against a wide range of insects that includes lepidopteran, hemipteran, and coleopteran pests (Figure 4.16).

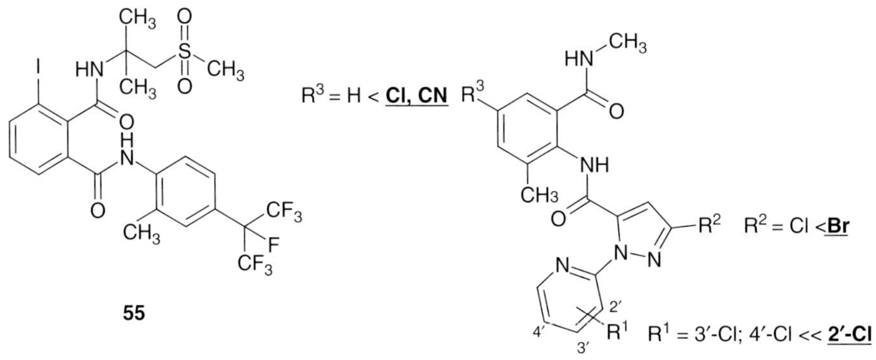

56a R^3 = Cl
56b R^3 = CN

Figure 4.16 Phthalic acid and anthranilic acid diamides such as flubendiamide (**55**), chlorantraniliprole (**56a**), and cyantraniliprole (**56b**).

4.4 Fungicides Containing Halogens

Today, modern specific fungicides control fungal plant pathogens more effectively, and at lower application rates, than the older "multisite" contact fungicides. Many of these agents have systemic properties and are therefore able to penetrate the plant tissue and to be further distributed via the xylem vessels into different plant parts that are not reachable directly by spray applications [146]. In this context, halogen substitution can have a remarkable effect on the physico-chemical behavior of fungicidally active compounds.

4.4.1 Sterol Biosynthesis Inhibitors (SBIs) and Demethylation Inhibitors (DMIs)

Conazole fungicides such as imidazoles and 1,2,4-triazoles represent one of the most important chemical groups of widely used agrochemicals [147, 148]. Most so-called DMIs [149] undergo systemic movement within plants. Almost 92% of the commercialized 1,2,4-triazoles are substituted with halogens, and about 68% of those commercialized so far possess chlorine substituents (F: ~8%, F/Cl: ~12%, Cl: ~68%, Cl/Br: ~4%). The chlorophenyl moiety is strongly preferred (Ph substitution pattern: 2,4,6-Cl_3; 3-Cl << 2,4-Cl_2 < 4-Cl), possibly because of the favorable physico-chemical properties such as the advantageous log *P*. Phenyl substituents – in most cases, halogens – adjust the lipophilicity of the product to a suitable value for systemic movement within the plant.

Details of the six most well-known and important halogenated DMIs – propiconazole (**57**) [150], tebuconazole (**58**) [151], cyproconazole (**59**) [152], difenoconazole (**60**) [153], tetraconazole (**61**) [154], and epoxiconazole (**62**) [155] – are listed in Table 4.8. Of these DMIs, the most successful – **58** and **62** – achieved sales of between US$ 250 and 400 million in 2009 alone.

Recently, a new fungicidal class of triazolinethiones was identified by a stepwise structural modification of the 1,2,4-triazole nucleus and its nucleophilic backbone [156]. The chlorine-containing prothiconazole (**63**; 2004, Proline®, Bayer CropScience) [157] was identified as an outstanding fungicide from this class (Figure 4.17). Apart from the 1,2,4-triazoline-5-thione ring system, prothiconazole (**63**) contains an *ortho*-chlorobenzyl residue and the innovative 1-chlorocyclopropyl moiety as a new lipophilic substituent. The commercial product **63** is a mixture of two active enantiomers, from which the (*S*)-(−)-enantiomer of **63** is significantly more active than the racemate [156].

Based on its broad spectrum of efficacy, excellent bioavailability, and long-lasting activity, prothiconazole (**63**) represents a systemic fungicide (log P_{OW} = 4.05) with protective and curative properties, and provides a very high standard of control of fungal diseases in cereals and other arable crops (usage rate: 200 g a.i. ha^{-1}).

Recently, Prosaro® (2011, Bayer CropScience) was registered in Western Canada, whereby a combination of tebuconazole (**58**) and prothioconazole (**63**) was made available for cereal disease control. Prosaro® will deliver the highest level of control

Table 4.8 Most important halogenated triazole fungicides (**57**–**62**).

Compound no.	Common name, trade name	Manufacturer (year introduced)	Structure
57	Propiconazole, Tilt®	Syngenta (1980)	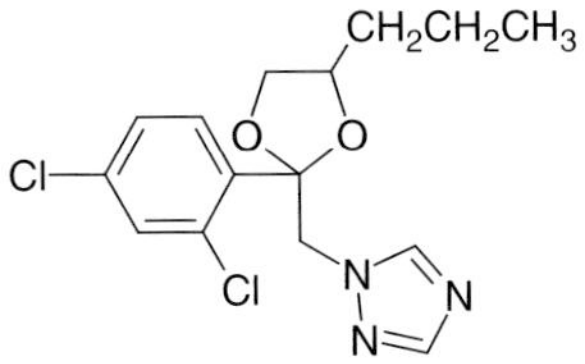
58	Tebuconazole, Folicur®	Bayer (1988)	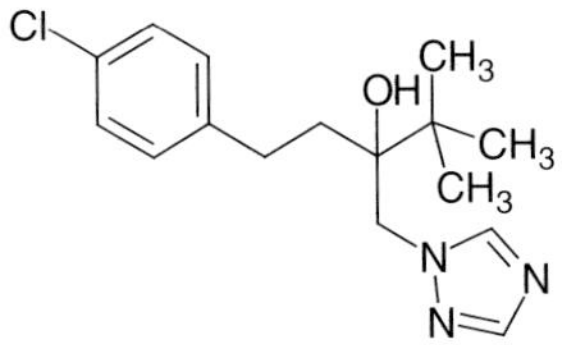
59	Cyproconazole, Sentinel®	(Sandoz) Bayer CropScience (1988)	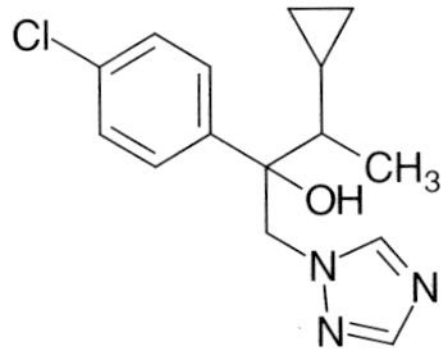
60	Difenoconazole, Score®	Syngenta (1989)	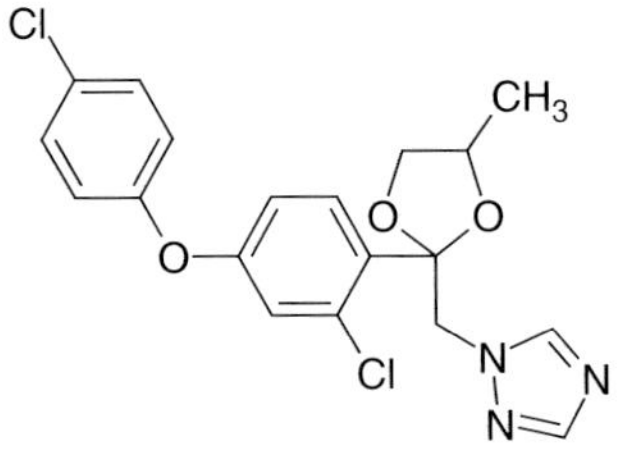
61	Tetraconazole, Eminent®	Montedison Enichem (1991)	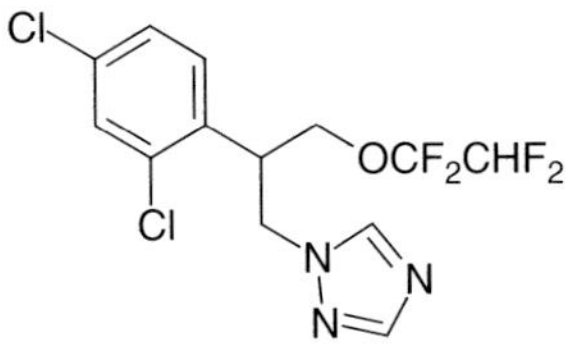
62	Epoxiconazole, Opus®	BASF (1992)	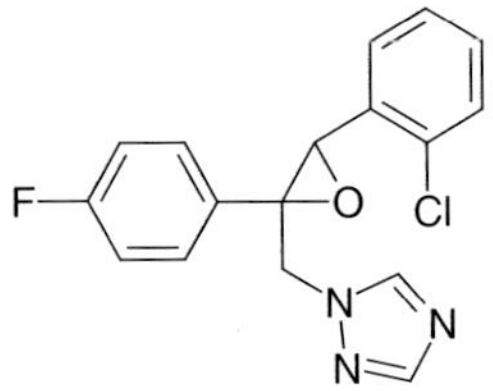

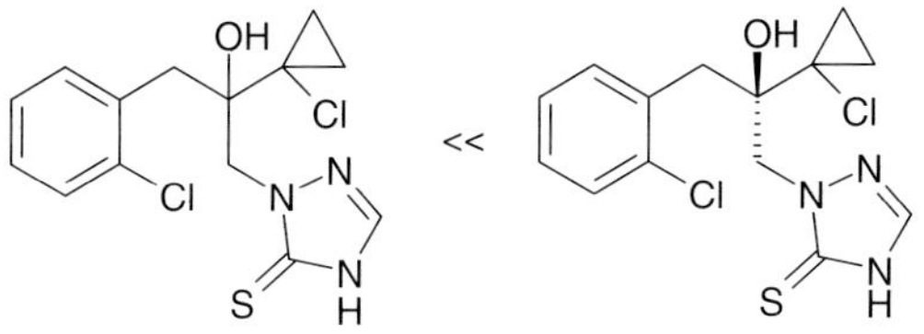

63 (S)-(–)-enantiomer

Figure 4.17 Prothioconazole (**63**) and its (S)-(−)-enantiomer.

and curative activity on leaf diseases, and the highest level of protection against *Fusarium* head blight (FHB) in wheat and barley.

4.4.2 Mitochondrial Respiratory Chain

4.4.2.1 Inhibitors of Succinate Dehydrogenase (SDH) – Complex II

Succinate dehydrogenase (SDH) is a membrane-bound enzyme that catalyzes the oxidation of succinic acid to fumaric acid. Following introduction of the non-halogenated carboxin (**64**; 1966, Vitavax®, Uniroyal) [158] and the iodine-containing benodanil (R = I) (**65**) [159], both of which are used for seed dressing with only particular activity against seedling diseases, a range of further halogen and/or halogen-substituted carboxamides with an *N*-phenyl-2-butene amide structure (R = F, Cl, CF_3, CHF_2) has been described. With the incorporation of halogen and/or halogen-substituted residues in combination with different heterocyclic moieties in the new generation of SDH inhibitors, an evolution of their biological profile and effects on plant physiology was apparent (Scheme 4.4, Table 4.9) [160].

For example, the two rice fungicides flutolanil (**66**) [161] and thifluzamide (**68**) [162] contain the systemic trifluoromethyl group (R = CF_3). Whereas, the chlorine-containing fungicide furametpyr (**67**) (R = Cl) addresses the rice market, boscalid (**69**) [163] is already more fused on specialty crops. Introduction of the novel 3-difluoromethyl-*N*-methyl-1*H*-pyrazol-4-yl moiety (R = CHF_2), as presented in isopyrazam (**70**) [164] and in various development candidates from different agrochemical companies such as bixafen (**71**) [165], fluxapyroxad (**72**) [166], or

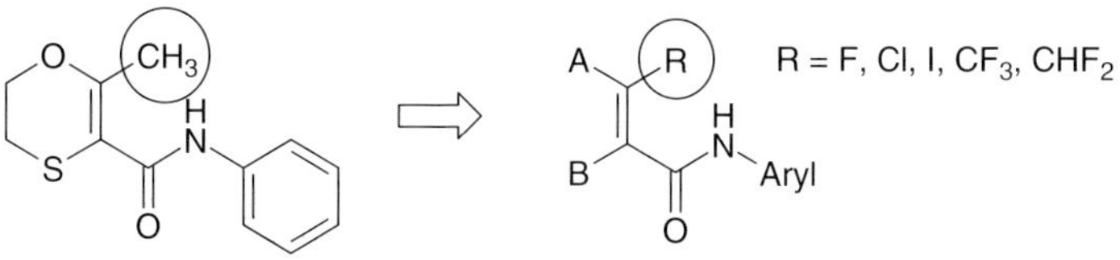

64 N-aryl-2-butene amides

Scheme 4.4 From carboxin (**64**) to halogenated inhibitors of succinate dehydrogenase (SDH) complex II (**65**–**75**; see Table 4.9).

Table 4.9 Halogenated inhibitors of succinate dehydrogenase (SDH) – complex II (**65–75**) (see Scheme 4.4 for the basic structures).

Compound no.	Common name, trade name	Manufacturer (year introduced)	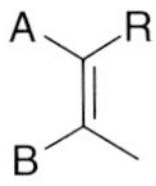	Aryl	Application rate[a] (crop) (g a.i. ha^{-1})
65	Benodanil, Benefit®	BASF (1974)	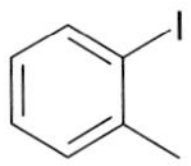	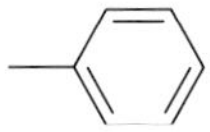	750–1000 (barley)
66	Flutolanil, Flutranil®	Nihon Nohyaku (1985)	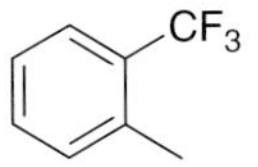	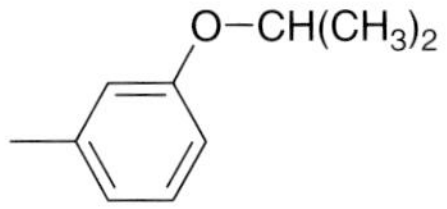	450–600 (rice)
67	Furametpyr, Limber®	Sumitomo (1996)	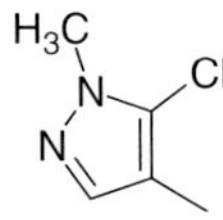	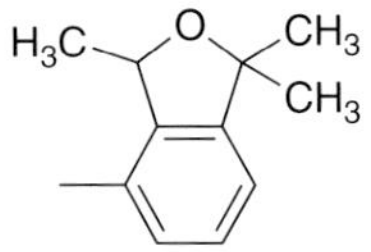	150–200 (rice)
68	Thifluzamide, Pulsor®	Monsanto (1997)	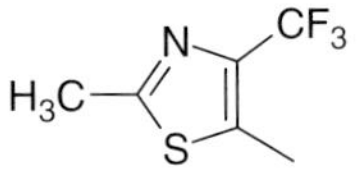	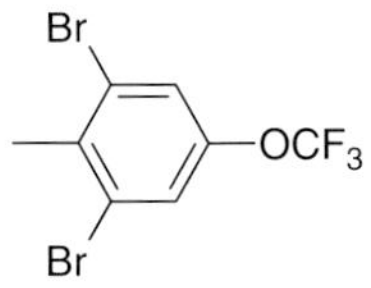	255–340 (rice)
69	Boscalid, Cantus®	BASF (2003)	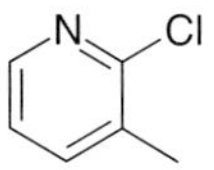	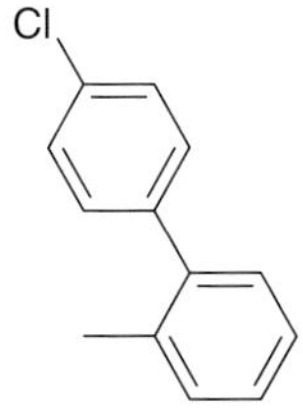	150–500[b](rice)
70	Isopyrazam[c] Bontima®	Syngenta (2010)	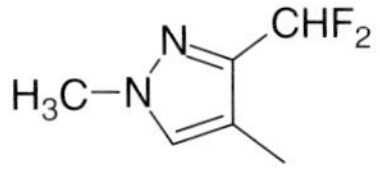	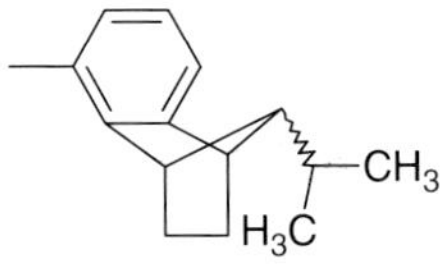	125 (cereals)
71	Bixafen Aviator®	Bayer CropScience (2011)	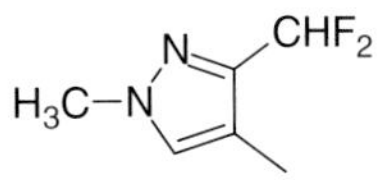	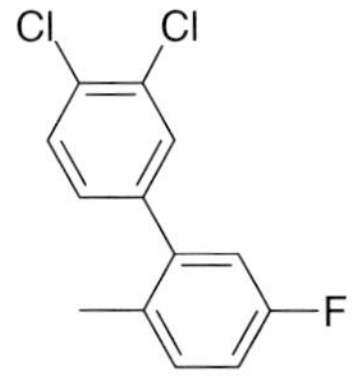	125 (cereals)

Table 4.9 (*continued*).

Compound no.	Common name, trade name	Manufacturer (year introduced)	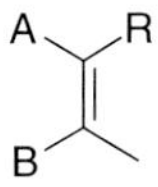	Aryl	Application rate[a] (crop) (g a.i. ha^{-1})
72	Fluxa-pyroxad Xemium®	BASF (2010)	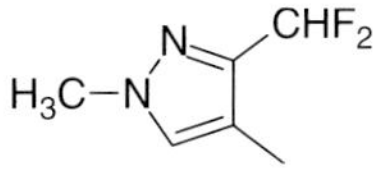	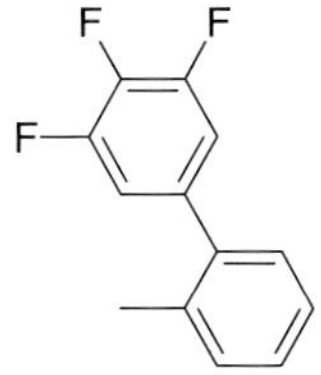	75 (cucurbits, vine)
73	Sedaxane Vibrance®	Syngenta (2011)	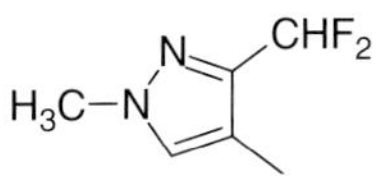	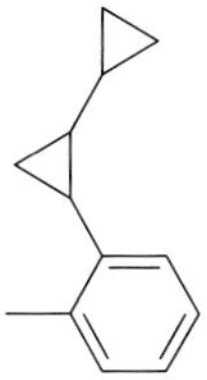	20[d] (canola)
74	Penthiopyrad Vertisan®	Mitsui (2011)	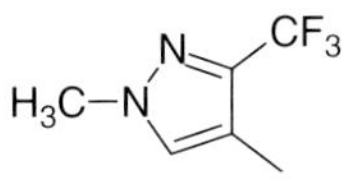	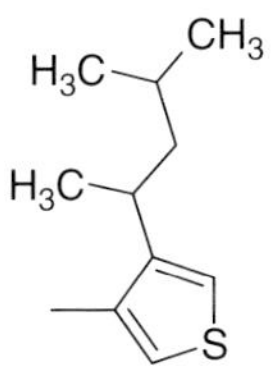	100–250 (cereals)
75	Penflufen Emesto®[e]	Bayer Crop-Science(2011)	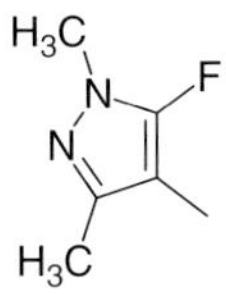	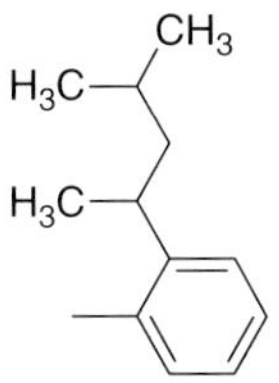	50 (potatoes)

[a]Seed treatment, systemic activity against *Rhizoctonia solani*.
[b]Broad-spectrum systemic fungicide.
[c]Technical isopyrazam in a mixture of 2 *syn*-isomers and 2 *anti*-isomers; ratios between 70 : 30 and 100 : 0.
[d]Vibrance®: 2.5-20 g per 100 g seed
[e]Emesto Fusion®: mixture of 390 g a.i. h^{-1} fluoxastrobin and 50 g a.i. h^{-1} penflufen

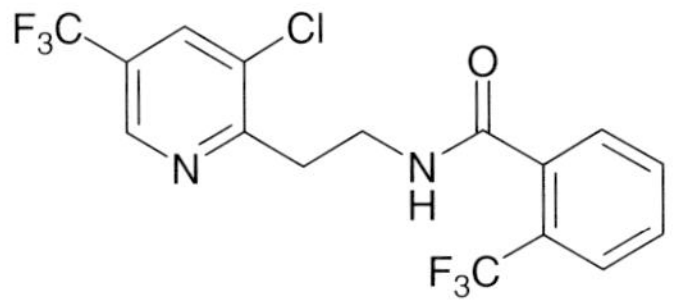

76

Figure 4.18 Pyridinyl-ethyl benzamide fluopyram (**76**).

sedaxane (**73**) [167], has either broadened their spectrum (cereals and soy bean) or improved the binding to the receptor compared with the halogen free precursors. As demonstrated for boscalid (**69**), and shown later for bixafen (**71**) or fluxapyroxad (**72**), additional halogen atoms (chlorine and/or fluorine) in the biphenyl part of the molecule can be favorable for long-lasting disease control (Table 4.9).

In this context, the fungicidal spectrum of penthiopyrad (**74**) [168] and penflufen (**75**) [169] containing trifluoromethyl- or fluoro-substituted *N*-methyl-pyrazole moieties ($R = F, CF_3$) and a 1,3-dimethylbutyl side chain in their het(aryl) parts as so-called mimics of biphenyl, has also been widened.

In the case of the novel pyridinyl-ethyl benzamide fungicide fluopyram (**76**) [170] (Figure 4.18), containing the 3-chloro-5-trifluoromethyl-2-pyridinyl moiety (as already outlined; see Section 4.1), a different cross-resistance pattern from other known SDH inhibitors (e.g., boscalid **69**) has recently been detected in *Corynespora cassiicola* and *Podosphaera xanthii*, the pathogens causing Corynespora leaf spot and powdery mildew disease on cucumber [171].

4.4.2.2 Inhibitors of Q_o Site of Cytochrome bc1 – Complex III

Over the past eight years, strobilurin fungicides have been among the most commercially successful class of agricultural fungicides [172]. Like the major lead structure strobilurin A [173], all strobilurins inhibit mitochondrial respiration by influencing the function of the so-called Q_o site of complex III (cytochrome bc_1 complex) [174–176], which is located in the inner mitochondrial membrane of fungi and other eukaryotes [177]. The binding site for the Q_o site inhibitors is distinct from the stigmastellin binding site within the membrane and below the peripheral helix αcd1 [178].

Despite being first introduced in 1996, the eight commercialized strobilurins are already the second largest group in the market, behind conazole fungicides (Section 4.4.1). Whereas, the first broad-spectrum systemic strobilurins such as azoxystrobin (1996, Amistar®, Syngenta) [179] and kresoxim-methyl (1996, Stroby®, BASF) [180] are not halogen-substituted, the incorporation of halogen atoms or halogenated substituents into the side chain began with trifloxystrobin (**77**; 1999, Flint®) [181], which contains a 3-trifluoromethylphenyl moiety in its oximether side chain and belongs to a new generation of strobilurin fungicides. Crystallographic studies of the binding of **77** have shown that this moiety interacts with a hydrophobic domain in the binding pocket (Phe128, Ile146, Ala277, Leu294), which presents a higher amino acid variability among organisms and plays a

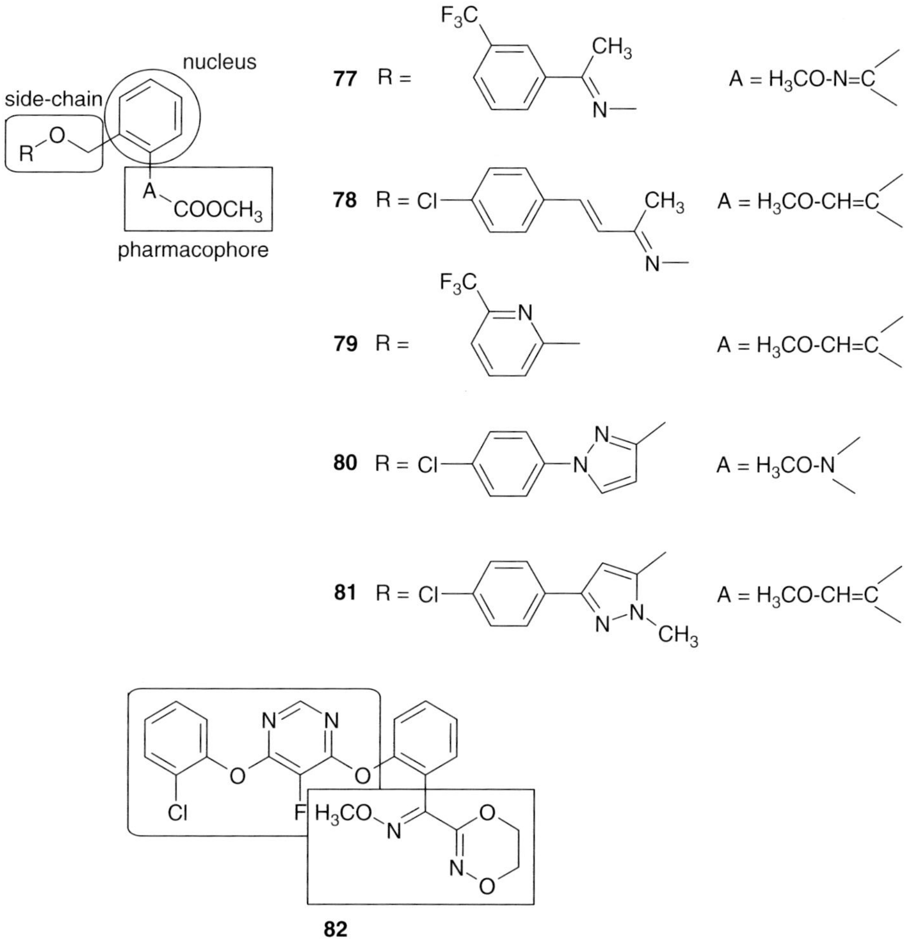

Figure 4.19 Commercialized strobilurine fungicides (**77**, **79**, **80**, and **82**) and the development products **78** and **81**.

role in species specificity toward β-methoxy-acrylates. During the preparation of oximethers, it has been shown that compounds with a fluorine-containing phenyl substituent such as trifluoromethyl showed a particularly strong systemic activity. Like kresoxim-methyl (vapor pressure: 2.3×10^{-3} mPa at 20 °C), **77** delivers disease control in the plant by virtue of a vapor action (Figure 4.19, Table 4.10) [182].

The low aqueous solubility (0.6 mg l^{-1}) and relatively high lipophilicity (log $P_{OW} = 4.5$) of **77** contribute to a high affinity for the waxy layer on the surface of the plant leaf for a long time, which in turn leads to the formation of a rain-resistant store of the active ingredient. A high level of humidity after a short drying phase aids retention of the fungicide and increases its redistribution. The special behavior of **77** on the surface of the plant, which is known as "*mesosystemic activity*," provides

Table 4.10 Physical Properties of Halogen-Containing Compounds of Stobilurin Type (**77–80, 82**).

Compound no.	Melting point (°C)	Vapor pressure (mPa at 20 °C)	Log P_{ow} (at 20 °C)	Solubility in water ($mg\,l^{-1}$ at 20 °C)
77	72.8–72.9	3.4×10^{-3a}	4.5	0.6
78	n.d.	n.d.	n.d.	n.d.
79	75.0	5.5×10^{-3}	3.6	3.1
80	63.7–65.2	2.6×10^{-5}	3.99[b]	1.9
81	n.d.	n.d.	n.d.	n.d.
82	103–108	6.0×10^{-7c}	2.86	2.56[d], 2.29[e]

[a] At 25 °C.
[b] At 22 °C.
[c] Extrapolated.
[d] Unbuffered.
[e] At pH 7.
n.d. = not described.

an excellent control of apple scab because of its inhibitory effects on the multiple stages of the life cycle of *Venturia inaequalis* [183].

Recently, the new development product enestrobin (**78**; Xiwojunzhi®, Shenyang Res. Inst. of Chem. Industry) [184], containing a 4-chlorophenyl unsaturated oxime ether side chain, has been presented. To date, field trial results have indicated that **78** is an especially active fungicide against crop diseases on cucumbers, such as downy mildew, powdery mildew, and gray mold; the agent is especially useful in plastic sheet-covered cucumber fields.

Picoxystrobin (**79**; 2002, Acanto®; Syngenta) [74] has a 6-CF_3-pyridin-2-yl moiety in its arylalkyl ether side chain, and was developed initially for disease control in cereals and apples.

BASF's second strobilurin, the *N*-(4-chloro-phenyl)-1*H*-pyrazol-3-yloxy-containing pyraclostrobin (**80**; 2002, Cabrio®; BASF) [185], derives its broader spectrum of activity from an introduction of the 4-chloro-phenyl-1*H*-pyrazol-3-yloxy moiety. Inspired by the azole structure, the new development product pyraoxystrobin (**81**; provisionally approved by ISO; Shenyang Res. Inst. of Chem. Industry) [186], demonstrated in field trials (used as 20% suspension concentrate; SC) an effective control of cucumber powdery mildew [187]. One of Bayer's research programs has focused on the variation of the toxophore moiety, which led to an incorporation of the carbocyclic acid moiety into a six-membered heterocycle.

The aryl ether structure of fluoxastrobin (**82**; 2005, Evito®; Bayer CropScience) [188] combines a methoximino 5,6-dihydro-1,4,2-dioxazin-2-yl toxophore with an optimally adjusted side chain bearing a 6-(2-chloro-phenoxy)-5-fluoro-pyrimidin-4-yloxy moiety as an essential element.

Under the provision that both **77** and **82** bind to their target in similar ways, it can be assumed that **82** has an advantage as no reorientation of the toxophore is necessary for binding to the target [189]. The excellent leaf systemicity serves as a basis for the rapid uptake and even acropetal distribution of **82** in the leaf. The SARs indicate that the fluorine atom has an important influence on the phytotoxicity and leaf systemicity of this fungicide and, in comparison with chlorine and hydrogen, fluorine is strongly preferred (Cl << H < F). Seed treatment with **82** provides both a very good broad-spectrum control and a long-lasting protection of young seedlings from seed and soil-borne pathogens.

Today, about 50% of all commercialized strobilurine fungicides are halogenated with chlorine and fluorine, or contain a "mixed" halogen substitution pattern such as chlorine and fluorine.

As a contribution to a resistance management strategy for strobilurins, fluoxastrobin (**82**) is developed as a co-formulation or recommended as a tank mix with fungicides from other chemical classes. For example, its combination (2004, Fandango®, Bayer CropScience) [190] with the chlorine-containing DMI-type fungicide prothioconazole (**63**) provided at least additive effects of **82** and extended the spectrum of activity towards all important seed and soil-borne pathogens [191].

4.4.2.3 NADH Inhibitors – Complex I

Agrochemical fungicides acting as NADH inhibitors with useful potency, spectrum, and toxicological properties, and which are sufficiently interesting for commercialization, are rare [192]. To date, only one compound, diflumetorim (**83**; 1997, Pyricut®, Ube Ind.), has been introduced into the market for use in ornamentals (Figure 4.20). The compound possess a trifluoromethoxy group that acidifies the NH binding in the amide moiety to improve the inhibitory properties.

4.4.3 Fungicides Acting on Signal Transduction

The non-systemic pyrrole fungicides fenpiclonil (**85**; 1988, Beret®, Syngenta) [193] (R^1, R^2 = Cl) and its difluoromethylenedioxy analog fludioxonil (**86**; 1993, Saphire®, Syngenta) [194] have been developed from the photo-unstable and chlorine-containing natural antibiotic pyrrolnitrin (**84**; Pyroace®, Fujisawa) that was first isolated from *Pseudomonas pyrrocinia* [195] (Figure 4.21, Table 4.11).

Especially, the introduction of the difluoromethylenedioxy moiety improved both biological activity and soil stability. Subsequent biochemical studies revealed

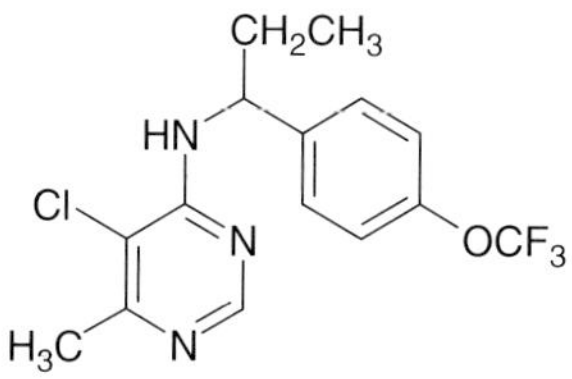

83

Figure 4.20 Diflumetorim (**83**).

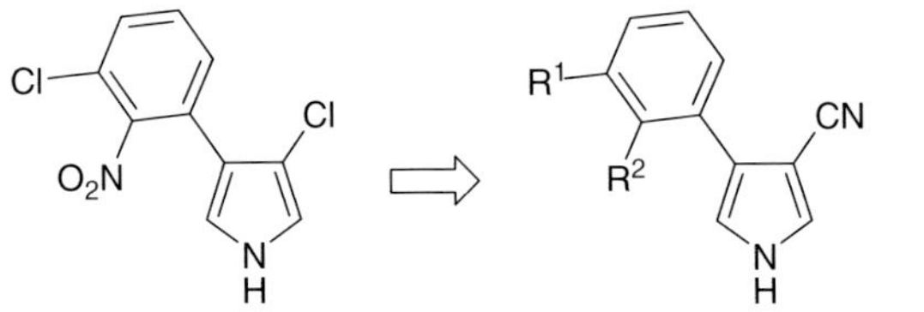

84

85 $R^1, R^2 = Cl$

86 $R^1 - R^2 = O\text{-}CF_2\text{-}O$

Figure 4.21 From natural antibiotic pyrrolnitrin (**84**) to the pyrrole fungicides fenpiclonil (**85**) and fludioxonil (**86**).

Table 4.11 Physical properties of pyrrolnitrin (**84**) and synthetic pyrrole fungicides (**85** and **86**).

Compound no.	Melting point (°C)	Vapor pressure (mPa at 25 °C)	Log P_{ow} (at 25 °C)	Solubility in water ($mg\,l^{-1}$ at 25 °C)	Light stability $t_{1/2}$ (h)
84	124.5	1.42×10^{-6a}	3.09	n.d.	n.d.
85	144.9–151.1	1.1×10^{-2}	3.86	4.8	48.0
86	199.8	3.9×10^{-4}	4.12	1.8	24.5

[a] Torr.
n.d. = not determined.

that the pyrrole fungicides inhibit a PK-III that is potentially involved in the osmosensing signal transduction pathway [196].

4.5 Plant Growth Regulators (PGRs) Containing Halogens

4.5.1 Reduction of Internode Elongation: Inhibition of Gibberellin Biosynthesis

Some of the triazoles, and especially their bioisosteric pyrimidine analogs such as the non-halogenated ancymidol (**87**; 1973, Arest®, Eli Lilly) [197], exhibit PGR activity in mono- and dicotyledonous species, and act by reducing internodal elongation through interaction with the gibberellin biosynthesis pathway [198]. Replacement of the 4-methoxy-phenyl group with a 4-trifluoromethoxy-phenyl moiety, and the cyclopropyl group with isopropyl, leads to flurprimidol (**88**; 1989, Cutless®, Dow AgroSciences) [199], which has several different physico-chemical properties, for example, log P and DT_{50} values or vapor pressure (Figure 4.22, Table 4.12).

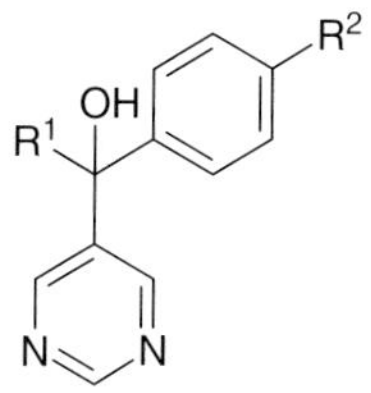

87 R^1 = cyclopropyl, $R^2 = OCH_3$

88 $R^1 = CH(CH_3)_2$, $R^2 = OCF_3$

Figure 4.22 Plant growth regulators ancymidol (**87**) and flurprimidol (**88**).

Table 4.12 Comparison of the physical properties of ancymidol (**87**) and flurprimidol (**88**).

Compound no.	Melting point (°C)	Vapor pressure (mPa at 25 °C)	Log P_{ow} (at 20 °C)	Solubility in water ($mg\,l^{-1}$ at 25 °C)	DT_{50}
87	110–111	<0.13[a]	1.9[b]	~650	>30 days[c]
88	93.5–97	4.85×10^{-2}	3.34	114[d]	~3 h

[a] At 50 °C.
[b] pH 6.5 at 25 °C.
[c] pH 5–9 at 25 °C.
[d] At 20 °C.

Whereas, **87** translocated only in the phloem of plants, compound **88** is both xylem- and phloem-mobile.

4.6 Herbicides Containing Halogens

Since the early 1950s, agrochemical companies have launched a series of new compounds that have consistently offered farmers progressive solutions for weed management in cereals. Today, more than 60% of proprietary cereal herbicides contain halogens, notably fluorine, chlorine, bromine, and iodine. The excellent efficiency, selectivity, and plant compatibility are the most prominent advantages for these halogen-containing commercial products.

4.6.1 Inhibitors of Carotenoid Biosynthesis: Phytoene Desaturase (PDS) Inhibitors

Most commercial so-called "bleaching herbicides" inhibit the synthesis of carotenoids by interfering with the carotenoid biosynthesis of photosynthetic pigments, chlorophylls, or carotinoids [200–202]. Subsequently, enzyme kinetics studies with several inhibitors have revealed a reversible binding to the enzyme, and non-competitive inhibition [203].

Table 4.13 Chemically different classes of phytoene desaturase inhibitors (**89–95**) (see Figure 4.23 for the basic structure).

Compound no.	Common name, trade name(s)	Manufacturer (year introduced)	R^1	X	R^2
89	Norflurazon, Zorial®, Telok®	Syngenta (1971)	H	N	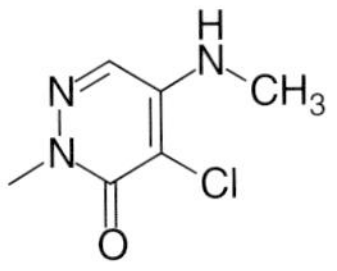
90	Fluridone, Brake®, Sonar®	Dow AgrowSciences (1981)	H	–	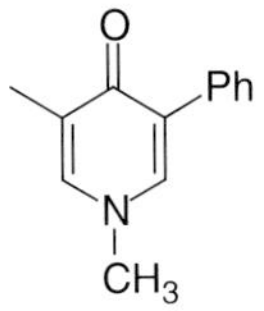
91	Flurochloridone, Rainbow®, Racer®	Makhteshim-Agan (1985)	H	N	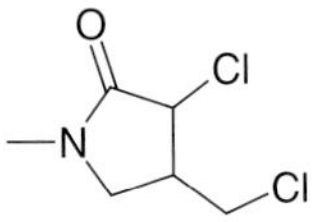
92	Diflufenican, Quartz®, Fenikan®	Bayer CropScience (1985)	H	O	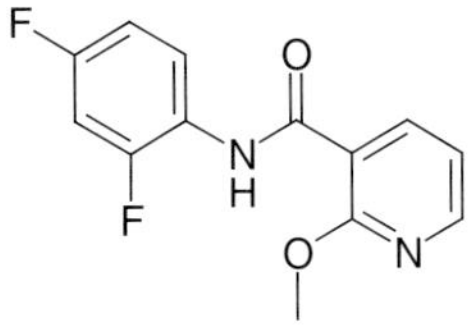
93	Flurtamone, Benchmark®, Bleacher®	Bayer CropScience (1997)	H	–	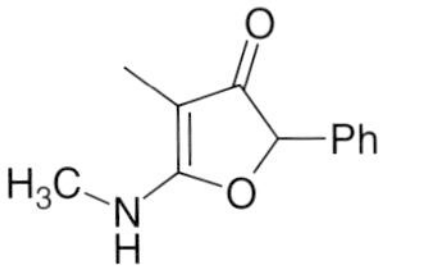
94	Picolinafen, Pico®, Sniper®	BASF (2001)	H	O	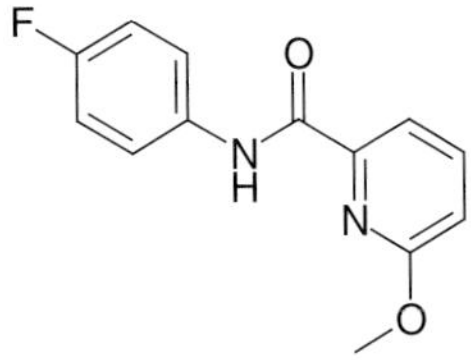
95	Beflubutamid, Benfluamid®	Ube Ind. (2003)	F	O	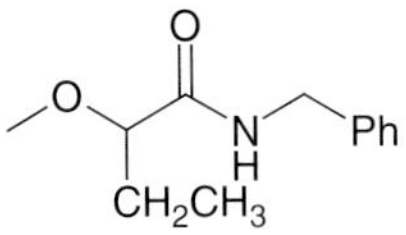

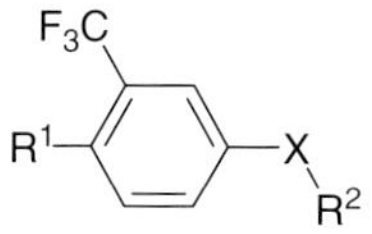

Figure 4.23 The 3-trifluoromethylphenyl moiety common to commercial products **89–95**.

A common fragment in all such commercial products (**89–95**; see Table 4.13) is the 3-trifluoromethyl-phenyl moiety (Figure 4.23), for example, norflurazon (**89**) [204], fluridone (**90**) [205], and fluorochloridone (**91**) [206]. Other commercial products are diflufenican (**92**) [207], flurtamone (**93**) [208], and picolinafen (**94**) [209], which contains a pyridine skeleton similar to that of **92**. A more recent addition to the list is the new selective herbicide for weed control in cereals, beflubutamid (**95**; 2003, Benfluamid®, Ube Ind.) [210], which contains as an exemption the 4-fluoro-3-trifluoromethyl-phenyl moiety ($R^1 = F$).

This moiety is a feature of either an arylanilide (X = N; log $P = 2.45, 3.36$; **89**, **91**, respectively), arylether (X = O; log $P = 4.90, 5.37, 4.28$; **92**, **94**, **95**, respectively) or of a substituted *N*-methyl-enaminone structure (X = C–C bond; log $P = 1.87$, n.d.; **90**, **93**, respectively) in a five- or six-membered heterocycle, respectively. It is assumed that the biochemical activity of these compounds is determined by the properties of the *meta*-trifluoromethyl-phenyl group, such as high lipophilicity (for X = O) and an electron-withdrawing nature. Furthermore, there are strict requirements for substitution at the five- or six-membered heterocycle of the inhibitor, especially at the position most distant from the carbonyl group.

4.6.2 Inhibitors of Acetolactate Synthase (ALS)

4.6.2.1 Sulfonylurea Herbicides

Sulfonylureas are extremely potent inhibitors of ALS [211], the key enzyme involved in the biosynthesis of branched amino acids such as leucine (Leu), isoleucine (Ile), or valine (Val) [211, 212].

Until now, approximately 48% of commercialized sulfonylureas or development products have been halogen-free, while almost 26% of the sulfonylureas have contained fluorine, and about 26% contained other halogens such as chlorine or iodine.

Sulfonylureas (Figure 4.24) can be further divided into two subclasses: (i) triazinylsulfonylurea herbicides (Y = N; Table 4.14); and (ii) pyrimidinylsulfonylurea herbicides (Y = CH; Table 4.15):

4.6.2.1.1 Halogen-containing triazinylsulfonylurea herbicides

Exchange of the *ortho*-chloro substituent (ring A, R^1) in the cercal-selective herbicide chlorsulfuron (**96**; 15–20 g a.i.ha^{-1}) with *ortho*-2-chloroethyl or *ortho*-3,3,3-trifluoropropyl leads to triasulfuron (**97**; wheat and barley, 5–10 g a.i. ha^{-1}) [213] and prosulfuron (**99**; maize, 10–40 g a.i. ha^{-1}; winter wheat, 20–30 g a.i. ha^{-1}) [214], respectively, which shows a selectivity shift (such changes facilitate patent applications). Maize-selective **99** is metabolized in maize

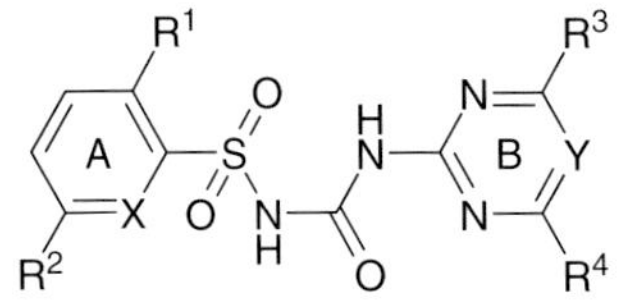

(i) Y = N Triazinylsulfonylurea herbicides
(ii) Y = CH Pyrimidinylsulfonylurea herbicides

Figure 4.24 Sulfonyl ureas.

Table 4.14 Halogen-containing triazinylsulfonylurea herbicides (**96–102**) (see Figure 4.23 for the basic structure).

Compound no.	Common name, trade name(s)	Manufacturer (year introduced)	R^1	R^2	X	R^3	R^4	Y
96	Chlorsulfuron, Glean®, Telar®	DuPont (1982)	Cl	H	CH	OCH_3	CH_3	N
97	Triasulfuron, Logran®, Amber®	Syngenta (1987)	$O(CH_2)_2Cl$	H	CH	OCH_3	CH_3	N
98	Triflusulfuron-methyl, Safari®, Debut®	DuPont (1992)	$COOCH_3$	H	CCH_3	$N(CH_3)_2$	OCH_2CF_3	N
99	Prosulfuron Peak®, Scoop®	Syngenta (1994)	$(CH_2)_2CF_3$	H	CH	OCH_3	CH_3	N
100	Iodosulfuron-methyl[a], Husar®, Hussar®	Bayer CropScience (2000)	$COOCH_3$	I	CH	OCH_3	CH_3	N
101	Trifloxysulfuron[a] Envoke®	Syngenta (2001)	OCH_2CF_3	H	CH	OCH_3	OCH_3	N
102	Tritosulfuron, Biathlon®	BASF (2005)	CF_3	H	CH	OCH_3	CF_3	N

[a] Sodium salt.

via an additional hydroxylation at the methyl group (ring B, R^4) of the triazine moiety (cf. Figure 4.25). A novel combination of substituents in ring B (R^3) and (R^4) is given in triflusulfuron-methyl (**98**; sugar beet, 10–30 g a.i. ha^{-1}) [215], which contains the *N,N*-dimethylamino and 3,3,3-trifluoroethyoxy group. Iodosulfuron-methyl-sodium (**100**) [216], the iodine derivative (ring A, $R^2 = I$) of metsulfuron-methyl (**103**; 1984, Gropper®, DuPont) [217], has a 10-fold faster soil degradation ($DT_{50} = 1–5$ days) than the non-halogenated sulfonylurea (**103**; $DT_{50} = 52$ days). Trifloxysulfuron-sodium (**101**; cotton, 5–7.5 g a.i. ha^{-1}) [218] and tritosulfuron (**102**; cereals, rice, and maize, 40–75 g a.i. ha^{-1}) demonstrate that the

Table 4.15 Halogen-containing pyrimidinylsulfonylurea herbicides (**105–109**) (see Figure 4.24 for the basic structure).

Compound no.	Common name, trade name(s)	Manufacturer (year introduced)	R^1	R^2	X	R^3	R^4	Y
105	Chlorimuron-ethyl, Classic®, Darban®	DuPont (1985)	$COOCH_2CH_3$	H	CH	OCH_3	Cl	CH
106	Flazasulfuron, Shibagen®	Ishihara (1989)	CF_3	H	N	OCH_3	OCH_3	CH
107	Flurpyrsulfuron-methyl[a], Lexus®, Oklar®	DuPont (1997)	$COOCH_3$	CF_3	N	OCH_3	OCH_3	CH
108	Primisulfuron-methyl Beacon®, Tell®	Syngenta (1998)	$COOCH_3$	H	CH	$OCHF_2$	$OCHF_2$	CH
109	Flucetosulfuron[b] Fluxo™	LG Chem	CH(OR)-$CHFCH_3$[c]	H	N	OCH_3	OCH_3	CH

[a] Sodium salt.
[b] Provisionally approved by ISO, development product.
[c] R = CO-CH_2-OCH_3.

trifluoroethoxy or trifluoromethyl groups are useful substituents (R^1) in ring A; in addition, the methyl (R^4) in ring B is exchangeable with the 3,3,3-trifluoroethyoxy (**98**) and trifluoromethyl group (**102**).

4.6.2.1.2 Halogen-containing pyrimidinylsulfonylurea herbicides

The first halogenated member of this subclass is chlorimuron-ethyl (**105**; soya beans and peanuts, 9–13 g a.i. ha^{-1}), which demonstrates the successful exchange of methoxy group with chlorine atom in ring B (Figure 4.24). A comparison of rimsulfuron (**104**; 1991, Titus®, DuPont) [219] containing a 3-ethylsulfonyl-pyridin-2-yl moiety with flazasulfuron (**106**) [220] showed that its 3-trifluoromethyl-pyridin-2-yl moiety had a marked impact on metabolism (Figure 4.25). The key transformation in tolerant turf grass is an unusual rearrangement and contraction of the sulfonylurea bridge, followed by hydrolysis and *O*-demethylation of a pyrimidyl methoxy (R^3) group. In contrast to **106**, flurpyrsulfuron-methyl sodium (**107**) contains a 3-methoxycarbonyl-6-trifluoromethyl-pyridin-2-yl moiety, which influences its metabolic pathway (Figure 4.25). Besides glutathione (GSH) conjugate formation (attack of GSH), *O*-demethylation is predominant in the detoxification of **107** in cereals (10 g a.i. ha^{-1}).

Primisulfuron-methyl (**108**) [221] is a maize-selective herbicide. Comparison with the unfluorinated triazine counterpart metsulfuron-methyl (**103**) indicates

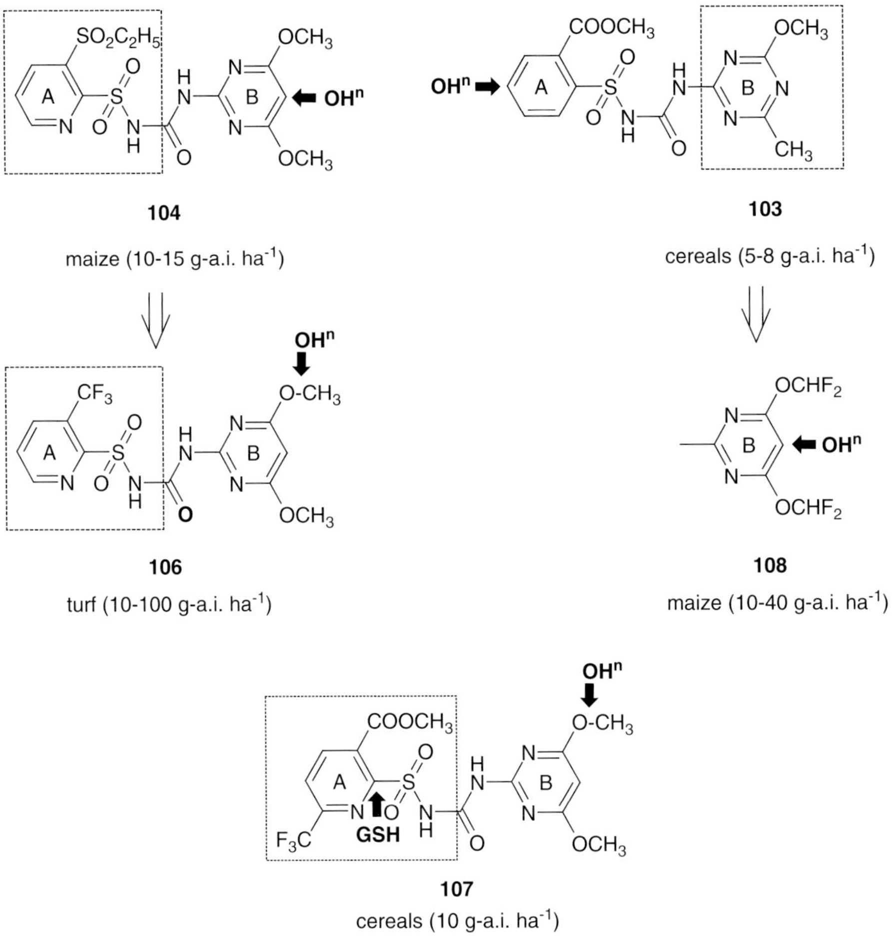

Figure 4.25 Influence of fluorine-containing substituents on selectivity and metabolism.

that crop safety for maize is achieved by a replacement of the triazine methoxy (R^3) and methyl (R^4) group in ring B with two difluoromethoxy groups (R^3, R^4).

In addition, **108** is deactivated in maize by hydroxylation of the phenyl ring A and pyrimidyl moiety B, followed by hydrolysis or further conjugation (Figure 4.25).

The systemic herbicide halosulfuron-methyl (**110**; maize, 18–35 g a.i. ha^{-1}) [222], metazosulfuron (**111**; provisionally approved by ISO, Nissan Chemical Industries) [223], imazosulfuron (**112**; Take-off®, Sumitomo; paddy rice, 75–95 g a.i. ha^{-1}) [224], and propyrisulfuron (**113**; provisionally approved by ISO, Sumitomo) [225] demonstrate the structural variability of ring A, for example, by the incorporation of further halogenated heterocyclic systems such as 3-chloro-1-methyl-1*H*-pyrazol-5-yl (**110** and **111**), 2-chloroimidazo [1,2-*a*]pyridine-3-yl (**112**) and 2-chloro-6-propylimidazo[1,2-b]pyridazin-3-yl (**113**) (Figure 4.26).

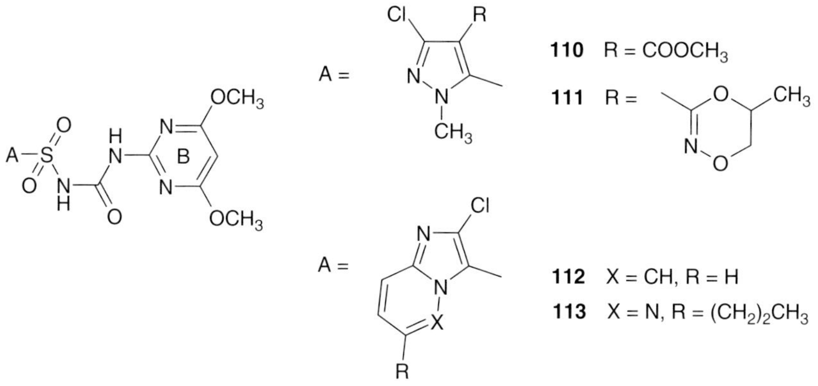

Figure 4.26 Halogen-containing pyrimidinylsulfonyl herbicides halosulfuron-methyl (**110**), metasulfuron (**111**; provisionally approved by ISO), imazosulfuron (**112**), and propyrisulfuron (**113**; provisionally approved by ISO).

4.6.2.2 **Sulfonylaminocarbonyl-Triazolone Herbicides (SACTs)**

The exchange of the *ortho*-$COOCH_3$ by the *ortho*-OCF_3 residue in the sulfonylaryl unit of propoxycarbazone sodium (**114**; 2001, Attribut®, Bayer CropScience) [226] led to the systemic herbicide flucarbazone sodium (**115**; 2000, Everest®, Bayer CropScience) [227], which provides excellent activity against grass weeds and several important broadleaf weeds when applied post-emergence to wheat (Figure 4.27).

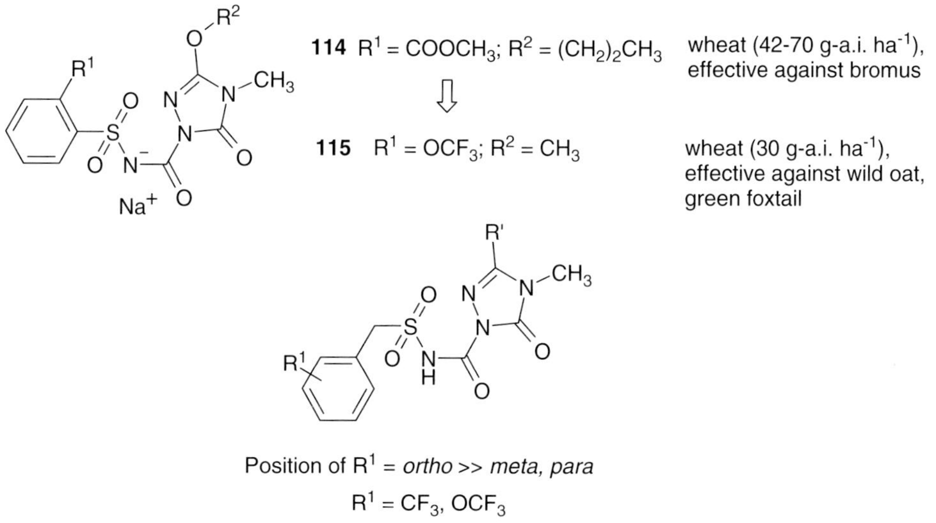

Figure 4.27 Triazolon herbicides and optimization in their sulfonylmethylaryl moiety.

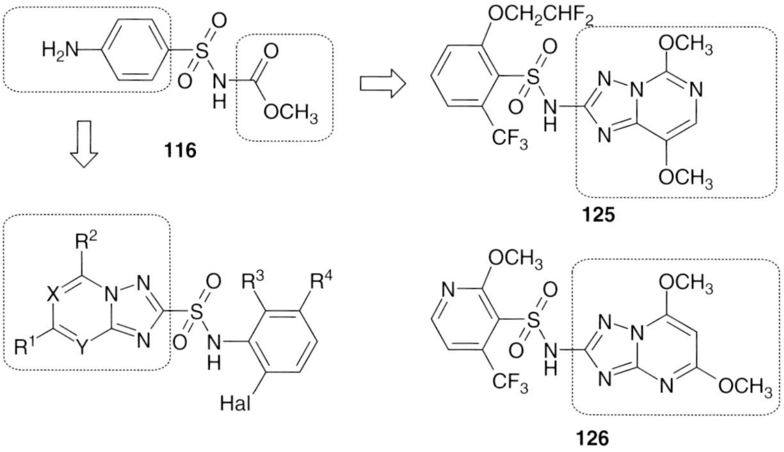

Figure 4.28 Asulam (**116**) and triazolopyrimidine herbicides (**117–121** (Table 4.16), **125**, **126**).

During optimization, the sulfonyl moiety was found to be more active than the corresponding sulfonylmethylaryl moiety. A particularly good activity and cereal selectivity was identified for the trifluoromethyl and trifluoromethoxy substitution.

4.6.2.3 **Triazolopyrimidine Herbicides**

The activity of the non-halogenated sulfonamide herbicide asulam (**116**; 1965, Asulox®, May & Baker; 1–10 kg a.i. ha^{-1}) [228] was remarkably improved by replacing the 4-aminophenyl ring with a halogenated triazolopyrimidine moiety and/or by replacement of the *N*-methoxycarbonyl group with a series of *ortho*-halogenated electron-deficient phenyl rings such as 2,6-difluoro-, 2,6-dichloro-, or 2-chloro-6-methoxycarbonyl-phenyl rings, forming the so-called "sulam" herbicides (Figure 4.28, Table 4.16).

With penoxsulam (**125**; 2004, Viper®, Dow AgroSciences) [229], however, the 4-amino-phenyl ring in **116** was replaced by the 2-(2,2-difluoroethoxy)-6-trifluoromethyl-phenyl ring, and the *N*-methoxycarbonyl group by a non-halogenated triazolopyrimidine moiety. All herbicides are active against broadleaf weeds after pre- and/or post-emergence application with different application rates versus crops: flumetsulam (**117**; 25–80 g a.i. ha^{-1}, soya/maize; systemic), metosulam (**118**; 5–30 g a.i. ha^{-1}, maize) [230], cloransulam-methyl (**119**; 40–50 g a.i. ha^{-1}, soya), diclosulam (**120**; 20–35 g a.i. ha^{-1}, soya beans/peanuts), florasulam (**121**; 5–10 g a.i. ha^{-1}, cereal/maize; systemic) [231], and pyroxsulam (**126**; 25–40 g a.i. ha^{-1}) with utility primarily in rice [232]. Because of the creation of a different set of particular halogenated basic triazolopyrimidine moieties, and the use of 2,6-dihalogenated anilines, a series of commercial valuable multi-outlet chemical intermediates was essential.

Table 4.16 Halogen-containing triazolopyrimidine herbicides (**117–121**) (see Figure 4.28 for the basic structure).

Compound no.	Common name, trade name(s)	Manufacturer (year introduced)	R^1	R^2	X	Y	R^3	R^4	Halogen
117	Flumetsulam, Broadstrike®, Preside®	Dow AgroSciences (1992)	CH_3	H	CH	N	F	H	F
118	Metosulam, Eclipse®, Uptake®	Dow AgroSciences (1993)	OCH_3	OCH_3	CH	N	Cl	CH_3	Cl
119	Chloransulam-methyl, Field Star®, First Rate®	Dow AgroSciences (1997)	F	OCH_2CH_3	N	CH	$COOCH_3$	H	Cl
120	Diclosulam, Spider®, Strongarm®	Dow AgroSciences (1997)	F	OCH_2CH_3	N	CH	Cl	H	Cl
121	Florasulam, Primus®, Boxer®	Dow AgroSciences (1999)	H	OCH_3	N	CF	F	H	F

4.6.3
Protoporphyrinogen IX Oxidase (PPO)

PPO inhibitors have a complex mechanism of action [233]. PPO, which is localized in the chloroplast and mitochondrial membranes, catalyzes the conversion of protoporphyrinogen IX into protoporphyrin IX. Many inhibitors mimic the hydrophobic region of protoporphyrinogen IX such that, over the past decade, a variety of different new PPO inhibitor classes containing halogen and/or halogen-substituted groups with even higher mimicry to protoporphyrinogen IX have been developed (see Figure 4.29):

- *Phenylpyrazole herbicides*, an example being the selective post-emergence cereal-selective herbicide pyraflufen-ethyl (**127**; 1999, Ecopart®, Nihon Nohyaku) [234].
- *N-Phenylphthalimide herbicides*, such as the cereal-selective herbicide cinidon-ethyl (**128**; 1998, Lotus®, BASF) [235], and the soya bean herbicides flumiclorac-pentyl (**129**; 1992, Resource®, Sumitomo) and flumioxazin (**130**; 1993, Sumisoya®, Sumitomo) [236].
- *Thiadiazole herbicides*, such as the post-emergence maize-selective herbicide fluthiacet-methyl (**131**; 1999, Action®, Kumiai) [237].
- *Oxadiazole herbicides*, for example, the pre- and post-emergence rice herbicide oxadiargyl (**132**; 1996, Raft®, Bayer CropScience) [238].

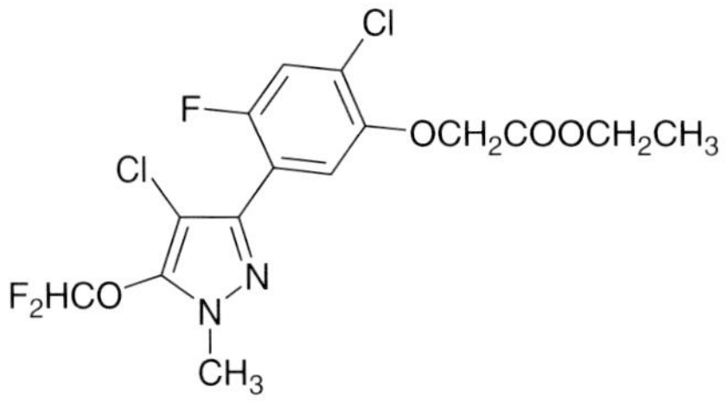

127

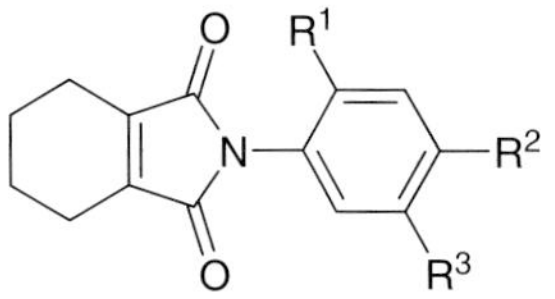

128 R^1 = H, R^2 = Cl, R^3 = 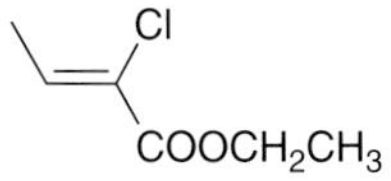

129 R^1 = F, R^2 = Cl, R^3= $OCH_2COO(CH_2)_4CH_3$

130 R^1 = F, R^2 -R^3 = 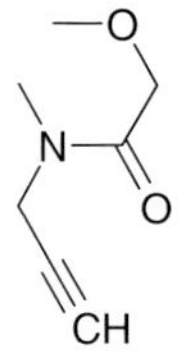

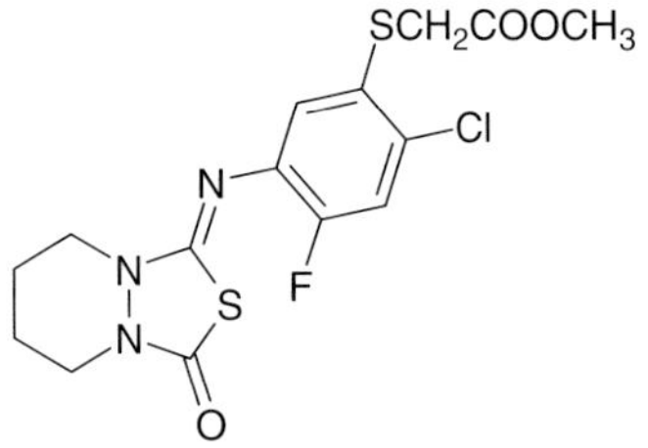

131

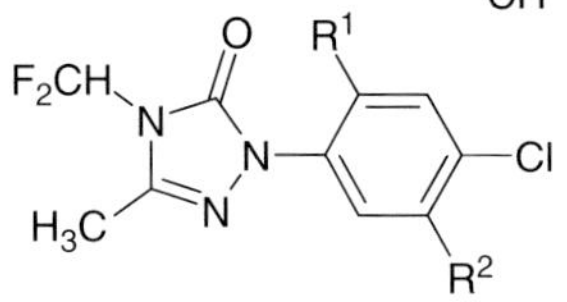

133 R^1 = Cl, R^2 = $NHSO_2CH_3$

134 R^1 = F, R^2 = $CH_2CHClCOOCH_2CH_3$

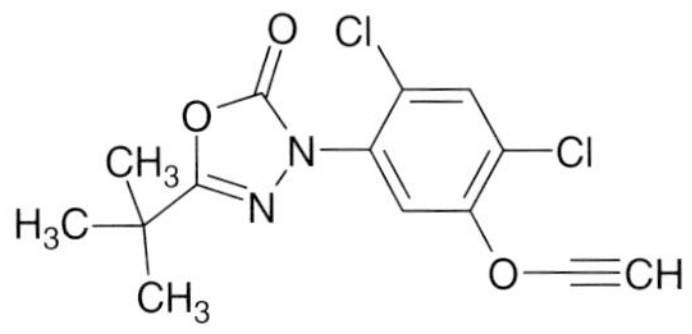

132

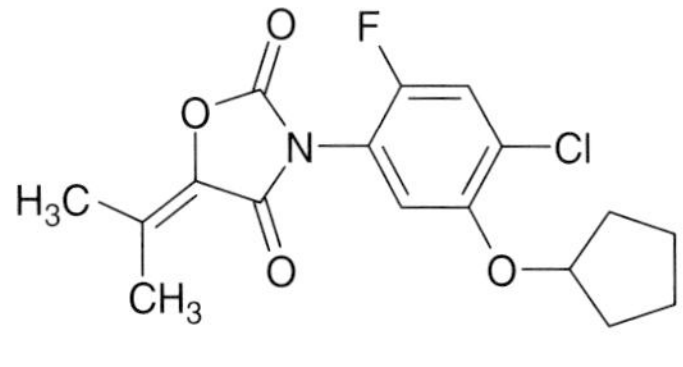

135

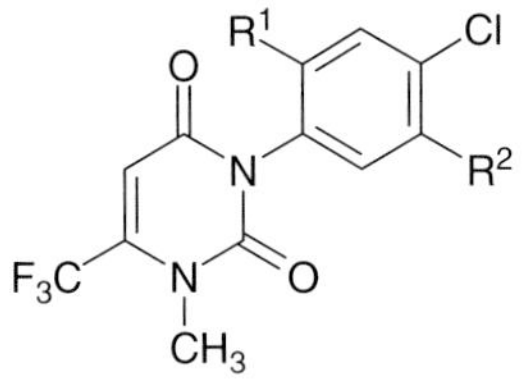

137 R^1 = H, R^2 = $COO(CH_3)_2COOCH_2CH{=}CH_2$

138 R^1 = F, R^2 = $CONHSO_2N(CH_3)CH(CH_3)_2$

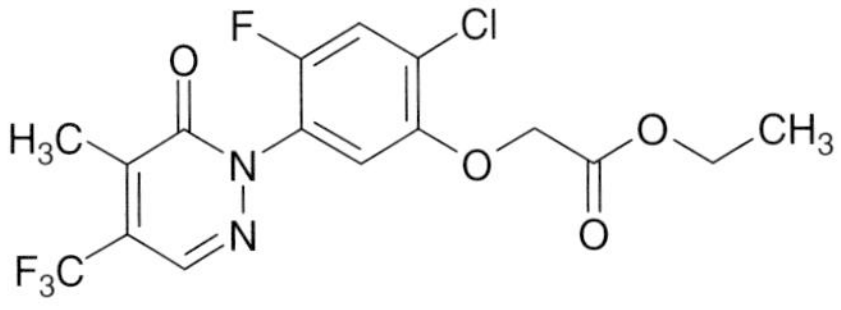

136

Figure 4.29 Different classes of halogen-substituted PPO inhibitors (**127**–**138**).

- *Triazolone herbicides*, such as sulfentrazone (**133**; 1995, Authority®, FMC) [239] and carfentrazone-ethyl (**134**; 1997, Aim®, FMC) [240], which are adsorbed via the roots (**133**) and by the foliage, with limited translocation in the phloem.
- *Oxazolinedione herbicides*, for example, pentoxazone (**135**; 1997, Wechser®, Kaken) [241].
- *Pyridazinone herbicides*, such as flufenpyr-ethyl (**136**; 2003, S-3153, Sumitomo) [242] described for controlling velvetleaf and morning glories in corn, soybean, and sugar cane.
- *Pyrimidinedione herbicides*, for example, butafenacil (**137**; 2000, Inspire®, Syngenta) [243] for use in vineyards, citrus, and non-crop land, and saflufenacil (**138**; 2010, Kixor®, BASF) for pre-emergence application for broadleaf weed control in maize and sorghum [244].

The various examples described above demonstrate that the introduction of halogens and/or halogen-substituted groups has had a dramatic effect on the metabolism of active ingredients through reaction at a location remote from the halogenated groups themselves. However, such effects cannot often be predicted as part of initial design of an active ingredient.

4.7 Summary and Outlook

In the search for a so-called "optimal product" in modern crop protection in terms of efficacy, environmental safety, user friendliness, and economic viability, the substitution of active ingredients with halogen atoms or halogen-containing substituents is an important tool. However, the introduction of such atoms or substituents into a molecule can lead to either an increase or a decrease in biological efficacy, depending on the MoA, the physico-chemical properties, or the target interaction of the compound. The metabolism of an active ingredient will be influenced by the substitution pattern in a given molecule, and also by its stability in the soil and/or in water. Finally, a shift into another crop protection area can be observed by the introduction of halogen atoms or halogen-containing substituents into different biologically active molecules. Because of the complex SARs within active ingredients, it is difficult to predict sites where halogens or halogen-substituted substituents will increase biological efficacy. Nonetheless, the technical availability of active ingredients containing halogens or halogen-substituted substituents has been greatly improved by an increase in access to new intermediates.

References

1. Leroux, F., Jeschke, P., and Schlosser, M. (2005) *Chem. Rev.*, **105**, 827–856.
2. Jeschke, P. (2010) *Pest Manag. Sci.*, **66**, 10–27.
3. Theodoridis, G. (2006) *Advances in Fluorine Science, Fluorine and the Environment: Agrochemicals, Archaeology, Green Chemistry and Water*, vol. 2, Elsevier B.V., pp. 121–175.

4. Jeschke, P. (2004) *ChemBioChem*, **5**, 570–589.
5. van der Waterbeemd, H., Clementi, S., Costantino, G., Carrupt, P.A., and Testa, B. (1993) in *3D QSAR in Drug Design: Theory, Methods and Applications* (ed. H. Kubinyi), Escom, Leiden, pp. 697–707.
6. Hiyama, T. (2000) *Organofluorine Compounds; Chemistry and Applications*, Springer, Berlin, and references cited therein.
7. Ma, J.A. and Cahard, D. (2004) *Chem. Rev.*, **104**, 6119–6146.
8. Audouard, C., Ma, J.A., and Cahard, D. (2006) *Advances in Organic Synthesis*, vol. 2, Bentham Science Publishers Ltd, pp. 431–461.
9. Quirmbach, M., Steiner, H., Meier, H., and Blaser, H.U. (2008) *Chem. Oggi/Chem. Today*, **26**, 12–13.
10. Shimizu, M. and Hiyama, T. (2005) *Angew. Chem. Int. Ed.*, **44**, 214–231.
11. Tomashenko, O.A. and Grushin, V.V. (2011) *Chem. Rev.*, **111** (8), 4475–4521.
12. Leroux, F., Manteau, B., Vors, J.P., and Pazenok, S. (2008) *Beilstein J. Org. Chem.*, **4**, 1–15.
13. Jeschke, P., Boston, E., and Leroux, F.R. (2007) *Mini-Rev. Med. Chem.*, **7**, 1027–1034.
14. Purser, S., Moore, P.R., Peter, R., Swallow, S., and Gouverneur, V. (2008) *Chem. Soc. Rev.*, **37**, 320–330.
15. Frohberger, P.E. (1978) *Pflanzenschutz-Nachr. Bayer (German Ed.)*, **31**, 11–24.
16. Yasui, K., Goto, T., Miyauchi, H., Yanagi, A., Feucht, D., and Fuersch, H. (1997) *Brighton Crop Prot. Conf. - Weeds*, **1**, 67–72.
17. Elbert, A., Overbeck, H., Iwaya, K., and Tsuboi, S. (1990) *Brighton Crop Prot. Conf. - Pests Dis.*, **1**, 21–28.
18. Kagabu, S. (1997) *Rev. Toxicol.*, **1**, 75–129.
19. Hassel, O. and Hvoslef, J. (1954) *Acta Chem. Scand.*, **8**, 873.
20. Adams, H., Cockroft, S.L., Guardigli, C., Hunter, C.A., Lawson, K.R., Perkins, J., Spey, S.E., Urch, C.J., and Ford, R. (2004) *ChemBioChem*, **5**, 657–665.
21. Price, S.L. and Stone, A.J. (2004) *ChemBioChem*, **5**, 1457–1470.
22. Auffinger, P., Hays, F.A., Westhof, E., and Ho, P.S. (2004) *Proc. Natl. Acad. Sci. USA*, **101**, 16789–16794.
23. Bondi, A. (1964) *J. Phys. Chem.*, **68**, 441–451.
24. Patani, G.A. and LaVoie, E.J. (1996) *Chem. Rev.*, **96**, 3147–3176.
25. Olesen, P.H. (2001) *Curr. Opin. Drug Disc. Dev.*, **4**, 471–478.
26. Basarab, G.S. and Boswell, G.A. Jr (1985) EP 137717A2 (DuPont, E. I. de Nemours and Co., USA); (*Chem. Abstr.*, **103**, 196096).
27. Northwood, P.J., Paul, J.A., Gibbard, R.A., and Noon, R.A. (1986) *Brighton Crop Prot. Conf. - Monogr.*, **33**, 233–238; (*Chem. Abstr.*, **104**, 143806).
28. Edwards, P.N. (1994) in *Organofluorine Chemistry: Principles and Commercial Applications* (eds R.E. Banks, B.E. Smart, and J.C. Tatlow), Plenum Press, New York, pp. 502–509.
29. Biffinger, J.C., Kim, H.W., and DiMagno, S.G. (2004) *ChemBioChem*, **5**, 622–627.
30. Smart, B.E. (1995) Cycloadditions forming three- and four-membered rings in *Chemistry of Organic Fluorine Compounds II* (eds M. Hudlicki and A.E. Pavlath), American Chemical Society, Washington, DC, pp. 767–796.
31. Berkowitz, D.B. and Bose, M. (2001) *J. Fluorine Chem.*, **112**, 13–33.
32. Zhou, P., Tian, F., Zou, J., and Shang, Z. (2010) *Mini-Rev. Med. Chem.*, **10**, 309–314.
33. Tohnishi, M., Nakao, H., Furuya, T., Seo, A., Kodama, H., Tsubata, K., Fujioka, S., Kodama, H., Hirooka, T., and Nishimatsu, T. (2005) Abstracts of Papers, 230th ACS National Meeting, Washington, D.C., 28 August–1 September 2005, AGRO-009.
34. Ebbinghaus-Kintscher, U., Luemmen, P., Lobitz, N., Schulte, T., Funke, C., Fischer, R., Masaki, T., Yasokawa, N., and Tohnishi, M. (2006) *Cell Calcium*, **39**, 21–33.
35. Ishaaya, I. and Casida, J.E. (1974) *Pestic. Biochem. Physiol.*, **4**, 484–490.
36. Grosscurt, A.C. (1977) *Brighton Crop Prot. Conf.- Pests Dis.*, **1**, 141–147.

37. Luteijn, J.M. and Tipker, J. (1986) *Pestic. Sci.*, **17**, 456–458.
38. Roberts, T. and Hutson, D. (eds) (1999) *Metabolic Pathway of Agrochemicals, Part 2: Insecticides and Fungicides*, The Royal Society of Chemistry, Cambridge, pp. 795–816.
39. Seebach, D. (1990) *Angew. Chem., Int. Ed. Engl.*, **29**, 1320–1367.
40. Pauling, L. (1960) *The Nature of the Chemical Bond*, Cornell University Press, Ithaca.
41. Borden, W.T. (1998) *Chem. Commun.*, 1919–1925.
42. Chorki, F., Grellepois, F., Ourévitch, M., Crousse, B., Charneau, S., Grellier, P., Charman, W.N., McIntosh, K.A., Pradines, B., Bonnet-Delpon, D., and Bégué, J.P. (2004) *J. Med. Chem.*, **5**, 637–654.
43. Huheey, J.E. (1965) *J. Phys. Chem.*, **69**, 3284–3289.
44. McClinton, M.A. and McClinton, D.A. (1992) *Tetrahedron*, **48**, 6555–6566.
45. Lien, E.J., Guo, Z.R., Li, R.L., and Su, C.T. (1982) *J. Pharm. Sci.*, **71**, 641–655.
46. Böhm, H.J., Banner, D., Bendels, S., Kansy, M., Kuhn, B., Müller, K., Obst-Sander, U., and Stahl, M. (2004) *ChemBioChem.*, **5**, 637–643.
47. Morgenthaler, M., Schweizer, E., Hoffmann-Roder, A., Benini, F., Martin, R.E., Jaeschke, G., Wagner, B., Fischer, H., Bendels, S., Zimmerli, D., Schneider, J., Diederich, F., Kansy, M., and Muller, L. (2007) *ChemMedChem*, **2**, 1100–1115.
48. Dunitz, J.D. (2004) *ChemBioChem*, **5**, 614–621.
49. Parsch, J. and Engels, J.W. (2002) *J. Am. Chem. Soc.*, **124**, 5664–5672.
50. Desiraju, G.R. (2002) *Acc. Chem. Res.*, **35**, 565–573.
51. Dunitz, J.D. and Tayler, R. (1997) *Chem. Eur. J.*, **3**, 89–98.
52. Ouvrad, C., Le Questel, J.Y., Berthelot, M., and Laurence, C. (2003) *Acta Crystallogr. B*, **59**, 512–526.
53. Corradi, E., Meille, S.V., Messina, M.T., Metrangolo, P., and Resnati, G. (2000) *Angew. Chem. Int. Ed.*, **112**, 1852–1856.
54. Lommerse, J.P.M., Price, S.L., and Taylor, R. (1997) *J. Comput. Chem.*, **18**, 757–774.
55. Banks, R.E. (1998) *J. Fluorine Chem.*, **87**, 1–17.
56. Smart, B.E. (1986) in *Molecular Structure and Energetics*, vol. 3 (eds J.F. Liebman and A. Greenberg), VCH Publishers, Deerfield Beach, pp. 141–191.
57. Prestwich, G.D. (1986) *Pestic. Sci.*, **37**, 430–440.
58. Bailey, W.A., Wilcut, J.W., Jordan, D.L., Swann, C.W., and Langston, V.B. (1999) *Weed Technol.*, **13**, 450–456.
59. Förster, H., Schmidt, R.R., Santel, H.J., and Andree, R. (1997) *Pflanzenschutz-Nachr. (German Ed.)*, **50**, 105–116.
60. Schiemenz, B. and Wessel, T. (2004) *Chem. Today*, **6**, 23–27.
61. Adams, D.J. and Clark, J.H. (1999) *Chem. Soc. Rev.*, **28**, 225–231.
62. Maienfisch, P. and Hall, R.G. (2004) *Chimia*, **58**, 93–99.
63. Hansch, C. and Leo, A. (1979) *Substituent Constants for Correlation Analysis in Chemistry and Biology*, John Wiley & Sons, Inc., New York.
64. Schmitt, M.H., van Almsick, A., and Willms, L. (2008) *Pflanzenschutz-Nachr. Bayer*, **61**, 7–14.
65. Schulte, A. and Köcher, H. (2008) *Pflanzenschutz-Nachr. Bayer*, **61**, 29–42.
66. Freigang, J., Laber, B., Lange, G., and Schulz, A. (2008) *Pflanzenschutz-Nachr. Bayer*, **61**, 15–28.
67. Gladysz, A. and Curran, D.P. (2002) *Fluorous Chemistry; Tetrahedron 58*, Elsevier, New York.
68. Hamaguchi, H., Hirooka, T. and Masaki, T. (2012) in *Modern Crop Protection Compounds* (eds W. Krämer, U. Schirmer, P. Jeschke and M. Witschel), Wiley-VCH Verlag GmbH, Weinheim, pp. 1389–1409.
69. Uehara, M., Watanabe, M., Kimura, M., Morimoto, M., and Yoshida, M. (2001) Eur. Pat. Appl. EP 1097932A1, (Nihon Nohyaku Co., Ltd., Japan); (*Chem. Abstr.*, **134**, 340516).
70. Zhang, J., Tang, X., Ishaaya, I., Cao, S., Wu, J., Yu, J., Li, H., and Qian, X.

(2010) *J. Agric. Food Chem.*, **58**, 2736–2740.

71. van de Waterbeemd, H., Constantino, G., Clementi, S., Cruciani, G., and Valigi, R. (1995) in *Chemometric Methods in Molecular Design*, vol. 2 (eds H. van der Waterbeemd), Wiley-VCH Verlag GmbH, Weinheim, New York, Basel, Cambridge, Tokyo, pp. 103–112.
72. Stetter, J. and Lieb, F. (2000) *Angew. Chem. Int. Ed.*, **39**, 1725–1744.
73. Colliot, F., Kukorowski, K.A., Hawkins, D.W., and Roberts, D.A. (1992) *Brighton Crop Prot. Conf. - Pests Dis.*, **1**, 29–34.
74. Godwin, J.R., Bartlett, D.W., Clough, J.M., Godfrey, C.R.A., Harrison, E.G., and Maund, S. (2000) *Brighton Crop Prot. Conf. - Pests Dis.*, **2**, 533–540.
75. Huang, S., Zhang, Z., Xu, H., and Zeng, D. (2005) *Nongyao*, **44**, 81–83.
76. Staudinger, H. and Ruzicka, L. (1924) *Helv. Chim. Acta*, **7**, 177–201.
77. O'Reilly, A.O., Khambay, B.P.S., Williamson, M.S., Field, L.M., Wallace, B.A., and Davies, T.G.E. (2006) *Biochem. J.*, **396**, 255–263.
78. Soderlund, D.M. (2008) *Pest Manag. Sci.*, **64**, 610–616.
79. Tomlin, C.D. (2000) *The Pesticide Manual*, 12th edn, British Crop Protection Council, Kent.
80. Bardner, R., Fletcher, K.E., and Griffiths, D.C. (1979) *Brighton Crop Prot. Conf. - Pests Dis.*, **1**, 223–229.
81. Eliott, M., Farnham, A.W., Janes, N.F., Needham, D.H., and Pulman, D.A. (1974) *Nature*, **248**, 710–711.
82. Breese, M.H. and Highwood, D.P. (1977) *Brighton Crop Prot. Conf. - Pests Dis.*, **2**, 641–648.
83. Hammann, I. and Fuchs, R. (1981) *Pflanzenschutz-Nachr. (German Ed.)*, **34**, 121–151.
84. Naumann, K. (1998) *Pestic. Sci.*, **52**, 3–20.
85. Naumann, K. (1990) in *Chemistry of Plant Protection 4, Synthetic Pyrethroid Insecticides* (eds W.S. Bowers, W. Ebing, D. Martin, and R. Wegeler), Springer, p. 43.
86. Robson, M.J., Cheetham, R., Fettes, D.J., Crosby, J., and Griffiths, D.C. (1984) *Brighton Crop Prot. Conf. - Pests Dis.*, **3**, 853–857.
87. Laskowski, D. (1991) *Rev. Environ. Contam. Toxicol.*, **174**, 49.
88. Doel, H.J.H., Crossman, A.R., and Bourdouxhe, L.A. (1984) *Meded. Fac. Landbouwwet, Univ. Gent*, **49**, 929–937.
89. Plummer, E.L., Cardis, A.B., Martinez, A.J., VanSaun, W.A., Palmere, R.M., Pincus, D.S., and Stewart, R.R. (1983) *Pestic. Sci.*, **14**, 560–570.
90. Gouger, R.J. and Neville, A.J. (1986) *Brighton Crop Prot. Conf. - Pests Dis.*, **3**, 1143–1150.
91. Pinochet, G. (1991) *Phytoma*, **428**, 54–57.
92. Ou, Z., Yang, S., Li, G., and Wang, J. (2010) CN 101747198A, (Guiyang Bestchem Co., Ltd., China); (*Chem. Abstr.*, **153**, 174709).
93. Mowlam, M.D., Highwood, D.P., Dowson, R.J., and Hattori, J. (1977) *Brighton Crop Prot. Conf. - Pests Dis.*, **2**, 649–656.
94. Agranat, I., Caner, H., and Caldwell, J. (2002) *Nat. Rev.*, **1**, 753–768.
95. Nakayama, I., Ohno, N., Aketa, K., Suzuki, Y., Kato, T., and Yoshioka, H. (1979) *Adv. Pestic. Sci.*, **2**, 174–181 (4th Plenary Lecture Symposium Paper International Congress Pesticide Chemistry).
96. Whitney, W.K., Keith, W., and Wettstein, K. (1979) *Brighton Crop Prot. Conf. - Pests Dis.*, **2**, 387–394.
97. Jin, W.G., Sun, G.H., and Xu, Z.M. (1996) *Brighton Crop Prot. Conf. - Pests Dis.*, **2**, 455–460.
98. Grosscurt, A.C., van Hes, R., and Wellinga, K. (1979) *J. Agric. Food. Chem.*, **27**, 406–409.
99. Salgado, V.L. (1990) *Pestic. Sci.*, **28**, 389–411.
100. Harder, H.H., Riley, S.L., McCann, S.F., and Irving, S.N. (1996) *Brighton Crop Prot. Conf. - Pests Dis.*, **2**, 449–454.
101. Bloomquist, J.R. (2001) in *Biochemical Sites Important of Insecticide Action and Resistance* (ed. I. Ishaaya), Springer, Berlin, p. 17.
102. Bloomquist, J.R. (1996) *Annu. Rev. Entomol.*, **41**, 163–190; (*Chem. Abstr.*, **124**, 79297).

103. Meinke, P.T. (2001) *J. Med. Chem.*, **44**, 641–659.
104. Dryden, M.W., Denenberg, T.M., and Bunch, S. (2000) *Vet. Parasitol.*, **93**, 69–75; (*Chem. Abstr.*, **134**, 26483).
105. Caboni, P., Sammelson, R.E., and Casida, J.E. (2003) *J. Agric. Food Chem.*, **51**, 7055–7061.
106. Wang, Z., Li, Y., Guo, T., and Song, Y. (2003) CN 1398515A (Faming Zhuanli Shenqing Gongkai Shuomingshu, China); (*Chem. Abstr.*, **141**, 35027).
107. Sheets, J.J. (2012) in *Modern Crop Protection Compounds* (eds W. Krämer, U. Schirmer, P. Jeschke and M. Witschel), Wiley-VCH Verlag GmbH, Weinheim, pp. 999–1011.
108. Mulder, R. and Gijswijk, M.J. (1973) *Pestic. Sci.*, **4**, 737.
109. Clarke, B.S. and Jewess, P.J. (1990) *Pestic. Sci.*, **28**, 377–388.
110. Scheltes, P., Hofman, T.W., and Grosscurt, A.C. (1988) *Brighton Crop Prot. Conf. - Pests Dis.*, **2**, 559–666.
111. Neumann, R. and Guyer, W. (1983) Proceedings of 10th International Congress of Plant Protection Conference, Vol. 1, pp. 445–451; (*Chem. Abstr.*, **102**, 57756).
112. Anderson, M., Fisher, J.P., Robinson, J., and Debray, P.H. (1986) *Brighton Crop Prot. Conf. - Pests Dis.*, **1**, 89–96.
113. Kim, K.S., Chung, B.J., and Kim, H.K. (2000) *Brighton Crop Prot. Conf. - Pests Dis.*, **2**, 41–46.
114. Komblas, K.N. and Hunter, R.C. (1986) *Brighton Crop Prot. Conf. - Pests Dis.*, **3**, 907–914.
115. Schenker, R. and Moyses, E.W. (1994) *Brighton Crop Prot. Conf. - Pests Dis.*, **3**, 1013–1021.
116. Ishaaya, I., Yablonski, S., Mendelson, Z., Mansour, Y., and Horowitz, A.R. (1996) *Brighton Crop Prot. Conf. - Pests Dis.*, **3**, 1013–1020.
117. Sbragia, R.J., Johnson, G.W., Karr, L.L., Edwards, J.M., and Schneider, B.M. (1998) WO 9819542A1 (Dow AgrowSciences, LLC, USA); (*Chem. Abstr.*, **129**, 13497).
118. Ishaaya, I. (1990) in *Pesticides and Alternatives* (ed. J.E. Casida), Elsevier, Amsterdam, p. 365.
119. Hammann, I. and Sirrenberg, W. (1980) *Pflanzenschutz-Nachr. (German Ed.)*, **33**, 1–34.
120. Blaß, W. (1998) *Pflanzenschutz-Nachr. (German Ed.)*, **51**, 79–96.
121. Ishida, T., Suzuki, J., Tsukidate, Y., and Mori, Y. (1994) *Brighton Crop Prot. Conf. - Pests Dis.*, **1**, 37–44.
122. Suzuki, J., Ishida, T., Kikuchi, Y., Ito, Y., Morikawa, C., Tsukidate, Y., Tanji, I., Ota, Y., and Toda, K. (2002) *Nihon Noyaku Gakkaishi (J. Pestic. Sci.)*, **27**, 1–8.
123. Pap, L., Hajimichael, J., Bleicher, E., Botar, S., and Szekely, I. (1994) *Brighton Crop Prot. Conf. - Pests Dis.*, **1**, 75–82.
124. Pap, L., Hajimichael, J., and Bleicher, E. (1996) *J. Environ. Sci. Health, Part B: Pestic. Food Contam. Agric. Wastes*, **31**, 521–526.
125. Bryan, K.M.G., Geering, Q.A., and Reid, J. (1981) *Brighton Crop Prot. Conf. - Pests Dis.*, **1**, 67–74.
126. Eberle, M., Farooq, S., Jeanguenat, A., Mousset, D., Steiger, A., Trah, S., Zambach, W., and Rindlisbacher, A. (2003) *Chimia*, **57**, 705–709.
127. Sparks, T.C., and DeAmicis, C.V. (2012) in *Modern Crop Protection Compounds* (eds W. Krämer, U. Schirmer, P. Jeschke and M. Witschel), Wiley-VCH Verlag GmbH, Weinheim, pp. 1078–1108.
128. Konno, T., Kuiyama, K., and Hamaguchi, H. (1990) *Brighton Crop Prot. Conf. - Pests Dis.*, **1**, 71–78.
129. Longhurst, C., Bacci, L., Buendia, J., Hatton, C.J., Petitprez, J., and Tsakonas, P. (1992) *Brighton Crop Prot. Conf. - Pests Dis.*, **1**, 51–58.
130. Hirata, K., Kudo, M., Miyake, T., Kawamura, Y., and Ogura, T. (1988) *Brighton Crop Prot. Conf. - Pests Dis.*, **1**, 41–48.
131. Kyomura, N., Fukuchi, T., Kohyama, Y., and Motojima, S. (1990) *Brighton Crop Prot. Conf. - Pests Dis.*, **1**, 55–62.
132. Hopkins, W.L. (1993) *Agric. Chem. New Compd. Rev.*, **11**, 98.
133. Jovell, J.B., Wright, D.P., Gard, I.E., Miller, T.P., and Treacy, M.F. Jr (1990) *Brighton Crop Prot. Conf. - Pests Dis.*, **1**, 43–48.

134. Kuhn, D. and Armes, N. (2012) in *Modern Crop Protection Compounds* (eds W. Krämer, U. Schirmer, P. Jeschke and M. Witschel), Wiley-VCH Verlag GmbH, Weinheim, pp. 1070–1078.
135. Hunt, D.A. (1994) in *Advances in the Chemistry of Insect Control III* (ed. G. Briggs), Cambridge University Press, Cambridge, pp. 127–140.
136. Kim, S.S. and Yoo, S.S. (2001) *Appl. Entomol. Zool.*, **36**, 509–514.
137. Kim, S.S. and Yoo, S.S. (2002) *BioControl*, **47**, 563–573.
138. Black, B.C., Hollingworth, R.M., Ahammadsahib, K.I., Kukel, C.D., and Donovan, S. (1994) *Pestic. Biochem. Physiol.*, **50**, 115–128.
139. Nishimatsu, T., Hirooka, T., Kodama, H., Tohnishi, M., and Seo, A. (2005) Congress Proceedings – BCPC International Congress: Crop Science and Technology, Glasgow, 31 October–2 November 2005, Vol. 1, pp. 57–64.
140. Tohnishi, M., Nakao, H., Furuya, T., Seo, A., Kodama, H., Tsubata, K., Fujioka, S., Kodama, H., Hirooka, T., and Nishimatsu, T. (2005) *J. Pestic. Sci. (Tokyo)*, **30**, 354–360.
141. Nauen, R. (2006) *Pest Manag. Sci.*, **62**, 690–692.
142. Lahm, G.P., Selby, T.P., Freudenberger, J.H., Stevenson, T.M., Myers, B.J., Seburyamo, G., Smith, B.K., Flexner, L., Clark, C.E., and Cordova, D. (2005) *Bioorg. Med. Chem. Lett.*, **15**, 4898–4906.
143. Lahm, G.P., Cordova, D., and Barry, J.D. (2009) *Bioorg. Med. Chem.*, **17**, 4127–4133.
144. Cordova, D., Benner, E.A., Sacher, M.D., Rauh, J.J., Sopa, J.S., Lahm, G.P., Selby, T.P., Stevenson, T.M., Flexner, L., Gutteridge, S., Rhoades, D.F., Wu, L., Smith, R.M., and Tao, Y. (2006) *Pestic. Biochem. Physiol.*, **84**, 196–214.
145. Selby, T.P., Lahm, G.P., Stevenson, T.M., Hughes, K.A., Annan, I.B., Cordova, D., Bellin, C.A., Benner, E.A., Wing, K.D., Barry, J.D., Currie, M.J., and Pahutski, T.F. (2010) Abstract of Papers, *239th ACS National Meeting, San Francisco, CA, USA, 21–25 March 2010, AGRO-3*, American Chemical Society, Washington, DC.
146. Kuck, K.H., Leadbeater, A. and Gisi, U. (2012) in *Modern Crop Protection Compounds* (eds W. Krämer and U. Schirmer, P. Jeschke and M. Witschel), Wiley-VCH Verlag GmbH, Weinheim, pp. 539–557.
147. Lamb, D., Kelly, D., and Kelly, S. (1999) *Drug Resist. Updates*, **2**, 390–402.
148. Berg, D., Buechel, K.H., Kraemer, W., Plempel, M., and Scheinflug, H. (1988) in *Sterol Biosynthesis Inhibitors* (eds D. Berg and M. Plempel), Ellis Horwood, Chichester, pp. 168–184; (*Chem. Abstr.*, **111**, 36288).
149. Buchenauer, H., Edgington, L.V., and Grossmann, F. (1973) *Pestic. Sci.*, **4**, 343–348.
150. Urech, P.A., Schwinn, F.J., Speich, J., and Staub, T. (1979) *Brighton Crop Prot. Conf. - Pests Dis.*, **2**, 508–515.
151. Berg, D., Born, L., Büchel, K.H., Holmwood, G., and Kaulen, J. (1987) *Pflanzenschutz-Nachr. (German Ed.)*, **2**, 11–132.
152. Gisi, U., Rimbach, E., Binder, H., Altwegg, P., and Hugelshofer, U. (1986) *Brighton Crop Prot. Conf. - Pests Dis.*, **2**, 857–864.
153. Ruess, W., Riebli, P., Herzog, J., Speich, J., and James, J.R. (1988) *Brighton Crop Prot. Conf. - Pests Dis.*, **2**, 543–550.
154. Garavaglia, C., Mirenna, L., Puppin, O., and Spagni, E. (1988) *Brighton Crop Prot. Conf. - Pests Dis.*, **1**, 49–56.
155. Ammermann, E., Loecher, F., Lorenz, G., Janssen, B., Karbach, S., and Meyer, N. (1990) *Brighton Crop Prot. Conf. - Pests Dis.*, **2**, 407–414.
156. Jautelat, M., Elbe, H.L., Benet-Buchholz, J., and Etzel, W. (2004) *Pflanzenschutz-Nachr. (English Ed.)*, **57**, 145–162.
157. Mauler-Machnik, A., Rosslenbroich, H.J., Dutzmann, S., Applegate, J., and Jautelat, M. (2002) *Brighton Crop Prot. Conf. - Pests Dis.*, **1**, 389–394.
158. von Schmeling, B. and Kulka, M. (1966) *Science*, **152**, 659–660.

159. Frost, A.J.P. and Hampel, M. (1976) in *Proceedings of 4th European and Mediterranean Cereal Rusts Conference* (ed. A. Broennimann), Federal Research Station for Agronomy, Zurich, pp. 99–101.
160. Rheinheimer, J. (2007) in *Modern Crop Protection Compounds* (eds W. Kramer and U. Schirmer), Wiley-VCH Verlag GmbH, Weinheim, pp. 496–505.
161. Araki, F. and Yabutani, K. (1981) *Proc. Brighton Crop. Prot. Conf. Pests Dis.*, **1**, 3–9.
162. O'Reilly, P., Kobayashi, S., Yamane, S., Phillips, W.G., Raymond, P., and Castanho, B. (1992) *Brighton Crop Prot. Conf. - Pests Dis.*, **1**, 427–434.
163. Stammler, G., Brix, H.D., Glaettli, A., Semar, M., and Schoefl, U. (2007) Proceedings of the 16th International Plant Protection Congress, Glasgow, 2007, Vol. 1, pp. 40–45.
164. Ehrenfreund, J., Tobler, H., and Walter, H. (2004) PCT Int. Appl., WO 2004/035589A1 (Syngenta A.-G., Switzerland); (*Chem. Abstr.*, **140**, 375164).
165. Dunkel, R., Rieck, H., Elbe, H.L., Wachendorff-Neumann, U., and Kuck, K.H. (2003) PCT Int. Appl., WO 2003/070705A1 (Bayer CropScience A.-G., Germany); (*Chem. Abstr.*, **139**, 180060).
166. Gewehr, M., Dietz, J., Grote, T., Blettner, C., Grammenos, W., Huenger, U., Mueller, B., Schieweck, F., Schwoegler, A., Lohmann, J.K., Rheinheimer, J., Schaefer, P., Strathmann, S., and Stierl, R. (2006) PCT Int. Appl. WO 2006/087343A1 (BASG A.-G., Germany); (*Chem. Abstr.*, **145**, 271772).
167. Ehrenfreund, J., Tobler, H., and Walter, H. (2003) PCT Int. Appl. WO 2003/074491A1 (Syngenta A.-G., Switzerland); (*Chem. Abstr.*, **139**, 246007).
168. Yanase, Y., Yoshikawa, Y., Kishi, J., and Katsuta, H. (2007) in *Pesticide Chemistry* (eds H. Ohkawa, H. Miyagawa, and P.W. Lee), Wiley-VCH Verlag GmbH & Co. KGaA, Weinheim, pp. 295–303.
169. Elbe, H.L., Rieck, H., Dunkel, R., Zhu-Ohlbach, Q., Mauler-Machnik, A., Wachendorff-Neumann, U., and Kuck, K.H. (2003) PCT Int. Appl. WO 2003/010149A1 (Bayer CropScience A.-G., Germany); (*Chem. Abstr.*, **138**, 137308).
170. Coqueron, P.Y., Lhermitte, F., Perrin-Janet, G., and Dufour, P. (2006) Paul. Eur. Pat. Appl. EP Pat. 1674455A1 (Bayer CropScience S.A., France); (*Chem. Abstr.*, **145**, 83234).
171. Ishii, H., Miyamoto, T., Ushio, S., and Kakishima, M. (2011) *Pest. Manag. Sci.*, **67**, 474–482.
172. Sauter, H. (2007) in *Modern Crop Protection Compounds* (eds W. Kramer and U. Schirmer), Wiley-VCH Verlag GmbH, Weinheim, pp. 457–495.
173. Anke, T., Schramm, G., Schwalge, B., Steffan, B., and Steglich, W. (1984) *Liebigs Ann. Chem.*, **9**, 1616–1625.
174. Becker, W.F., von Jagow, G., Ange, T., and Steglich, W. (1981) *FEBS Lett.*, **132**, 329–333.
175. Yu, C.A., Xia, J.X., Kachurin, A.M., Yu, A.M., Xia, D., Kim, H., and Deisenhofer, J. (1996) *Biochim. Biophys. Acta*, **1275**, 47–53.
176. Zhang, Z., Huang, L., Shulmeister, V.M., Chi, Y.I., Kim, K.K., Hung, L.W., Crofts, A.R., Berry, E.A., and Kim, S.H. (1998) *Nature*, **392**, 677–684.
177. Barlett, D.W., Clough, J.M., Goldwin, J.R., Hall, A.A., Hamer, M., and Parr-Dobranski, B. (2002) *Pest Manag. Sci.*, **58**, 649–662.
178. Link, T.A., Iwata, M., Bjoerkman, J., van der Spoel, D., Stocker, A., and Iwata, S. (2003) in *Chemistry of Crop Protection: Progress and Prospects in Science and Regulation* (eds G. Voss and G. Ramos), Wiley-VCH Verlag GmbH, Weinheim, pp. 110–127.
179. Frank, J.A. and Sanders, P.L. (1994) *Brighton Crop Prot. Conf. - Pests Dis.*, **2**, 871–876.
180. Ammermann, E., Lorenz, G., Schelberger, K., Wenderoth, B., Sauter, H., and Rentzea, C. (1988) *Brighton Crop Prot. Conf. - Pests Dis.*, **1**, 403–410.

181. Margot, P., Huggenberger, F., Amrein, J., and Weiss, B. (1998) *Brighton Crop Prot. Conf. - Pests Dis.*, **2**, 375–382.
182. Ziegler, H., Benet-Buchholz, J., Etzel, W., and Gayer, H. (2003) *Pflanzenschutz-Nachr. (German Ed.)*, **56**, 213–230.
183. Häuser-Hahn, I., Pontzen, R., and Baur, P. (2003) *Pflanzenschutz-Nachr. (German Ed.)*, **56**, 246–258.
184. Zhang, L.X., CH Li, Z., Li, B., Sun, K., Zhang, Z.J., Zhan, F.K., and Wang, J. (2003) *Brighton Crop Prot. Conf. - Pests Dis.*, **1**, 93–98.
185. Ammermann, E., Lorenz, G., Schelberger, K., Mueller, B., Kirstgen, R., and Sauter, H. (2000) *Brighton Crop Prot. Conf. - Pests Dis.*, **2**, 541–548.
186. Liu, C., Li, M., Zhang, H., Li, L., Zhang, M., Guan, A., Hu, C., Li, Z., and Jia, Y. (2005) PCT Int. Appl. WO 080344 A1 (Shenyang Res. Inst. Chem. Ind., China); (*Chem. Abstr.*, **143**, 229848).
187. Wang, L., Li, B., Xiang, W., Shi, Y., and Liu, C. (2008) *Nongyao*, **47**, 378–380.
188. Dutzmann, S., Mauler-Machnik, A., Kerz-Möhlendick, F., Appelgate, J., and Heinemann, U. (2002) *Brighton Crop Prot. Conf. - Pests Dis.*, **1**, 365–370.
189. Heinemann, U., Benet-Buchholz, J., Etzel, W., and Schindler, M. (2004) *Pflanzenschutz-Nachr. (German Ed.)*, **57**, 299–318.
190. Suty-Heinze, A., Häuser-Hahn, I., and Kemper, K. (2004) *Pflanzenschutz-Nachr. (German Ed.)*, **57**, 451–472.
191. Dutzmann, S., Hayakawa, H., Oshima, A., and Suty-Heinze, A. (2004) *Pflanzenschutz-Nachr. (German Ed.)*, **57**, 415–435.
192. Walter, H. (2007) in *Modern Crop Protection Compounds* (eds W. Kramer and U. Schirmer), Wiley-VCH Verlag GmbH, Weinheim, pp. 528–538.
193. Nevill, D., Nyfeler, R., and Sozzi, D. (1988) *Brighton Crop Prot. Conf. - Pests Dis.*, **1**, 65–72.
194. Gehmann, K., Nyfeler, R., Leadbeater, A., Nevill, A.J., and Sozzi, D. (1990) *Brighton Crop Prot. Conf. - Pests Dis.*, **2**, 399–406.
195. Arima, K., Imanaka, H., Kousaka, M., Fukuda, A., and Tamura, G. (1965) *J. Antibiot. Ser. A (Tokyo)*, **18**, 201–204; (*Chem. Abstr.*, **64**, 26701).
196. Corran, A. (2012) in *Modern Crop Protection Compounds* (eds W. Krämer, U. Schirmer, P. Jeschke and M. Witschel), Wiley-VCH Verlag GmbH, Weinheim, pp. 715–721.
197. Sisler, H.D., Ragsdale, N.N., and Waterfield, W.F. (1984) *Pestic. Sci.*, **15**, 167–176.
198. Rademacher, W. (2000) *Annu. Rev. Plant Physiol. Plant Mol. Biol.*, **51**, 501–531.
199. Almond, J.A. and Dawkins, T.C.K. (1985) *Proc. Brighton Crop Prot. Conf. - Weeds*, **2**, 481–488.
200. Sandmann, G. and Bäger, P. (1997) *Rev. Toxicol. (Amsterdam)*, **1**, 1–10; (*Chem. Abstr.*, **128**, 137455).
201. Hamprecht, G. and Witschel, M. (2012) in *Modern Crop Protection Compounds* (eds W. Krämer and U. Schirmer, P. Jeschke and M. Witschel), Wiley-VCH Verlag GmbH, Weinheim, pp. 197–225.
202. Sandmann, G. and Bäger, P. (1992) in *Rational Approaches to Structure, Activity, and Ecotoxicology of Agrochemicals* (eds W. Draber and T. Fujita), CRC Press, Boca Raton, FL.
203. Kowalczyk-Schräder, S. and Sandmann, G. (1992) *Pestic. Biochem. Physiol.*, **42**, 7–12.
204. Devlin, R.M. and Karczmarczyk, S.J. (1975) Proceedings, Annual Meeting of the Northeastern Weed Science Society, vol.29, pp. 161–168.
205. Waldrep, T.W. and Taylor, M.H. (1976) *J. Agric. Food Chem.*, **24**, 1250–1251.
206. Forbes, G.R. and Mathews, P.R. (1985) *Proc. Brighton Crop Prot. Conf. - Weeds*, **3**, 797–804.
207. Cramp, M.C., Gilmour, J., Hatton, L.R., Hewett, R.H., Nolan, C.J., and Parnell, E.W. (1985) *Proc. Brighton Crop Prot. Conf. - Weeds*, **1**, 23–28.
208. Rogers, D.D., Kirby, B.W., Hulbert, J.C., Bledsoe, M.E., Hill, L.V., Omid, A., and Ward, C.E. (1987) *Proc. Brighton Crop Prot. Conf. - Weeds*, **1**, 69–75.

209. White, R.H. and Clayton, W.S. (1999) *Brighton Conf. Weeds*, **1**, 47–52.

210. Takamura, S., Okada, T., Fukuda, S., Akiyoshi, Y., Hoshide, F., Funaki, E., and Sakai, S. (1999) *Brighton Conf. Weeds*, **1**, 41–46.

211. Gutteridge, S. and Thompson, M.E. (2012) in *Modern Crop Protection Compounds* (eds W. Krämer, U. Schirmer, P. Jeschke and M. Witschel), Wiley-VCH Verlag GmbH, Weinheim, pp. 29–49.

212. Berger, B.M., Müller, M., and Eing, A. (2002) *Pest Manag. Sci.*, **58**, 724–735.

213. Amrein, J. and Gerber, H.R. (1985) *Proc. Brighton Conf. - Weeds*, **1**, 55–62.

214. Schulte, M., Kreuz, K., Nelgen, N., Hudetz, M., and Meyer, W. (1993) *Proc. Brighton Crop Prot. Conf. - Weeds*, **1**, 53–59.

215. Peeples, K.A., Moon, M., Lichtner, F.T., Wittenbach, V.A., Carski, T.H., Woodward, M.D., Graham, K., and Reinke, H. (1991) *Proc. Brighton Crop Prot. Conf. - Weeds*, **1**, 25–30.

216. Hacker, E., Bieringer, H., Willms, L., Ort, O., Koecher, H., Kehne, H., and Fischer, R.C. (1999) *Brighton Conf. - Weeds*, **1**, 15–22.

217. Doig, R.I., Carraro, G.A., and McKinley, N.D. (1983) Proceedings of 10th International Congress of Plant Protection Conference, 20–25 November 1983, Vol. 1, pp. 324–331.

218. Howard, S., Hudetz, M., and Allard, J.L. (2001) *Brighton Crop Prot. Conf. - Weeds*, **1**, 29–34.

219. Palm, H.L., Liang, P.H., Fuesler, T.P., Leek, G.L., Strachan, S.D., Wittenbach, V.A., and Swinchatt, M.L. (1989) *Proc. Brighton Crop Prot. Conf. - Weeds*, **1**, 23–28.

220. Hashizume, B. (1990) *Jpn. Pestic. Inf.*, **57**, 27–30; (*Chem. Abstr.*, **115**, 108450).

221. Maurer, W., Gerber, H.R., and Rufener, J. (1987) *Brighton Crop Prot. Conf. - Weeds*, **1**, 41–48.

222. Suzuki, K., Nawamaki, T., Watanabe, S., Yamamoto, S., Sato, T., Morimoto, K., and Wells, B.H. (1991) *Proc. Brighton Crop Prot. Conf. - Weeds*, **1**, 31–37.

223. Kita, H., Tamada, Y., Nakaya, Y., Yano, T., and Saeki, M. (2005) PCT Int. Appl., WO 2005/103044 A1 (Nissan Chemical Industries, Ltd., Japan); (*Chem. Abstr.*, **143**, 440432).

224. Barefoot, A.C., Strahan, J.C., Powley, C.R., Shalaby, L.M., and Klemens, F.K. (1995) *Proc. Brighton Crop Prot. Conf. - Weeds*, **2**, 707–712.

225. Tanaka, Y., Kajiwara, Y., Noguchi, M., Kajiwara, T., and Tabuchi, T. (2003) PCT Int. Appl., WO 2003/061388 A1 (Sumitomo Chemical Takeda Agro Company, Ltd., Japan); (*Chem. Abstr.*, **139**, 149639).

226. Müller, K.H. (2002) *Pflanzenschutz-Nachr. (German Ed.)*, **55**, 15–28.

227. Santel, H.J., Bowden, B.A., Sorensen, V.M., and Müller, K.H. (1990) *Brighton Crop Prot. Conf. - Pests Dis.*, **1**, 23–28.

228. Cottrell, H.J. and Heywood, B.J. (1965) *Nature (London)*, **207**, 655–656.

229. Larelle, D., Mann, R., Cavanna, S., Bernes, R., Duriatti, A., and Mavrotas, C. (2003) Congress Proceedings – BCPC International Congress: Crop Science and Technology, Glasgow, 10–12 November 2003, Vol. 1, pp. 75–80.

230. Snel, M., Watson, P., Gray, N.R., Kleschick, W.A., and Carson, C.M. (1993) *Meded. - Facul. Landbouwkd. Toegep. Biol. Wet. (Univ. Gent)*, **58**(3a), 845–852; (*Chem. Abstr.*, **120**, 291962).

231. Thompson, A.R., McReath, A.M., Carson, C.M., Ehr, R.J., and DeBoer, G.J. (1999) *Brighton Conf. Weeds*, **1**, 73–80.

232. Johnson, T.C., Martin, T.P., Mann, K.K., and Pobanz, M.A. (2009) *Bioorg. Med. Chem. Lett.*, **17**, 4230–4240.

233. Theodoridis, G. (2007) in *Modern Crop Protection Compounds* (eds W. Kramer and U. Schirmer), Wiley-VCH Verlag GmbH, Weinheim, pp. 153–186.

234. Miura, Y., Ohnishi, M., Mabuchi, T., and Yanai, I. (1993) *Brighton Crop Prot. Conf. - Weeds*, **1**, 35–40.

235. Nuyken, W., Landes, M., Grossmann, K., and Gerber, M. (1999) *Brighton Crop Prot. Conf. - Weeds*, **1**, 81–86.

236. Yoshida, R., Sakaki, M., Sato, R., Haga, T., Nagano, E., Oshio, H., and Kamoshita, K. (1991) *Brighton Crop Prot. Conf. - Weeds*, **1**, 69–75.

237. Lemon, R.G., Hoelewyn, T.A., Abrameit, A., and Gerik, T.J. (1993) *Proc. - Beltwide Cotton Conf.*, **1**, 605–606.

238. Dickmann, R., Melgarejo, J., Loubiere, P., and Montagnon, M. (1997) *Brighton Crop Prot. Conf. - Weeds*, **1**, 51–57.

239. Dayan, F.E., Green, H.M., Weete, J.D., Hancock, H.G., and Gary, H. (1996) *Weed Sci.*, **44**, 797–803.

240. van Saun, W.A., Bahr, J.T., Bourdouxhe, L.J., Gargantiel, F.J., Hotzman, F.W., Shires, S.W., Sladen, N.A., Tutt, S.F., and Wilson, K.R. (1993) *Brighton Crop Prot. Conf. - Weeds*, **1**, 19–28.

241. Hirai, K., Yano, T., Ugai, S., Yoshimura, T., and Hori, M. (2001) *Nippon Noyaku Gakkaishi*, **26**, 194–202.

242. Katayama, T., Kawamura, S., Sanemitsu, Y., and Mine, Y. (1997) PCT Int. Appl. WO 9707104A1 (Sumitomo Chemical Company, Ltd., Japan); (*Chem. Abstr.*, **126**, 238386).

243. Liu, C. and Zhang, X. (2002) *Nongyao*, **41**, 45–46.

244. Sikkema, P.H., Shropshire, C., and Soltani, N. (2008) *Crop. Prot.*, **27**, 1495–1497.

Part II
New Methods to Identify the Mode of Action of Active Ingredients

Modern Methods in Crop Protection Research, First Edition.
Edited by Peter Jeschke, Wolfgang Krämer, Ulrich Schirmer, and Matthias Witschel.

5
RNA Interference (RNAi) for Functional Genomics Studies and as a Tool for Crop Protection

Bernd Essigmann, Eric Paget, and Frédéric Schmitt

5.1
Introduction

Although molecular cell biology has long been dominated by a protein-centric view, the discovery of new types of gene expression regulation (e.g., small non-coding RNAs) challenges this perception. The first observation of RNA silencing was described in 1990 by Napoli *et al.* [1] and van der Krol *et al.* [2] in petunia flowers transformed with a chalcone synthase or dihydroflavonol-4 reductase gene, such that color variegation was associated with the co-suppression of both endogenous and introduced genes. It was proposed at this time that the RNA molecules might interact either with RNA or with DNA and interfere with the transcription process either directly or via DNA methylation. Since then, the results of numerous studies have identified RNA silencing mechanisms under different terms such as co-suppression or post-transcriptional gene silencing (PTGS) in plants, or "quelling" in fungi. The double-stranded nature of the RNA implicated was first described during the late 1990s by Fire *et al.* [3], and this phenomenon was named RNA interference (RNAi). Since then, much progress has been made in the understanding of RNA silencing pathways in plants and animals, such that RNAi is today not only used widely as a tool for gene validation but has also become an interesting means for crop protection. Details of the use of RNAi against major crop diseases and pests will be reviewed in this chapter.

5.2
RNA Silencing Pathways

In eukaryotic genomes, most of the genes are subject to regulation in a sequence-specific manner through non-coding small RNAs. This phenomenon is termed *RNA silencing*. Small RNAs (sRNAs) are heterogeneous in size, 18–25 nt in length, and have roles in the development, reproduction, maintenance of genome integrity, and metabolism. They are also involved in responses to environmental

Modern Methods in Crop Protection Research, First Edition.
Edited by Peter Jeschke, Wolfgang Krämer, Ulrich Schirmer, and Matthias Witschel.

factors, including nutrient uptake, biotic, and abiotic stresses. Based on their biogenesis and precursor structure, two distinct groups account for these responses:

- The small interfering RNA (siRNA) pathway, which is naturally triggered by viral infections, aberrant genome transcripts or specific loci.
- The microRNA (miRNA) pathway, which is triggered by genome-encoded miRNAs.

Both mechanisms lead either to mRNA degradation and/or to translation repression. Irrespective of their origin, RNA silencing activities all share four consensus steps:

- The generation of double-stranded RNA (dsRNA).
- Processing of dsRNA into 18–25 bp portions by RNAase III-type enzymes called *Dicer*.
- *O*-methylation at 3′ ends.
- Incorporation into the RNA-induced silencing complex (RISC) containing Argonaute (AGO) proteins and associated with partially or fully complementary target RNA or DNA. AGO proteins bear a PAZ (Piwi, Argonaute, Zwille homology) domain for sRNA binding and a PIWI domain responsible for endonucleotidic slicing activity of target RNAs.

Most of the RNA silencing studies performed in plants have been conducted in *Arabidopsis thaliana*. In this model plant, four Dicer like (DCL, named by homology to animal Dicer), ten AGO and six RNA-dependent RNA polymerase (RDR) genes were identified. In maize, five DCL, 18 AGO, and five RDR genes were very recently described [4] while previously, in 2008, eight 8 DCL, 19 AGO, and five RDR genes were identified in rice [5]. Depending on the biological process and the associated RNA silencing pathway, different Dicer and AGO would be recruited to generate the respective silencing effect (Table 5.1 and Figure 5.1). For further details, see Refs [6, 7].

Table 5.1 Summary of small RNA types and functions.

Small RNA type	Small RNA size (nucleotides)	Origin	Silencing pathway components	Biological effect
miRNA	20–25	miRNA genes	DCL1, AGO1	mRNA degradation, translational repression
hc-siRNA	24	Transposons, repeated DNA loci	RDR2, DCL3, AGO4/6	Cytosine methylation and/or histone modification at target sites
Nat-siRNA (lsiRNA)	21 (~40)	NAT transcripts	DCL4, AGO1	mRNA degradation
ta-siRNA RNAi)	21	TAS transcripts viral replication	RDR6, DCL4, AGO1	mRNA degradation

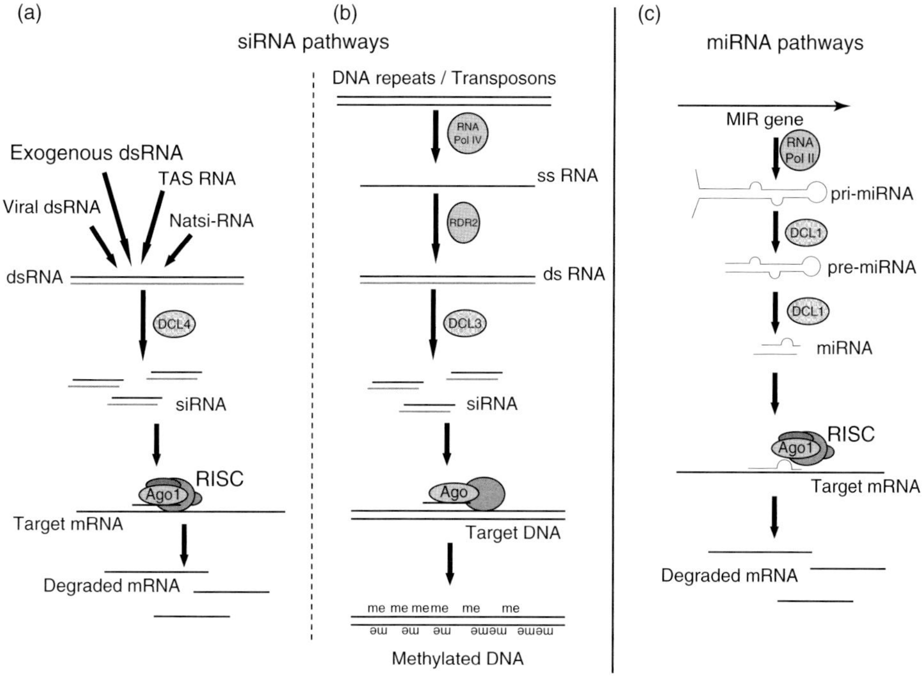

Figure 5.1 Schematic representation of RNA silencing pathways. (a) The siRNA pathway mediated by DCL4 and Ago1 leading to mRNA degradation (including RNAi); (b) The heterochromatic siRNA pathways effected by DCL3 and AGO4/6 directing DNA methylation and/or histone modifications; (c) The miRNA pathway acted by DCL1 and AGO1/10 and conducting to mRNA degradation.

5.2.1 The MicroRNA (miRNA) Pathway

MicroRNAs are encoded by intronic or intergenic genes and transcribed by RNA polymerase II. The primary transcript (pri-miRNA) forms an imperfect fold-back structure that is then processed into a stem-loop precursor known as a *pre-miRNA*. The pre-miRNAs are matured mostly in subnuclear bodies by DCL1, together with HYL1 (HYPONASTIC LEAVES 1) and SE (SERRATE), to form a sRNA duplex with an imperfectly based hairpin loop structure. The duplex is then methylated at its 3′ end by HEN1 (HUA ENHANCER 1), transported to the cytoplasm, and preferentially incorporated into AGO1 (or AGO10) containing the RISC complex to promote the slicing or translational repression of target mRNA based on sequence complementarity. Mature miRNAs are generally 20–25 nt long. The expression of miRNAs is highly regulated, and often subject to modification (induction or repression) by external stimuli that include nutrient availability as well as abiotic or biotic stresses, and they frequently exhibit tissue-specific expression. To date, more than 230 miRNA loci have been identified in *Arabidopsis*, representing around 80 miRNA families. In rice and maize, 491 and 170 miRNAs respectively have been described so far (*http://www.mirbase.org/*; 5 October 2011 [8–11].

5.2.2
The Small Interfering Pathway (siRNA)

siRNAs are perfectly 21–24 base-paired dsRNAs that arise from long dsRNA precursors. Different classes of siRNAs have been identified in plants.

- Heterochromatic small interfering RNAs (hc-siRNAs) are mainly derived from transposons and DNA repeated loci. Their dsRNA precursors are produced from the action of RDR2 (RNA-dependent RNA polymerase 2) copying single-stranded RNAs or from specific RNA polymerase IV, and processed into 24 nt long siRNAs (lsiRNAs) by DCL3. These siRNAs incorporate into AGO4 or AGO6 and mediate cytosine methylation and/or histone modifications at target sites.
- Natural antisense-transcript-derived small interfering RNAs (nat-siRNAs) are 21 bases in length, and are produced from overlapping sense and antisense transcripts and cleaved by DCL4. lsiRNAs of around 40 nt were recently discovered and are also generated by NAT transcripts. Nat-siRNAs play important functions in biotic and abiotic stress responses.
- Trans-acting small interfering RNAs (ta-siRNAs) are generated upon miRNA-guided cleavage of non-coding TAS transcripts converted to dsRNA by RDR6 and subsequently diced into active ta-siRNAs primarily by DCL4 (redundantly by DCL2). The ta-siRNAs are involved in PTGS of mRNA implicated in developmental phase changes and organ polarity. The effector molecules of the ta-siRNA pathway are also responsible for the so-called RNAi that initially was discovered as a mechanism against viral defense but which today is widely used for experimental gene knockdown.

5.3
RNAi as a Tool for Functional Genomics in Plants

For many years, the quality and quantity of crop plants have been steadily improved by using well-known conventional plant breeding methods that remain both time-consuming and laborious. Subsequently, genetic engineering has contributed to rapid and significant changes in crop improvement through novel genes and traits which can be effectively inserted into elite crops. After having successfully sequenced the genomes of several organisms, however, the aim of the research groups is to use these sequence databases to understand precisely how genes operate, as the study of genes and the role played by their resulting proteins in biochemical processes forms part of functional genomics. As in other kingdoms, a major challenge in the post-genome era of plant biology has been to determine the functions of all genes in a plant genome. The goal of functional genomics is to understand the relationship between an organism's genome and its phenotype [12] – a procedure which often requires not only high-throughput technologies to address several functions simultaneously but also the generation of numerous knockdown lines for which selection needs marker genes and several crosses to assess for the correct mutant/phenotype association.

Before the discovery of RNAi, the alteration of gene expression and functions was achieved by using transposons or T-DNAs insertion, chemical or radiation mutagenesis, and antisense RNA technology, the subsequent aim being to correlate plant gene functions with a new phenotype. These powerful tools presented certain limitations, however, such as addressing multigene families, multiple insertion sites, or lethality due to essential gene knockdown [13]. Using transient or stable expression systems, RNAi has a great potential as a tool for downregulating gene expression, and has boosted the present knowledge of gene regulation, gene function, and gene analysis [14].

Indeed, RNAi has several advantages for functional genomics, notably because it is dominant and can spread systematically [15, 16]. By targeting a specific sequence, it is possible to silence either a single chosen gene or a gene family, provided that the target sequence is a highly conserved domain among the family. Moreover, when studying the functions of essential genes, RNAi allows either different lines having variable levels of gene silencing with the same dsRNA insert to be obtained, or the use of inducible promoters that can regulate siRNA expression both time- and/or tissue-wise. Compared to gene knock-out, RNA silencing only downregulates gene expression, thus overcoming potential issues of lethality [13, 17–19].

However, in similar fashion to animals, it appears that in plants siRNA does not always target the correct gene (the "off-target silencing phenomenon") when sequences have partial homology for unintended genes; in fact, off-target silencing effects can occur all along the PTGS, or result from endogenous miRNA system perturbation [20]. Studies performed both *in silico* and *in vivo* have revealed that a great number of gene transcripts in *Arabidopsis* plants can have potential off-target effects when used as silencing trigger for PTGS. Thus, it was concluded that these effects can have a major impact when identifying the exact functional role of target genes in improving crop biosafety, and also provide give some hints to overcome such issues in the future.

As mentioned above, different techniques have been developed to induce dsRNA-mediated silencing in plants, though each methods has its own pros and cons [21, 22]. Most of the techniques require the construction of a vector to express – either transiently or stably – a dsRNA that often is transcribed as a single RNA with a spacer and inverted repeats that fold back to create a hairpin RNA (hpRNA). Many vectors have been based on the Gateway® system, and developed for use in different crops (monocotyledonous or dicotyledonous) or even tissues [23]). Artificial microRNAs (amiRNAs), based on endogenous miRNA backbones, are also effective in triggering gene silencing [24–26] and show some advantages over the hpRNA as they are more specific and less prone to silencing by other sRNAs.

Plant genetic transformation can be performed by various often-used techniques, including *Agrobacterium* strain infiltration, particle bombardment, or viral sequences that are used to infect the plant. Virus-induced gene silencing (VIGS), although transient, offers a great potential for large-scale reverse genetic studies [21, 27–29]. Although plants to which high-throughput techniques can be applied remain a major bottleneck, increasing numbers of species can now be transformed by agroinfiltration or by the use of VIGS vectors [30, 31].

As in animal systems, direct siRNA delivery into plants cells has also been proposed for large-scale RNAi screens [20]. For example, siRNAs have been directly introduced into the protoplasts of *Coptis japonica* [22], potato [32] or in rice, cotton, and pine cells by using laser-induced stress wave cell cultures [33], though these techniques are very rarely used.

Following the investigation of functional genomics in *Caenorhabditis elegans* [34], RNAi is today widely used also for functional genetics studies in plants. For example, the AGRIKOLA consortium (*Arabidopsis* Genomic RNAi Knock-out Line Analysis: *http://www.agrikola.org/index.php?o=/agrikola/main*; 5 October 2011), funded by the European Union, has utilized polymerase chain reaction (PCR) products to silence through RNAi each *Arabidopsis* gene, based on GSTs (Gene-specific Sequence Tags) previously developed by the CATMA consortium (Complete *Arabidopsis* Transcriptome MicroArray) [35]. Similarly, the objectives of the National Science Foundation (NSF)-funded *Medicago truncatula* RNAi database are to identify genes that are essential for symbiotic development. Based on a previous expressed sequence tag (EST) project, the aim of the research groups is to use RNAi to systematically silence the expression of approximately 1500 genes that are implicated in the development or functioning of the rhizobia-legume or mycorrhizal symbioses (*https://mtrnai.msi.umn.edu/*; 5 October 2011). In the *Arabidopsis* 2010 project, the amiRNAi Central Project team is developing a comprehensive resource for the analysis of *Arabidopsis* gene function through the creation of genome-wide collection of amiRNAs (ca. 17 699 clones to date; see *http://2010.cshl.edu/arabidopsis/2010/scripts/main2.pl*; 5 October 2011). Currently, numerous reports are available of RNAi being used for functional genomics studies; indeed, some previous reviews have already listed major contributions in model and agricultural plants [12, 13, 21]; also see examples in Table 5.2).

A few more examples can be given at this point. Functional genomics via RNAi has been used not only in large-scale experiments for agricultural crops but also on less-common plants for fundamental research into different metabolites pathways. For instance, although gene targeting via homologous recombination is possible, RNAi has been used in the moss *Physicomitrella patens*, to study families of genes [51]. In fruits, genes involved in ripening could be investigated, as has been shown in strawberry, where the agroinfiltration of a chalcone synthase RNAi construct impaired anthocyanin production and the flavonoid pathway [52]. In the medicinal plant *Salvia miltiorrhiza* (Chinese sage), phenylalanine ammonia-lyase (PAL) catalyzes the first step in the phenylpropanoid pathway, and is probably encoded by a multi-gene family. The suppression of PAL by RNAi led to the production of plants with dwarfism, altered leaves, a lower lignin content,

Table 5.2 Examples of plant gene function revealed by RNA interference.

Plant	Target gene	Reference(s)
Arabidopsis thaliana	AGAMOUS, CLAVATA3, APETALA1, PERIANTHIA	[36]
	Phytoene desaturase, ethylene signaling EIN2, Flowering repression FLC1, Chalcone synthase	[37]
Tobacco	Phytoene desaturase	[38]
	Polyphenol oxidase	[37]
Rice	RAD2/XPG nuclease OsGEN-L	[39]
	Heme oxidase, RAC GTPase, putative proteins	[40]
Cotton	Δ9 and Δ12 desaturase	[37]
	Myb transcription factor	[41]
Nicotiana benthamiana	N-gene response pathway: Rar1, Eds1, Npr1, Sgt1, Skp1, Csn8	[42, 43]
Tomato	Phytoene desaturase, constitutive triple response CTR1, ribulose bisphosphate carboxylase (tRbcS)	[44]
Soybean	Phytoene desaturase	[30, 45]
	Rust resistance Rpp4	[46]
Maize	Male sterility factor 45, dihydrol flavenol reductase, cytochrome P450	[47]
Potato	Starch synthase	[48]
Canola	Farnesyl transferase	[49]
Wheat	Phytoene desaturase, transmembrane protein	[50]

and an underdeveloped root system. The results of these studies confirmed that PAL plays a major role during the development of *S. miltiorrhiza*, and in the metabolism of both rosmarinic acid and salvianolic acid that are thought to be at the heart of the plant's antioxidant, antitumor, and antimicrobial properties [53]. In another study, an assessment was made of the possible physiological role of G-strand-specific single-stranded telomere binding proteins (GTBPs) via RNA-mediated gene knockdown. In this case, the RNAi-GTBP1 transgenic tobacco plants showed severe developmental abnormalities and genome instability, thus revealing the important role that GTBPs play in telomere structure and function [54].

DNA methylation is established in plants, through the RNA-directed DNA methylation (RdDm) pathway, that requires 24-nt siRNAs [55]. RNAi has been used in many studies to decipher the epigenetic regulation of gene expression, such as the activity of plant methyltransferase. For example, orthologs of DDM1 (decrease in DNA methylation), a chromatin remodeling factor required for the maintenance of DNA methylation, have been characterized in *Brassica rapa* [56]. It has also been shown, by using RNAi in hypomethylated knockdown plants, that the BrDDM1 genes regulate the DNA methylation of repetitive sequences such as transposons, in preferential fashion.

As will be seen below, although RNAi has also been employed in the battle against pathogens, its primary use has been in the discovery of resistance gene pathways and functions. For example, in the case of the soybean rust disease, some genes that play a role in plant resistance (Rpp2 or Rpp4 loci) towards *Phakopsora pachyrhizi* have been identified using VIGS [46, 57]). Similarly, a double agroinfiltration protocol has been developed for the functional assay of candidate genes in potato late blight resistance [58].

Reverse genetics through RNAi was also used to identify the function of those genes involved in RNAi metabolism itself [18], and miRNAs have also been shown clearly to play a role in tuning their own biogenesis and function [59]. One of the major miRNA functions is to control plant development, mainly by targeting transcription factors [60]. New approaches have been developed to study not only general metabolism genes but also endogenous miRNA in plants: "mimicries" are natural or artificial non-cleavable miRNA-targets that can sequester their associated miRNA and thus inhibit their activity on other targets [61]. Likewise, a new MIR VIGS system has been identified on *N. benthamiana* that allows not only VIGS amiRNA expression but also functional analyses of endogenous miRNAs [62]. More recently, the expression of endogenous miRNAs has been efficiently silenced in *A. thaliana* by using artificial miRNA, thus providing a new and powerful tool for the analysis of miRNA function in plants [63]. In contrast to sRNAs, less is known regarding the functions of long non-coding RNAs, although in the future these relatively new RNAs may become very relevant as new tools for biotechnological applications [64].

Although, according to numerous reports in the literature, RNAi is clearly a very useful tool for functional genomics applications, it must be borne in mind that, in almost every plant species studied using RNAi, variability in the silencing efficiency has been reported both at the mRNA or the phenotypic level, depending sometimes on the generation observed [12]. Important variables should be considered in the experimental design of reverse functional genomics, and great caution taken when interpreting the results obtained.

It has been seen that reverse genetics through RNAi is applicable to a large variety of plant species, from model plants as simple as moss to important agricultural crops such as maize, soybean, or wheat. An RNAi-associated loss of function can also sometimes improve plant performance, and RNAi has been used – and will continue to be used – in transgenic crop development for commercial uses [14, 20, 65].

5.4
RNAi as a Tool for Engineering Resistance against Fungi and Oomycetes

The use of the RNAi appears to be an attractive technology to downregulate essential genes in a plant pathogen or pest to engineer resistance. This technology has proved to be efficient for functional genomics studies in nematodes, insects, mammalian organisms, or plants. However, one of the first hurdles to be overcome is to identify an efficient transfection or transformation method to deliver into the organism the dsRNA or siRNA molecules that will trigger the RNAi effect. When considering plant biotechnologies applications, it is easy to imagine delivering a dsRNA expressed in a plant to a phytophageous insect, less obvious for sucking/piercing insects or nematodes (due to their feeding mode), and even more questionable when considering fungi or oomycetes, as little is known regarding their ability to absorb nucleotidic molecules. The questions are: "Does RNAi work at all in fungi and/or oomycetes?," and "Are the fungi able naturally to take up such molecules?"

The first evidence came from a pioneering study conducted by Baltz *et al.* in 2003 [66], in which it was shown that a tobacco plant expressing a hpRNAi construct against the β-tubulin gene of *Cercospora nicotianae* had strongly reduced symptoms upon infection. Different classes of resistant plants were obtained, and resistance was shown to correlate inversely with detection of the β-tubulin hairpin dsRNA in plants, which suggested an *in planta* dicing of the dsRNA taken up by the fungus into active siRNA molecules. Similar observations were made with rice plants engineered to become resistant against rice blast disease (*Magnaporthe grisea*). This was the first direct evidence of the use of RNAi to confer resistance against fungi, and the concept was later designated as host-induced gene silencing (HIGS) [67].

The second direct evidence for dsRNA uptake by fungal-like organisms was described in 2005 by Whisson *et al.* [68] in the oomycete *P. infestans*, by delivering *in vitro* dsRNA into protoplast to transiently silence a GFP marker gene or two endogenous genes. The study results confirmed that the external application of dsRNA could lead to a gene downregulation, and that RNAi could be used as a tool for functional genomics in *P. infestans*. Similar results were obtained *in vitro* on germinating spores of *Aspergillus nidulans* [69]. Subsequently, a phylogenetic analysis performed in 2006 by Nakayashiki *et al.* [70] provided more insight into the occurrence of RNA silencing components (Dicer, AGO, and RDR),and indicated that a wide range of fungi comprising ascomycetes, basidiomycetes, and zygomycetes possessed multiple components in their genomes. However, some ascomycetes (e.g., *Saccharomyces cerevisiae*) and basidiomycetes (e.g., *Ustilago maydis*) lack some or all of these components.

The *in vitro* delivery of dsRNA and target gene knockdown was also achieved on protoplasts of the basidiomycete *Moniliophthora perniciosa*, the causal agent of witches broom disease of cacao, providing here also a tool for functional genomics in this system [71].

The concept of controlling fungal diseases by HIGS seemed very attractive, and this point was confirmed by the filing of several patents issued between 2005 and 2009. Several fungi were targeted, including *Magnaporthe grisea* [72] *Phytophthora* sp. [73] *Sclerotinia sclerotorium*, *P. pachyrhizi* [74], and *Blumeria graminis* [75]. Further examples were described in 2010 on *Fusarium verticilloides* [76], demonstrating the silencing of a *gus* gene expressed by the fungi after transfer of *gus* dsRNA/siRNA expressed in the tobacco host plant. It was also possible to demonstrate the HIGS effect in wheat and barley against the obligate pathogen *B. graminis*, leading to a reduction in fungal infection [67].

The examples reported to date have demonstrated the feasibility of using RNAi technology to engineer resistance against fungi and oomycetes in plants; however, the available data also show that the resistance obtained is more likely partial than complete. A major challenge remains to identify the best genes to target, and these genes can vary among fungi depending on their life cycles and their biotrophic interactions with the plants. The use of *in vitro* dsRNA delivery systems, and also of *in planta* transient expression based on VIGS, will certainly accelerate discoveries in this field.

5.5
RNAi as a Tool for Engineering Insect Resistance

The worldwide economic damage to agricultural and horticultural crops and to orchards that is caused by insect pests today stands at US$ one hundred billion annually [77], and has been estimated as 10–20% of the major crops [78].

Consequently, the engineering of crop plants with an endogenous resistance to insect pests, to reduce not only yield loss but also pesticide utilization, has been one of the most important achievements of genetically modified (GM) technology [79]. In particular, the generation of crop plants expressing genes coding for insecticidal crystalline proteins from *Bacillus thuringiensis* (Bt) – the so-called Cry toxins – have achieved great success from both economical and ecological points of view [80]. Unfortunately, some pest species (e.g., sucking insects) are difficult to target with the Bt toxins and, as a result, resistant populations of targeted pests have evolved in the field. Recently, the fall armyworm *Spodoptera frugiperda* was found to be resistant to Cry1F corn grown in Puerto Rico. Both, Moar *et al.* [81] and Tabashnik and Carrière [82] reported the evolution of resistance in the field by populations of *Busseola fusca* to Cry1Ab corn grown in South Africa, and by populations of *Helicoverpa zea* to Cry1Ac and Cry2Ab cotton in the USA. In this context, additional or new technologies or modes of action would be helpful for durable pest control, of which RNAi technology is thought to play an important role.

The possibility of using plant-mediated RNAi to protect plants against insects has long been recognized but initially was considered unfeasible, as two of the components that were important for systemic RNAi response could not be identified in the genomes of the model insect *Drosophila melanogaster*. First, no homolog for encoding an RNA-dependent RNA polymerase (RDR) had been detected. RDR is necessary for the siRNA amplification that leads to persistent and systemic RNAi effects in *Caenorhabditis elegans*. In addition, the homologs of the *C. elegans* sid-1 (systemic RNAi deficient-1) gene, which functions as a channel for the uptake and release of dsRNA among cells, also appeared to be missing from the *Drosophila* genome [83]. Taken together, this information led to speculation as to why no systemic RNAi response could be established in *D. melanogaster*.

Homologs of the sid-1 gene have, however, been identified in the moths *Bombyx mori* and *Spodoptera exigua*, as well as in another model insect *Tribolium castaneum*, so that insect systemic RNAi was achieved first in *T. castaneum* (flour beetle), where multiple genes were targeted by the injection of specific dsRNA [84, 85]. Since then, it has been shown that RNAi effects can be produced in a wide range of insects.

A summary of successful RNAi experiments, by injection or feeding, in a number of lepidopteran species is provided by Terenius *et al.* [86]. Notably, among the first reports of lepidopteran RNAi made in 2002, one reported the knockdown of a pigment gene following dsRNA injection into *B. mori* embryos [87], while another targeted a pattern recognition protein, hemolin, in *Hyalophora cecropia* embryos by heritable RNAi [88].

As an example of sucking insects, Mutti *et al.* [89] showed that the injection of siCoo2-RNA into pea aphids led to Coo2 suppression, and caused a significant mortality of insects fed on host plants. RNAi was also successfully achieved by feeding sucking insects with sugar solutions containing dsRNA synthesized *in vitro* [90–92]. All of these results indicated that several insect pests from different orders can be effectively targeted by the oral delivery of dsRNA [93, 94]. This confirmation of whether RNAi effects could be induced in insects by orally delivered dsRNA is a prerequisite for utilization of RNAi for crop protection against insect pests.

In 2007, a breakthrough was achieved regarding plant-expressed dsRNAs for insect protection, when two research groups reported an enhanced resistance to insects by transgene-encoded RNAi in plants [95, 96]. In the first case, an hpRNA construct directed against a cytochrome P_{450} monooxygenase from the cotton bollworm (*Helicoverpa armigera*), termed *CYP6AE14*, was transferred into *planta*. This gene, CYP6AE14, is involved in detoxification of the major endogenous defense compound in cotton, the sesquiterpene aldehyde gossypol. Larvae fed on transgenic *A. thaliana* and tobacco plants producing the dsRNA, showed a suppressed expression of CYP6AE14 and a decreased tolerance to a gossypol-containing diet [95]. In generated transgenic cotton plants producing the dsCYP6AE14, the results have been verified by the expression of CYP6AE14 being suppressed in bollworms, while GM cotton showed an enhanced protection against bollworms by having a deleterious, but not lethal, effect on the worms [97]. The results confirmed that transgenic plants could provide sufficient levels of dsRNA to suppress gene expression in the midgut of the insect, so as to stunt its growth.

In the second example, candidate targets were screened by feeding larvae with an artificial diet supplemented with dsRNAs specific to a large number of essential insect genes from western corn rootworm (WCR, *Diabrotica virgifera*). In this screen, 14 targeted genes were identified that displayed a dramatic suppression of expression and additional larval stunting and mortality. In order to demonstrate the practical application of this technology, transgenic corn was engineered to express WCR-derived dsRNA directed against V-ATPase A. The generated corn challenged with WCR showed a suppression of mRNA in the insect and a reduction in feeding damage by WCR. Further feeding assays with the same V-ATPase A dsRNA caused mortality in the related coleopteran species, southern corn rootworm (SCR; *Diabrotica undecimpunctata howardii*) and Colorado potato beetle (CPB; *Leptinotarsa decemlineata*) [96]. The conclusions from these findings were that plant-expressed dsRNA could be delivered into insects and trigger systemic silencing, although certain challenges remained regarding the delivery, uptake and efficiency of ds/siRNA from plant-expressed ds/siRNA to ensure a systemic RNAi response in the insects [98].

5.6
RNAi as a Tool for Engineering Nematodes Resistance

Plant parasitic nematodes (PPNs) are still referred to as a "*hidden pest*," although comprehensive surveys undertaken some years ago estimated the damage to world agriculture to be worth US$ 125 billion annually [99]. The vast majority of this damage ought to be attributed to sedentary species such as root knot nematodes (RKNs, e.g., *Meloidogyne* spp.) and cyst nematodes (e.g., *Heterodera* spp. and *Globodera* spp.).

Both of these PPNs invade the host roots and release secretions into the plant cells, so as to induce physiological and morphological changes. In the case of the RKNs, the ultimate aim is to modify the plant cells into very specialized and metabolically active cells (galls or giant cells), from which the nematodes constantly obtain the nutrients that are necessary to support their development. The galls block water and nutrient flow to the plant, which results in a stunted growth, an impaired fruit production, and causes the foliage to yellow and wilt, with an increasing susceptibility to pathogen attack [100, 101].

Current nematode control strategies involve the application of nematicides, as well as good agricultural practice with crop rotation and resistant cultivars; however, each approach (as well as its combination) has limitations as the PPNs can remain dormant for many years. Especially nematicides which have been widely used to control both migratory and sedentary plant-parasitic nematodes are often associated with harmful environmental effects, and therefore their use has been contested in recent years. For example, *methyl bromide*, one of the most important chemical fumigants used to control nematodes and other pests, affects a wide range of organisms (including beneficial organisms), and was defined as a "chemical that contributes to the depletion of the Earth's ozone layer" [102]. As a result of the Montreal Protocol of 1991, methyl bromide was phased out in 2005 in developed countries, and will be phased out in developing countries by 2015. Other nematicides such as the neurotoxic acetylcholinesterase inhibitor fenamiphos and the carbamate nematicide carbofuran, will soon be phased out in the USA [103]. All of these different constraints will lead to a demand for the development of new strategies for nematode control, and an approach using RNAi could clearly be included among these.

The use of RNAi as a tool for nematode control was inspired by studies on *C. elegans* performed by Fire and Mello [3], who were awarded the 2006 Nobel Prize in Physiology or Medicine. The *in vitro* delivery of dsRNA to PPN proved to be more complex, however. The injection of dsRNA into the PPNs was clearly not feasible due to their small size (equally so for *C. elegans*); however, the problem of enhancing the uptake of dsRNA was overcome by using octopamine to stimulate dsRNA uptake from a soaking solution. Oral ingestion, as exemplified by the two cyst nematodes *Heterodera glycines* (soybean cyst nematode, SCN) and *Globoderapallida*, was monitored by employing a visual fluorescent marker, while the RNAi effect was confirmed by analyzing the transcript abundance and the silencing phenotypes on the development or sexual fate of target genes [104].

For the RKN *Meloidogyne incognita*, Rosso *et al.* introduced two other stimulation reagents (resorcinol and serotonin) to induce dsRNA uptake more effectively. Two genes expressed in the subventral esophageal glands – the calreticulin (Mi-crt) gene and the polygalacturonase (Mi-pg-1) gene – were targeted in this study. The incubation of nematodes in 1% resorcinol induced dsRNA uptake through the organism's alimentary track and led to an up to 92% depletion of Mi-crt transcripts. The time course of silencing showed different temporal patterns for Mi-crt and Mi-pg-1, while for the both genes the silencing effect was highly time-limited as no transcript depletion was detectable by three days after soaking [105]. Subsequently, the soaking strategy became a preferred method for studying gene function and identifying potential targets for parasite control, and led to an increasing number of reports (as summarized by Li *et al.* [103], Lilley *et al.* [106], Rosso *et al.* [107]). Thus, a large number of genes, expressed in a range of different tissues and cell types, have been successfully targeted for silencing in different PPNs in the intestine [104, 108] and female reproductive system [109], sperm formation [104, 110], and in both subventral and dorsal esophageal glands [105, 111–113]

From an agricultural point of view, the other very interesting application strategy was a host-delivered RNAi by transient or stable expression of the ds/siRNA *in planta*. These ds/siRNAs directed against essential genes of PPNs would, after being taken up by the nematode through ingestion, elicit a knockdown of the target gene in the nematode, resulting in a reduced vitality or reproductivity and triggering less crop damage.

One of the first reports to describe a successful reduction of root-knot formation via host-derived RNAi was made by Yadav *et al.* [114], who selected two genes of the RKN– one coding for a splicing factor and the other coding for an integrase – as targets. Subsequently, tobacco plants transformed with the corresponding hpRNA constructs revealed an almost complete resistance to RKN infection.

In the same year, Huang *et al.* showed that the ingestion of *16D10* dsRNA *in vitro* silenced the target parasitism gene in RKN and resulted in a reduced nematode infectivity [113]. Additionally, the *in vivo* expression of *16D10* dsRNA in *Arabidopsis* resulted in a resistance that was effective against the four major RKN species (*M. incognita, M. javanica, M. arenaria,* and *M. hapla*). In another study, Steeves *et al.* [110] used transgenic soybeans, transformed with an RNAi expression vector containing the major sperm protein gene from *H. glycines*, to significantly reduce the reproductive potential of this nematode; in this case, an almost 70% reduction in egg occurrence per gram root tissue was obtained. Additional examples showing reductions in the numbers of females or downregulation of the transcript level of the targeted gene in different plant species, have also been demonstrated [115–118].

Beside stable transformation, VIGS with the tobacco rattle virus (TRV) were also tested as a means of expressing dsRNA into plant cells and to mediate RKN gene silencing [119]. In one example, a knockdown of the targeted genes was observed in the progeny of the feeding nematodes, which suggested that a continuous ingestion of dsRNA triggers might be used for the functional analysis of genes involved in early development. However, the heterogeneity in RNAi efficiency between TRV-inoculated plants appeared to be a limitation in this approach [120].

In a study performed by Dalzell *et al.* [121], a gene-specific knockdown of FMRFamide-like peptide (FLP) transcripts using discrete 21 bp siRNAs was achieved. In contrast the "traditional" RNAi vectors usually contained inverted repeats of more than 100-bp sequences, which resulted in a population of at least several different species of siRNAs and the potential for "off-target" gene silencing. The data presented for the potato cyst nematode *G. pallida*, and *M. incognita* showed both knockdown at the transcript level and functional data derived from migration assay, indicating that siRNAs targeting certain areas of the FLP transcripts are potent and specific in the silencing of gene function. Very recently, Ibrahim *et al.* [122] also examined RNAi constructs targeted towards four different genes to determine their efficacy in reducing galls formed by RKN in soybean roots. These genes have a high similarity with essential SCN and *C. elegans* genes. When the transformed roots were challenged with RKN, two constructs targeted towards genes encoding tyrosine phosphatase (TP) and mitochondrial stress-70 protein precursor (MSP), respectively, showed a strong interference with gall formation, reducing their numbers by 92% and 94.7%, respectively, and confirming the potential of the RNAi approach.

When taking into consideration the application of RNAi technology within the different fields of agricultural disease or pest control, the current results obtained from plant-delivered RNAi experiments in regard to PPN control have been very encouraging. Although it is mandatory to perform a correct ecological risk assessment, this new technology is thought be environmentally friendly because of its high specificity. Nevertheless, off-target effects have been reported [123] and therefore a careful selection of a suitable target gene (and its region) is essential. For instance, it is known that siRNAs produced can interfere with other sRNA pathways such as the miRNA pathway [124], and that dsRNA can induce the innate immune response through interaction with Toll-like receptors, at least in vertebrates. Nonetheless, this issue will be addressed in the future as further data are acquired from RNA deep sequencing approaches and other functional genomic tools such as amiRNAs.

5.7
RNAi as a Tool for Engineering Virus Resistance

Historically, the phenomenon of cross-protection, where the inoculation of a plant with a mild viral strain protected it against subsequent infections by more severe viruses, in parallel with the pathogen-derived resistance (PDR) [125] concept in bacteria, led to the hypothesis that the introduction of viral sequences into plants may trigger resistance to homologous viruses. Indeed, it was found that the expression of recombinant virus-derived genes could confer resistance to the corresponding virus via PTGS (also known as co-suppression) [126–128]. The role of dsRNA in this resistance mechanism, as confirmed later [129], shed light on the discovery of RNAi in worms [3]. RNA silencing in plants is a natural defense system against foreign genetic elements such as viruses, transposons, or even transgenes, and recent progress has been made on the identification and mechanism of the plant's (and of other organisms) sRNA-directed viral immunity [130–133]. Both, RNA and DNA plant viruses activate RNA silencing through the formation of viral dsRNA and the accumulation of virus-derived small interfering RNAs (vsiRNAs) at a high level during infection [134]. Transgenic plants expressing a viral gene could then develop a resistance specific to that virus through RNAi targeting the expressed gene. However, as a defensive strategy, many plant viruses have evolved viral suppressors of RNA silencing (VSRs) to counteract antiviral silencing [131, 135]. These VSRs have different targets in the silencing pathway, and can hamper viral RNA recognition, dicing, RISC assembly, RNA targeting, and amplification [136]. The fact that almost all plant viruses encode RNA silencing suppressors reinforces the concept that RNA silencing evolved originally as an antiviral defense mechanism. It is noteworthy that such RNA silencing suppressors in their own right may be responsible for some viral symptoms, as they interact also on endogenous miRNA pathways [135, 137]. On the other hand, the silencing of these specific suppressors represents one means of rendering a plant resistant [129], as will be seen in some examples below.

Apart from resistance, an induction of the RNA silencing machinery by viruses led various research groups to develop VIGS. This technique utilizes the ability of viruses to carry and induce RNAi against foreign sequences, and is based on vectors derived from the genome sequence of different RNA virus combined with host or pathogen target gene sequences [27, 38]. In return, VIGS have become one of the main techniques used to investigate PTGS in plants, as described earlier in the method of choice for the functional genomics of plants [138]. VIGS efficiency relies on the capacity of viruses to infect the studied plants; vectors were first developed on model plants (tobacco, *N. benthamiana*, *A. thaliana*) using *tobacco mosaic virus* (TMV) or *Potato virus X* (PVX); later TRV-based VIGS vectors were used successfully in crops and vegetables such as tomato, pepper, and potato [139, 140]. Other viruses have been recombined to use VIGS in crops of economic interest, testing also different ways of inoculation (agroinfiltration, biolistic, spray, leaf rubbing; see Table 5.3).

More than being simply the development of a technique to assess plant gene function, viral sequence expression in plants has led to transgenic plants that are resistant to RNA or DNA viruses. Indeed, a major loss of plant productivity is

Table 5.3 Examples of viruses used as VIGS vectors.

Virus	Silenced hosts	Reference
Bean pod mottle virus	Soybean, *Phaseolus vulgaris*	[141]
Tobacco rattle virus	Soybean	[45]
Brome mosaic virus	Barley, rice, maize	[142]
Barley stripe mosaic virus	Barley, wheat	[143]
Pea early browning virus	*Pisum sativum*	[144]
African cassava mosaic virus	*Manihot esculenta*	[145]

due to viral diseases. In contrast to disease resistance against bacteria and fungi, transgenic virus-resistant plants have now been in commercial use for almost 15 years, based on the concept of PDR mediated either by viral proteins or by RNA co-suppression. Although, as noted above, the molecular mechanism of cross-protection was unknown at that time, RNA-mediated silencing was rapidly found to be the major component of virus inhibition in vegetables and fruits [146, 147]. *Papaya ringspot virus*-resistant papaya has proven to be a great success in the past, and indeed saved the papaya industry in Hawaii. A further example relates to *Potato virus Y* (PVY)-resistant GM potatoes (Newleaf Y®: *http://www.monsanto.com/newsviews/Pages/new-leaf-potato.aspx*; 5 October 2011), which were released onto the North American market in 1998 but a few years later were withdrawn because of a very poor public acceptance. Most of these crops were expressing the coding sequences of coat proteins (CPs), as well as replication or movement proteins. The discovery of RNA silencing rapidly allowed much more sophisticated approaches to be developed (e.g., use of hpRNA, amiRNA, etc.), as shown in the examples below. However, due mainly to market acceptance issues none of these second-generation plants has yet been released, despite the fact that the safety of virus-resistant transgenic plants has been extensively addressed over the past 15 years (for a review, see Ref. [146]).

Currently, model plants are still helping to provide an understanding of virus resistance. For example, tobacco plants were recently transformed with intron-hairpin RNA (ihpRNA) constructs expressing either the partial movement gene of the TMV or the partial replicase gene of the *cucumber mosaic virus* (CMV). Several of the obtained transgenic lines were immune to the respective virus, and showed that the resistance could be inherited and kept stable in T_4 progeny [148].

Potato was among the first plants developed to be resistant to viruses: transgenic lines had been produced that were resistant to PVY by expressing dsRNA derived from the three terminal parts of the CP gene [149]. Actually, these lines were also resistant to three other PVY isolates, as the sequence used was highly conserved. Recently, an opposite approach was undertaken by constructing a chimeric vector containing three partial gene sequences specific to three different viruses (PVX, PVY, *potato leaf roll* virus). Subsequently, 20% of the transgenic plants (accumulating specific siRNAs) were immune against all three viruses, confirming that a single transgene can effectively confer resistance to multiple viruses [150].

Often, RNA silencing solutions against viruses represents the "last hope" in crops where resistance genes have not been identified in the wild germplasm. The technology has, for example, been used to successfully combat viruses in plum trees with an ihpRNA containing the *plum pox virus* (PPV) CP sequence [151]. Similarly, sweet orange plants were rendered fully or partially resistant to *citrus psorosis virus* (CPsV) through the production of ihpRNA transcripts corresponding to viral CP, 54K, or 24K genes [152].

On occasion, there may be an urgent need to identify a cure to a particularly virulent viral disease. Cassava (*Manihot esculenta*) is an important subsistence food crop in the tropical regions of Africa, but is susceptible to 20 different viral diseases among which *cassava mosaic disease* (CMD) and *cassava brown streak disease* (CBSD) are the most threatening. A recent study showed that CBSD can be efficiently controlled using RNAi, even though the disease is caused by isolates of at least two phylogenetically distinct species of single-stranded RNA viruses. Although, these experiments were performed on the model plant *N. benthamiana*, some plants were also immune to very distant isolates of both viruses, and a positive correlation was found between the level of specific siRNAs and the level of virus resistance [153].

Economically important cereal crops could also benefit from RNA silencing for virus protection. This was first applied to barley (*Hordeum vulgare*) to combat *barley yellow dwarf virus* (BYDV); lines constitutively expressing the 5′ end of the virus gained complete immunity to BYDV [154]. Similarly, transgenic rice plants encoding the RNA-dependent RNA polymerase of *rice yellow mottle virus* (RYMV) proved to be resistant to RYMV strains from different African locations, and stable over at least three generations [155]. With regards to rice, *rice stripe virus* (RSV) is one of the most widespread and severe virus diseases with no known naturally occurring resistance genes. An RNAi approach was performed by the transformation of rice plant with three different RNAi vectors based either on the CP, or a special-disease protein (SP) or a chimeric sequence containing the both CP/SP genes [156]. The resistance assays showed that the chimeric CP/SP RNAi lines were more resistant than the single CP or SP lines, and stable at least to the T_2 progeny.

Few studies have attempted to induce RNAi-mediated transgenic virus resistance in maize, although *maize dwarf mosaic virus* (MDMV) resistance has been addressed by expressing hairpin constructs targeting either the replicase [157], the P1 protein (protease) [141] or the CP gene [158], and transgenic lines were obtained with an improved resistance. In the latter study [158], several T_2 transgenic lines were evaluated as being resistant to MDMV in field inoculation trials, the resistance being relative to the length of the inverted-repeat sequence, the copy number of T-DNA insertion, and the repeatability of integration sites.

In wheat, the use of RNAi to generate resistance to the *wheat streak mosaic virus* (WSMV) and *triticum mosaic virus* (TriMV) was evaluated, again by expressing their respective CP genes in a hairpin construct [159]; viral resistance was improved in up to 60% of the T_1 generation, while T_2 generation analyses revealed transgene silencing and deletion phenomena, but still some stable transgenic lines.

miRNAs are known to be important regulators of plant development, and have also been implemented to confer resistance to pathogens such as fungi and bacteria. It is unclear, however, as whether host miRNAs respond also to viral infection. In fact, the infection of *Brassica rapa* plants by *turnip mosaic virus* (TuMV), compared to other virus or fungi pathogens, specifically induced in the plant the upregulation of two miRNAs – bra-miR158 and bramiR1885 – which probably targeted the disease-resistant proteins [160]. The possibility of using amiRNAs to control viruses was confirmed when Niu and colleagues [161] successfully used *Arabidopsis* miR159a-based amiRNAs against the viral suppressor genes P69 and HC-Pro of the *turnip yellow mosaic virus* and TuMV, respectively. Moreover, a dimeric construct with these two amiRNAs also conferred resistance against these two viruses in inoculated *Arabidopsis* plants. Similarly, a different amiRNA was used to target the 2b viral suppressor of the CMV and confer resistance to CMV infection in transgenic tobacco [162]. Recently, using *A. thaliana* miR159a, miR167b, and miR171a precursors as backbones, amiRNAs were developed to target the silencing suppressor HC-Pro of PVY and the TGBp1/p25 (p25) of PVX. The resulting transgenic tobaccos became specifically resistant to PVY or PVX infections, and the level of resistance was correlated to amiRNA expression level. As seen previously, a high resistance to both PVY and PVX could further be obtained by expressing a dimeric amiRNA precursor [163].

Geminiviruses are single-stranded circular DNA viruses that cause economically significant diseases in a wide range of crop plants worldwide. RNA-mediated resistance toward DNA viruses is believed to function through a silencing of mRNA transcripts (PTGS) and through the methylation of viral DNA [164]. Recent studies have been conducted to determine how these viruses trigger and can suppress the induced PTGS, and also how the system might be used to better control these viruses in plants. For instance, in tomato and cassava, the expression of hpRNA targeting replication genes provided full resistance to the geminiviruses *tomato yellow leaf curl virus* (TYLCV) and *African cassava mosaic virus* (ACMV), respectively [165, 166]. Affecting rice production in Asia, Rice tungro is a viral disease caused by the simultaneous infection of *rice tungro bacilliform virus* (RTBV), a double-stranded DNA virus, and *rice tungro spherical virus* (RTSV), a single-stranded RNA virus. Transgenic rice plants were obtained that expressed the RTBV gene, both in sense and anti-sense orientation, resulting in dsRNA formation. Some transgenic lines showed extremely mild symptoms against RTBV [167]. To the present authors' knowledge, no transgenic crop that is resistant to DNA viruses has yet been released.

Viroids are plant infectious agents that do not code for proteins but rather recruit host enzymatic machineries for their replication and systemic movement. Plant tissues infected by viroids produce 21–24 nt viroid-derived small RNAs (vd-sRNAs). However, it is unclear whether viroid RNAs are both triggers and targets of RNA silencing (just like viruses), or whether they are directly involved in viroid pathogenesis. Deep sequencing to assess the composition and molecular nature of vd-sRNAs in plants infected by diverse viroids should help to provide an understanding of their functions [168]. However, siRNAs appear to effectively target the mature viroid RNA: indeed, transgenic tomato plants expressing a hpRNA construct derived from *potato spindle tuber viroid* (PSTVd) sequences, exhibited resistance to PSTVd infection through an accumulation of corresponding siRNAs in the plant [169].

The best means of delivering virus targeting-dsRNA into plant cells remains through stable transformation of an ihpRNA construct, although the main technical limitation relates to the fact that many important crop species are difficult or impossible to transform. To assess this issue, Tenllado and coworkers [170] investigated (and patented) different strategies based on exogenous RNA applications. Notably, these authors showed that both, *in-vitro*-transcribed dsRNA delivered by mechanical inoculation and hpRNA constructs transiently expressed after agroinfiltration, could activate RNAi in plants and interfere with viral infection in a sequence-specific manner [170]. This peculiar RNA silencing triggering method, appeared also to be effective against the inoculation of non-persistently transmitted viruses by aphids [171]. The same group also showed that the spraying of crude extracts of bacterially expressed dsRNAs protected *Nicotianae* sp. plants against infection by *pepper mild mottle virus* (PMMoV) and PPV [172]. As confirmed by other studies aimed at improving the technology [173], exogenous dsRNA application could be an effective and environmentally friendly tool to protect plants from viruses. Indeed, maize plants were protected against *sugarcane mosaic virus* (SCMV) when first sprayed with a bacterial crude extract containing dsRNA targeting the CP [174].

RNAi is an important natural pathway for virus resistance in plants. Today, a few virus-resistant genetically modified cultivars are available commercially, and their numbers are likely to increase in the future as the different mechanisms of resistance (including RNAi) are increasingly deciphered [146, 175]. This assumption can be confirmed in public databases, by the increasing number of field trials, and by patent files detailing RNAi-derived virus resistance. However, for the other plant diseases, more fundamental studies are still needed.

5.8 RNAi as a Tool for Engineering Bacteria Resistance

When considering the mechanisms that plants use to combat viruses, questions arise regarding the role of RNAi in protecting a plant against other invading organisms such as bacteria or fungi. To date, very few reports have been made on the use of specific sRNAs to combat bacterial diseases, although it is now clear that host endogenous sRNAs – including miRNAs and siRNAs – play important roles in the defense of plants against microorganisms [7, 176].

The present understanding of plant/microbe interactions is based on an evolutionary "zig-zag" model developed by Jones and Dangl [177]. Briefly, basal defense is triggered by the plant's perception of pathogen-associated molecular patterns (PAMPs). To adapt to this PAMPs-triggered immunity (PTI), the pathogens secrete effectors into the plant cells that suppress the PTI; host plants have thus evolved a second effector-triggered immunity (ETI) level to counteract adapted pathogens. The first evidence that bacterial attack could modify miRNA pattern was discovered in *Arabidopsis* plants challenged by the virulent *Pseudomonas syringae Pst*DC3000 strain [178]. The plant recognition of the bacterial Flg22 flagellin-derived PAMP induced miR393 expression which negatively regulate the auxin receptor genes, thus inhibiting *P. syringae* growth. In another study [179], a *P. syringae* infection of *Arabidopsis* plants induced miRNAs involved in the auxin signaling pathway (miR160, miR167), but downregulated other miRNAs.

Katiyar-Agarwal and coworkers [180, 181] identified a new *Arabidopsis* siRNA induced by *P. syringae Pst* carrying the effector *avrRpt2*. The natural antisense nat-siRNAATGB2 regulates plant immunity by cleaving *PPRL* mRNA, which is a negative regulator of *RPS2*, a member of the *R*-genes. Moreover, a new class of endogenous lsiRNA of 30–40 nt was found to be induced notably during bacterial infection. As these sRNA also target a negative regulator of PTI and ETI, it is suggested that plant defense is repressed under natural conditions but is rapidly switched on by RNAi mechanisms when facing infection [7, 182]. In another study [183], it also appeared that DNA methylation regulated by argonaute4 is implicated in basal plant immunity in *Arabidopsis*.

Similar to plant viruses, pathogen bacteria have evolved effector proteins named bacterial silencing repressors (BSRs) to suppress sRNA pathways. In *P. syringae Pst*, several type III secretion effectors have been identified that suppress the *Arabidopsis* RNA silencing machinery and enhance disease susceptibility [184]: AvrPtoB appears to downregulate pri-miR393 transcription, while AvrPto – which shares sequence similarities – downregulates mature miR393. A third effector, HopT1-1, may suppress the AGO1 slicing activity as well as the miRNA activity in translational inhibition. At present, it is unclear whether these effectors act directly on sRNA metabolism, or indirectly through other pathways. It is noteworthy, however, that some Solanaceae plants have evolved to counteract AVrPto effector by expressing the protein kinase Pto that competes with its natural target [185].

All of these results prove that sRNAs are important for regulating plant–pathogen interaction, but they also appear to play a part in nitrogen fixation by rhizobia or tumor formation by agrobacteria. Indeed, during *Agrobacterium* infection, siRNAs corresponding to transferred-DNA oncogenes accumulate in infected tissues, and plants expressing silencing suppressors or mutated in the silencing pathway are more susceptible [186]. Moreover, it was shown that *Arabidopsis* and tomato plants transformed with hpRNAs targeting the *iaaM* and *ipt* oncogenes, required for tumor formation, became highly resistant to crown gall disease [187]. This raises the question of natural miRNA or siRNA trafficking between plants and pathogens (including fungi) as means of resistance. So, might it be possible to directly target bacteria essential genes via HIGS to prevent further infection?

RNA silencing was also identified in bacteria and proven to be mainly responsible for defenses against phages and plasmids. However, the types of small non-coding RNAs that mediate RNA silencing and their related proteins differ in bacteria and eukaryotes [188]. As an increasing number of studies are focusing on bacterial sRNAs discovery and metabolism [189], including plant pathogens (*Xanthomonas oryzae* [190], *P. syringae* [191]), it cannot be excluded that silencing might be of interest in the perspective of plant resistance against bacteria.

5.9 RNAi as a Tool for Engineering Parasitic Weed Resistance

Apart from losses caused by pest and fungal diseases another serious menace to many food crop plants including maize, sorghum, rice, millet, and a range of legumes, especially in Africa, is caused by parasitic weeds. Within the different parasitic plants, the most economically damaging are witchweeds (*Striga* spp.), which account for an estimated 40% of crop damage annually in Africa with regards to grain yield loss (or more than US$ 7 billion). Broomrapes (*Orobanche* spp. and

Phelipanche spp.) are endemic in semiarid regions, especially in the Mediterranean and Middle East, while dodders (*Cuscuta* spp.) are obligate parasitic plants that include about 170 species worldwide [192].

Until now, the different control strategies for parasitic weed management, such as the use of herbicides or the development of resistant crops as well as agronomic practices, have been partly successful and can alleviate damage to a limited extent [193]. A significant success, which led to the awarding of 2009 World Food Prize to Gebisa Ejeta, has been reported for the breeding of sorghum for *Striga* resistance. However, this approach requires a complicated combination of separate recessive genes, each on different chromosomes; namely, genes involved in the control of the secretion of germination stimulants, genes involved in the inhibition of attachment structures, and also in the modulation of the crop vascular system to inhibit penetration [194, 195].

One possibility for a successful approach was to induce a suicidal germination in the *Striga* by treating the unplanted fields with germination promoters in the absence of any host plants, as *Striga* seeds that germinate without an available host plant die very quickly [196]. Unfortunately, however, the field treatments required are expensive and often are not feasible in underdeveloped regions where *Striga* is endemic and widespread.

Today's view of optimal parasitic weed control could be achieved by either the use of parasite resistance crops (from conventional breeding), or by crops that have been genetically engineered for resistance. So far, only a few crop varieties with stable resistance have been developed after decades of conventional plant breeding, and genetic resources for resistance genes are limited. A recent promising achievement was the identification and cloning of the first resistance gene to *Striga* in cowpea [197].

Among the genetically engineered crops, those that are herbicide-resistant represent the most obvious approach, but they are not necessarily the most effective for parasitic weed control because of the parasite lifestyle. Therefore, new approaches to tackle this problem are currently under investigation.

The downregulation of genes involved in strigolactone biosynthesis or other germination stimulants *in planta* might provide a resistance mechanism against parasitic weeds, as has been shown in the fast-neutron-mutagenized tomato mutant SL-ORT1. This mutant, which is unable to produce and secrete natural germination stimulants to the rhizosphere, was found to be highly resistant to various *Phelipanche* and *Orobanche* spp. [198]. However, because of the additional roles of strigolactones as arbuscular mycorrhizal fungus signaling molecules and in the regulation of plant architecture, the use of such a strategy for resistance against parasitic plants would require careful evaluation [199]. A more promising approach would be to engineer host plants with a double-stranded RNAi construct targeted against a gene that is crucial for growth, development or parasitic behavior of the parasitic plant, and capable of being translocated via the haustorium into the parasite.

The suggested possible use of this RNAi technology for genetic engineering of weed resistance was supported by findings that RNAs could freely move between plants and their parasitic host [200, 201]. Elsewhere it has been shown that, along with solutes, it is possible to transport viruses, proteins and also mRNA transcripts from the host to the parasite. As an example, pyrophosphate-dependent phosphofructokinase (LePFP) transcripts were found in the growing dodder stem up to 30 cm from the tomato–dodder connection [202].

Furthermore, RNAi signals have also been demonstrated to be trafficked between hosts and parasites. Transgenic *Triphysaria* plants expressing the beta-glucuronidase (GUS) reporter gene were attached to hairpin hp*GUS*-expressing lettuce. Transcript quantification indicated an up to 95% reduction in the steady-state message level of *GUS* mRNA in *Triphysaria* attached to *hpGUS* lettuce compared to control lettuce. These results revealed that the GUS silencing signal generated by the host roots was dislocated across the haustorium interface, and was functional in the parasite [203]. As these experiments have shown that hpRNA constructs engineered in the host plants can silence the expression of the corresponding gene in the parasitic plants genes, the future demand would be the identification of candidate targets which, when degraded, would lead to lethality or an inhibited development of the parasitic weeds.

One of the first reports of this RNAi approach to control or reduce parasitic weed impact on tomato plants was in selecting the mannose 6-phosphate reductase (M6PR) from *Orobanche aegyptiaca* as a potential target. It has been suggested that mannitol accumulation may be very important for *Orobanche* development, and that M6PR is a key enzyme in mannitol biosynthesis.

Hence, transgenic tomato plants bearing a gene construct containing a specific inverted-repeat fragment from *Orobanche* M6PR-mRNA were produced and M6PR-siRNA was detected in three independent transgenic tomato lines. Additionally, a quantitative RT-PCR analysis showed that the amount of endogenous M6PR mRNA in the tubercles and underground shoots of *O. aegyptiaca* grown on transgenic host plants was reduced by 60–80%. Furthermore, in connection with M6PR mRNA suppression, there was a significant decrease in the mannitol level and a significant increase in the percentage of dead *O. aegyptiaca* tubercles on the transgenic host plants [204].

Another promising target for an RNAi approach may be a recently characterized soluble acid invertase (PrSAI1) or cell wall invertase (PrCWI) from *Phelipanche ramosa*. Both enzymes are thought to be actively involved in growth, germination, and attachment to host roots, and therefore constitute good targets for the silencing strategy [205].

Another attempt to employ the same strategy, but performed on transgenic maize for *Striga* resistance, has not yet proved successful. In this case, five *Striga asiatica* genes were selected as targets for hairpin constructs, transformed into maize, and subsequently challenged with germinating seeds of *Striga*, but no reproducible control of the parasite has yet been observed [206].

Although, when taken together, these findings indicate that the RNAi technology has some potential for protecting crop plants from parasitic plants, it is clear that much more research is required to explore the use of RNAi as an alternative to current approaches for controlling parasitic weeds. One crucial issue will be the further identification of optimal targets for silencing to achieve sufficient efficiency.

Consequently, an understanding of the molecular basis of the interaction between hosts and parasitic weeds, and an identification of the genetic factors involved in haustorium formation and parasite development, will facilitate the recognition of potential targets [207].

5.10 RNAi Safety in Crop Plants

As an accompaniment to the emerging use of this new technology, aspects of safety for the environment and for health in humans should also be addressed. Today, it is well known that the phenomenon of RNAi is universal in plants and animals, with studies having shown that rice contains a huge number of short dsRNAs, some of which have similarities and matching segments to human genes [208]. Clearly, humans have long been exposed to these dsRNAs, including plant-derived RNAs, that match human genes. In general, RNA is recognized by the US Food and Drug Administration (FDA) as "Generally Regarded as Safe" (GRAS) and, as such, its consumption is not regulated.

5.11 Summary and Outlook

In this chapter, a review has been presented of the uses of RNAi as a tool in crop protection. Whilst all data reported to date have indicated that this technology has a considerable potential for combating plant pests and diseases, only limited examples of its use in a field context have been described, and further research is required before switching from laboratory and greenhouse uses of RNAi to industrial large-scale applications. Most of the experiments conducted to date have involved the targeting of plant pests, especially chewing insects and nematodes, although today an increasing number of reports demonstrate feasibility against other pests. Two of the major challenges for the successful use of RNAi are the availability of sequence information for relevant pests or diseases, and the identification of the appropriate target genes. An increasing understanding of RNAi, and more generally of the RNA silencing mechanism, will surely open the door to new generations of crop protection technologies which, due to their biological origin, may be perceived as being environmental friendly.

References

1. Napoli, C., Lemieux, C., and Jorgensen, R. (1990) *Plant Cell*, **2**, 279–289.
2. van der Krol, A.R., Mur, L.A., Beld, M., Mol, J.N.M., and Stuitje, A.R. (1990) *Plant Cell*, **2**, 291–299.
3. Fire, A., Xu, S., Montgomery, M.K., Kostas, S.A., Driver, S.E., and Mello, C.C. (1998) *Nature*, **391**, 806–811.
4. Qian, Y., Cheng, Y., Cheng, X., Jiang, H., Zhu, S., and Cheng, B. (2011) *Plant Cell Rep.*, **30**, 1347–1363.

5. Kapoor, M., Arora, R., Lama, T., Nijhawan, A., Khurana, J.P., Tyagi, A.K., and Kapoor, S. (2008) *BMC Genomics*, **451** (9), 1471–2164.
6. Ruiz-Ferrer, V. and Voinnet, O. (2009) *Annu. Rev. Plant Biol.*, **60**, 485–510.
7. Katiyar-Agarwal, S. and Jin, H. (2010) *Annu. Rev. Phytopathol.*, **48**, 225–246.
8. Kozomara, A. and Griffiths-Jones, S. (2011) *Nucleic Acids Res.*, **39** (Database Issue), D152–D157.
9. Griffiths-Jones, S., Saini, H.K., van Dongen, S., and Enright, A.J. (2008) *Nucleic Acids Res.*, **36** (Database Issue), D154–D158.
10. Griffiths-Jones, S., Grocock, R.J., van Dongen, S., Bateman, A., and Enright, A.J. (2006) *Nucleic Acids Res.*, **34** (Database Issue), D140–D144.
11. Griffiths-Jones, S. (2004) *Nucleic Acids Res.*, **32** (Database Issue), D109–D111.
12. Mcginnis, K.M. (2010) *Briefings Funct. Genomics*, **9** (2), 111–117.
13. Zhou, B.B., Li, W., and Chen, X.Y. (2008) *Forest. Stud. China*, **10** (4), 280–284.
14. Jagtap, U.B., Gurav, R.G., and Bapat, V.A. (2011) *Naturwissenschaften*, **98** (6), 473–492.
15. Dunoyer, P., Brosnan, C.A., Schott, G., Wang, Y., Jay, F., Alioua, A., Himber, C., and Voinnet, O. (2010) *EMBO J.*, **29** (10), 1699–1712.
16. Chitwood, D.H. and Timmermans, M.C.P. (2010) *Nature*, **467** (7314), 415–419.
17. Matthew, L. (2004) *Comp. Funct. Genomics*, **5** (3), 240–244.
18. Chen, X. (2010) *Plant J.*, **61** (6), 941–958.
19. Masclaux, F. and Galaud, J.P. (2011) *Methods Mol. Biol.*, **744**, 37–55.
20. Senthil-Kumar, M. and Mysore, K.S. (2011) *Methods Mol. Biol.*, **744**, 13–25.
21. Waterhouse, P.M. and Helliwell, C.A. (2003) *Nat. Rev. Genet.*, **4** (1), 29–38.
22. Sato, F. (2005) *Plant Biotechnol.*, **22** (5), 431–442.
23. Muranaka, T. (2011) *Methods Mol. Biol.*, **744**, 27–35.
24. Hirai, S. and Kodama, H. (2008) *Open Plant Sci. J.*, **2**, 21–30.
25. Warthmann, N., Chen, H., Ossowski, S., Weigel, D., and Herv, P. (2008) *PLoS ONE*, **3** (3), 1–10.
26. Sablok, G., Perez-Quintero, A.L., Hassan, M., Tatarinova, T.V., and Lopez, C. (2011) *Biochem. Biophys. Res. Commun.*, **406** (3), 315–319.
27. Baulcombe, D.C. (1999) *Curr. Opin. Plant Biol.*, **2** (2), 109–113.
28. Senthil-Kumar, M., Anand, A., Uppalapati, S.R., and Mysore, K.S. (2008) *CAB Rev.: Perspect. Agric., Vet. Sci., Nutr. Nat. Resour.* **3**, 1–11.
29. Vaghchhipawala, Z., Rojas, C.M., Senthil-Kumar, M., and Mysore, K.S. (2011) *Methods Mol. Biol.*, **678**, 65–76 (Plant Reverse Genetics).
30. Zhang, C., Bradshaw, J.D., Whitham, S.A., and Hill, J.H. (2010) *Plant Physiol.*, **153** (1), 52–65.
31. Barampuram, S. and Zhang, Z.J. (2011) *Methods Mol. Biol.*, **701**, 1–35 (Plant Chromosome Engineering).
32. Jung, H.I., Zhai, Z., and Vatamaniuk, O.K. (2011) *Methods Mol. Biol.*, **744**, 109–127.
33. Tang, W., Weidner, D.A., Hu, B.Y., Newton, R.J., and Hu, X.H. (2006) *Plant Sci.*, **171** (3), 375–386.
34. Kamath, R., Fraser, A.G., Dong, Y., Poulin, G., Durbin, R., Gotta, M., Kanapin, A., Le Bot, N., Moreno, S., Sohrmann, M., Welchman, D., Zipperlen, P., and Ahringer, J. (2003) *Nature*, **421** (6920), 231–237.
35. Hilson, P., Allemeersch, J., Altmann, T., Aubourg, S., Avon, A., Beynon, J., Bhalerao, R.P., Bitton, F., Caboche, M., Cannoot, B., Chardakov, V., Cognet-Holliger, C., Colot, V., Crowe, M., Darimont, C., Durinck, S., Eickhoff, H., Falcon De Longevialle, A., Farmer, E.E., Grant, M., Kuiper, M.T.R., Lehrach, H., Leon, C., Leyva, A., Lundeberg, J., Lurin, C., Moreau, Y., Nietfeld, W., Paz-Ares, J., Reymond, P., Rouze, P., Sandberg, G., Segura, M.D., Serizet, C., Tabrett, A., Taconnat, L., Thareau, V., Van Hummelen, P., Vercruysse, S., Vuylsteke, M., Weingartner, M., Weisbeek, P.J., Wirta, V., Wittink, F.R.A., Zabeau, M., and Small, I. (2004) *Genome Res.*, **14** (10b), 2176–2189. Available at:

http://www.agrikola.org/html/agrikola/Versatile_GSTs.pdf.

36. Chuang, C.F. and Meyerowitz, E.M. (2000) *Proc. Natl. Acad. Sci. USA*, **97** (9), 4985–4990.
37. Wesley, S.V., Helliwell, C.A., Smith, N.A., Wang, M., Rouse, D.T., Liu, Q., Gooding, P.S., Singh, S.P., Abbott, D., Stoutjesdijk, P.A., Robinson, S.P., Gleave, A.P., Green, A.G., and Waterhouse, P.M. (2001) *Plant J.*, **27** (6), 581–590.
38. Kumagai, M., Donson, J., Della-Cioppa, G., Harvey, D., Hanley, K., and Grill, L. (1995) *Proc. Natl. Acad. Sci. USA*, **92** (5), 1679–1683.
39. Moritoh, S., Miki, D., Akiyama, M., Kawahara, M., Izawa, T., Maki, H., and Shimamoto, K. (2005) *Plant Cell Physiol.*, **46** (5), 699–715.
40. Okano, Y., Miki, D., and Shimamoto, K. (2008) *Plant J.*, **53** (1), 65–77.
41. Machado, A., Wu, Y., Yang, Y., Llewellyn, D.J., and Dennis, E.S. (2009) *Plant J.*, **59** (1), 52–62.
42. Liu, Y., Schiff, M., Marathe, R., and Dinesh-Kumar, S.P. (2002) *Plant J.*, **30** (4), 415–429.
43. Liu, Y., Schiff, M., Serino, G., Deng, X.W., and Dinesh-Kumar, S.P. (2002) *Plant Cell*, **14** (7), 1483–1496.
44. Liu, Y., Schiff, M., and Dinesh-Kumar, S.P. (2002) *Plant J.*, **31** (6), 777–786.
45. Jeong, R.D., Hwang, S.Y., Kang, S.H., Choi, H.S., Park, J.W., and Kim, K.H. (2005) *Plant Pathol. J.*, **21** (2), 158–163.
46. Meyer, J.D.F., Silva, D.C.G., Yang, C., Pedley, K.F., Zhang, C., Van De Mortel, M., Hill, J.H., Shoemaker, R.C., Abdelnoor, R.V., Whitham, S.A., and Graham, M.A. (2009) *Plant Physiol.*, **150** (1), 295–307.
47. Cigan, A.M., Unger-Wallace, E., and Haug-Collet, K. (2005) *Plant J.*, **43** (6), 929–940.
48. Heilersig, B.H.J.B., Loonen, A.E.H.M., Janssen, E.M., Wolters, A.M.A., and Visser, R.G.F. (2006) *Mol. Genet. Genomics*, **275** (5), 437–449.
49. Wang, Y., Beaith, M., Chalifoux, M., Ying, J., Uchacz, T., Sarvas, C., Griffiths, R., Kuzma, M., Wan, J., and Huang, Y. (2009) *Mol. Plant*, **2** (1), 191–200.
50. Travella, S., Klimm, T.E., and Keller, B. (2006) *Plant Physiol.*, **142** (1), 6–20.
51. Cove, D.J., Perroud, P.F., Charron, A.J., Mcdaniel, S.F., Khandelwal, A., and Quatrano, R.S. (2009) *Emerging Model Org.*, **1**, 69–104.
52. Hoffmann, T., Kalinowski, G., and Schwab, W. (2006) *Plant J.*, **48** (5), 818–826.
53. Song, J. and Wang, Z. (2011) *J. Plant Res.*, **124** (1), 183–192.
54. Lee, Y.W. and Kim, W.T. (2010) *Plant Cell*, **22** (8), 2781–2795.
55. He, X.J., Chen, T., and Zhu, J.K. (2011) *Cell Res.*, **21** (3), 442–465.
56. Sasaki, T., Fujimoto, R., Kishitani, S., and Nishio, T. (2011) *Plant Cell Rep.*, **30** (1), 81–88.
57. Pandey, A., Yang, C., Zhang, C., Graham, M., Horstman, H., Lee, Y., Zabotina, O., Hill, J., Pedley, K., and Whitham, S.A. (2011) *Mol. Plant Microbe Interact.*, **24** (2), 194–206.
58. Bhaskar, P.B., Venkateshwaran, M., Wu, L., Ane, J.M., and Jiang, J. (2009) *PLoS ONE*, **4** (6), 1–8.
59. Mallory, A. and Bouche, N. (2008) *Trends Plant Sci.*, **13** (7), 359–367.
60. Jones-Rhoades, M.W., Bartel, D.P., and Bartel, B. (2006) *Annu. Rev. Plant Biol.*, **57**, 19–53.
61. Franco-Zorrilla, J.M., Valli, A., Todesco, M., Mateos, I., Puga, M.I., Rubio-Somoza, I., Leyva, A., Weigel, D., Garcia, J.A., and Paz-Ares, J. (2007) *Nat. Genet.*, **39** (8), 1033–1037.
62. Tang, Y., Wang, F., Zhao, J., Xie, K., Hong, Y., and Liu, Y. (2010) *Plant Physiol.*, **153** (2), 632–641.
63. Eamens, A., Agius, C., Smith, N.A., Waterhouse, P.M., and Wang, M.B. (2011) *Mol. Plant*, **4** (1), 157–170.
64. Jouannet, V. and Crespi, M. (2011) *Prog. Mol. Subcell. Biol.*, **51**, 179–200.
65. Eamens, A., Wang, M.B., Smith, N.A., and Waterhouse, P.M. (2008) *Plant Physiol.*, **147** (2), 456–468.
66. Baltz, R., Dumain, R., Ferullo, J.M., Peyrard, S., and Beffa, R. (2005) Method for modifying gene expression in a phytopathogenic fungi. Patent WO Pat. 2005/071091, filed Dec. 20, 2004 and issued June. 22, 2010.

67. Nowara, D., Gay, A., Lacomme, C., Shaw, J., Ridout, C., Douchkov, D., Hensel, G., Kumlehn, J., and Schweitzer, P. (2010) *Plant Cell*, **22** (9), 3130–3141.

68. Whisson, S.C., Avrova, A.O., van West, P., and Jones, J.T. (2005) *Mol. Plant Pathol.*, **6**, 153–163.

69. Kathri, M. and Rajam, M.V. (2007) *Med. Mycol.*, **45**, 211–220.

70. Nakayashiki, H., Kadotani, N., and Mayama, S. (2006) *J. Mol. Evol.*, **63**, 127–135.

71. Caribé dos Santos, A.C., Sena, J.A.L., Santos, S.C., Dias, C.V., Pirovani, C.P., Pungartnik, C., Valle, R.R., Cascardo, J.C.M., and Vincentz, M. (2009) *Fungal Genet. Biol.*, **46**, 825–836.

72. Van De Crean, M., Goh, P.Y., Logghe, M.G., Khu, Y.L., Mortier K., and Bogaert, T.A.O.E. (2006) Method for down regulating gene expression in fungi. Patent WO Pat. 2006070227, filed Oct. 10, 2005.

73. Niblett, C.L. (2006) Methods and materials for conferring resistance to pests and pathogens of plants. Patent WO Pat. 2006047495, filed Oct. 21, 2005.

74. Roberts, J.K., Pitkin J.W., and Adams, T.H. (2008) *In planta* RNAi control of fungi. Patent US Pat. 20080022423, filed Feb. 01, 2007.

75. Schweitzer, P., Nowara, D., and Douchkov, D. (2006) Method for increasing a fungal resistance of transgenic plants by host induced suppression of a pathogenic fungi gene expression. Patent WO Pat. 2006097465.filed Mar. 14, 2006.

76. Tinoco, M.L.P., Dias, B.B.A., Dall'Astta, R.C., Pamphile, J.A., and Aragão, F.J.L. (2010) *BMC Biol.*, **8** (27), 3–11.

77. Dhalmini, Z., Spillane, C., Moss, J.P., Ruane, J., Urquia, N., and Sonnino, A. (2005) Status of Research and Application of Crop Biotechnologies in Developing Countries – A Preliminary Assessment, Food and Agriculture Organization of the United (FAO), Rome, 62 p.

78. Ferry, N., Edwards, M.G., Gatehouse, J., Capell, T., Christou, P., and Gatehouse, A.M.R. (2006) *Transgenic Res.*, **15**, 13–19.

79. Kos, M., van Loon, J.J., Dicke, M., and Vet, L.E. (2009) *Trends Biotechnol.*, **27**, 621–627.

80. Raybould, A. and Quemada, H. (2010) *Food Sec.*, **2**, 247–259.

81. Moar, W., Roush, R., Shelton, A., Ferré, J., Macintosh, S., Leonard, B.R., and Abel, C. (2008) *Nat. Biotechnol.*, **26**, 1072–1074.

82. Tabashnik, B.E. and Carrière, Y. (2010) *Southwest. Entomol.*, **35** (3), 417–424.

83. Roignant, J., Carre, C., Mugat, B., Szymczak, D., Lepesant, J., and Antoniewski, C. (2003) *RNA*, **9**, 299–308.

84. Bucher, G., Scholten, J., and Klingler, M. (2002) *Curr. Biol.*, **12**, R85–R86.

85. Tomoyasu, Y. and Denell, R.E. (2004) *Dev. Genes Evol.*, **214**, 575–578.

86. Terenius, O., Papanicolaou, A., Garbutt, J.S., Eleftherianos, I., Huvenne, H., Kanginakudru, S., Albrechtsen, M., An, C., Aymeric, J.L., Barthel, A., Bebas, P., Bitra, K., Bravo, A., Chevalier, F., Collinge, D.P., Crava, C.M., de Maagd, R.A., Duvic, B., Erlandson, M., Faye, I., Felföldi, G., Fujiwara, H., Futahashi, R., Gandhe, A.S., Gatehouse, H.S., Gatehouse, L.N., Giebultowicz, J.M., Gómez, I., Grimmelikhuijzen, C.J., Groot, A.T., Hauser, F., Heckel, D.G., Hegedus, D.D., Hrycaj, S., Huang, L., Hull, J.J., Iatrou, K., Iga, M., Kanost, M.R., Kotwica, J., Li, C., Li, J., Liu, J., Lundmark, M., Matsumoto, S., Meyering-Vos, M., Millichap, P.J., Monteiro, A., Mrinal, N., Niimi, T., Nowara, D., Ohnishi, A., Oostra, V., Ozaki, K., Papakonstantinou, M., Popadic, A., Rajam, M.V., Saenko, S., Simpson, R.M., Soberón, M., Strand, M.R., Tomita, S., Toprak, U., Wang, P., Wee, C.W., Whyard, S., Zhang, W., Nagaraju, J., French-Constant, R.H., Herrero, S., Gordon, K., Swevers, L., and Smagghe, G. (2011) *J. Insect Physiol.*, **57**, 231–245.

87. Quan, G.X., Kanda, T., and Tamura, T. (2002) *Insect Mol. Biol.*, **11**, 217–222.

88. Bettencourt, R., Terenius, O., and Faye, I. (2002) *Insect Mol. Biol.*, **11**, 267–271.

89. Mutti, N.S., Louis, J., Pappan, L.K., Pappan, K., Begum, K., Chen, M.S.,

Park, Y., Dittmer, N., Marshall, J., Reese, J.C., and Reeck, G.R. (2008) *Proc. Natl. Acad. Sci. USA*, **105**, 9965–9969.
90. Bautista, M.A., Miyata, T., Miura, K., and Tanaka, T. (2009) *Insect Biochem. Mol. Biol.*, **39**, 38–46.
91. Walshe, D.P., Lehane, S.M., Lehane, M.J., and Haines, L.R. (2009) *Insect Mol. Biol.*, **18**, 11–19.
92. Zhou, X., Wheeler, M.M., Oi, F.M., and Scharf, M.E. (2008) *Insect Biochem. Mol. Biol.*, **38**, 805–815.
93. Price, D.R. and Gatehouse, J.A. (2008) *Trends Biotechnol.*, **26**, 393–400.
94. Huvenne, H. and Smagghe, G. (2010) *J. Insect Physiol.*, **56**, 227–235.
95. Mao, Y.B., Cai, W.J., Wang, J.W., Hong, G.J., Tao, X.Y., Wang, L.J., Huang, Y.P., and Chen, X.Y. (2007) *Nat. Biotechnol.*, **25**, 1307–1313.
96. Baum, J.A., Bogaert, T., Clinton, W., Heck, G.R., Feldmann, P., Ilagan, O., Johnson, S., Plaetinck, G., Munyikwa, T., Pleau, M., Vaughn, T., and Roberts, J. (2007) *Nat. Biotechnol.*, **25**, 1322–1326.
97. Mao, Y.B., Xue, X.Y., Hong, G.J., Tao, X.Y., Wang, L.J., and Chen, X.Y. (2011) *Transgenic Res.*, **20**, 665–673.
98. Gatehouse, J.A. and Price, D.R.G. (2011) in *Insect Biotechnology, Biologically-Inspired Systems*, vol. 2, Part 2 (ed. A. Vilcinskas), Springer, pp. 145–168.
99. Chitwood, D.J. (2003) *Pest Manag. Sci.*, **59**, 748–753.
100. Davis, E.L., Hussey, R.S., Baum, T.J., Bakker, J., and Schots, A. (2000) *Annu. Rev. Phytopathol.*, **38**, 365–396.
101. Gheysen, G. and Fenoll, C. (2002) *Annu. Rev. Phytopathol.*, **40**, 191–219.
102. Carpenter, J., Gianessi, L., and Lynch, L. (2000) The Economic Impact of the Scheduled Phase-Out of Methyl Bromide in the U.S., *http://www.ncfap.org/publications.html* (accessed 5 October 2011).
103. Li, J., Todd, T.C., Lee, J., and Trick, H.N. (2011) *Plant Biotechnol.*, **9**, 1–9.
104. Urwin, P.E., Lilley, C.J., and Atkinson, H.J. (2002) *Mol. Plant Microbe Interact.*, **15**, 747–752.
105. Rosso, M.N., Dubrana, M.P., Cimbolini, N., Jaubert, S., and Abad, P. (2005) *Mol. Plant Microbe Interact.*, **18**, 615–620.
106. Lilley, C.J., Bakhetia, M., Charlton, W.L., and Urwin, P.E. (2007) *Mol. Plant Pathol.*, **8**, 701–711.
107. Rosso, M.N., Jones, J.T., and Abad, P. (2009) *Annu. Rev. Phytopathol.*, **47**, 207–232.
108. Shingles, J., Lilley, C.J., Atkinson, H.J., and Urwin, P.E. (2007) *Exp. Parasitol.*, **115**, 114–120.
109. Lilley, C.J., Goodchild, S.A., Atkinson, H.J., and Urwin, P.E. (2005) *Int. J. Parasitol.*, **35**, 1577–1585.
110. Steeves, R.M., Todd, T.C., Essig, J.S., and Trick, H.N. (2006) *Funct. Plant Biol.*, **33**, 991–999.
111. Chen, Q., Rehman, S., Smant, G., and Jones, J.T. (2005) *Mol. Plant Microbe Interact.*, **18**, 621–625.
112. Huang, G., Allen, R., Davis, E.L., Baum, T.J., and Hussey, R.S. (2006) *Proc. Natl. Acad. Sci. USA*, **103**, 14302–14306.
113. Bakhetia, M., Urwin, P.E., and Atkinson, H.J. (2007) *Mol. Plant Microbe Interact.*, **20**, 306–312.
114. Yadav, B.C., Veluthambi, K., and Subramaniam, K. (2006) *Mol. Biochem. Parasitol.*, **148**, 219–222.
115. Fairbairn, D.J., Cavallaro, A.S., Bernard, M., Mahalinga-Iyer, J., Graham, M.W., and Botella, J.R. (2007) *Planta*, **226** (6), 1525–1533.
116. Sindhu, A.S., Maier, T.R., Mitchum, M.G., Hussey, R.S., Davis, E.L., and Baum, T.J. (2009) *J. Exp. Bot.*, **60**, 315–324.
117. Klink, V.P., Kim, K.H., Martins, V., Macdonald, M.H., Beard, H.S., Alkharouf, N.W., Lee, S.K., Park, S.C., and Matthews, B.F. (2009) *Planta*, **230**, 53–71.
118. Patel, N., Hamamouch, N., Li, C., Hewezi, T., Hussey, R.S., Baum, T.J., Mitchum, M.G., and Davis, E.L. (2010) *J. Exp. Bot.*, **61** (1), 235–248.
119. Valentine, T.A., Randall, E., Wypijewski, K., Chapman, S., Jones, K., and Oparka, K.J. (2007) *Plant Biotechnol. J.*, **5**, 827–834.

120. Dubreuil, G., Magliano, M., Dubrana, M.P., Lozano, J., Lecomte, P., Favery, B., Abad, P., and Rosso, M.N. (2009) *J. Exp. Bot.*, **60**, 4041–4050.
121. Dalzell, J.J., McMaster, S., Fleming, C.C., and Maule, A.G. (2010) *Int. J. Parasitol.*, **40**, 91–100.
122. Ibrahim, H.M.M., Alkharou, F.N.W., Meyer, S.L.F., Aly, M.A.M., Gamal El-Din, A.E.K.Y., Hussein, E.H.A., and Matthews, B.F. (2011) *Exp. Parasitol.*, **127** (1), 90–99.
123. Dalzell, J.J., McMaster, S., Johnston, M.J., Kerr, R., Fleming, C.C., and Maule, A.G. (2009) *Int. J. Parasitol.*, **39**, 1503–1516.
124. Brodersen, P. and Voinnet, O. (2009) *Nat. Rev. Mol. Cell Biol.*, **10**, 141–148.
125. Sanford, J.C. and Johnston, S.A. (1985) *J. Theor. Biol.*, **113**, 395–405.
126. Baulcombe, D.C. (1996) *Plant Cell.*, **8** (10), 1833–1844.
127. Baulcombe, D.C. (2004) *Nature*, **431** (7006), 356–363.
128. Lindbo, J. and Dougherty, W. (2005) *Annu. Rev. Phytopathol.*, **43**, 191–204.
129. Waterhouse, P.M., Graham, M.W., and Wang, M.B. (1998) *Proc. Natl. Acad. Sci. USA*, **95** (23), 13959–13964.
130. Soosaar, J., Burch-Smith, T., and Dinesh-Kumar, S. (2005) *Nat. Rev. Microbiol.*, **3** (10), 789–798.
131. Bucher, E. and Prins, M. (2006) in *Natural Resistance Mechanisms of Plants to Viruses* (eds G. Loebenstein and J.P. Carr), Springer, Netherlands, pp. 45–72.
132. Aliyari, R. and Ding, S. (2009) *Immunol. Rev.*, **227** (1), 176–188.
133. Fernandez-Calvino, L., Donaire, L., and Llave, C. (2011) in *Recent Advances in Plant Virology* (eds C. Caranta, M.A. Aranda, M.Tepfer, and J.J. Lopez-Moya), Caister Academic Press, pp. 121–135.
134. Llave, C. (2010) *Trends Plant Sci.*, **15** (12), 701–707.
135. Burgyan, J. and Havelda, Z. (2011) *Trends Plant Sci.*, **16** (5), 265–272.
136. Yang, J. and Yuan, Y. (2009) *Biochim. Biophys. Acta*, **1789** (9-10), 642–652.
137. Dunoyer, P., Lecellier, Ch., Parizotto, E., Himber, C., and Voinnet, O. (2004) *Plant Cell*, **16** (5), 1235–1250.
138. Shao, Y., Zhu, H.L., Tian, H.Q., Wang, X.G., Lin, X.J., Zhu, B.Z., Xie, Y.H., and Luo, Y.B. (2008) *Russ. J. Plant Physiol.*, **55** (2), 168–174.
139. Burch-Smith, T.M., Anderson, J.C., Martin, G.B., and Dinesh-Kumar, S.P. (2004) *Plant J.*, **39** (5), 734–746.
140. Zhu, X. and Dinesh-Kumar, S.P. (2009) in *RNA Interference, Methods for Plants and Animals* (eds T. Doran and C. Helliwell), CABI Publishing, Wallingford, pp. 26–49.
141. Zhang, Z.Y., Fu, F.L., Gou, L., Wang, H.G., and Li, W.C. (2010) *J. Plant Biol.*, **53** (4), 297–305.
142. Ding, X.S., Schneider, W.L., Chaluvadi, S.R., Mian, M.A.R., and Nelson, R.S. (2006) *Mol. Plant Microbe Interact.*, **19** (11), 1229–1239.
143. Scofield, S.R., Huang, L., Brandt, A.S., and Gill, B.S. (2005) *Plant Physiol.*, **138** (4), 2165–2173.
144. Constantin, G.D., Krath, B.N., Macfarlane, S.A., Nicolaisen, M., Johansen, I.E., and Lund, O.S. (2004) *Plant J.*, **40** (4), 622–631.
145. Fofana, I.B.F., Sangare, A., Collier, R., Taylor, C., and Fauquet, C.M. (2004) *Plant Mol. Biol.*, **56** (4), 613–624.
146. Fuchs, M. and Gonsalves, D. (2007) *Annu. Rev. Phytopathol.*, **45**, 173–202.
147. Collinge, D.B., Joergensen, H.J.L., Lund, O.S., and Lyngkjaer, M.F. (2010) *Annu. Rev. Phytopathol.*, **48**, 269–291.
148. Hu, Q., Niu, Y., Zhang, K., Liu, Y., and Zhou, X. (2011) *Virol. J.*, **8** (41), 1–11.
149. Missiou, A., Kalantidis, K., Boutla, A., Tzortzakaki, S., Tabler, M., and Tsagris, M. (2004) *Mol. Breed.*, **14** (2), 185–197.
150. Arif, M., Azhar, U., Arshad, M., Zafar, Y., Mansoor, S., and Asad, S. (2011) *Transgenic Res.*, **21** (2), 3 03–311.
151. Hily, J.M. and Liu, Z. (2007) *J. Theor. Biol.*, **113**, 123–147.
152. Reyes, C.A., De Francesco, A., Pena, E.J., Costa, N., Plata, M.I., Sendin, L., Castagnaro, A.P., and Garcia, M.L. (2011) *J. Biotechnol.*, **151** (1), 151–158.
153. Patil, B.L., Ogwok, E., Wagaba, H., Mohammed, I.U., Yadav, J.S., Bagewadi, B., Taylor, N.J., Kreuze, J.F., Maruthi, M.N., Alicai, T., and

Fauquet, C.M. (2011) *Mol. Plant Pathol.*, **12** (1), 31–41.
154. Wang, M.B., Abbott, D.C., and Waterhouse, P.M. (2000) *Mol. Plant Pathol.*, **1** (6), 347–356.
155. Pinto, Y.M., Kok, R.A., and Baulcombe, D.C. (1999) *Nat. Biotechnol.*, **17** (7), 702–707.
156. Ma, J., Song, Y., Wu, B., Jiang, M., Li, K., Zhu, C., and Wen, F. (2011) *Transgenic Res.*, **20** (6), 1367–1377.
157. Bai, Y., Yang, H., Qu, L., Zheng, J., Zhang, J.P., Wang, M., Xie, W., Zhou, X., and Wang, G. (2008) *Front. Agric. China*, **2**, 125–130.
158. Zhang, Z.Y., Yang, L., Zhou, S.F., Wang, H.G., Li, W.C., and Fu, F.L. (2011) *J. Biotechnol.*, **153** (3-4), 181–187.
159. Cruz, L.F. (2009) Resistance to wheat streak mosaic *virus* and *triticum mosaic virus* in wheat mediated by RNAi. Thesis, Kansas State University, *http://hdl.handle.net/2097/1653* (accessed 08 September 2011).
160. He, X.F., Fang, Y.Y., Feng, L., and Guo, H.S. (2008) *FEBS Lett.*, **582** (16), 2445–2452.
161. Niu, Q.W., Lin, S.S., Reyes, J.L., Chen, K.C., Wu, H.W., Yeh, S.D., and Chua, N.H. (2006) *Nat. Biotechnol.*, **24** (11), 1420–1428.
162. Qu, J., Ye, J., and Fang, R. (2007) *J. Virol.*, **81** (12), 6690–6699.
163. Ai, T., Zhang, L., Gao, Z., Zhu, C.X., and Guo, X. (2011) *Plant Biol.*, **13** (2), 304–316.
164. Rodriguez-Negrete, E.A., Carrillo-Tripp, J., and Rivera-Bustamante, R.F. (2009) *J. Virol.*, **83** (3), 1332–1340.
165. Fuentes, A., Ramos, P.L., Fiallo, E., Callard, D., Sanchez, Y., Peral, R., Rodriguez, R., and Pujol, M. (2006) *Transgenic Res.*, **15** (3), 291–304.
166. Vanderschuren, H., Alder, A., Zhang, P., and Gruissem, W. (2009) *Plant Mol. Biol.*, **70** (3), 265–272.
167. Tyagi, H., Rajasubramaniam, S., Rajam, M.V., and Dasgupta, I. (2008) *Transgenic Res.*, **17** (5), 897–904.
168. Navarro, B., Pantaleo, V., Gisel, A., Moxon, S., Dalmay, T., Bisztray, G., Di Serio, F., and Burgyan, J. (2009) *PLoS ONE*, **4** (11), 1–12.
169. Schwind, N., Zwiebel, M., Itaya, A., Ding, B., Wang, M., Krczal, G., and Wassenegger, M. (2009) *Mol. Plant Pathol.*, **10** (4), 459–469.
170. Tenllado, F., Llave, C., and Diaz-Ruiz, J.R. (2004) *Virus Res.*, **102** (1), 85–96.
171. Vargas, M., Martinez-Garcia, B., Diaz-Ruiz, J.R., and Tenllado, F. (2008) *Virol. J.*, **5** (42), 1–5.
172. Tenllado, F., Martinez-Garcia, B., Vargas, M., and Diaz-Ruiz, J.R. (2003) *BMC Biotechnol.*, **3**, 1–11.
173. Yin, G., Sun, Z., Liu, N., Zhang, L., Song, Y., Zhu, C., and Wen, F. (2009) *Appl. Microbiol. Biotechnol.*, **84** (2), 323–333.
174. Gan, D., Zhang, J., Jiang, H., Jiang, T., Zhu, S., and Cheng, B. (2010) *Plant Cell Rep.*, **29** (11), 1261–1268.
175. Auer, C. and Frederick, R. (2009) *Trends Biotechnol.*, **27** (11), 644–651.
176. Chellappan, P. and Jin, H. (2009) *Methods Mol. Biol.*, **495**, 121–132 (Plant Hormones).
177. Jones, J. and Dangl, J. (2006) *Nature*, **444** (7117), 323–329.
178. Navarro, L., Dunoyer, P., Jay, F., Arnold, B., Dharmasiri, N., Estelle, M., Voinnet, O., and Jones, J.D.G. (2006) *Science*, **312** (5772), 436–439.
179. Fahlgren, N., Howell, M.D., Kasschau, K.D., Chapman, E.J., Sullivan, C.M., Cumbie, J.S., Givan, S.A., Law, T.F., Grant, S.R., Dangl, J.L., and Carrington, J.C. (2007) *PLoS ONE*, **2** (2), 1–14.
180. Katiyar-Agarwal, S., Morgan, R., Dahlbeck, D., Borsani, O., Villegas, A., Zhu, J.K., Staskawicz, B.J., and Jin, H. (2006) *Proc. Natl. Acad. Sci. USA*, **103** (47), 18002–18007.
181. Katiyar-Agarwal, S., Gao, S., Vivian-Smith, A., and Jin, H. (2007) *Genes Dev.*, **21** (23), 3123–3134.
182. Jin, H. (2008) *FEBS Lett.*, **582** (18), 2679–2684.
183. Agorio, A. and Vera, P. (2007) *Plant Cell*, **19** (11), 3778–3790.
184. Navarro, L., Jay, F., Nomura, K., He, S.Y., and Voinnet, O. (2008) *Science*, **321** (5891), 964–967.
185. Xiang, T., Zong, N., Zou, Y., Wu, Y., Zhang, J., Xing, W., Li, Y., Tang, X.,

Zhu, L., Chai, J., and Zhou, J.M. (2008) *Curr. Biol.*, **18** (1), 74–80.
186. Dunoyer, P., Himber, C., and Voinnet, O. (2006) *Nat. Genet.*, **38** (2), 258–263.
187. Escobar, M.A., Civerolo, E.L., Summerfelt, K.R., and Dandekar, A.M. (2001) *Proc. Natl. Acad. Sci. USA*, **98** (23), 13437–13442.
188. Liu, J.M. and Camilli, A. (2010) *Curr. Opin. Microbiol.*, **13** (1), 18–23.
189. Liu, J.M. and Camilli, A. (2011) *Methods Mol. Biol.*, **733**, 63–79 (High-Throughput Next Generation Sequencing).
190. Liang, H., Zhao, Y., Zhang, J., Wang, X., Fang, R., and Jia, Y. (2011) *BMC Genomics*, **12** (87), 1–14.
191. Filiatrault, M., Stodghill, P., Bronstein, P., Moll, S., Lindeberg, M., Grills, G., Schweitzer, P., Wang, W., Schroth, G., Luo, S., Khrebtukova, I., Yang, Y., Thannhauser, T., Butcher, B., Cartinhour, S., and Schneider, D. (2010) *J. Bacteriol.*, **192** (9), 2359–2372.
192. Runo, S., Alakonya, A., Machukaa, J., and Sinhab, N. (2011) *Pest Manag. Sci.*, **67**, 129–136.
193. Aly, R. (2007) *In Vitro Cell Dev. Biol. Plant*, **43**, 304–317.
194. Gressel, J. (2010) *New Biotechnol.*, **27** (5), 522–527.
195. Ejeta, G., Rich, P.J., and Mohamed, A. (2007) in *Integrating New Technologies for Striga Control – Towards Ending the Witch Hunt* (eds G. Ejeta and J. Gressel), World Scientific, Singapore, pp. 87–98.
196. Egley, G.H., Eplee, R.E., and Norris, R.S. (1990) in *Witchweed Research and Control in the United States* (eds P.P. Sand, R.E. Eplee, and R.C. Westbrooks), WSSA, Champaign, IL, pp. 56–67.
197. Li, J.X. and Timko, M.P. (2009) *Science*, **325**, 1094.
198. Dor, E., Yoneyama, K., Wininger, S., Kapulnik, Y., Yoneyama, K., Koltai, H., Xie, X., and Hershenhorn, J. (2011). *Phytopathology*, **101**, 213–222.
199. Cardoso, C., Ruyter-Spira, C., and Bouwmeester, H.J. (2011) *Plant Sci.*, **180** (3), 414–420.
200. Roney, J.K., Khatibi, P.A., and Westwood, J.H. (2007) *Plant Physiol.*, **143**, 1037–1043.
201. Westwood, J.H., Roney, J.K., Khatibi, P.A., and Stromberg, V.K. (2009) *Pest Manag. Sci.*, **65**, 533–539.
202. David-Schwartz, R., Runo, S., Townsley, B., Machuka, J., and Sinha, N. (2008) *New Phytol.*, **179**, 1133–1141.
203. Tomilov, A.A., Tomilova, N.B., Wroblewski, T., Michelmore, R., and Yoder, J.I. (2008) *Plant J.*, **56**, 389–397.
204. Aly, R., Cholakh, H., Joel, D.M., Leibman, D., Steinitz, B., Zelcer, A., Naglis, A., Yarden, O., and Gal-On, A. (2009) *Plant Biotechnol. J.*, **7**, 487–498.
205. Draie, R., Peron, T., Pouvreau, J.B., Veronesi, C., Jegou, S., Delvault, P., Thoiron, S., and Simier, P. (2011) *Mol. Plant Pathol.*, **12**, 1364–3703.
206. de Framond, A., Rich, P.J., McMillan, J., and Ejeta, G. (2007) in *Integrating New Technologies for Striga Control* (eds G. Ejeta and J. Gressel), World Scientific Publishing Co., Singapore, pp. 185–196.
207. Yoder, I.J. and Scholes, D.J. (2010) *Curr. Opin. Plant Biol.*, **13**, 478–484.
208. Ivashuta, S.I., Petrick, J.S., Heisel, S.E., Zhang, Y., Guo, L., Reynolds, T.L., Rice, J.F., Allen, E., and Roberts, J.K. (2009) *Food Chem. Toxicol.*, **47**, 353–360.

6
Fast Identification of the Mode of Action of Herbicides by DNA Chips

Peter Eckes and Marco Busch

6.1
Introduction

Despite agrochemicals having played a major role in the large increase in agricultural productivity over the past 50 years, about 40% of the harvest still is lost due to pests or weed infestations. The primary method of weed control – at least, in industrialized countries – is the use of herbicides. Being by far the biggest segment of the crop protection market, herbicide sales have grown only moderately over the past 10 years, mainly because the market dynamics have been driven by the replacement of established products, but with the new herbicides showing only slightly better properties than their predecessors. Higher demands on the efficiency and spectrum of the new products, as well as stricter regulatory hurdles, make it increasingly difficult to bring new products to the market. This is highlighted by the fact that, in 2004, five of the six top-selling herbicides were originally launched between 30 and 60 years ago, and these five products still comprise more than 30% of herbicide sales worldwide.

In order to be successful in the future, a company must develop novel compounds for weed control which have superior agronomic properties and which can alter the market-landscape, or even create new market segments. Compounds with a novel herbicidal mode of action (MoA) should have the potential to fulfill these requirements, as they would open new market segments and thus trigger an above-average growth in herbicide sales.

Due to the high competitiveness of the crop protection market, research progress from the synthesis of new chemicals, to the promotion of lead compounds, to the project phase must be as streamlined as possible. It is imperative not only to eliminate those compounds with a weak efficacy or phytotoxicity but also those with non-desirable modes of action as early as possible from further evaluation, and then to concentrate on a few promising candidates. Besides a phenotypical inspection of the treated plants, the target site of a compound is usually determined with specific enzyme assays in test tubes or on microtiter plates, though this is a time-consuming, labor-, and cost-intensive process. Notably, it requires either the purification of a respective enzyme from plants or the preparation of proteins by

Modern Methods in Crop Protection Research, First Edition.
Edited by Peter Jeschke, Wolfgang Krämer, Ulrich Schirmer, and Matthias Witschel.

heterologous expression in, for example, bacteria or yeast. In addition, for each enzyme a specific assay must be developed in which the activity of that enzyme can be determined in the presence and absence of the compound under test. In order to identify the MoA of several compounds, each compound must – at least in theory – be tested against each enzyme, bearing in mind that for most enzymes no test tube assays are available. Consequently, it would be highly desirable to have available a method that could provide a clue as to the MoA of an herbicide, in a single experiment.

6.2 Gene Expression Profiling: A Method to Measure Changes of the Complete Transcriptome

The functionality of an organism is determined by the information contained within its genes. Genes are transcribed into messenger RNA (mRNA), which is subsequently translated into the different proteins which, as enzymes, are the ultimate effectors in the cell as they convert one metabolite into another. The controlled action of these enzymes is necessary for the coordinated interaction of the metabolic pathways that maintain the functionality of the organism (Figure 6.1). In this context, a key regulatory mechanism of living cells is the controlled expression of the respective genes. During development and differentiation, in addition to external perturbations, this network of expressed genes varies constantly to adapt to changes in environmental conditions. It has been well established that the

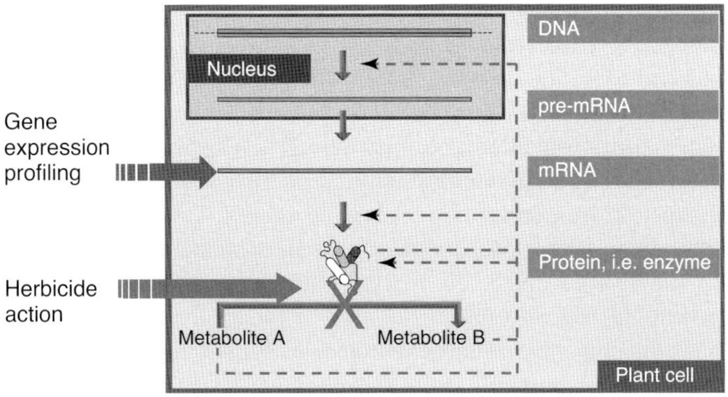

Figure 6.1 Schematic representation of general cellular processes. DNA as the storage of genetic information is localized in the nucleus and transcribed into messenger RNA (mRNA). The mRNA transports this information out of the nucleus into the cytoplasm of the cell, where it is translated into proteins. The proteins may be enzymes that catalyze a reaction from metabolite A to metabolite B. A herbicide blocks this reaction by inhibition of the enzyme activity. Proteins and metabolites exercise regulation on DNA transcription and RNA translation (dashed lines). Thus, the effects on cell processes exhibited by herbicides are reflected by changes in mRNA levels, which can be analyzed using gene expression profiling.

measurement of mRNA expression is a valuable tool for assessing the reactions of an organism to its environment, although ultimately the metabolic processes are mediated by the mRNA-encoded proteins.

When a plant is treated with a herbicide, its vital processes – such as photosynthesis, cell wall formation, or the biosynthesis of cellular components – are each affected, and this is reflected by changes in the *transcriptome*, which contains a set of all of the plant's mRNAs. Typically, the amount of mRNA of some genes will be increased, whereas that of some other genes will be decreased. This snapshot of the transcriptional status of a plant is referred to as the Gene Expression Profile (GEP).

The whole genomes of *Arabidopsis thaliana* [1] and rice [2–4] have been sequenced, and this information – together with technical advances in automation, miniaturization, and parallel synthesis of oligonucleotides – has been used to develop full-genome DNA microarrays for those plants [5, 6] which represent almost all genes of the respective species. In addition to the full-genome plant microarrays, there exist DNA chips for many different plants, such as corn, soybean, barley, or tomato. These chips do not represent the complete genome, but rather a large proportion of the expressed genes of the respective plants. Such DNA microarrays have been used to analyze the reaction of plants to biotic factors, including their defense against pathogens [7], seed development [8], nitrate assimilation [9], and fruit ripening [10], or to abiotic factors such as drought [11], cold [12], and heat [13]. In this way, it became possible to obtain new insights into the molecular mechanisms that regulate these processes.

The experiments described in this chapter employ the *Arabidopsis* ATH1 GeneChip microarray. This full-genome chip, which is manufactured by Affymetrix (*http://www.affymetrix.com/*), is about 1 × 1 cm in size and contains the nucleic acid sequences of about 24 000 genes (Figure 6.2). Short 25 mer nucleotide sequences for each gene have been synthesized on specific spots on the chip, while each gene is represented by 11 different oligonucleotides (gene probes), scattered randomly over the chip. The multitude of oligonucleotides for each gene and their random distribution increases the significance of the statistical analysis of the expression results.

Because the ATH1 chip represents almost all *Arabidopsis* genes, it can also detect changes in the transcriptome caused by the circadian clock of the plant, or by other environmental stimuli such as biotic or abiotic stresses. The effects of these stimuli on transcription can sometimes be much stronger than the changes caused by the action of herbicides. As this would mask the expression pattern produced by the herbicide, it is imperative to grow the plants under conditions which are as standardized as possible. Typically, plant growth chambers are required in which the light, temperature, and humidity can each be controlled. Moreover, all of the process steps – from the sowing and watering of *Arabidopsis*, the spraying of the compounds, and harvesting of the plants down to preparation of the mRNA – must be highly reproducible from experiment to experiment. By these means, it is possible to compare different expression profiles that have been produced over several years.

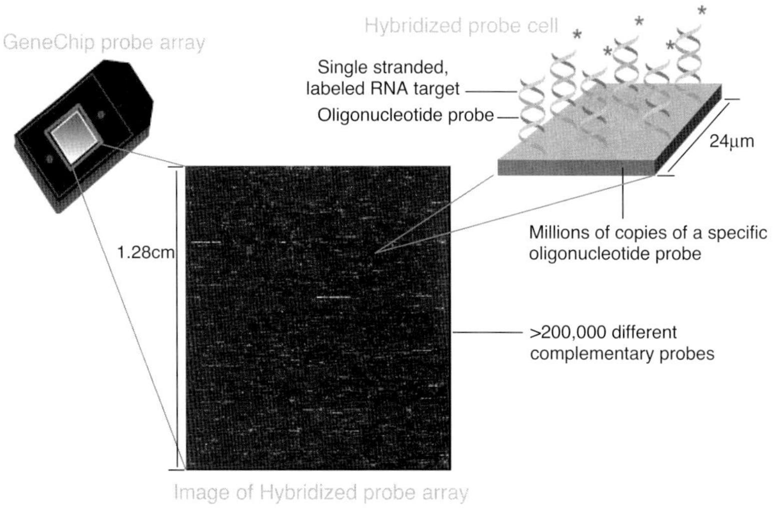

Figure 6.2 The GeneChip system. RNAs isolated from herbicide-treated plants and labeled with a fluorescent dye bind to their corresponding gene probes. Highly abundant RNAs produce bright signals, whereas rare RNAs produce only dim signals.

In a standard expression profiling experiment, the plants are harvested at 24 h after treatment with the chemical under test. RNA from the compound-treated plants and from control plants is then isolated separately, labeled with a specific dye, and incubated with the nucleotide sequences on the chip. Because of sequence homology, the individual RNAs bind to their corresponding gene probes. As the location of each *Arabidopsis* gene probe on the chip is known, and as the RNA is labeled with a fluorescent dye, the amount of bound RNA for each gene probe can be measured individually with a scanner. As a result, highly abundant RNAs will produce bright signals, while rare RNAs will produce only very dim signals, and the difference in brightness between the samples will determine whether the amount of RNA for a given gene has increased or decreased due to the herbicide treatment. As all *Arabidopsis* genes are located on a single DNA chip, it is possible to measure changes in RNA abundance for all genes in a single experiment. In this way, each herbicide will produce a distinctive gene expression pattern, which can be seen as a type of "fingerprint" for that herbicide.

6.3
Classification of the Mode of Action of an Herbicide

Since compounds that have the same MoA will affect the same metabolic processes, the expression profiles of plants treated with compounds having the same MoA should be very similar, and be clearly different from those of compounds with alternative modes of action. Under this assumption, a compendium of expression profiles from *Arabidopsis* plants treated with compounds/herbicides of known MoA has been established. This compendium represents about 40 herbicides from 11 known modes of action, including acetolactate synthase (ALS), protoporphyrinogen oxidase (PPO), photosystem I, photosystem II, or 5-enolpyruvylshikimi-3-phosphate-synthase (EPSPS). All expression profiles in the compendium are derived from *Arabidopsis* plants sprayed with two different concentrations of the respective compounds and harvested at 24 h after treatment. An analysis of the expression profiles by statistical methods such as hierarchical clustering [14] revealed that the assumption was correct. In fact, the profiles of compounds representing the same MoA were much more similar to each other than to any profile derived from a compound with an alternative MoA (Figure 6.3).

The expression profiles are stored in a database, termed the GEP Compendium, such that it becomes possible to classify compounds from the research pipeline with an unknown MoA into one of the known modes of action of the GEP Compendium, by conducting a single experiment. In this case, *Arabidopsis* plants are sprayed with the respective compound and the isolated and labeled RNA is then analyzed on the *Arabidopsis* chip. The resulting expression profile is then compared with those in the compendium by employing supervised learning algorithms such as Support Vector Machine (SVM) [15] or Analysis of Variance (ANOVA) [16]. When the new expression profile groups together with profiles of a specific MoA in the compendium, there is an utmost probability that the corresponding compound has the same MoA (Figure 6.4). If necessary, however, the MoA can be verified by classical methods such as enzyme assays or supplementation tests, if available.

In the meantime, many different compounds deriving from the research pipeline, and with an unknown MoA, have also been classified. In addition, it was possible to eliminate compounds with an unwanted MoA at a very early stage of the research process and then to concentrate on more promising substances. If compounds cannot be classified into an already existing MoA, the standard GEP Compendium approach can at least place them into specific unknown MoA groups.

6.4 Identification of Prodrugs by Gene Expression Profiling

An inherent problem in the MoA determination of herbicides by classical enzymatic assays is the evaluation of prodrugs. These are compounds which are not active *per se*, but which must be converted into an herbicidally active product inside the plant, for example, by cytochrome P_{450} enzymes [17] or esterases [18]. In conventional enzyme assays, the compounds are tested on purified target enzymes; however, because the prodrug is not converted into its active form the enzyme is not affected and the enzyme assay would not identify its MoA. Gene expression profiling is much closer to the "real situation," since whole plants are sprayed with a compound. Moreover, there is sufficient time for a potential prodrug to be taken up by the plant, to be converted into the active form, and to exert its effect on

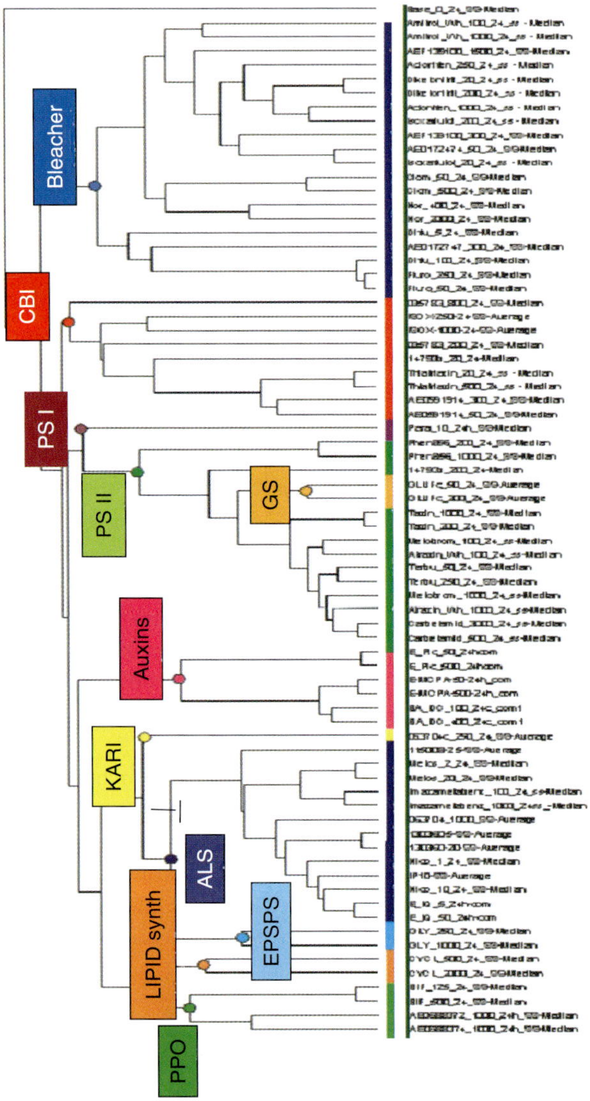

Figure 6.3 The herbicide GEP compendium. A hierarchical clustering of gene expression profiles of 40 compounds from 11 different modes of action is shown. The individual profiling experiments are listed in the lower part. Experiments clustering in the individually colored branches belong to the same MoA.

the target enzyme before the *Arabidopsis* plants are harvested for gene expression analysis. Therefore, gene expression profiling can be used to identify even the MoA of such prodrugs.

The active ingredient Compound A is an example of how gene expression profiling can be used to identify the MoA of a prodrug. Compound A has the ability

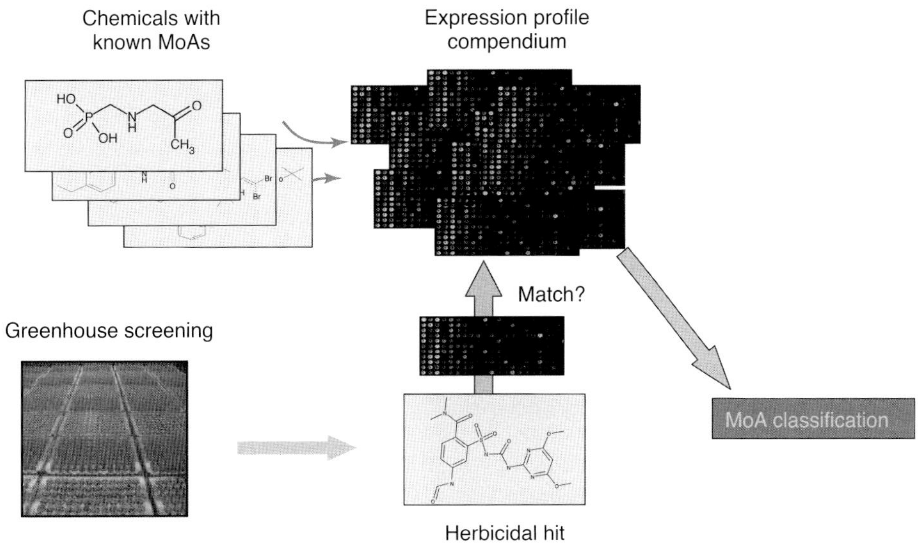

Figure 6.4 The gene expression profile compendium approach. A gene expression profile of a compound identified as herbicidally active in the greenhouse is compared with the already existing profiles in the database. If a similar profile is present, the new compound most probably has the same MoA.

to kill many different weeds, but unfortunately its MoA could not be identified by employing a collection of very diverse classical enzymatic assays. In a gene expression profiling experiment, however, it was possible to classify Compound A into the group of ALS inhibitors (Figure 6.5). The expression profile of Compound A-treated *Arabidopsis* plants was much more similar to the profiles of plants treated with other ALS inhibitors than to the profiles of plants treated with compounds with other modes of action. Further supporting evidence for ALS as the MoA derives from the fact that Compound A induces the genes for alternative oxidase (data not shown). It has been well established that ALS inhibitors increase the level of α-ketoacids such as pyruvate, which in turn leads to an increase in the alternative oxidase protein [19].

Final proof that Compound A affects ALS was obtained from supplementation experiments with the small plant *Lemna gibba*. ALS is the first common enzyme in the parallel pathways for synthesis of the branched-chain amino acids valine (Val), leucine (Leu), and isoleucine (Ile). The production of these amino acids is blocked by ALS inhibitors, but this inhibition can be overcome by the addition of micromolar concentrations of the branched-chain amino acids [20]. *Lemna gibba* plants treated with Compound A were only able to grow further when the growth medium was supplemented with Val, Leu, and Ile. Neither one branched-chain amino acid alone, nor any other of the 20 L-amino acids, could overcome the growth-inhibitory effects of the compound (Figure 6.6).

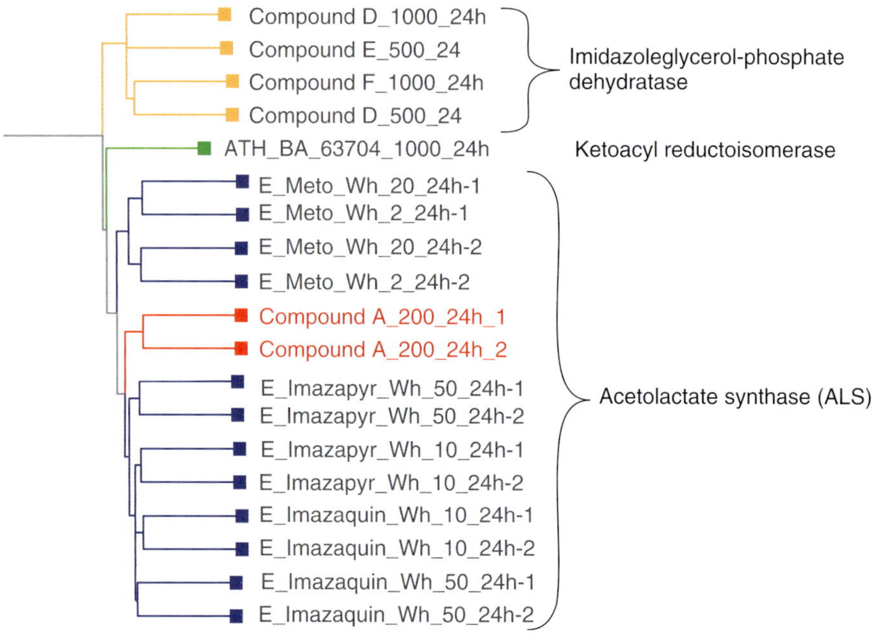

Figure 6.5 Classification of an herbicide. The expression profile of compound A clusters together with the profiles of known ALS inhibitors such as metosulam (Meto), imazapyr, or imazaquin.

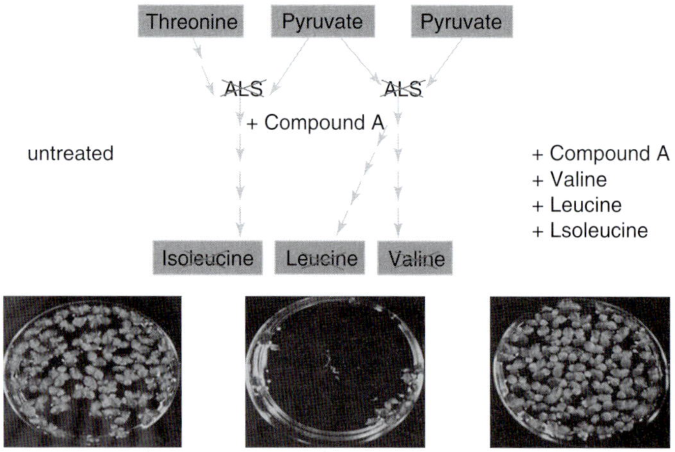

Figure 6.6 Supplementation assay of *Lemna gibba* plants. Plants grow normally in water without compound A (left). The addition of compound A prevents plants from growing (middle). Addition of the three branched-chain amino acids valine (Val), leucine (Leu) and isoleucine (Ile) overcomes the herbicidal effect of compound A and restores plant growth (right).

6.5
Analyzing the Affected Metabolic Pathways

In case the MoA of a compound could not be identified by the standard GEP Compendium approach, more detailed gene expression profiling studies – including several harvest time points and more compound concentrations – may be performed. This further in-depth analysis can provide some hints as to which genes are consistently upregulated or downregulated. If these genes belong to one or a few specific metabolic pathways, then there is a good chance that these pathways are affected by the compound, and that the actual MoA can be assigned to an enzyme within this pathway. In a first attempt to validate this assumption, the genes that are upregulated by the synthetic auxin DICAMBA were analyzed. It was expected that auxin-responsive genes would be over-represented among the upregulated genes and, indeed, it was observed that eight of the 13 highest upregulated genes belonged to the auxin-responsive genes group (Figure 6.7). An analysis of *Arabidopsis* genes, which are annotated as "auxin -related," revealed that about two-thirds of those genes are induced after treatment with DICAMBA. This provided yet another clear indication that DICAMBA effects gene expression, as would be expected for an auxin herbicide. It is well known that auxins induce ethylene production in plants by triggering expression of the genes for 1-aminocyclopropane-1-carboxylate (ACC) synthase (EC 4.4.1.14) [21]. In the present DICAMBA experiments, this gene was among the 13 highest upregulated genes. Some of the genes coding for the next enzyme in ethylene biosynthesis, namely ACC oxidase (EC 1.14.17.4), were also induced, and a recent study of the change in expression of *Arabidopsis* genes following the application of 2,4-dichlorophenoxyacetic acid (another synthetic auxin) described similar observations [22]. Most of the other highly upregulated genes encode stress-related proteins such as lipid transfer proteins, protein phosphatases 2C, or transcription factors involved in general stress response. The data show that, even without prior knowledge of the MoA of DICAMBA, the changes in the transcriptome would clearly have pointed to an auxin effect of that compound.

Another example of how gene expression profiling can help to identify the pathway that is affected by an herbicidal compound derives from the analysis of Compound B. About 60% of the photosynthesis-related genes are downregulated after treatment with this compound (data not shown). However, on examining the different metabolic pathways related to photosynthesis more specifically, it was noted that almost all genes responsible for the biosynthesis of chlorophyll were downregulated (Figure 6.8). This suggested that the target of Compound B might be located in the chlorophyll biosynthesis pathway, a key enzyme of which – and well-known herbicidal target – is PPO. In a PPO inhibition assay, a similar IC_{50} was obtained for Compound B as for Bifenox, a well-known inhibitor of PPO (Figure 6.9). In contrast to Compound B, the expression of chlorophyll biosynthetic genes remained unaffected after treating *Arabidopsis* with compounds that inhibited other herbicidal targets such as cellulose biosynthesis (Figure 6.8), acetyl-CoA carboxylase (ACCase), or hydroxyphenylpyruvate dioxygenase (4-HPPD)

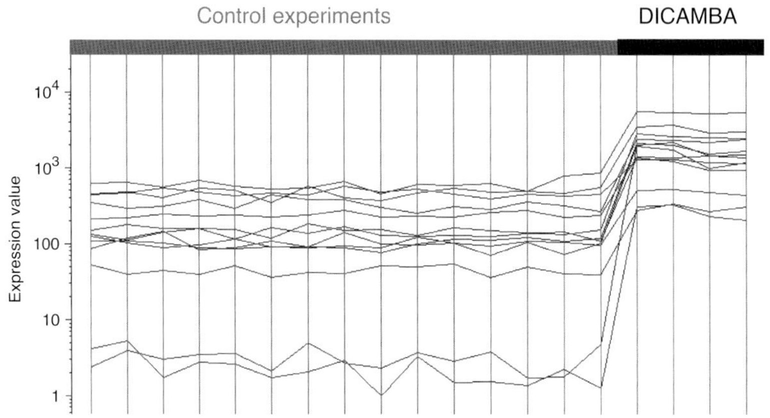

Probe name	ORF Number	Gene description
253908_at	At4g27260	Auxin-responsive GH3 family protein
249306_at	At5g41400	Zinc finger (C3HC4-type Ring finger) family protein
262912_at	At1g59740	Proton-dependent oligopeptide transport (POT) family protein
262099_s_at	At1g59500	Auxin-responsive GH3 family protein
245397_at	At4g14560	Auxin-responsive protein (IAA1)
266415_at	At2g38530	Nonspecific lipid transfer protein 2 (LTP2)
261114_at	At1g75390	bZIP transcription factor family protein
261368_at	At1g53070	Legume lectin family protein
253423_at	At4g32280	Auxin-responsive AUX/IAA family protein
257766_at	At3g23030	Auxin-responsive protein (IAA2)
248163_at	At5g54510	Auxin-responsive GH3 protein, putative (DFL-1)
255177_at	At4g08040	1-aminocyclopropane-1-carboxylate synthase, putative
255788_at	At2g33310	Auxin-responsive protein (IAA13)

Figure 6.7 Mode of action of DICAMBA. The gene expression of untreated control plants and DICAMBA-treated plants are compared. The expression values (scaled logarithmically) represent the level of expression of the genes highly upregulated by DICAMBA. The names of the genes are listed in the table. As expected for the action of a synthetic auxin, most of the genes are annotated as auxin-responsive. Each vertical line represents one experiment.

(data not shown). With gene expression profiling, it was possible to identify the affected pathway (chlorophyll biosynthesis) and to exclude other pathways from the analysis. Although PPO was among the downregulated genes, it proved impossible to pinpoint the actual target by gene expression profiling analysis alone. However, by deriving a GEP it was possible to reduce the number of potential target sites from the complete enzyme universe to the few enzymes involved in the chlorophyll biosynthetic pathway. These examples demonstrate the capability of gene expression profiling to reduce the number of potential targets from the complete proteome to only a few promising candidates.

This provides a starting point for more detailed biochemical, cellular, or molecular methods by which the actual target can be identified [23].

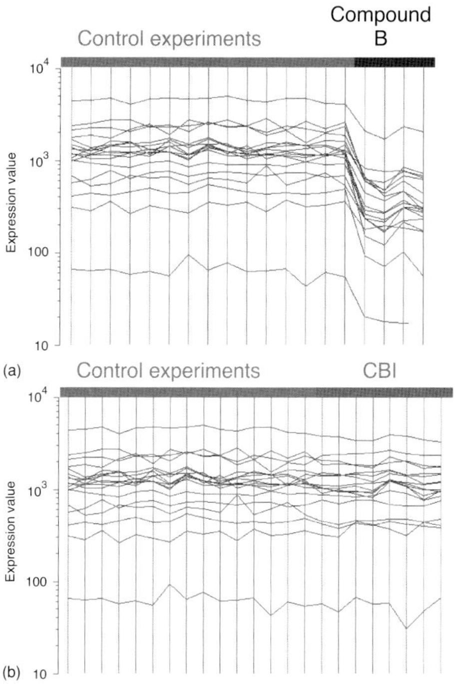

Gene name	Gene description
At1g03475	Coproporphyrinogen III oxidase
At1g03630	Protochlorophyllide reductase C
At1g08520	Magnesium-chelatase subunit chlD
At1g44446	Chlorophyll a oxygenase (CAO) / chlorophyll b synthase
At1g58290	Glutamyl-tRNA reductase 1
At2g26540	Uroporphyrinogen-III synthase family protein
At2g30390	Ferrochelatase II
At2g40490	Uroporphyrinogen decarboxylase, putative
At3g48730	Glutamate-1-semialdehyde aminotransferase 2 (GSA-AT 2)
At3g51820	Chlorophyll synthetase, putative
At3g56940	Magnesium-protoporphyrin IX monomethyl ester cyclase
At4g01690	Protoporphyrinogen oxidase (PPOX)
At4g18480	Magnesium-chelatase subunit chlI
At4g25080	Magnesium-protoporphyrin IX methyltransferase, putative
At4g27440	Protochlorophyllide reductase B
At5g08280	Hydroxymethylbilane synthase
At5g45930	Magnesium-chelatase subunit chlI
At5g54190	Protochlorophyllide reductase A

Figure 6.8 Mode of action of compound B: The gene expression of untreated control plants and compound B treated plants are compared. The expression values (scaled logarithmically) represent the level of expression of the highly downregulated genes. The names of the genes are listed in the table. Genes involved in chlorophyll biosynthesis are downregulated in plants treated with compound B (a), but not in plants treated with a cellulose biosynthesis inhibitor (CBI) (b). Each vertical line represents one experiment.

Compound	IC50 (M) on PPO
Bifenox	1.8×10^{-08}
Compound B	2.5×10^{-08}

Figure 6.9 *In vitro* PPO inhibition assay. The IC_{50} value of compound B is very similar to that of bifenox, a known PPO inhibitor.

6.6
Gene Expression Profiling: Part of a Toolbox for Mode of Action Determination

With gene expression profiling, it is possible to classify compounds into known modes of action or to identify pathway(s) affected by such compounds. However, it must be borne in mind that not only RNA levels but also the amount and stability of expressed proteins (proteome) and the concentration of metabolites (metabolome) within a given cellular context will determine gene activity (Figure 6.1). This makes it difficult – if not impossible – to precisely identify a new target solely by gene expression profiling. Recently, it has been possible to confirm the target of a herbicide by measuring the changes in the concentration of plant metabolites [24]. Further significant advances in the fields of proteomics and metabolomics [25–27] facilitated a thorough analysis of the changing pattern of proteins and metabolites of cells in a varying environment [28, 29], giving rise to the hope that these techniques could complement gene expression profiling for MoA analysis in the near future. The systematic analysis of the symptoms produced by different herbicidal compounds represents another important means of obtaining information regarding their MoA [30]. Finally, the target must be eliminated from the cell by employing molecular methods such as "gene knockouts" to unequivocally identify the target of a compound. This tool box of very diverse, but complementary, methods will surely lead to the identification of new modes of action that will serve as targets for herbicides with superior properties.

References

1. The Arabidopsis Genome Initiative (2000) *Nature*, **408**, 796–815.
2. Goff, S.A., Ricke, D., Lan, T.H., Presting, G., Wang, R., Dunn, M., Glazebrook, J., Sessions, A., Oeller, P., Varma, H., Hadley, D., Hutchison, D., Martin, C., Katagiri, F., Lange, B.M., Moughamer, T., Xia, Y., Budworth, P., Zhong, J., Miguel, T., Paszkowski, U., Zhang, S., Colbert, M., Sun, W.L., Chen, L., Cooper, B., Park, S., Wood, T.C., Mao, L., Quail, P., Wing, R., Dean, R., Yu, Y., Zharkikh, A., Shen, R., Sahasrabudhe, S., Thomas, A., Cannings, R., Gutin, A., Pruss, D., Reid, J., Tavtigian, S., Mitchell, J., Eldredge, G., Scholl, T., Miller, R.M., Bhatnagar, S., Adey, N., Rubano, T., Tusneem, N., Robinson, R., Feldhaus, J., Macalma, T., Oliphant, A., and Briggs, S. (2002) *Science*, **296**, 92–100.
3. Yu, J., Hu, S., Wang, J., Wong, G.K., Li, S., Liu, B., Deng, Y., Dai, L., Zhou, Y., Zhang, X., Cao, M., Liu, J., Sun, J., Tang, J., Chen, Y., Huang, X., Lin, W., Ye, C., Tong, W., Cong, L., Geng, J., Han, Y., Li, L., Li, W., Hu, G., Huang, X., Li, W., Li, J., Liu, Z., Li, L., Liu, J., Qi, Q., Liu, J., Li, L., Li, T., Wang, X., Lu, H., Wu, T., Zhu, M., Ni, P., Han, H., Dong, W., Ren, X., Feng, X., Cui, P., Li, X., Wang, H., Xu, X., Zhai, W., Xu, Z., Zhang, J., He, S., Zhang, J., Xu, J., Zhang, K., Zheng, X., Dong, J., Zeng, W., Tao, L., Ye, J., Tan, J., Ren, X., Chen, X., He, J., Liu, D., Tian, W., Tian, C., Xia, H., Bao, Q., Li, G., Gao, H., Cao, T., Wang, J., Zhao, W., Li, P., Chen, W., Wang, X., Zhang, Y., Hu, J., Wang, J., Liu, S., Yang, J., Zhang, G., Xiong, Y., Li, Z., Mao, L., Zhou, C., Zhu, Z., Chen, R., Hao, B., Zheng, W., Chen, S., Guo, W., Li, G., Liu, S., Tao, M., Wang, J., Zhu, L., Yuan, L., and Yang, H. (2002) *Science*, **296**, 79–92.
4. International Rice Genome Sequencing Project (2005) *Nature*, **436**, 793–800.
5. Zhu, T., Budworth, P., Chen, W., Provart, N., Chang, H.S., Guimil, S., Su, W., Estes, B., Zou, G., and Wang, X. (2003) *Plant Biotechnol. J.*, **1**, 59–70.
6. Redman, J.C., Haas, B.J., Tanimoto, G., and Town, C.D. (2004) *Plant J.*, **38**, 545–561.
7. Wan, J., Dunning, F.M., and Bent, A.F. (2002) *Funct. Integr. Genomics*, **2**, 259–273.
8. Ruuska, S.A., Girke, T., Benning, C., and Ohlrogge, J.B. (2002) *Plant Cell*, **14**, 1191–1206.
9. Wang, R., Okamoto, M., Xing, X., and Crawford, N.M. (2003) *Plant Physiol.*, **132**, 556–567.
10. Aharoni, A. and O'Connell, A.P. (2002) *J. Exp. Bot.*, **53**, 2073–2087.

11. Seki, M., Narusaka, M., Ishida, J., Nanjo, T., Fujita, M., Oono, Y., Kamiya, A., Nakajima, M., Enju, A., Sakurai, T., Satou, M., Akiyama, K., Taji, T., Yamaguchi-Shinozaki, K., Carninci, P., Kawai, J., Hayashizaki, Y., and Shinozaki, K. (2002) *Plant J.*, **31**, 279–292.
12. Kreps, J.A., Wu, Y., Chang, H.S., Zhu, T., Wang, X., and Harper, J.F. (2002) *Plant Physiol.*, **130**, 2129–2141.
13. Rizhsky, L., Liang, H., Shuman, J., Shulaev, V., Davletova, S., and Mittler, R. (2004) *Plant Physiol.*, **134**, 1683–1696.
14. Eisen, M.B., Spellman, P.T., Brown, P.O., and Botstein, D. (1998) *Proc. Natl. Acad. Sci. USA*, **95**, 14863–14868.
15. Cristianini, N. and Shawe-Taylor, J. (2000) *An Introduction to Support Vector Machines (and other Kernel-Based Learning Methods)*, Cambridge University Press, Cambridge.
16. Mardia, K., Kent, J., and Bibby, J. (1979) *Multivariate Analysis*, Academic Press, London.
17. Thies, F., Backhaus, T., Bossmann, B., and Grimme, L.H. (1996) *Plant Physiol.*, **112**, 361–370.
18. Yamoto, S., Fusaka, T., and Tanaka, Y. (2005) *J. Pestic. Sci.*, **30**, 384–389.
19. Gaston, S., Ribas-Carbo, M., Busquets, S., Berry, J.A., Zabalza, A., and Royuela, M. (2003) *Plant Physiol.*, **133**, 1351–1359.
20. Ray, T.B. (1984) *Plant Physiol.*, **75**, 827–831.
21. Yip, W.K., Moore, T., and Yang, S.F. (1992) *Proc. Natl. Acad. Sci. USA*, **89**, 2475–2479.
22. Raghavan, C., Ong, E.K., Dalling, M.J., and Stevenson, T.W. (2006) *Funct. Integr. Genomics*, **6**, 60–70.
23. di Bernardo, D., Thompson, M.J., Gardner, T.S., Chobot, S.E., Eastwood, E.L., Wojtovich, A.P., Elliott, S.J., Schaus, S.E., and Collins, J.J. (2005) *Nat. Biotechnol.*, **23**, 377–383.
24. Lange, B.M., Ketchum, R.E., and Croteau, R.B. (2001) *Plant Physiol.*, **127**, 305–314.
25. Weckwerth, W. (2003) *Annu. Rev. Plant Biol.*, **54**, 669–689.
26. Fernie, A.R., Trethewey, R.N., Krotzky, A.J., and Willmitzer, L. (2004) *Nat. Rev. Mol. Cell Biol.*, **5**, 763–769.
27. Tyers, M. and Mann, M. (2003) *Nature*, **422**, 193–197.
28. Cook, D., Fowler, S., Fiehn, O., and Thomashow, M.F. (2004) *Proc. Natl. Acad. Sci. USA*, **101**, 15243–15248.
29. Renaut, J., Hausman, J.F., and Wisniewski, M.E. (2006) *Physiol. Plantarum*, **126**, 97–109.
30. Grossmann, K. (2005) *Pest Manag. Sci.*, **61**, 423–431.

7
Modern Approaches for Elucidating the Mode of Action of Neuromuscular Insecticides

Daniel Cordova

7.1
Introduction

As the human population continues to rise and of arable land remains constant or even decreases there is an ever-increasing demand for higher crop yields. Today, insecticides play a significant role in meeting this demand, and the discovery of new insecticides – particularly those acting at new target sites – is vital due to the continuous development of insecticide resistance. The escalating costs of research and development of new insecticides has, however, led to an increasing challenge to commercialization of new insecticidal agents. Nevertheless, new insecticides have entered the marketplace over the past decade (Table 7.1; Figure 7.1), several of which have novel modes of action. Given the investment needed for pesticide discovery and development, coupled with ever-stringent regulatory hurdles, elucidation of an insecticide's mode of action (MoA) early in the development process can play an integral role in assessing the compound's potential value.

Three criteria should be satisfied in order to validate an insecticide's MoA:

- Action at the proposed target is consistent with poisoning symptoms.
- Action at the target protein occurs at concentrations relevant to toxicity rates in pest species.
- Potency at the target protein correlates well with toxicity for a series of analogs.

Resistance to the candidate insecticide, whether due to mutation, overexpression, or knockout of the target protein, can further validate its target site. Insecticides which are fast-acting typically perturb neuromuscular function, and comprise 90% of the global insecticide market [1]. Historically, mechanistic studies on such insecticides (i.e., organophosphates, carbamates, pyrethroids, cyclodienes, and neonicotinoids) have relied on a combination of enzyme kinetic studies, radioligand binding studies, and *in situ* electrophysiological studies using model insects. The use of such approaches continues to provide a wealth of target site information for new insecticidal chemistry, and will therefore be highlighted only briefly in this chapter as they relate to the latest commercial insecticides. The bulk

Modern Methods in Crop Protection Research, First Edition.
Edited by Peter Jeschke, Wolfgang Krämer, Ulrich Schirmer, and Matthias Witschel.

Table 7.1 The mode of action of insecticides registered between 2001 and 2011 (bold) or which are currently in development.

Target	Insecticide	Target site studies
Acetyl CoA carboxylase	**Spirodiclofen** **Spiromesifen** **Spirotetramat**	Microscopic observations, lipid monitoring, resistance
GABA-gated chloride channel	**Lepimectin**	Oocyte voltage-clamp, radioligand binding
Nicotinic acetylcholine receptor	**Clothianidin**	WCVC, oocyte voltage-clamp, radioligand binding
	Dinotefuran	WCVC, radioligand binding
	Spinetoram	WCVC, radioligand binding, genomics
	Sulfoxaflor	Oocyte voltage-clamp, radioligand binding
	Thiomethoxam	Oocyte voltage-clamp, radioligand binding
Ryanodine receptor	**Chlorantraniliprole** cyantraniliprole	Calcium imaging, radioligand binding
	Flubendiamide	Calcium imaging, radioligand binding
Sodium channel	**Metaflumizone**	WCVC, oocyte voltage-clamp
Homopteran feeding blocker	**Flonicamid**	WCVC, resistance, EPG
Unknown	**Pyridalyl**	Cell proliferation, resistance

WCVC = whole-cell voltage-clamp recording; EPG = electrical penetration graph.

of the chapter will focus on more recent approaches that have been added to the MoA "elucidation toolbox."

7.2 Biochemical and Electrophysiological Approaches

7.2.1 Biochemical Studies

Radioligand binding studies have historically played a critical role in defining an insecticide's MoA. Indeed, with the exception of improvements in cell harvesters and radiometric counting instrumentation, little has changed in how binding

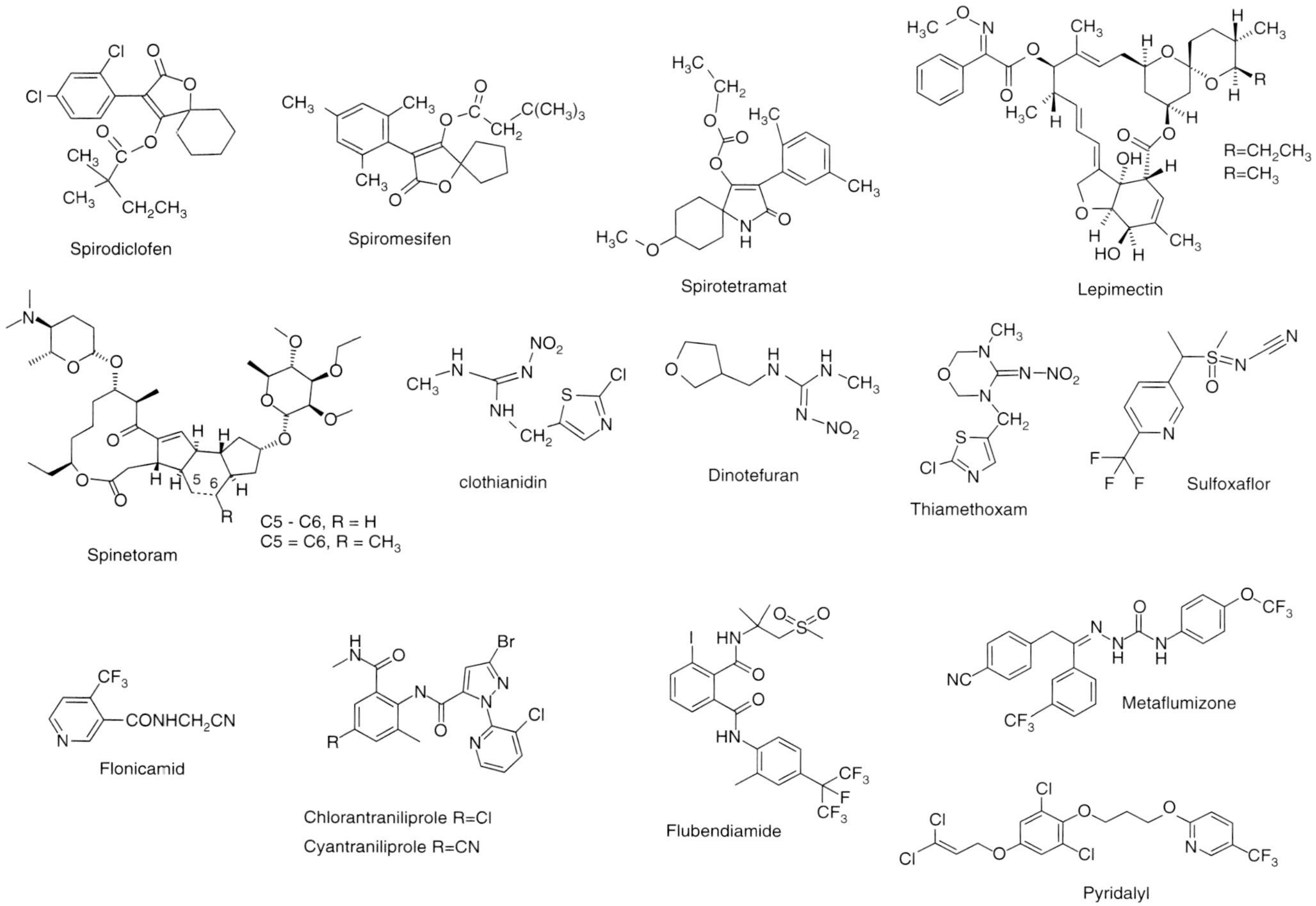

Figure 7.1 Chemical structures of the insecticides described in Table 7.1.

studies have been conducted over the past 10–15 years. Studies continue to rely on characterizing the interaction between membrane preparations, reference radioligands, and the chemistry of interest. Currently, numerous radioligands are commercially available for characterizing interactions between new candidate insecticides and neurotransmitter receptors and ion channels, as shown in Table 7.2. The vast majority of insect studies are conducted with tissues isolated from model insects such as cockroach (*Periplaneta americana*) [2–4], fruitfly (*Drosophila melanogaster*) [5–8], and house fly (*Musca domestica*) [9–11], as well as with relevant pest species such as the green peach aphid (*Myzus persicae*) and brown planthopper (*Nilaparvata lugens*) [12–15]. For investigations on nicotinic acetylcholine receptors (*n*AChRs), the displacement of [^{3}H]-imidacloprid from green peach aphid membranes is frequently used to characterize binding site interactions [12, 14, 15]. When using this preparation, a strong correlation has been observed between binding affinity and

Table 7.2 List of commercially available radioligands having utility for mode of action studies on neuromuscular targets.

Target	Radioligand	Vendors
Acetylcholine transport	[^{3}H]-Vesamicol	PerkinElmer, ARC, Moravek
Cl^- channel (GABA-gated)	[^{3}H]-BIDN	PerkinElmer
	[^{3}H]-EBOB	PerkinElmer, ARC
	[^{3}H]-GABA	PerkinElmer, ARC, Moravek, ViTrax
	[^{3}H]-Muscimol	PerkinElmer, ARC
Cl^- channel (glutamate-gated)	[^{3}H]-Ivermectin	PerkinElmer, ARC
Glutamate receptor	[^{3}H]-AMPA	PerkinElmer, ARC
	[^{3}H]-Kainic acid	PerkinElmer, ARC
	[^{3}H]-Quisqualic acid	PerkinElmer, ARC
*m*AChR	[^{3}H]-Quinuclidnyl benzilate	PerkinElmer, ARC, MP Biomedicals
*n*AChR	[^{3}H]-α-Bungarotoxin	PerkinElmer, ARC
	[^{125}I]-α-Bungarotoxin	PerkinElmer, ARC, MP Biomedicals
	[^{3}H]-Epibatidine	PerkinElmer, ARC
	[^{125}I]-Epibatidine	PerkinElmer, ARC
	[^{3}H]-Imidacloprid	PerkinElmer, ARC
RyR	[^{3}H]-Ryanodine	PerkinElmer, ARC
VGCC	[^{3}H]-Verapamil	PerkinElmer, ARC
VGSC	[^{3}H]-Batrachatoxin	PerkinElmer

insect toxicity for sulfoximines, a recently discovered class of sap-feeding insecticides produced by Dow AgroSciences [12, 16, 17]. Interestingly, the development candidate from this class, sulfoxaflor, exhibits a much weaker [^{3}H]-imidacloprid displacement potency relative to its toxicity to the green peach aphid.

In cases where insecticides fail to displace existing probes, a radiolabeled analog of the chemistry of interest is typically pursued. Custom radiosynthesis can be contracted with companies such as PerkinElmer, which was the case for the radiolabeled anthranilic diamides, [^{3}H]-DP-010 and [^{3}H]-DP-033. Chlorantraniliprole, the first anthranilic diamide to be commercialized by DuPont Crop Protection, activates insect ryanodine receptors (RyRs), but does not interact with the binding site for ryanodine [2, 3, 18]. Biochemical characterization studies conducted using tissue preparations from *P. americana* and the corn planthopper, *Perigrinus maidis,* have revealed that previously known RyR agents fail to interact with the binding site of anthranilic diamides.

The *n*AChR modulator, spinosad, is another example of an insecticide that binds to a unique site. Receptor binding studies have demonstrated that spinosyn A (the primary component of spinosad) failed to displace various *n*AChR radioligands, including [^{3}H]-imidacloprid, which indicated that spinosyns bind to a distinct site on the receptor [19]. Subsequent displacement studies using a radiolabeled spinosyn, [^{3}H]-dihydrospinosyn A, confirmed the nicotinic nature of the receptor and the lack of interaction with neonicotinoids [20].

Today, surface plasmon resonance (SPR) -based instruments are gaining acceptance for use in binding studies, due to their improved detection sensitivity and micro-fluidics. The SPR detection systems function on the principle of protein/ligand binding to a surface-bound recognition element (ligand/protein), resulting in a shift in the surface plasma wave [21]. More recent SPR instruments have been shown to provide comparable sensitivity to traditional radiolabeled bioassays [22]. Although SPR offers increasing utility for investigating ligand interactions with soluble proteins, membrane-bound receptors and ion channels remain a challenge as SPR-based assays require detergent solubilization or suspension of the target proteins in lipid membranes [23]. Accordingly, for neurotransmitter receptor and ion channel targets, high-throughput screening of expressed proteins is the more prevalent application of SPR.

7.2.2 Electrophysiological Studies on Native and Expressed Targets

7.2.2.1 Whole-Cell Voltage Clamp Studies

For investigating insecticide action on ligand-gated and voltage-gated channels, whole-cell voltage-clamp (WCVC) recording – a variation of the patch-clamp technique developed by Neher and Sakmann in 1976 [24] – remains a principal tool. In the whole-cell configuration, upon establishing a gigaohm resistance seal between the patch pipette tip and cell, the membrane patch under the tip is ruptured. This allows the voltage across the entire cell membrane to be clamped, and the cytosolic milieu to be dialyzed with the internal solution of the patch pipette.

The combination of voltage protocols and recording solutions allows for a detailed characterization of receptor and channel function. For further details on general WCVC methodology, see Chapter 6 in *Current Protocols in Neuroscience* (1997) [25].

Studies on insect ligand-gated and voltage-gated ion channels are frequently conducted using dissociated neurons isolated from cockroach and locust [26–31]. These preparations have been well characterized over the years, and therefore have proven invaluable for investigating numerous insecticides including neonicotinoids, spinosad, fipronil, pyrethroids, and indoxacarb [32–36].

Most studies on insect neuromuscular junction and muscle ion channels utilize the two-electrode voltage-clamp method rather than the WCVC technique, due to space clamp limitations and difficulties associated with obtaining stable fibers following enzymatic dissociation. Space clamp refers to the ability to maintain an adequate voltage-clamp as the distance from the patch electrode increases. Although a model organism, *Drosophila* is particularly well suited for voltage-clamp studies, due to the: (i) small size of the muscle fibers; (ii) the availability of ion channel mutants; and (iii) the wealth of published reports, particularly from the laboratory of Wu and colleagues [37, 38]. An excellent review on insect ion channels, in both neuronal and muscle preparations, is also available [39].

A more recent patch-clamp preparation utilizing single muscle fibers from honeybee (*Apis mellifera*) has been described by Collet and Belzunces [40]. In this case, the fibers were enzymatically isolated using a mixture of collagenase, pronase, papain, and trypsin. By using WCVC, the authors demonstrated the presence of glutamate-activated currents as well as voltage-gated calcium and potassium channel currents. Although these muscle fibers lack voltage-gated sodium channels (VGSCs), the pyrethroid, allethrin, suppressed action potential bursts due to its inhibitory effect on voltage-gated calcium channels [41]. With the recent sequencing of the *Apis mellifera* genome [42], and the increasing public scrutiny of insecticide actions on pollinators, studies on honeybee receptors and ion channels may prove particularly useful for monitoring insecticide mode of action and selectivity.

7.2.2.2 **Oocyte Expression Studies**

Since the early 1980s, *Xenopus laevis* oocytes have been exploited as an expression system for various neurotransmitter receptors and ion channels. Indeed, the use of two-electrode voltage-clamp on oocytes expressing insect targets remains a highly effective tool for target site elucidation today. The advantages of oocyte expression studies include: (i) a lack of endogenous expression of most insecticide targets; (ii) the ability to vary the combination of receptor/channel subunits and their relative ratios; (iii) suitability for mutation studies; and (iv) the ability to develop higher throughput voltage-clamp assays using commercially available systems. As with all approaches, *Xenopus* oocyte recording does have its disadvantages. For example, differences in subunit composition (relative to native receptors/channels) or the need to coexpress insect and vertebrate receptor subunits, as is the case for *n*AChRs, can mask subtle differences in the pharmacology and kinetics of the targets of interest. Furthermore, the sensitivity of expressed receptors can often be lower than that observed with native receptors, as is the case with imidacloprid (EC_{50} values of 2.7 and 0.3 μM for expressed Dα2β2 and native *Drosophila* *n*AChRs, respectively) [43–46]. Nevertheless, oocyte expression studies have proven highly successful for characterizing the target effects of neuronal insecticides.

An investigation of the MoA of sulfoxaflor involved the use of traditional radioligand binding studies coupled with voltage-clamp studies, using *Xenopus* oocytes [12]. Expressed chimeric *n*AChRs (Dα1β2 and Dα2β2) exhibit concentration-dependent inward currents in response to this insecticide, as shown in Figure 7.2. A comparison of *n*AChR currents in oocytes expressing Dα2β2 showed that sulfoxaflor behaves as a super-agonist, inducing currents greater than threefold that of acetylcholine (Table 7.3). Such super-agonist effects have previously been reported for the neonicotinoid, clothianidin, whereas other neonicotinoids – including imidacloprid, acetamiprid, and thiacloprid – behave as partial agonists [33, 47]. Interestingly, while imidacloprid (10 nM) antagonizes the acetylcholine response in oocytes expressing Dα1β2 *n*AChRs, this was not observed with sulfoxaflor.

The above-described studies utilized receptors from model insects and provided a wealth of knowledge regarding insecticide–target interactions. Although model systems are indispensable, the use of targets (either native or expressed) from commercially relevant pests can offer insight into how subunit composition influences insecticide binding.

The brown planthopper, *Nilaparvata lugens*, is a major pest of rice in Asia. Liu and colleagues have functionally co-expressed *n*AChR subunits Nlα1 and Nlα2 from *N. lugens* with a mammalian β2 subunit [48, 49]. Oocytes expressing three different hybrid *n*AChRs (Nlα1/β2, Nlα2/β2, and Nlα1/Nlα2/β2) produce imidacloprid-sensitive currents each with distinct pharmacological properties. Consequently, expression studies using *N. lugens* *n*AChR subunits may further delineate distinct differences in MoA among commercial neonicotinoids and new *n*AChR insecticides such as sulfoxaflor.

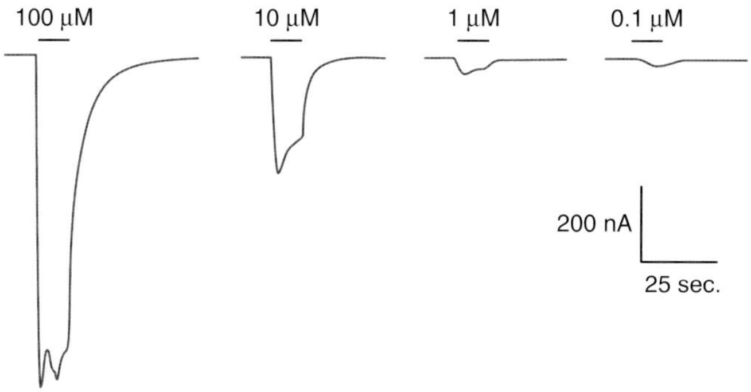

Figure 7.2 Dose-dependent activation of Dα2β2 receptors by increasing concentrations of sulfoxaflor. Sulfoxaflor was applied as indicated by the horizontal bar. Reprinted with permission from Ref. [12]; © 2011, Elsevier.

Table 7.3 Comparative green peach aphid (GPA) toxicity, maximal nAChR currents, and [^{3}H]-imidacloprid (IMI) binding data for sulfoxaflor and a set of neonicotinoid insecticides.

Compound	GPA toxicity[a]	I_{max} (% ACh response)[b]		[^{3}H]IMI binding	
	LC_{90} (ppm)	Dα2β2	Dα1β2	K_i(nM)[c]	n_H[d]
Sulfoxaflor	0.19(0.11–0.35)[e]	348 ± 48	27.9 ± 4.7	265 ± 49	0.9
Imidacloprid	0.24(0.17–0.36)[e]	32.8 ± 2.0	13.2 ± 4.0	5.1 ± 0.7	1.1
Clothianidin	1.2(0.6–2.3)	273 ± 49	19.3 ± 1.4	24.2 ± 2.3	0.8
Acetamiprid	0.35(0.2–0.7)[e]	20.5 ± 2.3	11.5 ± 1.5	19.2 ± 5.7	0.9
Thiacloprid	2.8(1.7–4.7)	12.2 ± 1.3	4.0 ± 0.6	5.4 ± 1.4	1.2
Dinotefuran	6.8(2.9–29.3)[e]	26.5 ± 4.4	8.6 ± 2.6	2223 ± 712	1.2
Nitenpyram	8.2(2.5–60)	47.2 ± 3.1	23.4 ± 4.0	7.3 ± 2.8	0.8

[a] LC_{90} value (with 95% fiducial limits).
[b] Mean current induced by 100 μM of each compound expressed as percentage of an initial response to 100 μM ACh (±SEM, $n \geq 4$ replicates per value except nitenpyram on Dα1/β2, where $n = 3$).
[c] Apparent inhibition constants (K_is) are expressed as mean ± SEM ($n \geq 3$ independent experiments for each value).
[d] Hill slope (n_H) is the mean for each compound.
[e] From Ref. [6]. Table reprinted from Ref. [12] with permission from Elsevier.

7.2.3 Automated Two-Electrode Voltage-Clamp TEVC Recording Platforms

Commercial automated systems for two-electrode voltage-clamp (TEVC) recording from *Xenopus* oocytes include Roboocyte® from Multi Channel Systems (Figure 7.3) and OpusXpress from Molecular Devices. While these systems have a particular utility for compound screening, the ability to evaluate insecticide leads against targets with modified amino acid residues or subunit combinations enables detailed target site characterization. Schnizler *et al.* [50] and Goldin [51] have provided excellent overviews of the Roboocyte® system. The workstation includes a head assembly which combines an automated pressure manifold for cDNA injection, a TEVC amplifier capable of recording up to ±32 μA of current with a resolution of 1 nA, and an eight- or 16-valve continuous-flow perfusion system. User-generated sequence protocols (Roboocyte Scripting Language) facilitate changes to the cDNA injection parameters, voltage-step protocols, and sequential compound application. Data processing and analysis software allows the user to perform baseline and leak current subtraction, data averaging, and curve-fit analysis.

The OpusXpress system differs significantly from Roboocyte® in two major aspects. First, cDNA/RNA injections must be made manually and second, recording is conducted on eight oocytes in parallel [52, 53]. While this system has the capability of testing over 200 different conditions daily (either distinct compounds, variations in target proteins or a combination of the two), it requires greater manual

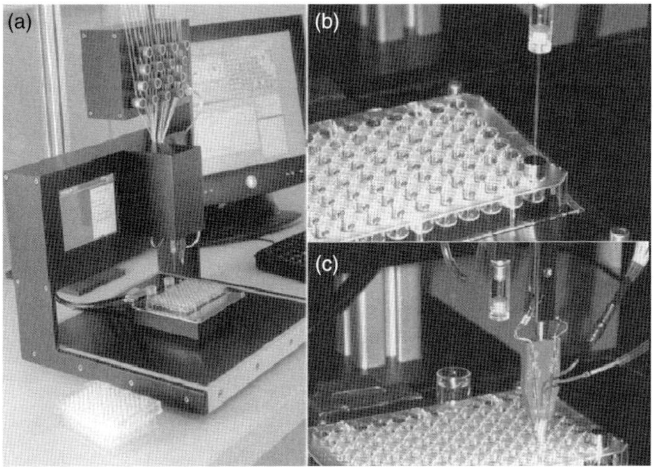

Figure 7.3 The Roboocyte automated voltage-clamp system from Multichannel Systems. (a) The instrument consists of a single head that moves vertically for both injection and recording with oocytes located in a 96-well plate; (b) Close-up view of an injection needle; (c) Close-up view of the recording head which contains both voltage and current electrodes and a perfusion needle. (Adapted from Goldin A.L., *Expression and Analysis of Recombinant Ion Channels: Structural Studies to Pharmacological Screening* (eds. J.J. Clare and D.J. Trezise). 1–25, (2006). Copyright Wiley-VCH Verlag GmbH & Co. KGaA. Reproduced with permission.)

involvement over Roboocyte®. In July 2010, the Molecular Devices Corporation announced that it was ceasing production of OpusXpress; consequently Roboocyte® would be the system of choice for laboratories seeking to implement an automated oocyte recording system.

7.3 Fluorescence-Based Approaches for Mode of Action Elucidation

7.3.1 Calcium-Sensitive Probes

Over the past two decades, the use of fluorescence probes as reporters for physiological function has expanded the repertoire of tools available for target site elucidation. Real-time changes in fluorescence are recorded using microscopic imaging systems and plate-based platforms such as FlexStation® and FLIPR® (Fluorometric Imaging Plate Reader) (both from Molecular Devices). Fluorescent probes allow the monitoring of chemically induced changes in cellular calcium concentration, membrane potential, intracellular pH, and mitochondrial function. Calcium, in particular, is an excellent reporter for insecticide action as it is involved in cell signaling, muscle contraction, neurotransmitter release, and fertilization [54]. The key mechanisms involved in calcium homeostasis in insect neurons are depicted in Figure 7.4; these include voltage-gated calcium channels, RyRs, inositol trisphosphate receptors, and calcium ATPase.

Currently, numerous fluorescent probes are available for calcium imaging with the dual-wavelength (ratio) dyes, Fura-2 and Indo-1, and the single-wavelength dyes, Fluo-3 and Fluo-4, having the greatest utility. All are available in membrane-permeant (acetoxymethylester; AM) forms which allows for easy dye loading into cells. Fura-2 AM is by far the dye of choice for ratio imaging [55, 56]. Since its development by Roger Tsien and colleagues during the 1980s, Fura-2 AM has been cited in thousands of papers [57]. Imaging with ratiometric dyes offers the advantage of allowing the conversion of fluorescence ratio values to quantitative calcium concentration while negating any effects of uneven dye loading, dye leakage, photobleaching, and variations in cell thickness. Calcium imaging offers several advantages over conventional patch-clamp techniques:

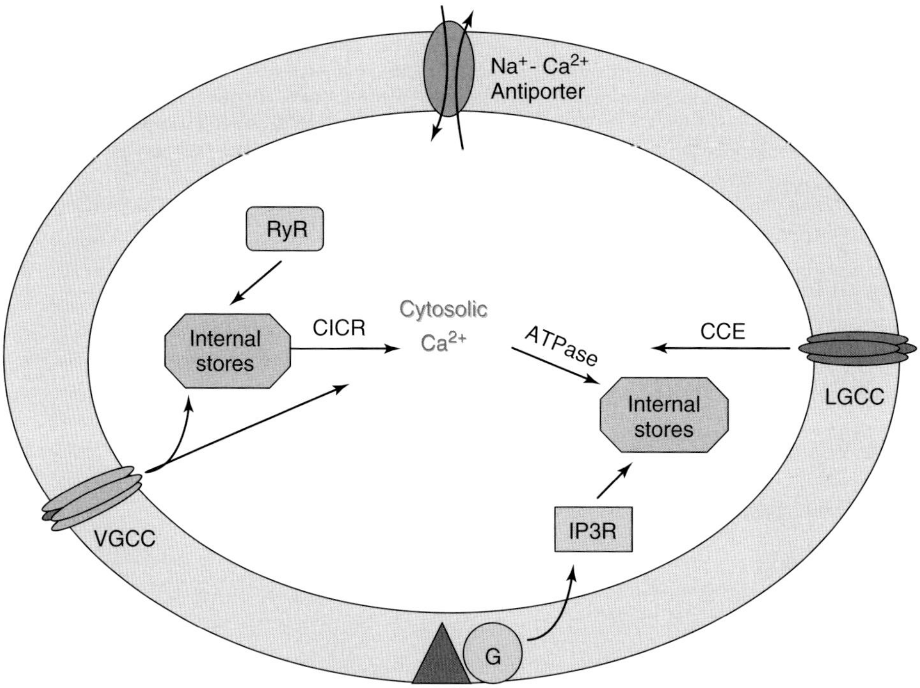

Figure 7.4 Diagram of the major effectors involved in calcium (Ca^{2+}) homeostasis of insect neurons. External Ca^{2+} enters the cell via a voltage-gated Ca^{2+} channel (VGCC), ligand-gated calcium channel (LGCC), or capacitative Ca^{2+} entry (CCE), a mechanism by which depletion of internal Ca^{2+} stores triggers Ca^{2+} entry through non-voltage-gated channels. The release of internal Ca^{2+} stores can be activated via inositol trisphosphate receptor (IP3R) coupled to G-proteins (G), ryanodine receptors (RyR), or Ca^{2+}-induced Ca^{2+} release (CICR), a mechanism in which Ca^{2+} entry triggers store release. The recovery of basal cytosolic Ca^{2+} levels occurs via a Na^+/Ca^{2+} antiporter, refilling of internal Ca^{2+} stores driven by an ATPase-dependent pump, and sequestration in Ca^{2+}-binding proteins (not shown).

- Ease of use, as cells of varying size can be recorded without the need for microelectrodes.
- Simultaneous recording from as many as 100 cells in a given experiment.
- An ability to detect action on voltage- and ligand-gated targets, as well as targets that modulate intracellular calcium stores.

Of course, as with any technique there are significant limitations with calcium imaging:

- The cells are not clamped at a desired voltage, and therefore voltage-protocols can not be employed.
- Calcium signals may reflect cumulative effects from multiple targets.
- The signal kinetics are relatively slow.
- The target of interest may not impact intracellular calcium concentration.

Numerous insectide targets have been investigated using calcium imaging, including RyRs and *n*AChRs (see below), muscarinic acetylcholine receptors [58–60], octopamine receptors [61–63], ligand-gated chloride channels [64], and voltage-gated ion channels [65–67]. As with patch-clamp studies, the saline composition (standard versus calcium-free) and the use of pharmacological agents can provide insight into the action of insecticides at neuronal targets. Compound-induced changes in cytosolic calcium may correspond to a direct action on calcium signaling mechanism, or may reflect a secondary or "down-stream" effect.

Calcium imaging studies have proven critical for elucidating RyR modulation as the MoA of the newly commercialized anthranilic diamide insecticide, chlorantraniliprole, and the phthalic diamide insecticide, flubendiamide. By using Fura-2-loaded embryonic neurons from *P. americana* brains, anthranilic diamides were shown to induce a dose-dependent increase in intracellular calcium [3, 18, 68, 69]. The application of anthranilic diamides in calcium-free saline or in cells pre-treated with ryanodine revealed that the calcium response corresponded to the release of ryanodine-sensitive calcium stores. Similarly, flubendiamide was found to release internal calcium stores in neurons isolated from *Heliothis virescens* [70]. Imaging studies performed on cells expressing recombinant insect receptors provided a genetic validation for RyR as the target of the diamides insecticides. Further, comparative studies with cells expressing mammalian RyRs proved critical for revealing the insect selectivity of these new insecticides.

An example of using calcium imaging to detect secondary effects of VGSCs is demonstrated with the *N*-isobutylamide, piperovatine. *N*-Isobutylamides are botanical insecticides which have been shown previously to induce repetitive nerve discharge via an action on VGSCs [71, 72]. Piperovatine was found to induce a strong increase in cytosolic calcium in *P. americana* neurons, with some cells exhibiting repetitive calcium spikes during saline washout [67]. Similar effects are observed with pyrethroids (unpublished results), and are consistent with persistent sodium channel activation. Further, piperovatine-induced calcium response is completely blocked by pre-treating with the sodium channel blocker, tetrodotoxin [67].

A second example of a target with a "down-stream" impact on cytosolic calcium is the *n*AChR. Numerous studies have been conducted in which insect *n*AChR

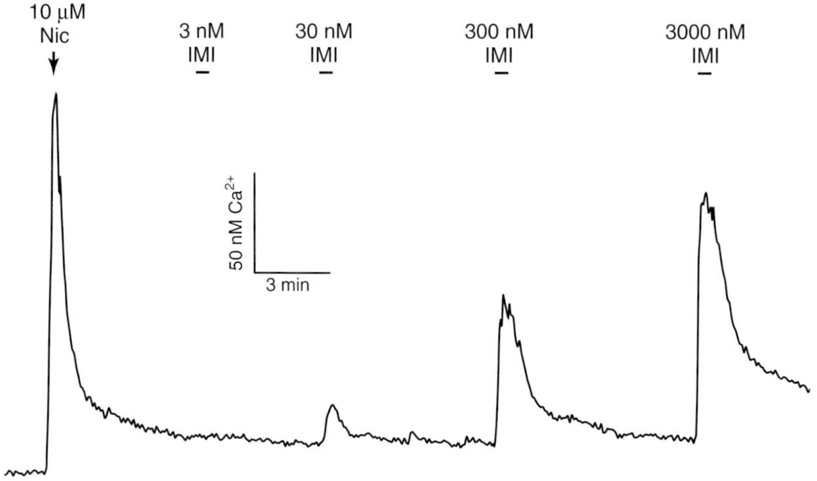

Figure 7.5 Ca^{2+} responses of *P. americana* embryonic neurons challenged with imidacloprid show sensitivity comparable with electrophysiological approaches. Ca^{2+}-imaging studies were conducted on neurons loaded with Fura-2 AM. Shown is a typical response of a neuron that was continuously perfused with saline and challenged with either nicotine (Nic, 15 s) or imidacloprid (IMI, 30 s).

function was investigated using calcium imaging [45, 73–78]. Although the predominant source of *n*AChR-induced calcium influx results from the activation of voltage-gated calcium channels, a significant portion of the signal can be attributed directly to receptor-mediated calcium entry. Interestingly, sensitivity to the neonicotinoid, imidacloprid, is comparable to that reported in various patch-clamp studies, with concentrations as low as 30 nM inducing significant calcium mobilization, as shown in Figure 7.5 [10, 33, 43, 44, 79]. While calcium imaging has limitations with regard to the level of detail that can be obtained, it can provide considerable information on insecticide–target interaction for multiple targets.

7.3.2
Voltage-Sensitive Probes

Voltage-sensitive fluorescent probes can be categorized as fast-responding (aminonaphthylethenylpyridinium (ANEP) and RH (dialkylaminophenylpolyenylpyridinium; originally synthesized by Rina Hildelsheim dyes) or slow-responding (carbocyanine and oxonol dyes). The principal advantage of fast voltage-sensitive probes over traditional electrophysiological recording methods is their ability to characterize patterns of neuronal activity with high temporal and spatial resolution [80]. With the exception of signal kinetics, the disadvantages relative to electrophysiological recording are similar to those described above for calcium imaging. On reviewing the literature, voltage-sensitive probes were used primarily to investigate spatiotemporal patterns of odor-evoked neural activity in insect antennal lobes [81–84]. The slow voltage-sensitive probes offer utility for reporting on mitochondrial membrane potential, and in characterizing the action of *Bacillus thuringiensis* toxins [85–87]. Thus, voltage-sensitive probes have a limited utility for investigating insecticides that act on neuromuscular targets.

7.4 Genomic Approaches for Target Site Elucidation

7.4.1 Chemical-to-Gene Screening

With the availability of complete genomes from multiple insect species, and reduced costs associated with sequencing, genomic approaches represent an effective alternative when target site identification proves elusive using traditional techniques. Historically, genetic mapping has been used to identify amino acid mutations that confer insecticide resistance, as was the case for the cyclodienes and nodulisporic acid [88–91]. Chemical-to-gene screening (a forward genetic screen) involves the mutagenesis of model organisms, such as *Drosophila* and *Caenorhabditis elegans*, to induce resistance to the insecticidal chemical of interest. The resistant organisms are then genetically sequenced to identify the molecular target of the chemistry, as depicted in Figure 7.6 [92]. This screening approach has served as

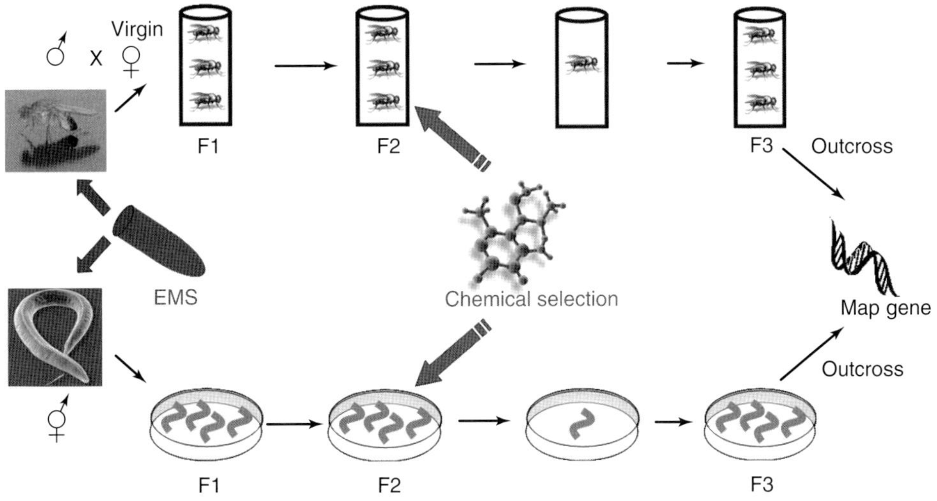

Figure 7.6 The chemical-to-gene screening process. Male *Drosophila melanogaster* or hermaphrodite *Caenorhabditis elegans* are mutagenized by exposure to ethyl methanesulfonate (EMS). *Drosophila* F2 offspring are obtained by crossing with marker virgin female flies, whereas *C. elegans* are self-fertilizing. The F2 offspring are challenged with the chemical of interest to select for resistance. These mutants are outcrossed to remove mutations unrelated to resistance and genetic mapping techniques are subsequently employed to identify the gene(s) of interest. The photographs of *Drosophila* and *C. elegans* were provided by André Karwath and A. J. Cann, respectively.

a successful platform for companies such as Cambria Pharmaceuticals (formerly Cambria Biosciences; Woburn, MA) and DevGen (Gent, Belgium), and has led to the identification of target proteins for several insecticides and anthelmintics. Based on their short life-cycle, ease of culture, and amenability to mutations, *Drosophila* and *C. elegans* are ideal for chemical-to-gene screening.

As discussed previously, spinosad interacts with *n*AChRs at a novel binding site. Most recently, a forward genetic screen using *Drosophila* identified a key *n*AChR subunit that confers sensitivity to spinosyns [93, 94]. In this case, *Drosophila* were mutagenized with ethyl methane sulfonate (EMS) and selected under pressure with spinosyn A to establish lines of resistant flies. By using complementation studies, a span of 29 genes associated with resistance were defined, including *n*AcRa-30D, the gene encoding the Dα6 subunit of the *n*AChR. Co-expression of the *Drosophila* Dα6 and Dα5 with the *C. elegans* chaperone protein ric-3, (resistance to cholinesterase) produced functional *n*AChRs which exhibited sensitivity to spinosyns, but not to imidacloprid (Figure 7.7).

Chemical-to-gene screening has also been instrumental in elucidating the MoA of spiroindolines, a class of lepidopteran insecticides recently discovered by Syngenta Crop Protection [95, 96]. An optimization program led to the identification of highly potent lepidopteran insecticides exemplified by SYN876. In 2010, Earley and colleagues reported that *C. elegans* treated with spiroindolines exhibited a "coiling" phenotype (see Figure 7.8b) consistent with a neurotoxic action [97]. Conventional biochemical and physiological studies ruled out targets of commercialized insecticides, thus suggesting a novel mechanism. Subsequent

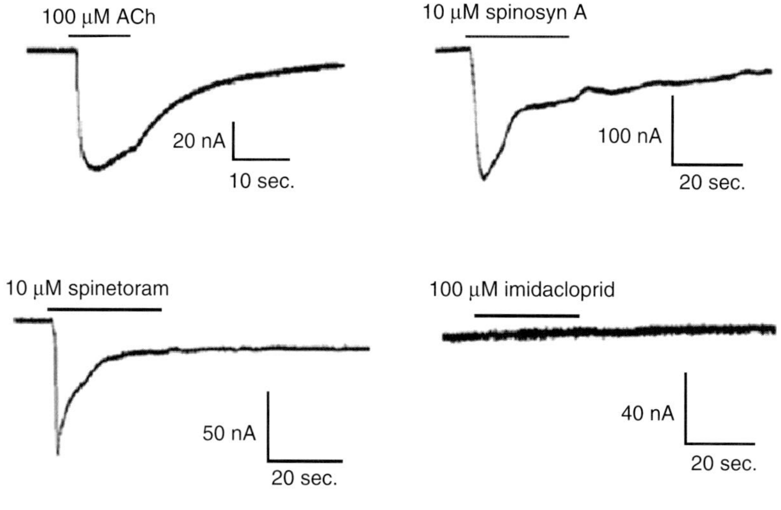

Figure 7.7 Agonist activation of Dα6/Dα5/ric-3. In addition to nicotine, spinosyn A, acetylcholine and spinetoram were also capable of eliciting currents when applied to oocytes expressing Dα6/Da5/ric-3. Imidacloprid (100 μM) did not elicit consistent currents when applied to oocytes expressing Dα6/Dα5/ric-3. Reprinted with permission from Ref. [93]; © 2010, Elsevier.

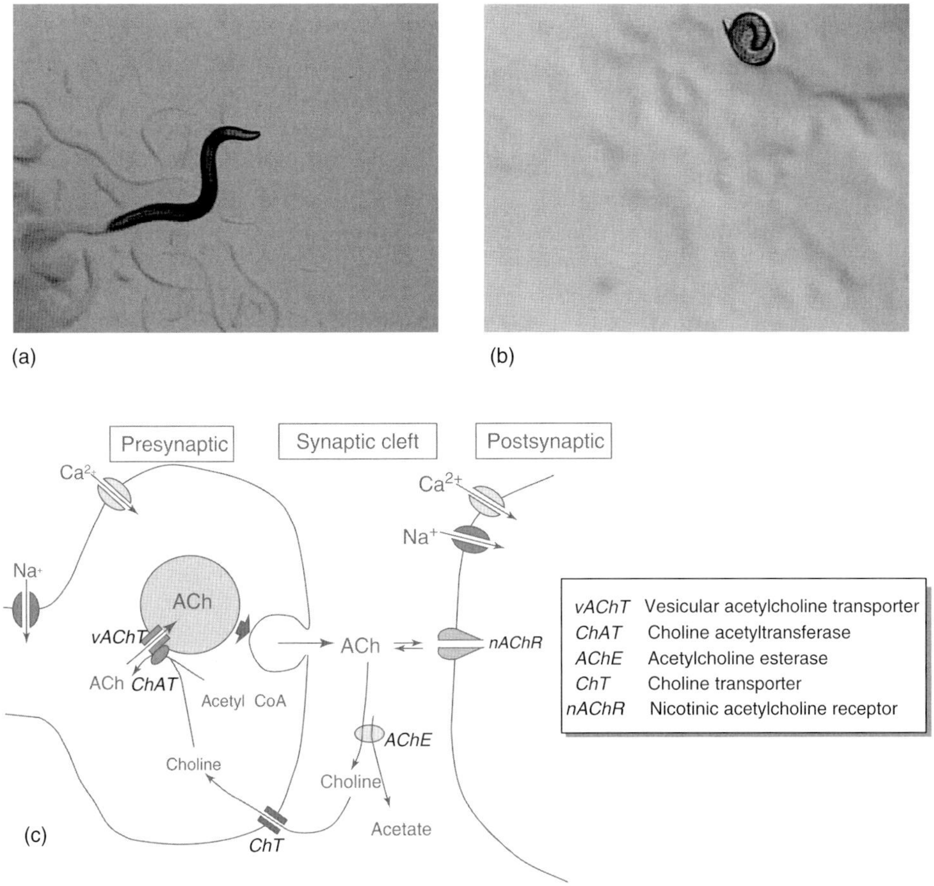

Figure 7.8 The spiroindoline, SYN876, inhibits the vesicular acetylcholine transporter (vAChT) resulting in a "coiling phenotype" in the nematode, *C. elegans*. Photomicrographs of (a) an untreated and (b) SYN-876-treated *C. elegans*; (c) Diagram of acetylcholine synthesis, transport, and release from the presynaptic neuron. Inhibition of the vAChT results in a depletion of vesicular acetylcholine and subsequent block in cholinergic transmission. This represents a novel mode of action, as commercial insecticides interfere with cholinergic transmission by inhibiting acetylcholinesterase (organophosphates and carbamates) or binding to postsynaptic nicotinic acetylcholine receptors (neonicotinoids, nereistoxin analogs, and spinosyns). Figure kindly provided by Fergus Earley, Syngenta.

mutagenesis studies pursued in collaboration with Cambria Biosciences produced spiroindoline-resistant *C. elegans* with dominant mutations mapping to chromosome IV and ultimately linked to a gene encoding vesicular acetylcholine transport (VAChT; Figure 7.8c). Follow-up studies involving the functional expression of *Drosophila* VAChT confirmed the inhibition of VAChT by spiroindolines.

The combination of genomics and target-based resistance has also been successfully applied to investigations of acaricidal MoA. The two-spotted spider mite,

Tetranychus urticae, is a haploid-diploid organism in which females and males arise from fertilized and unfertilized eggs, respectively [98, 99]. The hemizygous nature of male mites allows for the evaluation of the maternal or paternal basis of inherited resistance. Bifenazate, a miticide discovered by Uniroyal Chemical during the 1990s, is categorized as having an unknown MoA by the Insecticide Resistance Action Committee (IRAC). Although Uniroyal Chemical had proposed the putative target of bifenazate to be γ-aminobutyric acid (GABA) receptor as the putative target of bifenazate [100], resistance-based studies suggested that the toxicologically relevant target was mitochondrial rather than neuronal. Using a laboratory-selected bifenazate-resistant strain of *T. urticae* Koch, Van Leeuwen *et al.* [101] demonstrated a complete maternal inheritance associated with bifenazate resistance, which indicated that it was encoded by the mitochondrial genome. A reduction in ATP content observed with bifenazate-treated *T. urticae* further supported a mitochondrial target. Subsequent genomic studies with *T. urticae* and the citrus red mite, *Panonychus citri*, revealed that strong bifenazate resistance was associated with mutations in the cytochrome b Q_o-pocket [102–104].

7.4.2
Double-Stranded RNA Interference

In addition to forward-genetics, reverse-genetic approaches such as RNA interference (RNAi) offer useful tools for MoA elucidation, as well as discovery of novel targets. RNAi is the process by which double-stranded RNA (dsRNA) is introduced into the organism by feeding, soaking, or injection. The dsRNA is cleaved into short nucleotide fragments that are then incorporated into the RNA-inducing silencing complex (RISC); this eventually results in a silencing of homologous messenger RNA. As with chemical-to-gene screening, *Drosophila* and *C. elegans* are both highly amenable to this approach [105, 106]. In the case of *C. elegans*, dsRNA is introduced into the nematodes by allowing them to feed on *Escherichia coli* expressing the RNA of interest. For *Drosophila*, such studies can be conducted at the cellular level by soaking the cells with dsRNA, or at the organism level via direct injection into the fly larvae. In addition to model organisms, successful RNAi-induced gene silencing studies have been reported with insects from other orders, via feeding or injection [107–109].

Under situations where a putative target site has been identified, resistance to the chemistry following gene silencing can provide further genetic validation. Such studies can be conducted at both the cellular and organism level. For example, Boina and Bloomquist [110] employed RNAi studies in *C. elegans* to verify that the nematicidal activity of anion transport blockers such as 4,4″-diisothiocyantostilbene-2,2'-disulfonic acid (DIDS) and (5-nitro-2-(3-phenylpropylamino) benzoic acid (NPPB) resulted from an inhibition of the voltage-gated chloride channel encoded by ceclc-2 [110].

In addition to target validation, RNAi studies can also prove useful for identifying novel targets for exploitation. For this, a whole-genome-based screening approach can be used to identify novel insecticide targets, independent of prior knowledge regarding the protein's function. A genome-wide screen in *C. elegans* was described in detail by Kamath and Ahringer [111]. These authors constructed a bacterial library allowing RNAi of approximately 86% of *C. elegans* genes, and have made this library publicly available through MRC Gene Services (*http://www.hgmp.mrc.ac.uk*). Here, the nematodes were fed on the bacterial library to induce gene silencing, and progeny scored for loss-of-function phenotypes (i.e., lethal, uncoordinated, sluggish, paralyzed, etc.). Follow-up genetic studies are then conducted to verify that the phenotype is associated with mutation of the target gene. An example of the focused RNAi approach is seen with the levamisole-sensitive *n*AChR in *C. elegans*, where Gottschalk *et al.* [112] identified a subunit of calcineurin A, TAX-6, which was found to negatively regulate *n*AChR activity. In another study, ACR-16 was found to be essential to the levamisole-resistant *n*AChR [113]. Non-neuronal targets can also be readily validated as having insecticidal properties, as has been shown through the silencing of TcCHS2, a gene for chitin synthase in *Tribolium castaneum* [109].

7.4.3 Metabolomics

Metabolomics, the comprehensive study of metabolite profiles generated in response to a stimulus, has been used for herbicidal mode of action studies [114, 115]. Nonetheless, it has largely been unexploited for identifying insecticide targets [116]. One of the main challenges with fast-acting insecticides is to distinguish metabolic changes associated with direct action on a target from secondary effects. Recent studies have explored the use of mass spectrometry to generate distinct metabolome profiles in the water flea, *Daphnia magna*, and the earthworm, *Eisenia fetida*, following exposure to insecticides that affect the nervous system, such as fenvalerate, DDT, and endosulfan [117, 118]. The *Daphnia* metabolite profiles were distinct following exposure to fenvalerate versus other toxicants such as cadmium, and the oxidative phosphorylation uncoupler, dinitrophenol. Despite the use of metabolomics for detecting environmental toxicants, significant advancements are required before this approach proves useful for elucidating the insecticidal MoA.

7.5 Conclusion

Elucidating the target site of insecticides early stage in discovery and development can have an integral role for the prioritization of candidate leads. In order to be most effective, the research scientist relies upon a combination of investigative tools (some of which were detailed in this chapter), a historical knowledge of structurally related chemistry, and deductive reasoning. It is hoped that the topics described in this chapter, in conjunction with methods detailed in the numerous studies cited, will aid those exploring the interaction between candidate insecticides and targets associated with neuromuscular function. It should be noted that one topic not included in this chapter has been analytical methods which, while highly valuable – particularly for studying prospective pro-insecticides – fall beyond the scope of the chapter.

References

1. Nauen, R. and Bretschneider, T. (2002) *Pestic. Outlook*, **6**, 241–245.
2. Cordova, D., Rauh, J.J., Benner, E.A., Schroeder, M.E., Sopa, J.S., Lahm, G.P., Pahutski, T.F., Long, J.K., Holyoke, C.W., Smith, R.M., Barry, J.D., and Dung, M.H. (2010) Isoxazoline insecticides: a novel class of GABA-gated chloride channel blockers. Abstracts of Papers 12th IUPAC International Congress of Pesticide Chemistry, 4–10 July 2010, Melbourne, Australia.
3. Cordova, D., Benner, E.A., Sacher, M.D., Rauh, J.J., Sopa, J.S., Lahm, G.P., Selby, T.P., Stevenson, T.M., Flexner, L., Gutteridge, S., Rhoades, D.F., Wu, L., Smith, R.M., and Tao, Y. (2006) *Pestic. Biochem. Physiol.*, **84**, 196–214.
4. Orr, N., Shaffner, A.J., and Watson, G.B. (1997) *Pestic. Biochem. Physiol.*, **58**, 183–192.
5. Honda, H., Tomizawa, M., and Casida, J.E. (2007) *J. Agric. Food Chem.*, **55**, 2276–2281.
6. Honda, H., Tomizawa, M., and Casida, J.E. (2006) *J. Agric. Food Chem.*, **54**, 3365–3371.
7. Tomizawa, M., Millar, N.S., and Casida, J.E. (2005) *Insect Biochem. Mol. Biol.*, **35**, 1347–1355.
8. Cole, L.M. and Casida, J.E. (1992) *Pestic. Biochem. Physiol.*, **44**, 1–8.
9. Ozoe, Y., Asahi, M., Ozoe, F., Nakahira, K., and Mita, T. (2010) *Biochem. Biophys. Res. Commun.*, **391**, 744–749.
10. Nauen, R., Ebbinghaus-Kintscher, U., Elbert, A., Jeschke, P., and Tietjen, K. (2001) in *Biochemical Sites of Insecticide Action and Resistance* (ed. I. Ishaaya), Springer-Verlag, Heidelberg, pp. 77–105.
11. Liu, M.Y. and Casida, J.E. (1993) *Pestic. Biochem. Physiol.*, **46**, 40–46.
12. Watson, G.B., Loso, M.R., Babcock, J.M., Hasler, J.M., Letherer, T.J., Young, C.D., Zhu, Y., Casida, J.E., and Sparks, T.C. (2011) *Insect Biochem. Mol. Biol.*, **41**, 432–439.
13. Xu, X., Bao, H., Shao, X., Zhang, Y., Yao, X., Liu, Z., and Li, Z. (2010) *Insect Mol. Biol.*, **19**, 1–8.
14. Kayser, H., Lee, C., Decock, A., Bauer, M., Haettenschwiler, J., and Maienfisch, P. (2004) *Pest Manag. Sci.*, **60**, 945–958.
15. Lind, R.J., Clough, M.S., Reynolds, S.E., and Earley, F.G.P. (1998) *Pestic. Biochem. Physiol.*, **62**, 3–14.
16. Babcock, J.M., Gerwick, C.B., Huang, J.X., Loso, M.R., Nakamura, G., Nolting, S.P., Rogers, R.B., Sparks, T.C., Thomas, J., Watson, G.B., and Zhu, Y. (2011) *Pest Manag. Sci.*, **67**, 328–334.
17. Zhu, Y., Loso, M.R., Watson, G.B., Sparks, T.C., Rogers, R.B., Huang, J.X., Gerwick, B.C., Babcock, J.M., Kelley, D., Hegde, V.B., Nugent, B.M., Renga, J.M., Denholm, I., Gorman, K., DeBoer, G.J., Hasler, J., Meade, T., and Thomas, J.D. (2011) *J. Agr. Food Chem.*, **59**, 2950–2957.
18. Gutteridge, S., Caspar, T., Cordova, D., Tao, Y., Wu, L., and Smith, R.M. (2003) Nucleic acids encoding ryanodine receptors. US Patent US Pat. 7,205,147. filed Sep. 23, 2003 and issued Apr. 17, 2007.
19. Orr, N., Shaffner, A.J., Richey, K., and Crouse, G.D. (2009) *Pestic. Biochem. Physiol.*, **95**, 1–5.
20. Orr, N., Watson, G.B., Gustafson, G.D., Hasler, J.M., Chaoxian, G., Chouinard, S., Cook, K.R., and Salgado, V.L. (2010) Novel assays utilizing nicotinic acetylcholine receptor subunits. US Patent Appl. U.S. Pat. 20100212029, filed Dec. 22, 2009.

21. Homola, J. (2003) *Anal. Bioanal. Chem.*, **377**, 528–539.
22. Paulo, J.A. and Hawrot, E. (2009) *Anal. Biochem.*, **389**, 86–88.
23. Maynard, J.A., Lindquist, N.C., Sutherland, J.N., Lesuffleur, A., Warrington, A.E., Rodriquez, M., and Oh, S.H. (2009) *Biotechnol. J.*, **4**, 1542–1558.
24. Neher, E. and Sakmann, B. (1976) *Nature*, **260**, 799–802.
25. Chanda, V.B. and Crooks, G.P. (eds) (1997) *Current Protocols in Neuroscience*, John Wiley & Sons, Inc., Hoboken, NJ.
26. Chong, Y., Hayes, J.L., Sollod, B., Wen, S., Wilson, D.T., Hains, P.G., Hodgson, W.C., Broady, K.W., King, G.F., and Nicholson, G.M. (2007) *Biochem. Pharmacol.*, **74**, 623–638.
27. Heidel, E. and Pfluger, H.J. (2006) *Eur. J. Neurosci.*, **23**, 1189–1206.
28. Defaix, A. and Lapied, B. (2005) *Invert. Neurosci.*, **5**, 135–146.
29. Zhao, X., Salgado, V.L., Yeh, J.Z., and Narahashi, T. (2004) *NeuroToxicology*, **25**, 967–980.
30. Salgado, V.L. and Saar, R. (2004) *J. Insect Physiol.*, **50**, 867–879.
31. Brone, B., Tytgat, J., Wang, D.C., and Van Kerkhove, E. (2003) *J. Insect Physiol.*, **49**, 171–182.
32. Jannsen, D., Derst, C., Buckinx, R., Van den Eynden, J., Rigo, J.M., and Van Kerkhove, E. (2007) *J. Neurophysiol.*, **97**, 2642–2650.
33. Tan, J., Galligan, J.J., and Hollingworth, R.M. (2007) *NeuroToxicology*, **28**, 829–842.
34. Zhao, X., Ikeda, T., Salgado, V.L., Yeh, J.Z., and Narahashi, T. (2005) *NeuroToxicology*, **26**, 455–465.
35. Zhao, X., Yeh, J.Z., Salgado, V.L., and Narahashi, T. (2005b) *J. Pharm. Exp. Ther.*, **314**, 363–373.
36. Lapied, B., Grolleau, F., and Sattelle, D.B. (2001) *Br. J. Pharmacol.*, **132**, 587–595.
37. Singh, S. and Wu, C.F. (1999) *Int. Rev. Neurobiol.*, **43**, 191–220.
38. Singh, A. and Singh, S. (1999) *J. Neurosci.*, **19**, 6838–6843.
39. Wicher, D., Walther, C., and Wicher, C. (2001) *Prog. Neurobiol.*, **64**, 431–525.
40. Collet, C. and Belzunces, L. (2007) *J. Exp. Biol.*, **210**, 454–464.
41. Collet, C. (2009) *Pflugers Arch. - Eur. J. Physiol.*, **458**, 601–612.
42. The Honeybee Sequencing Consortium (2006) *Nature*, **443**, 931–949.
43. Nishiwaki, H., Nakagwa, Y., Kuwamura, M., Sato, K., Akamatsu, M., Matsuda, K., Komai, K., and Miyagawa, H. (2003) *Pest Manag. Sci.*, **59**, 1023–1030.
44. Brown, L.A., Ihara, M., Buckingham, S.D., Matsuda, K., and Sattelle, D.B. (2006) *J. Neurochem.*, **99**, 608–615.
45. Jepson, J.E.C., Brown, L.A., and Sattelle, D.B. (2006) *Invert. Neruosci.*, **6**, 33–40.
46. Ihara, M., Matsuda, K., Otake, M., Kuwamura, M., Shimomura, M., Komai, K., Akamatsu, M., Raymond, V., and Sattelle, D.B. (2003) *Neuropharmacology*, **45**, 133–144.
47. Ihara, M., Matsuda, K., Shimomura, M., Sattelle, D.B., and Komai, K. (2004) *Biosci. Biotechnol. Biochem.*, **68**, 761–763.
48. Liu, Z., Han, Z., Zhang, Y., Song, F., Ya, X., Liu, S., Gu, J., and Millar, N.S. (2009) *J. Neurochem.*, **108**, 498–506.
49. Liu, Z., Cao, G., Li, J., Bao, H., and Zhang, Y. (2009) *J. Neurochem.*, **110**, 1707–1714.
50. Schnizler, K., Kuster, M., Methfessel, C., and Fejtl, M. (2003) *Recept. Channels*, **9**, 41–48.
51. Goldin, A.L. (2006) in *Expression and Analysis of Recombinant Ion Channels: From Structural Studies to Pharmacological Screening* (eds J.J. Clare and D.J. Trezise), Wiley-VCH Verlag GmbH & Co. KGaA, Weinheim, Germany, pp. 1–25.
52. Leisgen, C., Kuester, M., and Methfessel, C. (2007) in *Patch Clamp Methods and Protocols* (eds P. Molnar and K. Hickman), Humana Press, Totawa, New Jersey, pp. 87–109.
53. Papke, R.L. and Stokes, C. (2010) *Methods*, **51**, 121–133.

54. Berridge, M.J., Lipp, P., and Bootman, M.D. (2000) *Nat. Rev. Mol. Cell Biol.*, **1**, 11–21.
55. O'Connor, N. and Silver, R.B. (2007) *Methods Cell Biol.*, **81**, 415–433.
56. Takahashi, A., Camacho, P., Lechleiter, J.D., and Herman, B. (1999) *Physiol. Rev.*, **79**, 1089–1125.
57. Grynkiewicz, G., Poenie, M., and Tsien, R.Y. (1985) *J. Biol. Chem.*, **260**, 3440–3448.
58. Raymond Delpech, V. and Sattelle, D.B. (2004) *Cell Calcium*, **35**, 131–139.
59. Cordova, D., Raymond Delpech, V., Sattelle, D.B., and Rauh, J.J. (2003) *Invert. Neurosci.*, **5**, 19–28.
60. Millar, N.S., Baylis, H.A., Reaper, C., Bunting, R., Mason, W.T., and Sattelle, D.B. (1995) *J. Exp. Biol.*, **198**, 1843–1850.
61. Hoff, M., Balfanz, S., Ehling, P., Gensch, T., and Baumann, A. (2011) *FASEB J.*, **25**, 2484–2491.
62. Huang, J., Hamasaki, T., and Ozoe, Y. (2010) *Arch. Insect Biochem. Physiol.*, **73**, 74–86.
63. Bischof, L.J. and Enan, E.E. (2004) *Insect Biochem. Mol. Biol.*, **34**, 511–521.
64. Vomel, M. and Wegener, C. (2007) *Dev. Neurobiol.*, **67**, 792–808.
65. Ueda, A. and Wu, C.F. (2006) *J. Neurosci.*, **26**, 6238–6248.
66. Berke, B.A., Lee, J., Peng, I.F., and Wu, C.F. (2006) *Neuroscience*, **142**, 629–644.
67. McFerren, M.A., Cordova, D., Rodriquez, E., and Rauh, J.J. (2002) *J. Ethnopharm.*, **83**, 201–207.
68. Lahm, G.P., Stevenson, T.M., Selby, T.P., Freudenberger, J.H., Cordova, D., Flexner, L., Bellin, C.A., Dubas, C.M., Smith, B.K., Hughes, K.A., Hollingshaus, J.G., Clark, C.E., and Benner, E.A. (2007) *Bioorg. Med. Chem. Lett.*, **17**, 6274–6279.
69. Lahm, G.P., Selby, T.P., Freudenberger, J.H., Stevenson, T.M., Myers, B.J., Seburyamo, G., Smith, B.K., Flexner, L., Clark, C.E., and Cordova, D. (2005) *Bioorg. Med. Chem. Lett.*, **15**, 4898–4906.
70. Ebbinghaus-Kintscher, U., Luemmen, P., Lobitz, N., Schulte, T., Funke, C., Fischer, R., Masaki, T., Yasokawa, N., and Tohnishi, M. (2006) *Cell Calcium*, **39**, 21–33.
71. Ottea, J.A., Payne, G.T., and Soderlund, D.M. (1992) in *Molecular Mechanisms of Insecticide Resistance* (ed. C. Mullins), American Chemical Society, Washington, DC, pp. 276–287.
72. Kubo, I., Klocke, J.A., and Matsumoto, T. (1984) in *Pesticide Synthesis Through Rational Approaches* (ed. P. Magee), American Chemical Society, Washington, DC, pp. 163–172.
73. Campusano, J.M., Su, H., Jiang, S.A., Sicaeros, B., and O'Dowd, D.K. (2007) *Dev. Neurobiol.*, **67**, 1520–1532.
74. Wegener, C., Hamasaka, Y., and Nassel, D.R. (2003) *J. Neurophysiol.*, **91**, 912–923.
75. Courjaret, R., Grolleau, F., and Lapied, B. (2003) *Eur. J. Neurosci.*, **17**, 2023–2034.
76. Vermehnren, A. and Trimmer, B.A. (2005) *J. Neurobiol.*, **62**, 289–298.
77. Oertner, T.G., Single, S., and Borst, A. (1999) *Neurosci. Lett.*, **274**, 95–98.
78. Oliveira, E.E., Pippow, A., Salgado, V.L., Buschges, A., Schmidt J., and Kloppenburg, P. (2010) *J. Neurophysiol.*, **103**, 2770–2782.
79. Buckingham, S.D., Lapied, B., Le Corronc, H., Grolleau, F., and Sattelle, D.B. (1997) *J. Exp. Biol.*, **200**, 2685–2692.
80. Chemla, S. and Chavane, F. (2010) *J. Physiol.*, **104**, 40–50.
81. Hiroyuki, A. and Kanzaki, R. (2004) *J. Exp. Biol.*, **207**, 633–644.
82. Hill, E.S., Okada, K., and Kanzaki, R. (2003) *J. Exp. Biol.*, **206**, 345–352.
83. Ai, H., Okada, K., Hill, E.S., and Kanzaki, R. (1998) *Neurosci. Lett.*, **258**, 135–138.
84. Okada, K., Kanzaki, R., and Kawachi, K. (1996) *Neurosci. Lett.*, **209**, 197–200.
85. Rodrigo-Simon, A., Caccia, S., and Ferre, J. (2008) *Appl. Environ. Microbiol.*, **74**, 1710–1716.
86. Leonardi, M.G., Caccia, S., Gonzalez-Cabrera, J., Ferre, J., and Giordana, B. (2007) *J. Membr. Biol.*, **214**, 157–164.
87. Gonzalez-Cabrera, J., Farinos, G.P., Caccia, S., Diaz-Mendoza, M., Castanera, P., Leonardi, M.G.,

Giordana, B., and Ferre, J. (2006) *Appl. Environ. Microbiol.*, **72**, 2594–2600.

88. Ffrench-Constant, R.H., Daborn, P.J., and Le Goff, G. (2004) *Trends Genet.*, **20**, 163–170.

89. Ffrench-Constant, R.H., Rocheleau, T.A., Steichen, J.C., and Chalmers, A.E. (1993) *Nature*, **363**, 449–461.

90. Ffrench-Constant, R.H. and Roush, R.T. (1991) *Genet. Res.*, **57**, 17–21.

91. Kane, N.S., Hirschberg, B., Qian, S., Hunt, D., Thomas, B., Brochu, R., Ludmerer, S.W., Zheng, Y., Smith, M., Arena, J.P., Cohen, C.J., Schmatz, D., Warmke, J., and Cully, D.F. (2000) *Proc. Natl. Acad. Sci. USA*, **97**, 13949–13954.

92. Jones, K., Buckingham, S.D., and Sattelle, D.B. (2005) *Nat. Rev.*, **4**, 321–330.

93. Watson, G.B., Chouinard, S.W., Cook, K.R., Geng, C., Gifford, J.M., Gustafson, G.D., Hasler, J.M., Larrinua, I.M., Letherer, T.J., Mitchell, J.C., Pak, W.L., Salgado, V.L., Sparks, T.C., and Stilwell, G.E. (2010) *Insect Biochem. Mol. Biol.*, **40**, 376–384.

94. Orr, N., Watson, G.B., Hasler, J., Michael, J., Geng, C., Cook, K.R., Salgado, V.L., and Chouinard, S. (2006) Sequences of *Drosophila melanogaster* nicotinic receptor alpha-6 and alpha-7 subunits for bioassay. WO Patent Appl. 2006091672, filed Aug. 31, 2006.

95. Cassayre, J., Hughes, D.J., Roberts, R.S., Worthington, P.A., Cederbaum, F., Maienfisch, P., and Molleyres, L.P. (2010) Spiroindolines: discovery of a novel class of insecticides. Abstracts of Papers 239th National Meeting of the American Chemical Society, 21–25 March 2010, San Francisco.

96. Cassayre, J., Maienfisch, P., Roberts, R.S., Worthington, P.A., Hughes, D.J., Molleyres, L.P., Cederbaum, F., Hillesheim, E., Sluder, A., Earley, F., and Shah, S. (2010) The discovery of spiroindolines – a new class of insecticides with a novel mode of action. Abstracts of Papers 12th IUPAC International Congress of Pesticide Chemistry, 4–10 July 2010, Melbourne, Australia.

97. Sluder, A., Clover, R., Shah, S., She, M., Hirst, L., Cutler, P., Flury, T., Stanger, C., Flemming, A., and Earley, F. (2010) Spiroindolines reveal a novel target protein for insecticide action. Abstracts of Papers 239th National Meeting of the American Chemical Society, 21–25 March 2010, San Francisco.

98. Bull, J.J. (1979) *Heredity*, **43**, 361–381.

99. Oliver, J.H. Jr (1971) *Am. Zool.*, **11**, 283–299.

100. Dekeyser, M.A. (2005) *Pest Manag. Sci.*, **61**, 103–110.

101. Van Leeuwen, T., Tirry, L., and Nauen, R. (2006) *Insect Biochem. Mol. Biol.*, **36**, 869–877.

102. Van Leeuwen, T., Van Nieuwenhuyse, P., Vanholme, B., Dermauw, W., Nauen, R., and Tirry, L. (2011) *Insect Mol. Biol.*, **20**, 135–140.

103. Van Nieuwenhuyse, P., Van Leeuwen, T., Khajehali, J., Vanholme, B., and Tirry, L. (2009) *Pest Manag. Sci.*, **65**, 404–412.

104. Van Leeuwen, T., Vanholme, B., Van Pottelberge, S., Van Nieuwenhuyse, P., Nauen, R., Tirry, L., and Denholm, I. (2008) *Proc. Natl. Acad. Sci.*, **105**, 5980–5985.

105. Busch, M., Villalba, F., Schulte, T., and Menke, U. (2005) *Pflanzenschutz. - Nachr. Bayer*, **58**, 34–50.

106. Kamath, R.S., Fraser, A.G., Dong, Y., Poulin, G., Durbin, R., Gotta, M., Kanapin, A., Le Bot, N., Moreno, S., Sohrmann, M., Welchman, D.P., Zipperlen, P., and Ahringer, J. (2003) *Nature*, **421**, 231–237.

107. Bai, H., Zhu, F., Shah, K., and Palli, S.R. (2011) *BMC Genom.*, **12**, 388–398.

108. Pitino, M., Coleman, A.D., Maffei, M.E., Ridout, C.J., and Hogenhout, S.A. (2011) *PLoS ONE*, **6**(10), e25709.

109. Alves, A.P., Lorenzen, M.D., Beeman, R.W., Foster, J.E., and Siegfried, B.D. (2010) *J. Insect Sci.*, **10**, 162–177.

110. Boina, D.J. and Bloomquist, J.R. (2010) *Pestic. Biochem. Physiol.*, **97**, 161–166.

111. Kamath, R.S. and Ahringer, J. (2003) *Methods*, **30**, 313–321.

112. Gottschalk, A., Almedom, R.B., Schedletzky, T., Anderson, S.D., Yates,

J.R. III, and Schafer, W.R. (2005) *EMBO J.*, **24**, 2566–2578.

113. Fracis, M.M., Evans, S.P., Jensen, M., Madsen, D.M., Mancuso, J., Norman, K.R., and Maricq, A.V. (2005) *Neuron*, **46**, 581–594.

114. Aranibar, N., Singh, B.J., Stockton, G.W., and Ott, K.H. (2001) *Biochem. Biophys. Res. Commun.*, **286**, 150–155.

115. Alliferis, K.A. and Chrysayi-Tokousbalides, M. (2006) *J. Agric. Food Chem.*, **54**, 1687–1692.

116. Aliferis, K.A. and Jabaji, S. (2011) *Pestic. Biochem. Physiol.*, **100**, 105–117.

117. Taylor, N.S., Weber, R.J.M., White, T.A., and Viant, M.R. (2010) *Toxicol. Sci.*, **118**, 307–317.

118. McKelvie, J.R., Yuk, J., Xu, Y., Simpson, A.J., and Simpson, M.J. (2009) *Metabolomics*, **5**, 84–94.

8
New Targets for Fungicides

Klaus Tietjen and Peter H. Schreier

8.1
Introduction: Current Fungicide Targets

The general term *fungicide* will be used throughout this chapter to describe active ingredients against plant diseases caused by fungi or *oomycetes* (though the latter are descendants of *brown algae*, and not of fungi) [1]. Bactericides are not accounted for here. Furthermore, an active ingredient is considered as a fungicide only when it can be used in praxis at a reasonable dose, for example, below 1 kg ha^{-1}, to effectively treat a plant disease. With this understanding, a plethora of compounds which are described as fungicides but are active only on target level or on artificial media, can be taken aside. Such a focus avoids blurring the industrial problems of identifying new fungicide targets, which are the primary molecular interaction partners in a cell. With our underlying understanding of a fungicide, the number of actual targets is unquestionably much lower than the number of conceivable potential targets.

The fungicide market has, for almost 15 years, comprised only two specific targets that well exceed a 5% market share, namely sterol C-14 demethylase and cytochrome c reductase (Figure 8.1), beyond long-established multisite fungicides which lack a specific molecular target. All other targets linger at a lower importance and radical innovation rarely happened. So, what are the reasons for such slow modernization?

It will be shown in this chapter that there is no lack of novel targets *per se*. However, the economical needs for compounds which are active against a broader evolutionary spectrum of diverse fungal and oomycetes species greatly narrows the success rate in identifying new suitable fungicides with a new mode of action (MoA). In comparison to herbicides or insecticides, economically viable market sizes for fungicides are achievable only when a relatively wide evolutionary species range is captured (see Figure 1.2 in Ref. [1]). A wide species range implies a high degree of molecular and physiological target diversity, especially with regards to the differences between fungi and oomycetes accounting for restricted applicability of many targets. However, even within fungi or oomycetes many active ingredients are also of limited value against different species.

Modern Methods in Crop Protection Research, First Edition.
Edited by Peter Jeschke, Wolfgang Krämer, Ulrich Schirmer, and Matthias Witschel.

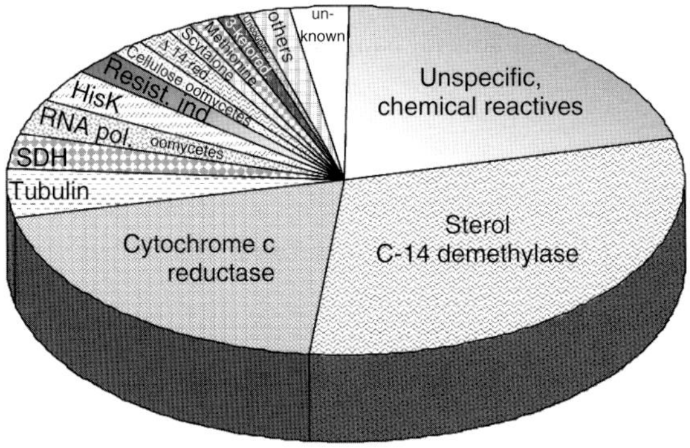

Figure 8.1 Targets of current fungicides (world market 2010). Targets down to 0.5% of the market share are shown. Abbreviations: SDH = succinate dehydrogenase; RNA pol· = RNA polymerase I; HisK = histidine kinase/MAP kinase (osmosensing); resist. ind· = inducer of plant resistance; cellulose = cellulose biosynthesis; Δ14red· = sterol Δ14 reductase; scytalone = scytalone reductase and scytalone dehydratase; methionine = methionine biosynthesis; 3-ketored· = sterol 3-ketoreductase; "others" include actin, spectrin (oomycetes), ATP synthase, trehalase, chitin synthase, adenosine deaminase, squalene epoxidase, and biologicals; "unknown" includes more than 20 different active ingredients. Source: Bayer CropScience Business Intelligence.

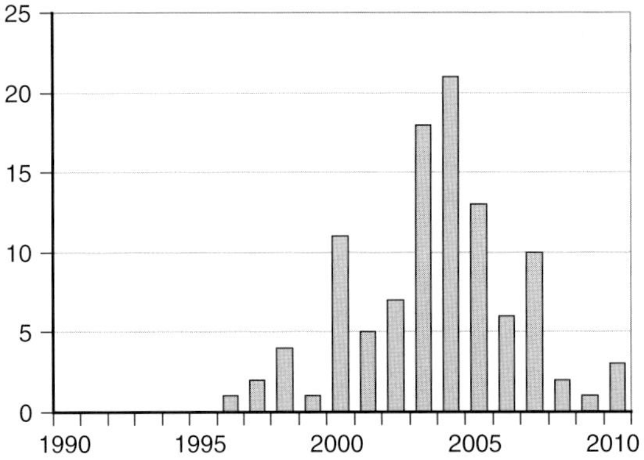

Figure 8.2 Numbers of patents claiming targets for fungicides. The search terms applied were "fungicide target" AND "screening." Patents not claiming specific targets were omitted. Although these search terms are neither necessary nor sufficient for an exhaustive result, the time curve provides evidence for the peak characteristic. For target details, see Table 8.1. Data are from a search in *Chemical Abstracts*, using the program SciFinder as a tool.

So, from where has innovation been derived during the past years, and how have modern molecular biology techniques contributed to such progress? Moreover, what might be expected for the future? Hopefully, some answers to these points will be provided in this chapter.

8.2
A Retrospective Look at the Discovery of Targets for Fungicides

Chemically reactive (electrophile) multi-site inhibitors, which represent the oldest class of fungicides, were first identified simply through empirical testing of simple low-cost chemicals [2]. Nonetheless, multi-site fungicides are expected to be useful agents for many years to come, as their unspecific MoA means that the development of resistance via simple mutations is impossible.

Although sterol C-14 demethylase inhibitors were first discovered during the 1960s, their MoA was not recognized during the first chemical optimization cycles [3]. Such inhibitors – azoles – exhibit a high-level and broad biological activity, and sterol C-14 demethylase inhibitors have subsequently become the basic "load carrier" for treatment of many fungal diseases in plants. There is, however, one important limitation in that, because oomycetes are unable to synthesize sterols of their own, they are insensitive to these inhibitors.

Conversely, cytochrome c reductase inhibitors are result of rational optimization with regards to their mode of action. Fungicidal activity of the natural products myxothiazole, oudemansin and strobilurin A, and their modes of action, were first discovered during the late 1970s [4]. Their chemical optimization led to successful development of strobilurin-type fungicides during the mid-1990s. Despite encountering certain problems of resistance, cytochrome c reductase inhibitors are today the second-important "load carrier" for treatment not only of fungal diseases but also of diseases caused by oomycetes.

Except for strobilurins, all other commercially available fungicide targets were apparently identified only after chemical optimization of the respective fungicide classes.

New fungicides targeting succinate dehydrogenase have been launched much more recently, since 2010 [5]. These new products were developed intentionally to achieve a higher target activity as well as a higher biological efficacy than the older products dating back to the 1960s. It appears that the new succinate dehydrogenase inhibitors are today's only example where a later optimization of the compounds' properties has enabled a boost in significance of a fungicide target. In future, the new succinate dehydrogenase-inhibiting fungicides may achieve significant sales, as this promising fungicide class is not affected by current resistance problems as experienced by sterol C14 demethylase and cytochrome c reductase inhibitors.

8.3
New Sources for New Fungicide Targets in the Future?

"Chemistry first" is the sustained basis for discovery of novel fungicides. Driven by competition between agrochemical companies, patent-busting activities are the major driver in identification and development of new compounds with increasing biological performance. A systematic biochemical check can reveal, whether novel compounds hit already-known targets. Whilst in positive cases use of target-based biochemical assays for optimization is standard [1], much fungicide innovation still derives without any knowledge of the molecular target.

A promising biological activity of a new chemical class creates great interest in elucidating the molecular target. However, the effort needed to elucidate a novel target for a given compound still can hardly be underestimated. For example, only after many years of research was cellulose synthase recently discovered to be target of carboxylic acid amides [6] active against oomycetes. Also in future, additional targets for already existing products are unlikely to be discovered on a frequent basis.

Although the more recent strategy of novel "target first" appears to be promising, it still awaits a demonstration of value.

8.4 Methods to Identify a Novel Target for a Given Compound

When for a new fungicidally active compound a validation of its activity on known targets is shown to be negative, the search for a target can include a series of diverse technologies, which are briefly detailed in the following subsections.

8.4.1 Microscopy and Cellular Imaging

Since early years of fungicide research, modes of action of fungicides were investigated by employing microscopic methods. Whilst metabolic disturbances may not produce microscopically visible phenotypes, certain other mechanisms such as interference with cell wall biosynthesis can clearly be observed. Consequently, a series of studies pinpointed cellulose synthase as a possible target for the oomyceticide iprovalicarb, based on valineamide carbamate chemistry [7–9]. Subsequently, Delvos showed that fluorescently labeled antibodies could locate against the proposed target protein, so as to manifest cellulose synthase dislocation induced by the oomyceticide [9]. However, microscopy cannot generally be used to identify a molecular target unambiguously. For example, the target of the oomyceticide fluopicolide was narrowed down to affect spectrin localization, although direct interaction with a protein has not yet been identified [10].

For other modes of action intracellular fluorescent reporters, like, for example, green fluorescent protein fused to tubulin or actin can be useful [11, 12].

8.4.2 Cultivation on Selective Media

When an active compound inhibits an enzyme within a biochemical pathway, metabolic disturbance may not manifest as a meaningful microscopic phenotype. In fact, in such a case the target may be identified by omission and/or supplementation of the culture medium with different intermediate or metabolic end products.

By using this technique, cystathionine beta-lyase was discovered as target for pyrimidineamine fungicides [13].

8.4.3
Incorporation of Isotopically Labeled Precursors and Metabolomics

In order to identify those biochemical pathways which are affected by a compound, one possibility is to follow incorporation of isotopically labeled biochemical precursors. Acylalanines, an older but very important class of oomyceticides, have been shown to inhibit incorporation of [^{3}H]-uridine into RNA [14]. Subsequently, inhibition of RNA polymerase I has been tentatively assumed as MoA of these compounds, though no further confirmation has been reported. Whilst generally useful, however, isotopically labeled precursors have never been employed as a strong basis for MoA elucidation.

Today's major progress in analytical methods has enabled identification of metabolites without use of isotope labeling. Typically, a target may be identified by analytical identification of metabolites being blocked and intermediates created, although in practice this approach has not become important beyond application to inhibitors of sterol biosynthesis. In case of the latter compounds, gas chromatography/mass spectrometry (GC-MS) or liquid chromatography-dual mass spectrometry (LC-MS-MS) are standard techniques used to define the target. In other cases, physiological counterbalance of inhibition may represent a prevalent mechanism that limits the power of metabolomics. In addition, many targets – for example, tubulin – may not have any direct influence on metabolism.

8.4.4
Affinity Methods

Affinity methods to capture a target for an active compound with an unknown mode of action seem to be very promising. The use of radioactively labeled compounds has appeared straightforward in particular, although this has not yet led to identification of a fungicide target. Most other affinity methods require derivatives of active compounds, which retain a high level of activity when a chemical linker arm is joined to the molecule. Whilst, at present, the ability to overcome this obstacle is rare, the recent identification of aurora kinase as a novel fungicidal target has demonstrated the power of affinity methods, at least in principle [15].

8.4.5
Resistance Mutant Screening

The elucidation of unknown targets by screening for resistance mutants has, in some cases, revealed a novel and very interesting target. For example, in 2002 Zhang and coauthors identified the osmo-sensing mitogen-activated protein

(MAP) kinase pathway as target for phenylpyrrole fungicides [16], although the primary interacting protein still remains unknown. As noted above, Blum *et al.* [6] more recently identified cellulose synthase of *Phytophthora* as target for carboxylic acid amide oomyceticides via a resistance mutation. Such technology may also be sufficiently effective to elucidate a target before a compound reaches the market: for example, glucosaminyl-phosphatidylinositol acyltransferase which is active in biosynthesis of glycosylphosphatidylinositol (GPI)-anchored cell wall mannoproteins was identified in this way as a target for novel fungicides such as 1-(4-butylbenzyl)isoquinoline [17, 18].

8.4.6
Gene Expression Profiling and Proteomics

Treatment of cells with biologically active compounds is expected to result in diverse physiological reactions and adaptations, and also in modifications of gene expression. Although treatment of fungal cells with a fungicide gives rise to extensive gene expression changes [19], the cells do not simply react by undergoing a proximate change of gene expression for the target of a given compound. Consequently, pinpointing of a biochemical pathway (e.g., sterol biosynthesis) is sometimes possible with a degree of confidence, despite the identification of a single target being almost impossible [20, 21]. To date, all targets identified via gene expression analysis have already been recognized by other means beforehand, there being no single example of a *de novo* identification of a certain target via gene expression analysis alone. Nevertheless, gene expression profiling may be helpful for focusing further target search on certain pathways. Indeed, gene expression profiling might well be utilized to categorize different unknown fungicidal compounds as a form of "fingerprinting."

Although, arguably, gene expression levels might not necessarily predict protein levels, it has been shown in human cells that gene expression and protein levels globally are indeed correlated [22]. In a proteomic study with fungicides, Hoehamer *et al.* [23] identified proteome changes that were related to changes that had already been recognized from gene expression studies. These findings indicate that proteome studies would, likely, not provide any better conclusions than would gene expression studies.

8.5
Methods of Identifying Novel Targets without Pre-Existing Inhibitors

A challenging approach to development of new fungicide targets may begin with identification of a novel target and subsequent identification of specific inhibitors.

8.5.1 Biochemical Ideas to Generate Novel Fungicide Targets

For many years, biochemists have conducted extensive surveys of biochemistry textbooks and of scientific literature to discover novel fungicide targets. An example of this occurred when Pillonel [24] reported cyclin-dependent protein kinases (cdks) as novel fungicide targets. In all eukaryotes cdks serve as essential enzymes in cell division, and it seemed likely that cdk-inhibiting compounds – which had already been explored for cancer therapy – might also exhibit fungicidal activity. In this situation, the success of these compounds was strongly supported by existence of a broad variety of cdk inhibitors already known from pharmaceutical research. In absence of a plethora of known inhibitors, however, such a rational approach has proved very difficult, and a breakthrough for *de novo* design of biologically active fungicides has not yet been demonstrated 10 years after introduction of target-based high-throughput screening [1].

8.5.2 Genomics and Proteomics

The emergence of genome sequencing technology by the late 1990s raised great expectations for the identification of novel antifungal targets [25]. A run for patents began that claimed the use of specified gene products (proteins) to discover new fungicides using target-based high-throughput screening (Figure 8.2). Although the targets claimed in the patents read like a gene inventory (Table 8.1), the rush ended when it became apparent that only a very few targets could be validated by compounds with sufficient biological activity to control disease in plants. Until now, it cannot be said that any market product or developmental candidate has arisen directly from these efforts. So what might be the reason(s) for this unexpected failure?

Typically, novel targets were validated by a knockout mutation. When the knockout was not viable, the gene product was considered to be essential and thus thought to be a potential target [26]. In many cases, avirulence genes like effector protein genes that are not essential but which are obligatory for virulence were considered as novel fungicide targets [27]. Also proteome studies might reveal fungal proteins that are potentially involved in the development of infection [28], and these could be regarded as potential fungicide targets in the same right as targets identified using the techniques described above.

Cloning, functional over-expression of the potential target protein, and development of a biochemical high-throughput assay has been, in many cases, either non-trivial or not feasible. Not all biochemical high-throughput assays have yielded high potential biochemical hits, but this may reflect the difference between screening library configurations and target site demands. For example, inhibitors of carbohydrate phosphate-metabolizing enzymes typically require highly polar physico-chemical properties similar to the substrate, that are by far not "agro-like" as described by Drewes *et al.* [1]. However, even when biochemical hits were

Table 8.1 Fungicide targets claimed in patents between 1996 and 2010. Target names and functions are sometimes tentative, as some descriptions in the patents are insufficient. The numbers of patents form the basis for Figure 8.2.

Target	Function	Company	Patent	Year of publication
Transcription factor MoPPF3 (BNI1)	Cytoskeleton and protein trafficking	China Agricultural University, PRC	CN 102021186	2011
Chitin synthase	Cell wall biosynthesis	China Agricultural University, PRC	CN 102021185	2011
Phosphopantetheinyl transferase	Fatty acid biosynthesis	F2g Ltd, UK	WO 2010139952	2010
Carbonic anhydrases	Basic metabolism	Union Life Sciences Ltd., UK	WO 2010061185	2010
γ-Glutamyltransferase	Basic metabolism	Wisconsin Alumni Research Foundation, USA	US 20100099122	2010
Flocculation suppression protein	Cell aggregation	Shanghai Institutes for Biological Sciences, Chinese Academy of Sciences, PRC	CN 101362798	2009
Histidinol dehydrogenase	Amino acid biosynthesis	National Institute of Pharmaceutical Education and Research (NIPER), India	IN 2007DE01117	2008

Histone deacetylase	Transcription regulation	Methylgene Inc., Canada	WO 2008021944	2008
FKBP12-associated protein 1, FAP1 (ROS resistant protein)	Transcription factor	Peking University First Hospital, PRC	CN 101062942	2007
Nonhistone protein 6, MNH6	Transcription regulation	Zhejiang University, PRC	CN 101058814	2007
Autophagy protein 5, ATG5	Cytoskeleton and protein trafficking	Zhejiang University, PRC	CN 101050463	2007
ER vesicle protein MgCON3	Cytoskeleton and protein trafficking	China Agricultural University, PRC	CN 101020711	2007
Protein kinase mgATG1	Cytoskeleton and protein trafficking	Zhejiang University, PRC	CN 1995353	2007
Transcription factor, MgCON1	Transcription factor, control of conidiospore generation	China Agricultural University, PRC	CN 1951959	2007
Alpha-mannosyltransferase MgPPF5	Glycoprotein biosynthesis	China Agricultural University, PRC	CN 1951958	2007
Transcription factor, MgPTH12	Control of maturation and pathogenicity of fungal appressorium	China Agricultural University, PRC	CN 1951957	2007
RNA splicing factor mgcon2	Transcription regulation, control of secondary conidiophore differentiation	China Agricultural University, PRC	CN 1952149	2007
UDP-galactopyranose mutase	Glycoprotein biosynthesis	Universiteit Leiden, Netherlands	WO 2007011221	2007
Isopentenyl pyrophosphate isomerase	Isoprenoid (sterol) biosynthesis	Bayer CropScience AG, Germany	WO 2006128593	2006
Transcription factor KIN17	Transcription factor, pathogenicity-related	China Agricultural University, PRC	CN 1821409	2006

(continued overleaf)

Table 8.1 (continued)

Target	Function	Company	Patent	Year of publication
Transcription factor rmlA	Transcription factor, involved in cell wall stress response	Stichting voor de Technische Wetenschappen, Netherlands; Universiteit Leiden	WO 2006071118	2006
EPA6 and EPA7	Adhesins involved in biofilm formation	Institut Pasteur, France	WO 2006059228	2006
Many	Genes induced by a heat shock	Schering Corporation, USA	US 20060088859	2006
Thymidylate kinase	Nucleotide metabolism	Bayer CropScience AG, Germany	EP 1612276	2006
Thioredoxin reductase	Redox equilibrium	Bayer CropScience AG, Germany	WO 2005098030	2005
Histidinol dehydrogenase	Amino acid biosynthesis, essential for pathogenicity	Icoria, Inc., USA	WO 2005089156	2005
Two-component histidine kinase	Signaling, osmotic stress	Riken Corp., Japan	WO 2005085416	2005
Adenylosuccinate synthase	Nucleotide biosynthesis	Icoria, Inc., USA	WO 2005071411	2005
Cyclic nucleotide phosphodiesterase	Signaling			
Mitochondrial translocase TOM20	Protein trafficking	BASF AG, Germany	WO 20050728	2005
Mitochondrial translocase TIM17				
Pyruvate kinase	Basic metabolism	Bayer CropScience AG, Germany	WO 2005054457	2005
Protein kinase AUT3	Involved in autophagocytosis	BASF AG, Germany	WO 2005042734	2005

Amidophosphoribosyl-transferase	Nucleotide biosynthesis	Icoria, Inc., USA	WO 2005029034	2005
Cutinase transcription factor 1 (CTF1)	Transcription regulation			
Promoter region of the YMR325W gene (unknown function)	Reporter for sterol biosynthesis	Rosetta Inpharmatics, Inc., USA	WO 2005012559	2005
Mannosyltransferase ANP1	Glycoprotein biosynthesis	Paradigm Genetics, Inc., USA	WO 2005012544	2005
Mannosyltransferase PMT2				
Uridylyl glucose pyrophosphorylase	Nucleotide metabolism	BASF AG, Germany	WO 2005005629	2005
Acetolactate synthase	Amino acid biosynthesis	Paradigm Genetics, Inc., USA	WO 2005002521	2005
Potassium transport protein	Ion translocation			
Guanylate kinase	Nucleotide metabolism	Bayer CropScience AG, Germany	EP 1493821	2005
Ornithine carbamoyltransferase	Amino acid biosynthesis	Paradigm Genetics, Inc., USA	WO 2004104176	2004
Ketol-acid reductoisomerase	Amino acid biosynthesis			
Pathogenicity-conferring gene PCGI	Essential for pathogenicity			
Syntaxin VAM3	Protein trafficking	Max-Planck-Gesellschaft, Germany	WO 2004089398	2004
rab GTPase YPT7				
tRNA splicing endonuclease	Protein biosynthesis	PTC Therapeutics, Inc., USA	WO 2004087070 WO 2004087884	2004
Porphobilinogen deaminase	Porphyrin biosynthesis	Paradigm Genetics, Inc., USA	WO 2004083396	2004
Bifunctional purine biosynthetic protein	Nucleotide metabolism			
Transcription factor	Gene regulation			

(continued overleaf)

Table 8.1 *(continued)*

Target	Function	Company	Patent	Year of publication
Wiskott-Aldrich syndrome protein CaWAL1	Cytoskeleton and protein trafficking, prevent the transition of yeast to hyphal growth habit	Hans-Knöll-Institut, Germany	DE 10309866	2004
Pyrroline-5-carboxylate reductase	Amino acid metabolism	Paradigm Genetics, Inc., USA	WO 2004078992	2004
Dihydrofolate reductase	Nucleotide metabolism	McMaster University, Canada	WO 2004069255	2004
310 diverse essential *Candida* genes	Essential for growth	Elitra Pharmaceuticals, Inc., USA	WO 2004056965	2004
ATP(CTP):tRNA nucleotidyltransferase (CCA1)	Protein biosynthesis	Oxford Glycosciences (UK) Ltd, UK	WO 2004053486	2004
Biotin protein ligase 1 BPL1	Protein biosynthesis	Oxford Glycosciences (UK) Ltd, UK	WO 2004053485	2004
1-Phosphotidylinositol-4-phosphate 5-kinase MSS4	Signaling	Oxford Glycosciences (UK) Ltd, UK	WO 2004053150	2004
mRNA guanylyltransferase CGT1	Transcription	Oxford Glycosciences (UK) Ltd, UK	WO 2004053149	2004
Phosphatidylinositol transfer protein SEC14	Protein trafficking	Oxford Glycosciences (UK) Ltd, UK	WO 2004053146	2004
tRNA ligase TRL1	Protein biosynthesis	Oxford Glycosciences (UK) Ltd, UK	WO 2004053145	2004
Trehalose-6-phosphate synthase	Carbohydrate metabolism	Paradigm Genetics, Inc., USA	WO 2004044148	2004
Osmosensing histidine kinase os-1	Osmotic stress signal receptor	Sumitomo Chemical Company, Limited, Japan	EP 1415996	2004
GTP cyclohydrolase II	Flavin biosynthesis	BASF AG, Germany	WO 2004022776	2004

Transcription factor UPC2	Modulation of drug resistance	McGill University, Canada	WO 2004014349	2004
Transcription factor RDS2				
Transcription factor STB5				
Ribose-5-phosphate isomerase	Basic metabolism	Bayer CropScience AG, Germany	EP 1394265	2004
Proteasome	Protein degradation	BASF AG, Germany	DE 10232773	2004
Spindle pole body-associated protein CIK1	Cell cycle regulation	K.K. Gni, Japan	WO 2004006866	2004
Chitin synthase 2	Cell wall biosynthesis	Frank *et al.*	US 20030228645	2003
Putrescine aminopropyltransferase	Polyamine biosynthesis for stress tolerance	Heiniger *et al.*	WO 2004042348	2003
S-Adenosylmethionine decarboxylase SPE2	Polyamine biosynthesis for stress tolerance	Mahanty *et al.*	WO 2004042348	2003
Eleven diverse essential *Candida* genes	Essential for growth	Bristol-Myers Squibb Co., USA	WO 2003091418	2003
Interaction of Gβ with Cdc24p	Essential for mating	Nem *et al.*	US 20030175712	2003
Transcription factor	Transcription factor	Achillion Pharmaceuticals, Inc., USA	WO 2003055448	2003
361 diverse *Cryptococcus* genes	Diverse	Elitra Pharmaceuticals, Inc., USA	WO 2003052076	2003
Asparagine synthase	Amino acid biosynthesis	Paradigm Genetics, Inc., USA; Icoria, Inc., USA	WO 2003050310	2003
5-Aminolevulinate synthase	Porphyrin biosynthesis			
Histidinol-phosphatase				
3-Isopropylmalate dehydratase				
Threonine synthase				
Inosine monophosphate dehydrogenase	Nucleotide biosynthesis	Bayer CropScience AG, Germany	WO 2003054221	2003
α-Aminoadipate reductase	Amino acid biosynthesis	Paradigm Genetics, Inc., USA	WO 2003046130	2003
Homocitrate synthase				

(continued overleaf)

Table 8.1 *(continued)*

Target	Function	Company	Patent	Year of publication
Acetoacetyl-CoA thiolase	Sterol biosynthesis	Bayer CropScience AG, Germany	WO 2003048349	2003
Glyoxylate cycle enzymes	Basic metabolism	Whitehead Institute for Biomedical Research, USA	US 20030082669	2003
MAP kinase	Signaling	BASF AG, Germany	DE 10150677	2003
Gpr1 G-protein-coupled receptor	Signaling	K.U. Leuven Research & Development, Belgium	WO 2003025218	2003
Glyoxal oxidase	Basic metabolism	Bayer CropScience AG, Germany	WO 2003023028	2003
α-1,3-Glucan synthase	Cell wall biosynthesis	Stichting, Netherlands	WO 2003020922	2003
Glutamine:fructose-6-phosphate amidotransferase				
Fructose-1,6-bisphosphate aldolase	Basic metabolism	Bayer CropScience AG, Germany	WO 2003020957	2003
Homoaconitase	Amino acid biosynthesis	BASF AG, Germany	WO 2003000880	2003
Protein kinase KIN28	Transcription regulation	Millennium Pharmaceuticals, Inc., USA	WO 2002094993	2002
>600 diverse	Essential gene	Elitra Pharmaceuticals, Inc., USA	WO 2002086090	2002
RNA triphosphatase	mRNA processing	Shuman, Stewart	US 6107040	2002
Cofilin	Cytoskeleton and trafficking	Cytokinetics, USA	WO 2002042451	2002
Toxic peptide synthase	Pathogenicity factor biosynthesis	Syngenta, Switzerland	WO 2002042444	2002
Essential for Mitotic Growth 1 (Emg1)	Ribosomal RNA methyltransferase, cell cycle regulation	Regents of the University of Michigan, USA	US 6383753	2002
Emg-1-nuclear interacting protein-1 ENIP1				

26 diverse	Essential proteins	Anadys Pharmaceuticals, Inc., USA	WO 2002002055	2002
Protein kinase AG007 Protein kinase AG008 Protein kinase AG009 Protein kinase AG0010	Essential proteins	Syngenta, Switzerland	US 6291665	2001
GTP-binding protein AG001 GTP-binding protein AG002 GTP-binding protein AG003 GTP-binding protein AG004	Essential proteins	Syngenta, Switzerland	US 6291660	2001
YDR141C, YDR091C, YOL022C, YOL026C, YOL034W, and YOL077C	Diverse essential genes	Rosetta Inpharmatics, Inc., USA	US 6221597	2001
YKR081C, YFR003C, YGR277C, YGR278W, YKR071C, YKR079C, and YKR083C	Diverse essential genes	Rosetta Inpharmatics, Inc., USA	WO 2000071161	2001
Sugar phosphatases	Basic metabolism	K.U. Leuven Research & Development, Belgium	WO 2001016357	2001
Punchless 1, PLS1	Cytoskeleton and trafficking	Aventis CropScience SA, France	WO 2000077036	2000
Many	Essential genes	Rosetta Inpharmatics, Inc., USA	WO 2000071161	2000
Mannosyltransferase CaKRE5 Ion transporter CaALR1 Rho guanyl-nucleotide exchange factor CaCDC24	Essential genes	Mycota Biosciences Inc., Canada	WO 2000068420	2000
RNA triphosphatase	mRNA processing	Sloan-Kettering Institute for Cancer Research, USA	WO 2000063433	2000

(continued overleaf)

Table 8.1 *(continued)*

Target	Function	Company	Patent	Year of publication
Many diverse	Essential proteins	Rosetta Inpharmatics, Inc., USA	WO 2000058457	2000
Cleavage/polyadenylation factor CstF64	Transcription regulation	Isis Innovation Ltd., UK; Aranda Fernandez, Agustin	WO 2000042204	2000
17 diverse	Essential genes	Millennium Pharmaceuticals, Inc., USA	WO 2000039342	2000
Inositolphosphoryl ceramide synthase	Sphingolipid biosynthesis	AstraZeneca UK Limited, UK	WO 2000029590	2000
Diverse	Essential genes	Novartis A.-G., Switzerland	WO 2000022133	2000
α-1,3 Mannosyltransferase	Protein glycosylation	Cornell Research Foundation, Inc., USA	WO 2000020568	2000
Many diverse	Essential proteins	Janssen Pharmaceutica N.V., Belgium	WO 2000009695	2000
Histone acetyltransferase TATA-box-binding associated transcription factor TAF-145	Transcription regulation	Scriptgen Pharmaceuticals, Inc., USA	WO 9940199	1999
Protein elongation factor 2 EF2	Protein biosynthesis	Merck & Co., Inc., USA	WO 9857176	1998
Histidine kinase OS-1	Osmotic stress signal receptor	Mycotox, Inc., USA	WO 9844148	1998
Transcription factor IIB yTFIIB	Transcription regulation	Children's Hospital Medical Center, Philadelphia, USA	WO 9839355	1998
Transcription factor Skn7	Transcription regulation	Medical Research Council, UK	WO 9838331	1998
Profilin	Cytoskeleton and trafficking	Bristol-Myers Squibb Co., USA	WO 9731104	1997
Endoglucanase egl1	Secreted, cell wall degradation	BASF AG, Germany	WO 9712911	1997
Phosphatidylinositol *N*-acetylglucosaminyltransferase CLY4	Glycoprotein biosynthesis	Bristol-Myers Squibb Co., USA	EP 735138	1996

PRC = People's Republic of China.

identified, the multifaceted problems of transferring target activity into sufficient activity in plants were found to be substantial [29]. In addition to "agro-like" physico-chemical properties, metabolic stability is needed such that compounds which are easily decomposed by enzymatic hydrolysis or oxidation, or which undergo conjugation (e.g., small esters, primary alcohols, or phenols), are excluded. Ultimately, none of the target patents listed in Table 8.1 has yet led to a recognizable fungicide developmental candidate.

8.6 Non-Protein Targets

RNA molecules can adopt interesting conformations, and may serve as targets to low-molecular-weight compounds. For example, the antibiotic pyrithiamine is known to bind to RNA involved in thiamine biosynthesis, and thus demonstrates the basic value of RNA as a drug target [30]. Since, in history of fungicide discovery, RNA-targeting compounds rather than protein-targeting compounds never have been identified by chance, RNA targets are unlikely to achieve a similar relevance as protein targets.

Alternatively, RNA can be targeted in highly specific fashion by other RNA molecules, which may in turn lead to RNA interference (RNAi) that can be used to knock out the function of a gene in a cell. Recently, RNAi has been considered as a mechanism for novel fungicides, although the delivery of active RNA molecules other than by transgenic plants or microorganisms may be economically out of reach [31].

8.7 Resistance Inducers

Some compounds in the fungicide market are not directly active on their target pathogens, but rather serve as activators of the plant's own defense mechanisms. In that case, MoA studies must be conducted in plants, and not in fungi. Phosphonates such as fosetyl-Al have been used since the late 1970s, especially against oomycetes diseases. In fact, after many diverse biochemical explanations of the compound's MoA [32], gene expression profiling in plants has shown that fosetyl-Al acts indirectly via an activation of a series of plant defense genes [33–35]. Recently, a deeper understanding of the MoA of another resistance inducer, acibenzolar S-Me, was developed by Jaskiewicz *et al.* [36]. These authors showed that acibenzolar S-Me actually primes the plant defense response through an epigenetic chromatin modification, thus eluding the potential penalty of a permanent defense activation. Such an elucidation of the MoA was only possible after development of chromatin immunoprecipitation techniques and a growing understanding of epigenetics in biology. In case of other resistance inducers, such as hairpin protein [37], laminarin [38], or chitin fragments [39], different – though partially overlapping – responses have been shown, and it is likely that several mechanisms involved in the induction of resistance are in place.

8.8 Beneficial Side Effects of Commercial Fungicides

The market success of plant protection compounds is not determined solely by their activities on target organisms. Clearly, it is not simply a coincidence that in crop plants the two most successful fungicide targets are linked to beneficial side effects of their inhibitors. Sterol demethylase inhibitors typically exhibit a degree of activity against other cytochrome P_{450} enzymes. In crop plants, such enzymes involved in plant hormone metabolism may be affected beneficially. For example, interference with biosynthesis of gibberellins, brassinosteroids or abscisic acid may have a favorable influence on the growth phenotype, or may confer drought resistance [40–43]. Likewise, inhibitors of cytochrome c reductase in plants may bring about certain abiotic stress resistance and greening effects [44–46]. From a commercial aspect it is likely that these side effects may have some influence on success of individual products.

8.9 Concluding Remarks

In spite of great technological progress having been made, the targeted discovery of novel fungicides remains an immense challenge. In fact, it is not discovery of a target that is the greatest hurdle, but rather the restrictions posed on new active compounds by the obligatory physico-chemical properties required for sufficient bioavailability to ensure biological activity. Complexity accrues with the major attrition rate caused by inacceptable toxicological or environmental properties of novel compounds, and it is not surprising that development of novel antimycotics encounters similar problems [47]. Nonetheless, molecular biological, biochemical, and chemical advancements continue to provide hope [27], with only the future proving progression in development of novel fungicide targets.

References

1. Drewes, M., Tietjen, K., and Sparks, T.C. (2012) High throughput screening in agrochemical research, in *Modern Crop Protection Compounds* (eds P. Jeschke, W. Krämer, U. Schirmer, and M. Witschel), Wiley-VCH Verlag GmbH, Weinheim, pp. 1–00.
2. Morton, V. and Staub, T. (2008) A Short History of Fungicides. Online, APSnet Features, 2008, doi: 10.1094/APSnetFeature-2008-0308. Available at: *http://www.apsnet.org/publications/apsnetfeatures/Pages/Fungicides.aspx* (accessed June 2012).
3. Büchel, K.H. (1986) The history of azole chemistry, in *Fungicide Chemistry*, ACS Symposium Series, Vol. 304 (eds M. Green and D.A. Spilker), American Chemical Society, Washington, DC, pp. 1–23.

4. Becker, W.F., Von Jagow, G., Anke, T., and Steglich, W. (1981) *FEBS Lett.*, **132** (2), 329–333.
5. Leroux, P. (2010) *Phytoma*, **631**, 8–11.
6. Blum, M., Boehler, M., Randall, E., Young, V., Csukai, M., Kraus, S., Moulin, F., Scalliet, G., Avrova, A.O., and Whisson, S.C. (2010) *Mol. Plant Pathol.*, **11** (2), 227–243.
7. Jende, G. (2001) Die Zellwand des Oomyceten *Phytophthora infestans* als Wirkort von Fungiziden. Dissertation, Institut für Pflanzenkrankheiten der Rheinischen Friedrich-Wilhelms Universität Bonn, Germany.
8. Mehl, A. (2006) Untersuchungen zum Wirkungsmechanismus von Iprovalicarb. Dissertation, Institut für Phytomedizin der Universität Hohenheim, Germany.
9. Delvos, B. (2009) Untersuchungen der Effekte von Iprovalicarb und Dimethomorph auf die Zellwand von *Phytophthora infestans*. Dissertation, Mathematisch-Naturwissenschaftlichen Fakultät der Heinrich-Heine-Universität Düsseldorf, Germany.
10. Toquin, V., Barja, F., Sirven, C., Gamet, S., Latorse, M.P., Zundel, J.L., Schmitt, F., and Beffa, R. (2006) *Pflanzenschutz - Nachr. Bayer (English Edition)*, **59** (2-3), 171–184.
11. Takano, Y., Oshiro, E., and Okuno, T. (2001) *Fungal Genet. Biol.*, **34** (2), 107–121.
12. Delgado-Alvarez, D.L., Callejas-Negrete, O.A., Gomez, N., Freitag, M., Roberson, R.W., Smith, L.G., and Mourino-Perez, R.R. (2010) *Fungal Genet. Biol.*, **47** (7), 573–586.
13. Masner, P., Muster, P., and Schmid, J. (1994) *Pestic. Sci.*, **42** (3), 163–166.
14. Davidse, L.C., Gerritsma, O.C.M., Ideler, J., Pie, K., and Velthuis, G.C.M. (1988) *Crop Prot.*, **7** (6), 347–355.
15. Tueckmantel, S., Greul, J.N., Janning, P., Brockmeyer, A., Gruetter, C., Simard, J., Gutbrod, O., Beck, M.E., Tietjen, K., Rauh, D., and Schreier, P.H. (2011) *ACS Chem. Biol.*, **6** (9), 926–933.
16. Zhang, Y., Lamm, R., Pillonel, C., Lam, S., and Xu, J.R. (2002) *Appl. Environ. Microbiol.*, **68** (2), 532–538.
17. Tsukahara, K., Hata, K., Nakamoto, K., Sagane, K., Watanabe, N.A., Kuromitsu, J., Kai, J., Tsuchiya, M., Ohba, F., Jigami, Y., Yoshimatsu, K., and Nagasu, T. (2003) *Mol. Microbiol.*, **48** (4), 1029–1042.
18. Umemura, M., Okamoto, M., Nakayama, K.I., Sagane, K., Tsukahara, K., Hata, K., and Jigam, Y. (2003) *J. Biol. Chem.*, **278** (26), 23639–23647.
19. Liu, T.T., Lee, R.E.B., Barker, K.S., Lee, R.E., Wei, L., Homayouni, R., and Rogers, P.D. (2005) *Antimicrob. Agents Chemother.*, **49** (6), 2226–2236.
20. Becher, R., Weihmann, F., Deising, H.B., and Wirsel, S.G.R. (2011) *BMC Genom.*, **12**, 52.
21. Florio, A.R., Ferrari, S., Carolis, E.D., Torelli, R., Fadda, G., Sanguinetti, M., Sanglard, D., and Posteraro, B. (2011) *BMC Microbiol.*, **11**, 97.
22. Schwanhäusser, B., Busse, D., Li, N., Dittmar, G., Schuchhardt, J., Wolf, J., Chen, W., and Selbach, M. (2011) *Nature*, **473**, 337–342.
23. Hoehamer, C.F., Cummings, E.D., Hilliard, G.M., and Rogers, P.D. (2010) *Antimicrob. Agents Chemother.*, **54** (5), 1655–1664.
24. Pillonel, C. (2005) *Pest Manag. Sci.*, **61** (11), 1069–1076.
25. De Backer, M.D., Dijck, P.V., and Luyten, W.H.M.L. (2002) *Am. J. PharmacoGenom.*, **2** (2), 113–127.
26. Beffa, R. (2004) *Pflanzenschutz - Nachr. Bayer (English Edition)*, **57** (1), 46–61.
27. Kamoun, S., Dong, S., Hamada, W., Huitema, E., Kinney, D., Morgan, W.R., Styer, A., Testa, A., and Torto, T.A. (2002) *Can. J. Plant Pathol.*, **24** (1), 6–9.
28. Garrido, C., Cantoral, J.M., Carbu, M., and Gonzalez-Rodriguez, V.E. (2010) *Curr. Proteomics*, **7** (4), 306–315.
29. Dunbar, S.J. and Corran, A.J. (2007) in *Pesticide Chemistry* (eds H. Ohkawa, H. Miyagawa, and P.W. Lee), Wiley-VCH Verlag GmbH, Weinheim, pp. 65–75.
30. Sudarsan, N., Cohen-Chalamish, S., Nakamura, S., Emilsson, G.M., and Breaker, R.R. (2005) *Chem. Biol.*, **12** (12), 1325–1335.
31. Schumann, U., Ayliffe, M., and Kazan, K. (2010) *Front. Biol.*, **5** (6), 478–494.

32. Hillebrand, S., Zundel, J.L., and Tietjen, K. (2012) Fungicides with unknown mode of action, in *Modern Crop Protection Compounds*, vol. 2 (eds W. Krämer, U. Schirmer, P. Jeschke, and M. Witschel), Wiley-VCH Verlag GmbH, Weinheim, Chapter 1.
33. Molina, A., Hunt, M.D., and Ryals, J.A. (1998) *Plant Cell*, **10**, 1903–1914.
34. Chuang, H.W., Hsieh, T.F., Duval, M., and Thomas, T.L. (2003) *Genomics of Plants and Fungi*, Mycology Series, Vol. 18, Marcel Dekker, New York, pp. 237–253.
35. Latorse, M.P., Mauprivez, L., Sirven, C., Gautier, P., and Beffa, R. (2010) Comparison of fosetyl-Al and another phosphonate on plant downy mildew protection and on Arabidopsis thaliana gene expression, in *Proceedings, 6th International Workshop of Grapevine Downy and Powdery Mildew, Bordeaux, France, 4–9 July 2010* (eds A. Callonec, F. Delmotte, B. Emmett, D. Gadoury, C. Gessler, D. Gubler, K.H. Kassemeyer, P. Magarey, M. Raynal, and R. Seem), INRA, Paris, France, pp. 158. Available at: *https://colloque.inra.fr/gdpm_2010_bordeaux/content/download/876/12286/version/1/file/ProceedingsGDPM2010r.pdf* (accessed June 2012).
36. Jaskiewicz, M., Conrath, U., and Peterhaensel, C. (2011) *EMBO Rep.*, **12** (1), 50–55.
37. Livaja, M., Zeidler, D., Rad, U., and Durner, J. (2008) *Immunobiology*, **213** (3–4), 161–171.
38. Aziz, A., Poinssot, B., Daire, X., Adrian, M., Bezier, A., Lambert, B., Joubert, J.M., and Pugin, A. (2003) *Mol. Plant-Microbe Interact.*, **16** (12), 1118–1128.
39. Wan, J., Zhang, X.C., Neece, D., Ramonell, K.M., Clough, S., Kim, S.Y., Stacey, M.G., and Stacey, G. (2008) *Plant Cell*, **20** (2), 471–481.
40. Fletcher, R.A., Hofstra, G., and Gao, J. (1986) *Plant Cell Physiol.*, **27** (2), 367–371.
41. Fletcher, R.A., Gilley, A., Sankhla, N., and Davis, T.D. (2000) *Horticultural Rev.*, **24**, 55–138.
42. Kitahata, N., Saito, S., Miyazawa, Y., Umezawa, T., Shimada, Y., Min, Y.K., Mizutani, M., Hirai, N., Shinozaki, K., Yoshida, S., and Asami, T. (2005) *Bioorg. Med. Chem.*, **13** (14), 4491–4498.
43. Goertz, A., Oerke, E.C., Puhl, T., and Steiner, U. (2008) *J. Appl. Bot. Food Q.*, **82** (1), 60–68.
44. Wu, Y.X. and Tiedemann, A. (2001) *Pestic. Biochem. Physiol.*, **71** (1), 1–10.
45. Clark, D.C. (2003) *Pflanzenschutz - Nachr. Bayer (English Edition)*, **56** (2), 281–296.
46. Zhang, Y.J., Zhang, X., Chen, C.J., Zhou, M.G., and Wang, H.C. (2010) *Pestic. Biochem. Physiol.*, **98** (2), 151–157.
47. Van Minnebruggen, G., Francois, I.E.J.A., Cammue, B.P.A., Thevissen, K., Vroome, V., Borgers, M., and Shroot, B. (2010) *Open Mycol. J.*, **4**, 22–32.

Part III
New Methods to Improve the Bioavailability of Active Ingredients

Modern Methods in Crop Protection Research, First Edition.
Edited by Peter Jeschke, Wolfgang Krämer, Ulrich Schirmer, and Matthias Witschel.

9
New Formulation Developments

Rolf Pontzen and Arnoldus W.P. Vermeer

9.1
Introduction

In the modern agrochemical market, innovation is not restricted to the discovery of new active ingredients; the application form of these substances is also a method for differentiation. During the past decades it has been shown that the different properties of the end product can be modified by choosing the most appropriate formulation type. Examples of these properties are related to the handling of the products, their dilutability, and mixability with further products, fertilizers, and so on, to their biological performance and, above all, to their impact on workers, bystanders, and the environment.

Over the past decades, several comprehensive reviews detailing trends in formulation technology have been published [1–7]. On closer examination, these show that whilst each review provides an extensive coverage of the subject, the focuses of the articles have changed over time. During the early 1970s, the formulation additives available were limited, and the development of a stable formulation and its application at the farm level were the first concerns of the formulation scientist. Later, the optimization of biological performance became most important, and this resulted in the development of new formulation types such as encapsulated and oil-based dispersions. Today, however, the protection of the environment is becoming increasingly the focal point of product development processes. The latter point not only concerns the use of non-classified formulation additives, but also relates to formulations that can be characterized as, for example, drift-reducing systems.

Next to realizing the changing focus of these reviews, it is interesting to more closely examine the accuracy of the predictions that were made. Some 20 years ago [1], the introduction of a significant number of products based on living organisms was forecast; yet, to date only a limited number of these products have reached the market. An exception can be made for organic farming, where the use of beneficials such as bacteria, spores, and nematodes has been established.

Although the need for a reduction in solvents and dusty formulations was correctly recognized 20 years ago, the solutions that were proposed to these problems have not always been correct. For example, emulsion in water (EW) and

Modern Methods in Crop Protection Research, First Edition.
Edited by Peter Jeschke, Wolfgang Krämer, Ulrich Schirmer, and Matthias Witschel.

water-dispersible granule (WG) formulations have not become the dominant types. Furthermore, the tendency to reduce complexity for the end-user by reducing the possible number of mixing partners – either in the form of formulations that contain multiple active ingredients, or by preparing combinations of active ingredients and adjuvants – has not yet resulted in these products having a significant market share. Outside Europe, in particular, tank-mixing remains very much a common practice. In Europe, registration procedures and discussions related to resistance management have been a hurdle for insecticide mixtures in particular. It is clear from these considerations that the prediction of trends on a scientific basis is difficult, and that the choice of a certain formulation during the development process depends also on numerous non-scientific aspects, including registration, public opinion, safety of use, cost, and the preferred method of application.

On examining the development of crop protection agents more closely, it must be borne in mind that the most important function of the formulation is – and will remain so in the future – to guarantee the homogeneous distribution of a small amount of active ingredient over a large area. Since the potential efficacy of most modern active ingredients has increased significantly, and doses of less than 10–100 g ha^{-1} are more the rule than the exception, this has become even more important. To illustrate this distribution problem, the following examples are given. For seed treatment, a dose of 1 g of formulated product per million seeds is very common, whereas spraying insecticides onto fully grown plant will result in the treatment of several hectares of leaf surface with only 200 ml of formulated product. These aspects indicate that today, formulation technology must be seen as an enabling technology that adds value and attractiveness to the crop protection industry, while at the same time improving both operator and bystander safety, reducing the environmental impact, and increasing food safety [5].

As mentioned above, discussions concerning trends in formulation technology are not straightforward, as numerous aspects that cannot be influenced by the formulation chemist will contribute to development decisions. For this reason, attention in this chapter is focused on the trends that have affected the crop protection market over the past decade, rather than looking into the future. Clearly, such an evaluation might allow some conclusions to be drawn with regards to future trends.

The turnover (ex company) of the most important formulation types are given for the period 2000–2010 in Figure 9.1. Here, the curves shown represent about 90% of the overall turnover as obtained from AgroWin. It can be seen that, over this period, the use of solvent-based products (emulsifiable concentrate; EC) decreased by about 16%, whereas that of wettable powders (WPs) decreased globally by even more than 30%. These figures clearly confirm the above-mentioned tendencies to develop formulations that are safe for people and the environment. However, it must be mentioned that, despite a decreasing market penetration of the ECs, this formulation type still dominates the market on a worldwide basis. The decrease of 30% in the turnover of granules (GRs) is most likely explained by the fact that developments in the application techniques led to the use of GRs being rather inconvenient. The formulation type that has grown most significantly over

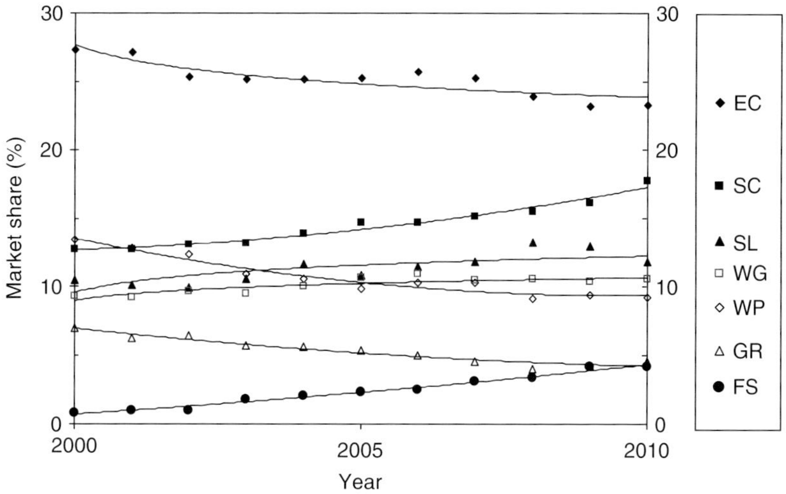

Figure 9.1 Market shares of the most important formulation types. The formulation types were selected to obtain at least 90% market share. The lines shown serve only as a guide for the eye. Data extracted from AgroWin.

the past decade has been the suspension concentrates (SCs) (>40%), the main advantages of these water-based flowables being a lack of solvent- and dust-related problems, low cost, ease of handling and dosing during application, and a relative straightforward production. A disadvantage of the SCs relates to their biological performance whenever systemic active ingredients are involved (this point is discussed in Section 9.1.4). The increased use of water-based flowables for seed treatment (FS) follows this line of reasoning. A slight increase in use was found for WGs and soluble liquids (SLs), albeit only to a minor extent.

Differentiating these results between the different indications, insecticides, fungicides, and herbicides, shows an interesting picture (see Table 9.1). The importance of the different formulation types varies strongly between the three indications. In the case of insecticides, the most dominant formulation type is still EC, which has been used predominantly for the formulation of pyrethroids, organophosphates, and carbamates. Because of the toxicological profile of these substances, however, it is expected that the relevance of these products will be further reduced over the next years, and this is reflected by the very strong growth rates of SC and WG types of formulation in this area. A further reason for the growth of the latter two formulation types is the fact that most of the recently introduced insecticides show a high molecular weight, and consequently have a reduced solubility in most polar and nonpolar solvents.

Due to the importance of numerous multisite fungicides such as mancozeb, propineb, or chlorothalonil, fungicides is the only indication where WPs – although decreasing significantly – remain a relevant formulation type. Furthermore, it can be seen that within fungicides ECs are still gaining in importance, which can be explained by the importance of this formulation type for cereal applications where

Table 9.1 Market share (MS) of the most important formulation types in 2010 for the different indications; insecticides, fungicides, and herbicides. Formulation types were selected to obtain at least 90% MS. The relative changes in MS for the different formulations are based on MS in 2000 are also shown.

	Insecticides (2010) (% MS[a])	Relative change to 2000 (%)	Fungicides (2010) (% MS[a])	Relative change to 2000 (%)	Herbicides (2010) (% MS[a])	Relative change to 2000 (%)
EC	34	−17	19	+45	26	−14
WP	6	−45	22	−40	3	−25
SC	18	+100	28	+50	16	+15
WG	8	+300	12	+71	13	−3
SL	4	−20	2	−50	27	+28
GR	10	−33	1	−80	3	−25

[a] Data without flowables for seed treatment.
Data extracted from AgroWin.

a high penetration rate and optimized retention are crucial. Again, the use of SCs and WGs is growing.

Finally, it can be seen that herbicides follow the general trends for different formulation types relatively well, but with one exception. Whereas, for insecticides and fungicides a reduction in the turnover of SLs was found, this is the formulation type with the highest market share within the herbicide category. Evidently, this can be related to the formulations containing the different glyphosate salts.

From these findings, it can be concluded that the current trends observed within the crop protection market are clearly determined to a large extent by the demands of society, with the safety of people working with pesticides, of bystanders, and also of the environment, being treated with the highest priority. However, it is also clear that these demands can only be fulfilled when a technical solution is available. The latter point depends mainly on the physical chemical properties of the active ingredient, as is reflected by the differences between the three indications summarized in Table 9.1. Depending on (among others) the melting point, log *P*-value, acidity, molecular weight, and chemical stability of these materials, certain formulation types will not easily be assessed, and neither do they fit into the working mechanism of the corresponding active ingredient. This leads to the fact that the ultimate product is mostly a compromise between different properties, and that the choice of a final formulation type is based on well-weighted arguments.

A further aspect to be considered during the development process is the biological performance of a product. As indicated above, SCs and WGs are increasingly the formulation type employed during recent times and, especially in the case of systemic pesticides, their uptake into the plant, insect, and/or fungus is a predominant step for good efficacy. Evidently, substances that are in a dissolved state are better available for uptake than solids. It may also be clear that active substances, when applied as a crystalline material, do not penetrate the leaf cuticle spontaneously. Consequently, bearing in mind the increasing importance of these types of product, the need to optimize their efficacy has also became essential.

Whenever there is a need to improve the efficacy of an active substance, the use of penetration-enhancing additives has become common practice. Typically, this can be achieved either as a tank-mix, or the adjuvants can be incorporated into the formulation. In particular, this need has been intensively investigated over the past decade in the case of SCs and WGs. With regards to the end user, the main advantage of in-can formulations compared to tank-mixes relates to the greater accuracy of the dosing, the lack of risk of under- or overdosing of one of the mix partners, and a reduction of logistics. From the producer's point of view, in addition to an improved reliability, the main advantage is an optimized registration for in-can products. Especially in Europe, tank-mix partners must form part of the registration dossier, which in turn will cause a significant increase in the complexity of the registration process. Indeed, this has resulted in the development of oil-based suspension concentrates (ODs), suspo-emulsions (SEs), and adjuvanted SCs (as discussed in the following subsections of the chapter). At this point, it should be emphasized that formulations with an optimized biological profile are mainly developed as stand-alone products. However, whether or not these products can be mixed with other materials, and whether they still perform as intended, must be investigated case by case.

The drivers for the choices to be made during the development process will be discussed in the following sections, after which an extensive discussion will be provided, presenting the different formulation concepts, together with details regarding the scientific background of these products. As formulations may form colloidal systems where different incompatible phases are in equilibrium, differentiation will also be made between actives that are dissolved, those that are present in a solid state within a liquid medium (suspensions), and particles that are present in a solid matrix. Doing so, attention will be focused on the most relevant formulation types (see the FAO Manual or Croplife International for further details related to the less-relevant types). Finally, the way in which formulations can affect biological performance will be discussed, focusing on the reasons why this aspect has become so important over the past 10 years. In this case, the properties of different formulations will be discussed, as will the methods by which such properties are determined.

9.2
Drivers for Formulation Type Decisions

During the development of commercial crop protection products, a number of factors must be taken into account for selecting the optimal formulation type. Once these parameters have been determined and prioritized, the selection process can be initiated, with different formulation ingredients (including surfactants and adjuvants) being combined to produce a stable formulation with a shelf life that, under varying climatic conditions, will be at least two years.

The mode of action (MoA) of the active substance is a first factor to be considered. Whenever a systemic activity is required, the pesticide should preferably be formulated in a dissolved form to assure a high bioavailability. Only in the case that acceptable solvents are not available will alternatives be taken into account. For these situations, an adjuvanted formulation should be considered to compensate for the loss in efficacy due to the physical state of the active ingredient. For contact activity, a property such as rainfastness becomes relevant, as the presence of the pesticide for long periods of time on the outer leaf surface is needed to assure a longlasting effect. Formulations that consist of crystalline materials are often chosen for this purpose. Whenever the spray residue containing these crystals has dried sufficiently on the leaf surface, the crystals may become aligned between the waxy crystals of the plant surface, this being the best guarantee of a good contact efficacy.

In addition to the MoA, the target pest, disease, or weed species will each also have an effect on the formulation type, as will the nature of the crop to be protected. For example, the treatment of weeds on hard surfaces may require products other than those needed for the treatment of scales in a fully grown citrus tree. Differences between crops with regards to plant compatibility may also require the use of different formulation types. For example, the leaves of grape vines are very sensitive to solvents so that spray solutions containing a suspension will be preferred, whereas in cereals an EC will be more effective.

Whether or not the preferred formulation is possible depends heavily on the physico-chemical properties of the active substance, including its molecular weight, melting point, lipophilicity (log*P*), and water solubility. Typically, substances with a low molecular weight, a low melting point, and a high log *P* are the best candidates for EC formulations, whereas substances with a high melting point and a low water solubility can be transferred into SC and WG formulations. A summary of these preferences is provided in Figure 9.2. In addition to the physico-chemical properties of the active substance, its chemical stability is also relevant. Notably, pesticides that are unstable against hydrolysis should not be formulated in the presence of

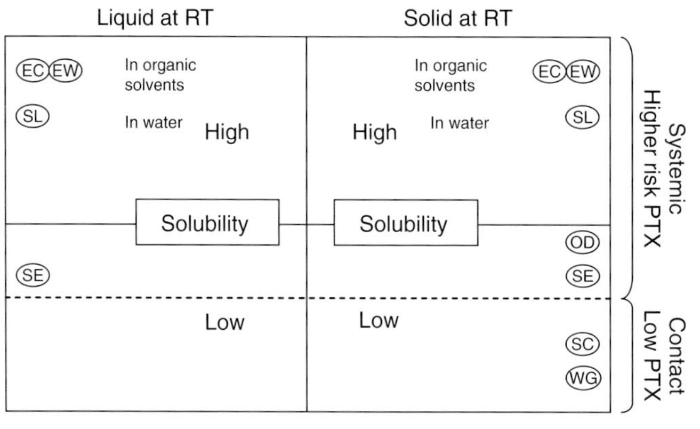

Figure 9.2 Guidelines for the choice of the most relevant formulation type on the basis of their physical chemical and biological profiles. RT: room temperature; PTX: phytotoxicity.

water, or in the presence of other polar solvents. Substances that are unstable to ultraviolet light should be formulated preferentially in crystalline fashion; the same applies to active substances that are highly volatile.

However, it is not only the biological and technological aspects that influence the choice of formulation type. Once the (eco)toxicological profile of an active ingredient is known, an appropriate formulation design can contribute to minimizing any risks that might be associated with this substance. It is for just such a reason that an active ingredient with skin-irritating properties should not be formulated as an EC. Another example is related to drift and bystander exposure, where it is well known that the formulation type may have a major impact. In such a case, EC, EW, or OD formulations are to be preferred.

Finally, market demands and cost of the formulation per treated hectare will each affect the final formulation type.

9.3 Description of Formulation Types, Their Properties, and Problems during Development

9.3.1 Pesticides Dissolved in a Liquid Continuous Phase

During recent years, formulations that contain the active ingredient in a dissolved form have dominated the market based on several advantages. In general, the production of these products is cheap, and does not require expensive equipment; furthermore, the handling, dosing, and application of the formulation by the end-user is straightforward, the packaging material is not difficult to clean, and the disposal of any empty containers is not problematic. Finally, these types of formulation guarantee a good biological performance. Agrochemicals that fit very well into this type of formulation are those with a high solubility (preferably $>30\%$) in a solvent, or are liquid at room temperature. Typical formulations in this group include SLs, ECs, dispersible concentrates (DCs), emulsions (both EW and WO), micro-emulsions (MEs), and capsule formulations (capsule suspension; CS). Although encapsulated products fit into this category of formulation, the properties are somewhat different and production is much more difficult (as will be discussed later).

Among the above-mentioned formulation types, the SL is the most simple to prepare. By dissolving the active agent in either water or a water-miscible solvent, a solution is obtained which merely requires dilution in the spray tank. However, due to the restricted number of active ingredients which are highly soluble in these polar solvents and are stable against hydrolysis, the market share of SLs remains limited. Since SL formulations are based on water(like) solvents, the surface tension of these products is far from optimal, and this often results in a very poor wetting and attachment of the spray droplets to the leaf surface. To improve the biological efficacy – and, in particular, the retention of the spray – surfactants may be added

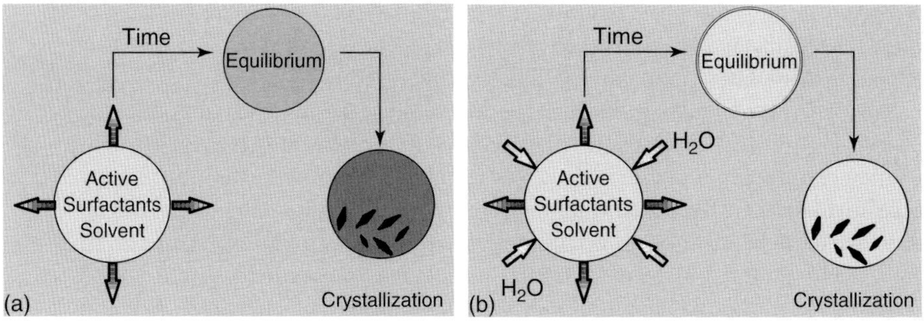

Figure 9.3 Different crystallization scenarios of (a) soluble liquid (SL) and (b) emulsifiable concentrate (EC) after dilution of the concentrate in the spray solution. Light gray = oil; dark gray = water.

to the formulation. The main risk with SLs relates to crystallization phenomena, both in the concentrate and in the spray solution. As most formulations contain a high dose per liter, storage at low temperatures may lead to an oversaturation in the concentrate and, as a result, crystals may be formed. Especially for SLs based on water-miscible solvents, this will cause problems when emptying the containers, as the crystallized actives often fail to dissolve in small amounts of pure water. Crystallization may also occur during preparation of the spray solution. Whenever the concentration of the active ingredient (a.i.) in the spray reaches a value that exceeds its water solubility, crystals will be formed, as shown schematically in Figure 9.3. The majority of active ingredients that are formulated as SLs are in the form of the free acid/base or salt, although some neutral active ingredients are also known (e.g., 2,4-D, MCPA, dicamba, paraquat, glyphosate, glyfosinate, or imidacloprid).

Currently, ECs remain the most widespread formulation type, based partly on their straightforward production, which involves simply dissolving the active ingredient in a nonpolar/oil solvent in combination with emulsifiers or an emulsifier blend [8, 9]. When preparing the spray solution, the concentrate is spontaneously emulsified, yielding a clear blue to milky white emulsion. This property depends mainly on the size of the emulsion droplets, and also therefore on the emulsifier (concentration) used. As the water solubility of the most hydrophilic parts of the emulsifiers depends on the temperature and the salt concentration, the water quality/hardness of the spray liquid is important for the stability of the spray solution. In order to obtain a stable emulsion in the spray solution, surfactant blends with a certain hydrophilic/lipophilic balance (HLB) value are used [8]. Surfactants with an HLB value of between 8 and 18 normally produce stable spray solutions. Due to the presence of emulsifiers in the ECs, the retention and uptake in plant tissues are generally acceptable, although on occasion adjuvants may be added to further improve the products. As with SLs, the main formulation risk with ECs is the occurrence of crystallization, both in the concentrate and the spray liquid. For ECs, crystallization at a low temperature in the concentrate is even more critical since, once the container has been emptied the remaining residue is not water-soluble. In spray solutions, crystallization may occur in both the water phase and the emulsion droplets. Whenever the oil phase is in equilibrium with water, small amounts of oil will dissolve in the continuous phase, small amounts of water will dissolve in the oil, and the active ingredient will then distribute between the two phases. However, this can result in an oversaturation of the oil phase and, as a result, in crystal formation (see Figure 9.3) that may cause blockage of the filters in the spray equipment and should therefore be prevented. A further disadvantage of ECs is the fact that they contain solvents which may be responsible for plant incompatibility or an increased dermal toxicity of the active ingredient; they may also represent a possible fire hazard and increase the cost of the product. As mentioned above, pyrethroids and organophosphates are typical active ingredients in ECs, although many cereal fungicides such as prothioconazole (Proline®), pyraclostrobin and metconazole (Twinline®), and isopyrazam and cyprodinil (Bontima®), are also formulated as ECs.

A DC is best described as an EC based on a water-soluble solvent (or an SL with an active ingredient that is not water-soluble); consequently, the active ingredient is intended to crystallize upon dilution in the spray solution. In order to obtain a stable spray solution, it is essential that the crystals remain small; by applying crystallization inhibitors (mostly polymers) it can be ensured that many small crystals are produced and that, over time, no significant crystal growth occurs. Flint®, which is based on trifloxystrobin, is a product that may be formulated as a DC.

Both oil-in-water (EW) and water in oil (WO) emulsions are produced by a forced emulsion process. An a. i. can be dissolved in the emulsified phase, or it may serve as the oil phase in an EW. In contrast to the previous formulation types, where the concentrate consists of a continuous phase, EW and WO emulsions already incorporate an emulsion in the concentrate. While these types of emulsion are not thermodynamically stable, and therefore will at some point in time coagulate, storage stability remains a subject of concern. However, by using polymers that surround the emulsion droplet as an emulsifier, it is possible to obtain a kinetically stabilized system. Stabilization against sedimentation or creaming of the oil phase can be achieved with polysaccharides such as xanthan gum. Examples of active ingredients used for this formulation type are cyfluthrin and spiroxamine.

Microemulsions (MEs) are defined as *"thermodynamically stable emulsions, which are formed spontaneously upon dilution of the oil surfactant mixture in water, because of their very low interfacial tension"*. They form a transparent solution because of the size of the emulsion droplets, which is smaller than 100 nm. Generally, MEs are prepared using a mixture of different type of surfactant, one with a very low HLB that is soluble in the oil phase, and a second one with a high HLB that is soluble in water. The total concentration of surfactant in a ME is usually on the order of 30%, which is high compared to that of emulsions (<5%). For this reason, MEs generally have a good biological profile, but may have some risk of phytotoxicity. Examples of MEs include Chinmix® containing beta-cypermethrin, and Flex-me® containing fomesafen.

One special formulation type in which the active ingredient is mostly in a dissolved state is the CS containing microcapsules. These formulations are best described as EWs coated with an impermeable wall rather than a surfactant or polymer layer (as with EWs). This method, whereby the active ingredient is dissolved in the encapsulated emulsion droplets, involves a rather complex technology and often requires a polymerization reaction at the interface. Once the capsules have been prepared they must be stabilized against sedimentation by the addition of dispersing agents and thickeners (as is the case for SCs); they will also require the addition of adjuvants to improve retention and wetting of the leaf surface. Although, this technology is clearly more expensive when compared to the above-mentioned formulation types, the reason for choosing this more complex process may be to reduce the acute toxicity of the formulation (in case of active ingredients with a high acute toxicity, such as organophosphates), to produce a controlled or retarded release of the active ingredient, or to be able to combine chemically incompatible active ingredients (or an active ingredient and adjuvants). Several encapsulation materials are available, of which melamine, gelatin, polyurethane, and acrylate walls are examples [10, 11].

The most frequent encapsulation method used in crop protection is by interfacial polymerization. In this case, an oil-soluble monomer (e.g., toluene di-isocyanate) is dissolved in the emulsified oil phase, where it can react with a reactive amine (e.g., ethylene diamine) in the surrounding water phase. The subsequent polymerization reaction takes place at the water–oil interface, which results in the creation of the capsule wall. Subsequently, by varying the thickness of the wall, an optimal release profile can be established. Examples of such products are Nemacur® and Mocab®. The release of the active ingredient from the capsule may be either gradual (due to diffusion through the capsule wall) or abrupt (by rupture of the wall material). The latter effect may be due either to mechanical stress when the spray hits the leaf surface, or to the drying process either on the leaf or inside the insect. Often, combinations of different release mechanisms can be incorporated into a single product so as to combine knockdown activity with residual efficacy, as has been achieved with Karate Zeon®.

9.3.2 Crystalline Pesticides in a Liquid Continuous Phase

One of the most important group of formulations involves the active ingredient being present as a solid, but dispersed in a liquid phase. Dispersions in water are generally described as SCs, and those in a water-immiscible solvent or oil, as ODs. Mixed ready-to-use formulations of a SC together with an EW are defined as SEs. According to the FAO definition, the oil-phase in a SE should be (or should contain) a second dissolved active ingredient. However, numerous cases are known where the oil phase in the SE consists of an adjuvant system to improve the biological efficacy of the crystalline active substance. Alternatively, this type of formulation can be seen as adjuvanted SC. As can be concluded from the data listed in Table 9.1, the SCs are the formulations that have demonstrated a growing importance over the past decade.

The typical properties of an active ingredient that allow the development of SC are a high melting point (preferably >80 °C) and a chemical stability in water so as to prevent hydrolysis of the active ingredient. Furthermore, a low water solubility is required in the case of SC, and of OD a low solubility in the solvent or oil, used in this particular case (evidently the latter situation is required to prevent crystal growth during storage). The main benefits of this technology are that highly concentrated products are possible, the dosing, handling and application of containers is easier, and the products are safer for the people working with them. Finally, both the production costs and the costs for the further ingredients are relatively low compared to solid and solvent-based formulations. Today, as farmers often prefer liquid solvent-free products, SCs are growing in popularity. However, one clear disadvantage of SCs in the broader sense is that these formulations can be very sensitive to minor changes in the recipe, and that consequently a greater expertise is required to develop a storage-stable product. Whenever SCs are prepared from systemic active ingredients, the biological performance is often lower than that of an EC containing the same active ingredient. To overcome this difficulty, several optimized SC concepts have been developed over the past decade. In the case of ODs, the continuous phase has been replaced with an oil/adjuvant system to improve the biological performance (see Section 9.1.4). In the case of adjuvanted SC or SE, this is achieved by incorporating adjuvants into the product. As most penetration-enhancers act on the waxy cuticle of the leaf, this requires a lipophilic adjuvant to be incorporated into the hydrophilic phase of the SC, though this might in turn lead to problems of incompatibility.

Typically, SCs are produced by premixing the crystalline active ingredient in an aqueous suspension with a wetting agent and dispersing agents. This pre-milling is performed with a colloid mill, and is also important for obtaining a good wetting of the active ingredient. Next, the slurry is passed several times through a bead mill to further comminute the crystals into particles smaller than a few micrometers in size. These bead mills are filled with glass or ceramic beads, and the efficiency of the milling process depends (among other factors) on the density of the beads, their degree of filling, the milling speed, and the throughput and viscosity of the slurry (for an extensive overview of this technology, see Refs [12, 13]). During the grinding process the surface area of the crystalline material is increased exponentially. Since the active ingredients used for SC are in general more lipophilic, wetting of the surface is needed to prevent reaggregation of the small particles during processing. Typically, the wetting agent used consists of small surfactants that have a high affinity for the surface, a high diffusion coefficient, and therefore cover the new created surface area more or less instantaneously. During and after the milling process these wetting agents are gradually replaced from the crystal surface by the dispersing agents used. The latter are surfactants with higher molecular weights, and often are even polymers with a high affinity for the surface and numerous binding sites to improve the irreversibility of the adsorption process. This process, which is referred to as *surface fractionation*, has been described extensively [14–16].

A schematic of possible dispersing systems is shown in Figure 9.4. Of the two stabilizing mechanisms generally available, one is based on steric stabilization, and

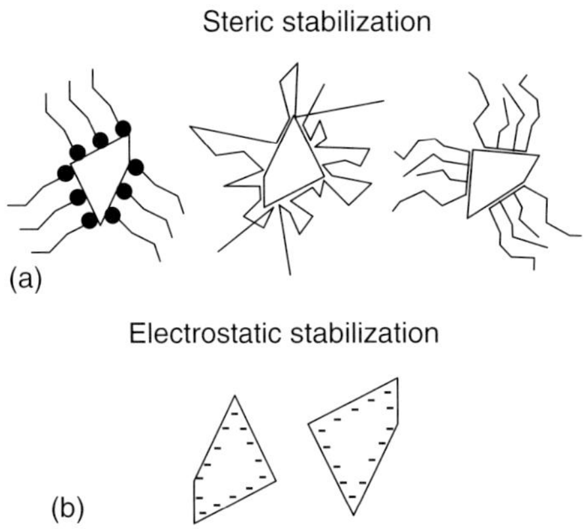

Figure 9.4 Schematic representation of the different dispersing systems used to disperse solids in a liquid phase. (a) Left to right: Using nonionic surfactant, which typically has only one anchoring point with the surface, polymers, and comb polymers; (b) Using surfactants that induce a (negative) charge on the surface and thus introduce electrostatic repulsion.

the other on electrostatic stabilization [17–19]. In the case of steric stabilization, the water-soluble part of the dispersing agent is mostly based on ethoxylated chains. As the water-solubility of these chains depends on their capacity to order water molecules surrounding the chain, the stabilizing effect will depend heavily on factors such as temperature and salt concentration. At higher temperatures and salt concentrations, these molecules lose their water solubility and consequently will induce a flocculation process. More recently, alkyl polyglucosides have been introduced as an alternative surfactant, but although these molecules do not depend very heavily on changes in temperature and salt concentration, their application to crop protection products is still limited. In order to prevent desorption of the surfactant molecules during storage, the affinity of the lipophilic part of the surfactant should match the active ingredient crystal surface as well as possible. Evidently, this affinity will differ for the different crystal phases of the active ingredient material, and will also be affected by the crystal polymorph [20, 21]. To overcome these aspects, mixtures of dispersing agents can be used or, in the case of polymers, polymers with different anchoring groups can be applied. A second effect where desorption of the surfactant molecules plays a role relates to crystal growth or Ostwald ripening during storage. With an increasing water solubility of the active ingredient molecules, which can be induced due to the presence of surfactant micelles in solution, the larger crystals will tend to grow at the cost of their smaller counterparts. Evidently this will lead to not only a reduction in storage stability but also to an increased sedimentation/caking; ultimately, it may even result in blockage of the filter during spray application.

In a final step, all further additives can be added such as biocides, antifreeze, and thickeners. Especially for SCs, this last step is crucial, because SCs must be storable for at least two years in an intrinsically "unstable" form. Due to the density difference between the active ingredient and the continuous phase, the particles will sediment, and to prevent this thickeners will be needed. Because of the presence of water, however, freezing of the product can cause instabilities in the dispersing and thickening system, resulting in its destabilization. Evidently bacterial growth should also be prevented. Some examples of SCs include rynaxypyr (Coragen®), sulcotrione (Mikado®), and epoxyconazol (Opus®).

In an effort to optimize SCs in terms of their biological performance, OD formulations have been the subject of much investigation. The main difference between ODs and SCs is that the continuous phase no longer consists of water but rather of a water-insoluble solvent that serves either as a penetration enhancer or as a carrier for adjuvants [22]. In addition to biological performance [23], the chemical stability of the active ingredient may serve as a reason for this type of formulation, and this is the case for many sulfonylurea OD formulations. Depending on the properties of the active ingredient, the behavior of the OD in the spray solution may be quite different; for a water-soluble active ingredient the OD will form an emulsion in the spray with the active ingredient being present dissolved in water. In the case of a water-insoluble active ingredient, however, the spray solution will consist of emulsion droplets formed by the solvent/oil phase, while in the water are suspended crystals of the active ingredient. On the leaf surface, following the evaporation of water, the active ingredient will be brought into contact with the oil phase, which assures the biological performance [22] (see also Figure 9.8). Whilst the productions of OD and SC formulations are comparable, the higher viscosity of the oil phase compared to that of water means that more time and a greater energy input are often required. Furthermore, these formulations require complicated thickener systems, since the polysaccharide-based thickeners commonly used in SCs are incompatible with solvent/oil-based systems. As an alternative, clays such as bentonite are often applied, though these have the disadvantage that they must be activated by small amounts of polar solvent, which often leads to problems in production scale-up processes. Typical examples of OD formulations have been prepared with the insecticides thiacloprid (Biscaya®, Proteus®) and spirotetramat (Movento®). An example of a herbicide OD is Atlantis®, based on mesosulfuron.

An alternative route to developing biological optimized formulations is to incorporate adjuvants into SC-type formulations. The majority of such products are known commercially as SCs, and occasionally also as SEs. In the latter case the adjuvant (or an adjuvant dissolved in an oil phase) is emulsified as in an EW in the continuous phase of the SC. Compared to ODs, the main difference from a formulation point of view relates to the type of adjuvant that can be used. In ODs (and SEs) the adjuvants are more lipophilic, whereas those that can be incorporated into SCs must be more hydrophilic. However, as the leaf cuticle is also more lipophilic, this brings limitations to the choice of adjuvant in these systems. A further aspect to be considered during the formulation development process, which relates to the use of hydrophilic surfactants, is the fact that these surfactants are used in high concentrations, and consequently they will form micelles in the concentrate which will in turn have an effect on the dissolved amount of active ingredient. Thus, Ostwald ripening and other crystal growth phenomena must be regarded with care.

Some example of adjuvanted SCs are based on spirotetramat (Movento®) or, in the case of SEs, on thiacloprid (Calypso®).

A final formulation to be mentioned in this category relates to the use of SCs for seed treatment. As can be seen in Figure 9.1, the market presence of FS formulations has been growing significantly during the past decade. The main differences compared to SCs relate to the dispersed phase in FS since, in SCs mostly one or two actives are suspended in the continuous phase. In FS formulations this number may be up to four, and pigments are also often added. Consequently, selection of the correct dispersing agent system is most relevant for FS. Under normal conditions the use of adjuvants is not required for FS formulations.

9.3.3
Pesticides in a Solid Matrix

A third group of formulations includes those in which the active ingredient is distributed in, or on, a solid matrix; typical formulations in this group are WPs, WGs, and GRs. Agrochemicals that are suited to these formulations in general have higher melting points, although an exception can be made for absorptive GRs, where the active ingredient can be dissolved in a solvent before being absorbed into the GR. Often, the possibilities of improving the biological efficacy of these type of formulation are very limited.

Although WPs have in the past been important within the agricultural industry, their market share is rapidly diminishing, due mainly to their intrinsic dust properties, and this trend is expected to continue even further. WPs are mostly produced from solid active ingredients with high melting points, with the crystalline materials being comminuted via a dry mechanical milling process, using a hammer or pin mill, or by air milling. During production, care must be taken to avoid the occurrence of dust explosions. In addition to the active ingredient, WPs usually contain solid surfactants as wetting and dispersing agents, and an inert filler to improve their storage stability.

As a more modern alternative to WPs, WGs have been developed. Although frequent mention has been made [1, 7] that these types of formulation would also be a preferred alternative to SCs, market observations have not confirmed this suggestion. One reason for this observation is that the dosing of WG is carried out by weight rather than by volume, which is inconvenient at the farm level. One major advantage of WGs is their convenience in packaging (disposal) and handling. Typically, WGs are non-dusty, free-flowing particles that are several millimeters in size and which, when diluted in the spray solution slowly fall apart, although the individual particles may be easily redispersed upon agitation. The ability to pack WGs into cartons or even paper bags is also a major advantage compared to liquid products.

The technology required to produce WGs is rather complex, and involves a large initial capital investment. Furthermore, as the running costs of WGs are among the highest of the different formulation types, only those products with a high active ingredient content and very large sales volumes are suited to WG development. WGs can be produced using various technologies that include extrusion granulation, fluid bed granulation, spray-drying, high-speed mixing agglomeration, and pan granulation. The latter two processes are used less frequently nowadays, mainly due to difficulties in controlling the process and because of large variabilities in product quality. Because of the high capital investment required, the choice between these technologies often depends on the availability of equipment. A further factor that determines the technology used relates to the physico-chemical properties of the active ingredient. Compared to extrusion, spray-drying and fluid bed granulation each require high temperatures (>70 °C) during processing, whereas extrusion can be conducted at about 40 °C. A second aspect relates to the comminution of the active ingredient which, for spray-drying, is performed using a wet milling step (as described in Section 9.3.2), while for other techniques dry milling is sufficient. The latter approach is preferred for active ingredients that are less stable against hydrolysis.

Granulation via extrusion involves several steps. The active ingredient is first milled by air-milling, and then mixed with other components of the recipe, and wetted. This stage of the process can be achieved batch-wise in, for example, a Lodiger-type of mixer or in a continuous process using a high-speed mixer (e.g., Schugi mixer). The resultant wet mass or paste, which contains 10–20% of water, is then pressed through an extrusion screen to produce elongated worm-like granules (see Figure 9.5). Following extrusion, the water must be removed from the granules by drying. One major advantage of extrusion as a granulation technology is that it can be carried out either as a batch or as a continuous process although, as WGs are related to large-volume products, the use of a continuous process will reduce production complexity. Extrusion is a technically difficult production, with the compaction step in particular requiring close attention; indeed, if this step is not sufficiently well controlled then dispersion in the spray solution may be problematic (see Figure 9.6).

Fluid bed granulation is a spray-drying technology where the active ingredient/additive powder is fluidized by hot air blown though a screen in the base of

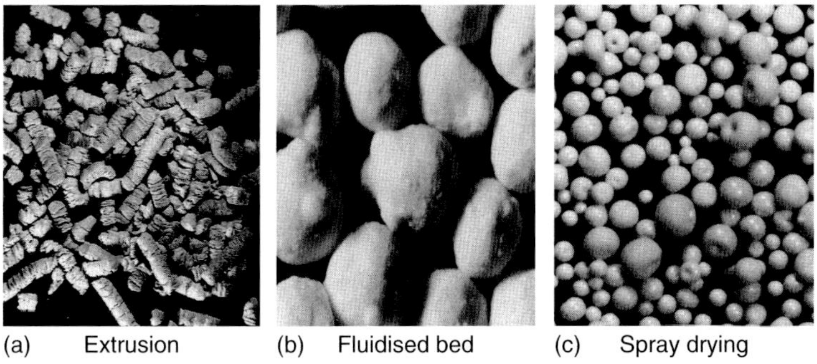

Figure 9.5 Photographs of the three most relevant water-dispersible granule (WG) types of formulation, indicating the differences in granule form and size. The average granule sizes (d_{50}) were (a) extrusion WG : ~700 μm; (b) fluidized bed WG : ~500 μm; and (c) spray-dried WG : ~250 μm.

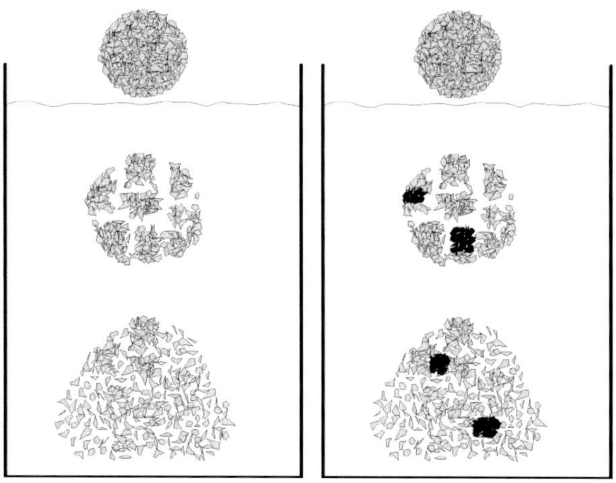

Figure 9.6 Schematic representation of the redispersion process of water-dispersible granule (WG). (a) The granule is brought into contact with water and slowly falls apart in isolated crystals; (b) An inadequate redispersion due to suboptimal production conditions.

the equipment. The liquid water/binder solution is sprayed on top of the fluidized powder, so as to induce an agglomeration of the WGs. During this process, the WG particles are dried by the hot fluidizing air in a single-step process. In general, fluidized bed is a batch technology, and the WGs produced in this way generally have good dispersing properties. A typical fluidized bed WG is shown in Figure 9.5.

A final WG production process relates to spray-drying, which involves spraying a slurry or solution into the top of a spray tower, against a flow of hot air. As the droplets pass to the bottom of the tower, the liquid contained within them evaporates, causing the remaining material to form a spherical particle with an average size of a few hundred micrometers (see Figure 9.5). A combination of fluidized bed technology at the bottom of the spray tower allows a further agglomeration of these particles into larger, less dusty WGs. Similar to the fluidized bed process, spray-drying produces WGs with very good redispersion properties. Unfortunately, the disadvantages of the process are the complexity and high operating costs, as well as the high energy input required to evaporate large amounts of liquid.

As noted above, the three technologies discussed each lead to products with different properties, including size, shape, and redispersion kinetics in water. Moreover, each of these techniques requires significant investment and has high running costs; consequently, these technologies are most appropriate for producing highly concentrated formulations and for active ingredients that require a high dose rate per hectare. In order to dilute the cost pro hectare, however, active ingredient contents between 50% and 80% are very common. In relation to the requirement for high active ingredient concentrations in WGs, it is very difficult to build in adjuvants to improve the biological performance. In fact, the possible active ingredient loadings in adjuvanted WGs would need to be of the order of

30%, and this would prove problematic for this technology from a commercial viewpoint.

During recent years, granular formulations (GRs) have shown a strongly decreasing relevance in the crop market. These formulations are not intended to be diluted in a spray liquid, but rather to be distributed directly in the field. In general, they contain lesser amounts of active ingredient, because a homogeneous distribution of GR is possible only when larger amounts are distributed per hectare. The GRs may be up to several millimeters in size, they should be dust-free, and they should disintegrate in the soil to release the active ingredient Two types of GR are commonly used: the first type is based on a coating process where small particles are adsorbed onto a carrier (e.g., sand) by using a "sticker"; and the second type is based on a porous carrier (e.g., silica, clay, or organic material) into which an active ingredient solution is absorbed. In the latter case, the solution may be loaded with surfactants so as to improve the soil activity of the active ingredient

9.4 Bioavailability Optimization

The design of a stable formulation, which can be easily diluted with water and evenly distributed by spraying, is often insufficient for the best biological performance of a modern agrochemical. In order specifically to increase the bioavailability of an agrochemical, further formulation ingredients known as *adjuvants* must be added. An adjuvant is defined as any substance in a formulation, or added to the spray liquid, that modifies the activity of a plant protectant or the spray characteristics, without having its own biological activity [24]. Bioavailability optimization refers to the quest for a well-designed distribution behavior of the active ingredient on or in the plant which is required for an optimum activity against the target organisms, whether fungal pathogens, insects, mites, or weeds.

The bioavailability of an agrochemical is affected by a vast number of factors which depend not only on the formulation but also on the intrinsic properties of the active ingredient, such as molecular weight and physico-chemical properties. In order to understand the field performance of a plant protectant, the complete sequence of steps affecting the efficacy must be considered, starting with spray formation at the nozzles, spray retention, spray deposit formation, deposit properties, penetration of the active ingredient into the plant pathogen, the plant pest organism or into the leaf, followed by redistribution in the plant tissue and long distance translocation within the plant. The first of these steps – spray formation, retention, and deposit formation – are solely affected by the formulation. Most important for the efficacy of a plant protectant is its release and bioavailability from the spray deposit, sorption into the cuticle, and penetration into the leaf tissue. These steps are greatly affected by both the formulation components and the physico-chemical properties of the active ingredient. The local, intercellular redistribution of the agrochemical within the leaf tissue, the plant pathogen or the insect/mite, and also the long distance translocation in the vascular tissue of xylem and phloem, is solely dependent on the physico-chemical properties of the active ingredient, and cannot be directly manipulated by the formulation.

Finally, a formulation with a well-designed bioavailability must insure the specific needs of agrochemicals. In the case of fungicides and insecticides, this includes a longlasting and well-balanced protective and/or curative/systemic performance, whilst in the case of a (leaf) herbicide it must include a high penetration and translocation, with as little as possible of the active ingredient being left on the leaf surface.

The most important factors affecting the bioavailability of agrochemicals, which can be manipulated by changes to the formulation, are highlighted in the following subsections.

9.4.1 Spray Formation and Retention

Following spray droplet formation at the nozzles of the spray boom, all droplets should reach the plant surface, ideally without any loss by drift caused by side winds or low retention.

It is not only the nozzle type, but also the adjuvants and other formulation components that can affect the droplet size spectra of the spray [25, 26]. Usually, any adjuvant or formulation that reduces the length of the liquid sheet below the nozzle causes larger droplets, as it is often the case with emulsions. A longer liquid sheet due to an increased viscosity, usually at a high concentration of water-soluble adjuvants (polymers), causes a reduction in the average droplet size and increases the risk of drift losses. A typical foliar spray has a mean droplet diameter of 150–300 μm, and the spray drift reduction is best if the proportion of small droplets (<100 μm diameter) present is low. Unfortunately, spray retention on difficult-to-wet leaves requires the opposite effect, with retention being increased in the case of small spray droplets. In order to improve both spray drift and retention, the formulation must be optimized for a good adhesion of especially the larger droplets of the size spectrum in the spray.

Spray retention describes the fraction of spray droplets that remains ("sticks") onto the treated leaf surface after impact. After being retained, these droplets adhere to the surface without changing the contact area and evaporate over this initial area; alternatively, they may continue to spread after impact until they are completely evaporated. The *final coverage* – the leaf area covered by spray deposit after evaporation – is the product of spray retention and droplet spreading.

Retention depends heavily on the leaf surface properties, with smooth leaf surfaces without crystalline waxes (e.g., easily wettable apple or vine leaves) always showing a high retention, while difficult-to wet plants characterized by the presence of crystalline surface waxes show a reduced retention up to complete repellency, depending on the density of the crystalline waxes. Many important crops belong to the latter group, including cereals and cabbages, early stages of corn and soybeans, all of which have crystalline (epicuticular) waxes on their leaf surfaces. The reduced retention on these crops can be fully compensated by the addition of specific

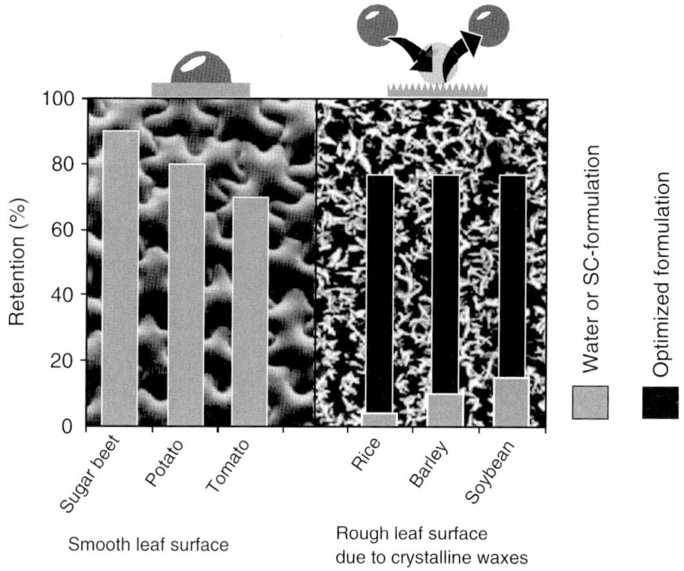

Figure 9.7 Spray retention as affected by leaf surface structure and formulation.

surfactants, either added as tank-mix or already included in the formulation recipe (Figure 9.7) [27].

Two different mechanisms exist by which surfactants can improve retention:

- The first mechanism is based on surfactants that can quickly decrease the surface tension of the freshly formed spray droplets at the nozzle. These surfactants must diffuse rapidly from the droplet interior to the surface, which leads to a low dynamic surface tension. This diffusion-controlled process is most efficient for adjuvants with a low molecular size and with a high critical micelle concentration (cmc); this causes a high concentration of molecules in the spray droplet before micelle formation. Next to the surfactant properties, by far the most important factor is the surfactant concentration in the spray solution, as coverage of the droplet surface with surfactant molecules is directly correlated with retention enhancement. Therefore, the absolute concentration of an adjuvant in the spray solution is decisive for best retention, and not the surfactant application rate per hectare. An optimal retention can usually be achieved with an adjuvant concentration of $0.5–1.0\,g\,l^{-1}$ in the spray solution [27].
- The second mechanism is based on polymers which change the viscosity of the spray solution. These provide the droplets with an internal structure, reduce the droplet deformation by relaxation after the impact, and enlarge the time for "wetting." Best-suited in this role are polymers with a high, so-called "extensional" viscosity – that is, a low shear viscosity under the application conditions, but a high viscosity of the expanding and retracting droplet. Polyethylene glycols, polyvinyl alcohols, or cellulose derivatives are examples of polymers used for the improvement of spray retention (in the range of $0.2–2\,g\,l^{-1}$).

It must be considered that, under field conditions, the retention can be higher than would be expected from greenhouse trials. The wettability of leaves also depends on numerous environmental factors affecting the surface wax morphology, such as water stress, radiation, temperature, and relative humidity [28]. Even more important is the presence of contaminants such as dust, soil particles, or microorganisms on the leaf surface, as this leads to a higher spray retention on difficult-to-wet plants in comparison to glasshouse-grown plants with an intact surface wax layer.

9.4.2
Spray Deposit Formation and Properties

Spray deposit formation and properties have a major impact on the bioavailability and performance of agrochemicals, although these phenomena are poorly understood. Some effects are solely formulation-based, such as spray droplet spreading (which is mainly affected by static surface tension and leaf surface structure) or hygroscopy, which can be adapted by humectants. Other effects are also dependent on the physico-chemical properties of the active ingredient, such as crystallization behavior and rain fastness (both of which are influenced by solubility), losses of the agrochemical by hydrolysis, evaporation, or UV-irradiation.

For the best performance of a plant protection agent, the spray deposit on the leaves must be stable over time and must release the active ingredient timely and well-dosed for initial penetration and – in the case of longlasting efficacy of insecticides and fungicides – to provide a protective control of the crop plants against pests and diseases. Unfavorable physico-chemical properties may lead to losses of the active ingredient from the deposit: a high water solubility can reduce rain fastness, while a vapor pressure exceeding 10^{-2} mPa may increase the risk of vapor loss [29].

A vast number of factors have been identified that influence the formation and properties of spray deposits. Environmental factors such as humidity, wind, and temperature affect the evaporation time of water and solvents (if present in the formulation), leading to a limited or prolonged spreading time of the spray droplets. Application techniques such as nozzle type, pressure, ground speed of the sprayer, or water application rate, also play very important roles in deposit formation. Especially, the droplet size spectrum of the spray is decisive not only for drift control and retention, but also for the deposit formation and properties. Larger droplets show a greater tendency for ring formation, and the uniformity in the distribution of the dissolved or particulate active ingredient across the droplet contact area can be strongly affected [30]. Deposit structure is also affected by biological factors such as leaf orientation, the presence of trichomes, or crystalline waxes on the leaf surface. Finally, the formulation type and the included formulation chemicals are of utmost importance for the deposit structure. Some examples of different spray deposits and other influential factors are illustrated in Figure 9.8.

The spray deposit must release the agrochemical in a diffusible, solubilized form; consequently, the most important point is the physical state of the active

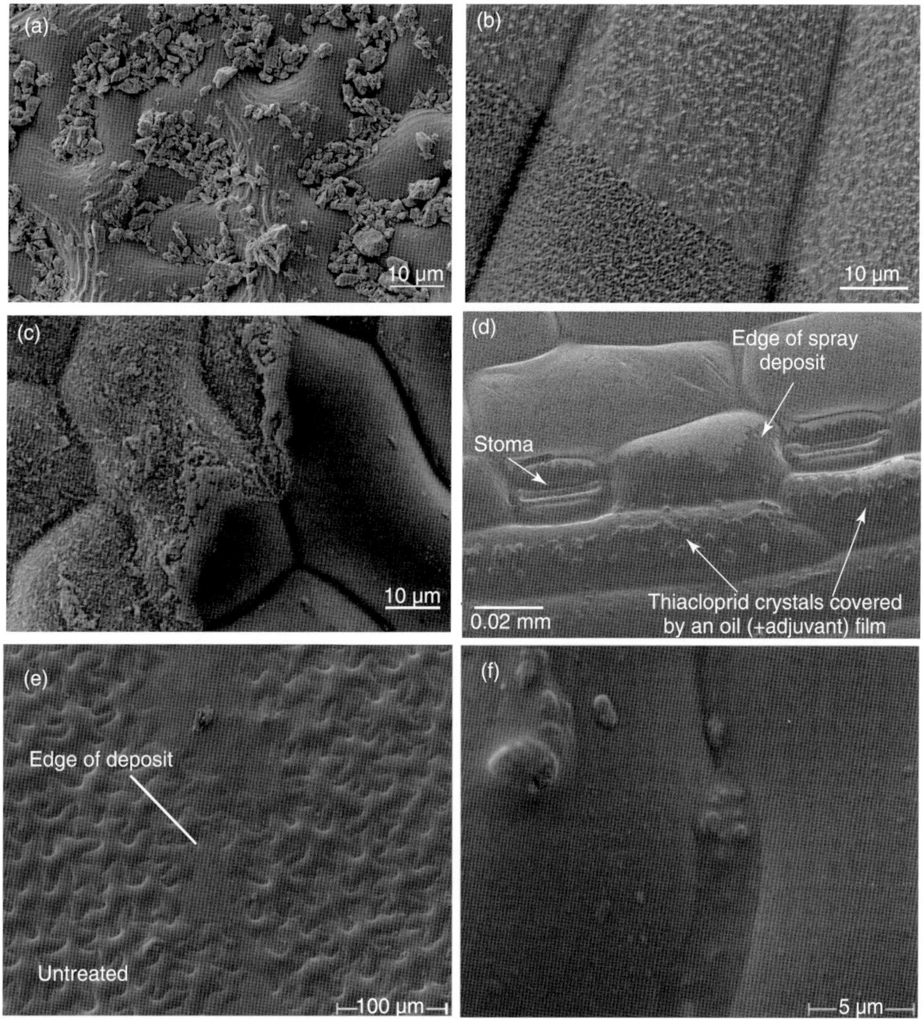

Figure 9.8 Examples of spray deposits. Scanning electron microscopy images of leaf surfaces after cryofixation. (a) WG-formulation on apple leaf (mancozeb, Dithane® WG75); (b) EC-formulation on barley leaf (fluoxastrobin and prothioconazole, Fandango® EC 200); (c) SC-formulation on cotton leaf (flubendiamide, Belt® SC 480); (d) OD-formulation on barley leaf (thiacloprid, Biscaya® OD 150); (e,f) Adjuvanted SC formulation on apple leaf (spirotetramat, Movento® SC 100), showing slight ring formation (e). The center of the deposit at higher magnification shows active ingredient-particles covered by an adjuvant film (f).

ingredient. A solid deposit with a crystalline active ingredient – formed by WG- and non-adjuvanted SC-formulations (Figure 9.8a,c) – causes a significantly reduced bioavailability in comparison to a liquid-like deposit, with the active ingredient in an amorphous state or even largely solubilized and having an intimate contact to the cuticle surface. Most EC-formulations form such liquid-like deposits; adjuvanted SC- and OD-formulations generate spray deposits with active ingredient particles covered by a thin liquid film of adjuvants or, in the case of an OD-formulation, a film of oil which may contain further solubilized, and penetration-enhancing additives (Figure 9.8b,d–f).

The higher the melting point of an agrochemical, the lower is the solubility and dissolution rate from crystalline deposits, as the melting point is related to the crystal lattice energy, which in turn leads to a limited penetration [31].

9.4.3 Cuticular Penetration

The aerial parts of a plant are covered by the cuticle, a lipophilic composite membrane which is synthesized by the epidermis of the leaves, fruits, and flowers. The main function of the cuticle is to protect the plant against water loss; however, this function unfortunately leads to the creation of a very strong barrier against the penetration of agrochemicals. The principal component of the cuticular membrane is *cutin*, an insoluble amorphous polyester of crosslinked hydroxy-fatty acids and hydroxyepoxy-fatty acids and mainly aliphatic waxes [32]. The waxes are deposited on the cuticle surface (epicuticular waxes), and are also embedded in the upper layer of the cuticle, forming the so-called "cuticular proper." The intracuticular waxes in the cutin matrix are mainly responsible for the rate-limiting barrier of cuticles for foliar uptake (Figure 9.9). This has been clearly demonstrated with isolated cuticles, by comparing permeabilities prior to and after extraction of the waxes; typically, the permeability of organic compounds was increases in the latter case by up to 9200-fold [33].

The penetration through the cuticle is a passive, diffusion-controlled process:

$$\frac{dM}{dt} \propto K \times D \times (C_o - C_i)$$

The steady-state flow rate of an organic solute or an agrochemical is a function of the partition coefficient K (partition spray deposit/cuticle and cuticle/epidermal cell wall), a function of diffusion coefficient D, and the concentration difference between outer and inner cuticle surface (= driving force). The partition coefficient is strongly affected by the lipophilicity of the agrochemical, and diffusion is mainly determined by the molecular size and shape (this is an approximate description of the factors of influence; for further details, see Refs [35] and [36]).

The most crucial point for understanding cuticular penetration is the uncertainty about the concentration of the agrochemical in the spray droplet during the initial evaporation phase, and afterwards in the deposit. The concentration C_0 changes during the process of deposit formation, and is mostly not clearly known for the final deposit structure. Therefore, the prediction of penetration behavior for an

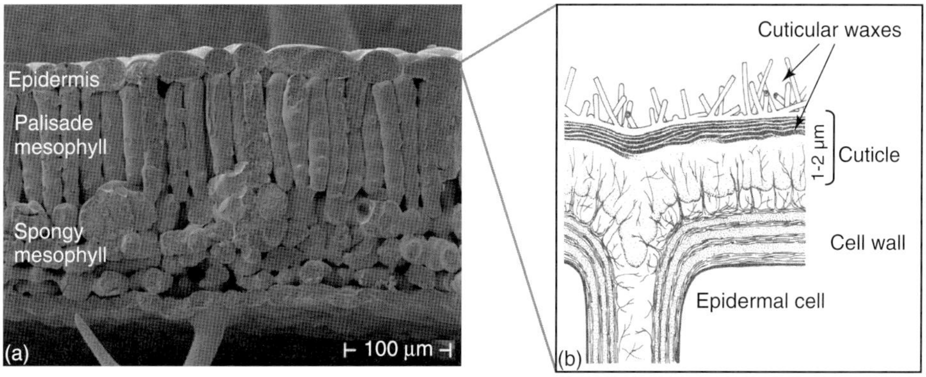

Figure 9.9 The cuticle as a barrier for leaf uptake of agrochemicals. (a) Cross-section of a *Solanum nigrum* leaf. Scanning electron microscopy (SEM) image after freeze fracture and slight freeze etching. The epidermal cells are covered by the thin cuticular membrane; (b) Internal structure of the cuticle. Adapted from Ref. [34].

agrochemical is extremely difficult. The availability of an active ingredient from the drying spray droplet depends heavily on its physico-chemical properties, and can be further manipulated by the formulation. In this case, the solubility and lipophilicity of the active ingredient are the most important parameters.

Agrochemicals with a low lipophilicity ($\log P < 2$) and a high water solubility show a limited uptake into the cuticle and the leaf tissue, due to an unfavorable distribution coefficient water/cuticle and low permeability; however, once having crossed the cuticle further systemic redistribution of the active ingredient in the plant is excellent. Many herbicides belong to this category of compounds. In this case, adjuvants have the most pronounced potential to improve penetration.

With increasing lipophilicity the charging of the cuticle and other lipophilic compartments becomes more pronounced, as long as the solubility is sufficiently high. These agrochemicals show only a limited systemic translocation within the plant, because the active ingredient is always trapped in the lipophilic phases. Cuticular uptake and systemic redistribution are further reduced in the case of compounds with very limited solubility in water and organic solvents.

For water-soluble agrochemicals (mostly herbicides), it remains unclear as to whether they pass the cuticle via a physically distinct pathway, the so-called "aqueous pores" [37, 38]. The importance of this pathway for the penetration of hydrophilic solutes has still not yet been completely resolved [39].

If the leaf uptake of an active ingredient is limited but essential for good biological performance, then an optimization process for improved foliar penetration of the active ingredient will be needed. For this purpose, isolated cuticle membranes represent a very valuable tool for the detailed and systematic analysis of the penetration behavior of agrochemicals, in combination with formulation components and specific adjuvants.

9.4.3.1 Cuticular Penetration Test

For this test, apple leaf cuticles are isolated from leaves taken from trees in an orchard, as described by Schönherr and Riederer [40]. Only the astomatous cuticular membranes of the upper leaf surface lacking stomatal pores are obtained. Discs with diameters of 18 mm are punched out of the leaves and infiltrated with an enzymatic solution of pectinase and cellulase. The cuticular membranes are then separated from the digested leaf cell broth, cleaned by gently washing with water, and dried. After storage for about four weeks, the permeabilities of the cuticles reach a constant level, and the cuticular membranes are ready for use in the penetration test.

When the cuticular membranes are applied to the diffusion vessels the correct orientation is important, with the inner surface of the cuticle facing towards the inner side of the diffusion vessel (Figure 9.10a). A spray droplet of 1–5 μl is applied with a pipette to the outer surface of the cuticle, after which the diffusion vessel is turned around and carefully filled with the acceptor solution. The choice of acceptor medium depends on the problem to be solved, but water (buffered to pH 5.5) is used to simulate the apoplast as a natural desorption medium at the inner surface of the cuticle. If more lipophilic agrochemicals are tested, however, an acceptor containing water mixed with an organic solvent is often better suited. In this case, the distribution coefficient of the active ingredient between the cuticle and acceptor is changed, leading to a stronger desorption of the agrochemical.

The diffusion vessels filled with the acceptor and stirrer are transferred to a thermostatically controlled stainless steel block which ensures not only a well-defined temperature but also a constant humidity at the cuticle surface with the spray

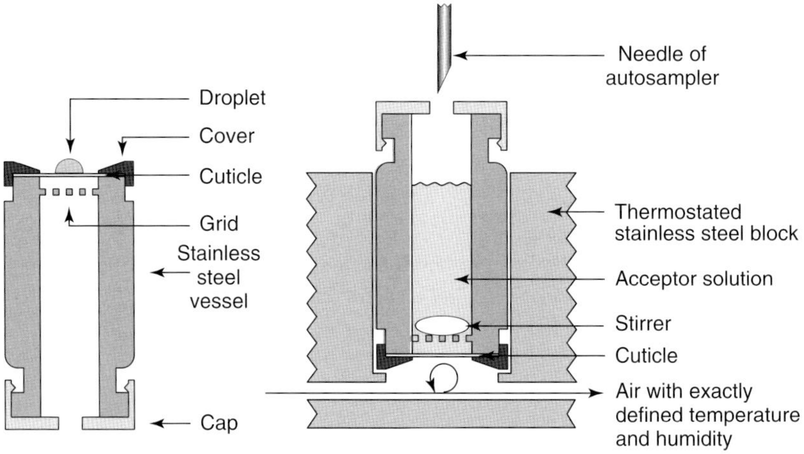

Figure 9.10 The cuticle penetration test. (a) Cuticular membranes are fixed to one side of a diffusion vessel and a single droplet of test solution (1–5 μl) is applied with a pipette; (b) The diffusion vessel is then turned around, filled with acceptor solution and fixed in a thermostatically controlled stainless steel block with exactly controlled temperature and humidity. Aliquots of acceptor solution are removed regularly by an autosampler and the concentration of analyte is estimated using HPLC.

deposit (Figure 9.10b). Aliquots of the acceptor are removed regularly using an auto sampler, and the active ingredient content is estimated using high-performance liquid chromatography (HPLC). All data points are finally combined to produce the penetration kinetic. As the degree of variation in the penetration barrier of the cuticles is high, at least 10 measurements must be taken.

This cuticle penetration test is a further developed and adapted version of the simulation of foliar uptake (SOFU) technique, as originally described by Schönherr and Baur [41]. It is well suited for performing systematic and mechanistic studies on the effects of formulations, adjuvants, and solvents on the penetration of agrochemicals. Moreover, it can be used not only for screening many different adjuvant types but also to analyze the optimal ratio of adjuvant (or combination of adjuvants) to the active ingredient in the formulation, and to examine any effects of temperature or humidity on penetration. Also, by specifically varying the test conditions, estimates can be provided of whether the low penetration of an active ingredient is dependent on a limited surface availability from the deposit, or on the cuticular penetration barrier. This background knowledge is important when optimizing a formulation, as adjuvants that improve surface availability differ from those that decrease the penetration barrier.

9.4.3.2 Effect of Formulation on Cuticular Penetration

The cuticular penetration of agrochemicals can be influenced considerably by adjuvants, whether integrated into the formulation recipe or added as a tank mix to the spray solution. Besides oils, methylated seed oils (MSOs), humectants, and ammonium salts, the most important adjuvants are surfactants (surface active agents). Among the surfactants, the nonionic types are the most effective penetration enhancers (see Ref. [24] for further information on different types of surfactant). The most widely used surfactants are alcohol ethoxylates or alkoxylates, alkylphenol ethoxylates, organosilicones (e.g., trisiloxane ethoxylates), crop oil ethoxylates, and ethylene oxide/propylene oxide polymers (EO-PO block polymers). The effect of adjuvants on cuticular penetration is an extremely complex process, and there is no one simple or unifying mechanism involved; different surfactants exert different influences on different agrochemicals in different target species [42, 43]. Consequently, it is impossible to predict the effect of a given surfactant on a certain agrochemical. Nonetheless, the MoA of adjuvants can be grouped into two different categories, related to an increased penetration by either indirect or direct effects.

Indirect effects of adjuvants are mainly caused by changing the spray deposit properties, in a manner that the availability of the agrochemical from the deposit is improved. This may be effected by:

- a better solubility of the active ingredient in the deposit;
- the avoidance of active ingredient crystallization in the deposit;
- in case of active ingredient precipitation, precipitation in an amorphous state;
- a longlasting resolubilization of the active ingredient by high humidity/dew, facilitated by humectants; and
- better or less coverage/spreading.

Alternatively, an indirect effect may be caused by an improved/reduced cuticular uptake of the active ingredient by changing the distribution coefficient deposit/cuticle.

A *direct effect* of adjuvants on cuticular penetration is a reduction of the penetration barrier of the cuticle by swelling agents or plasticizers [44]. Typically, MSOs and some alcohol ethoxylates (e.g., Genapol® C050) are well-known swelling agents that reduce the size selectivity of cuticular membranes and are especially active as penetration enhancers at low temperatures [45]. The accelerator action of such adjuvants is linked to their own penetration behavior in comparison to the penetration kinetic of the active ingredient. Both kinetics must be more or less synchronized, and depend on the mobilities in the cuticular membrane as well as the driving forces. Therefore, the same accelerator may not work equally well with different actives, while its activity will also depend on the plant species and temperature [45].

The penetration behavior of an active ingredient is heavily dependent on its physico-chemical properties, although it can be specifically adapted for use in different crops by the development of specific formulation types (Figure 9.11).

Some examples will now be provided, for all three indications, of how bioavailability is affected by the formulation. In the case of tebuconazole, the SC 430 formulation shows a very limited penetration, and is best suited to applications in crops that are very sensitive to any phytotoxic side effects of tebuconazole (e.g., stunting) and/or the effects of the formulation components of the EW 250 (both, the solvent and the included surfactant can be phytotoxic). The SC 430 formulation provides a predominantly preventive MoA while the addition of a tank-mix adjuvant can significantly improve the penetration of tebuconazole SC 430. This approach is regularly taken in areas where farmers are used to working with tank-mix adjuvants. The EW 250 was developed mainly for cereal use; indeed, this formulation was optimized for both high penetration and excellent retention, as the leaves of cereal crops are always covered by crystalline waxes. This formulation offers the active ingredient in a completely solubilized form, which enables a rapid distribution/uptake into the cuticle, while the included alcohol alkoxylate with plasticizer properties efficiently increases the penetration (Figure 9.11a). The penetration kinetic shows the typical shape produced by a formulation containing a plasticizer adjuvant, with an initially high penetration rate that then levels off. Such behavior can be explained by the self-penetration of the adjuvant leading to a temporary reduction in the penetration barrier of the cuticle.

The penetration behavior of different formulation types of spirotetramat, as estimated by the cuticle penetration test, are shown in Figure 9.11b. As expected, the SC 240 with no included adjuvants has the lowest penetration rate, whereas an adjuvant (e.g., ethoxylated crop oil) – whether added as a tank-mix or as a built-in additive (e.g., SC 100) – significantly increases the penetration. The highest penetration, however, is facilitated by the OD 150 formulation, which is based on a crop oil as liquid phase containing the adjuvant. As the oil neither evaporates nor penetrates, it forms a liquid film on the spirotetramat particles and provides the active ingredient in an available form for long periods of time (see also Figure 9.8f).

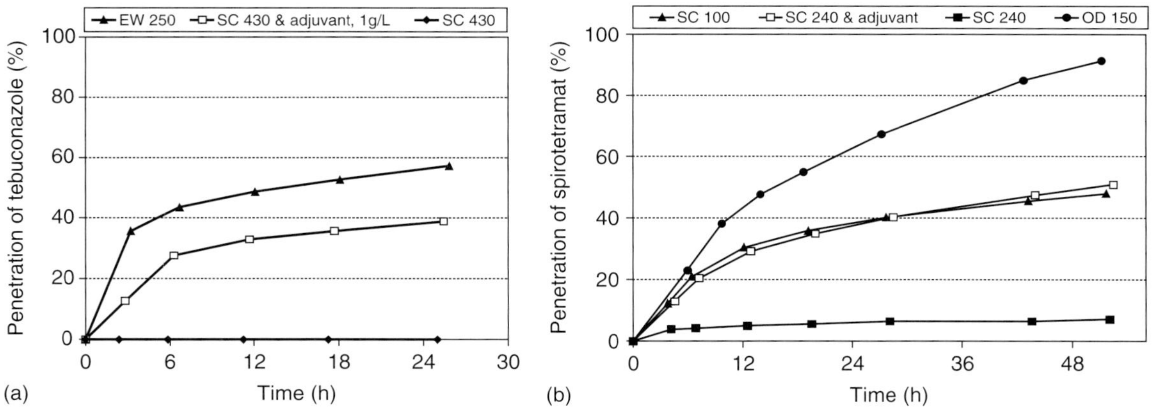

Figure 9.11 Cuticular penetration as affected by formulation type and adjuvants. (a) Tebuconazole (Folicur®, 0.5 g a·i·l^{-1}): the very limited penetration of SC 430 can be significantly improved by a tank-mix with adjuvant (alcohol alkoxylate); EW 250 has the highest penetration rate because of the included adjuvant and a completely dissolved active ingredient; (b) Spirotetramat (Movento®, 0.3 g a·i·l^{-1}): the adjuvanted SC 100 with improved penetration similar to the SC 240 tank-mixed with the same adjuvant (ethoxylated crop oil); the OD 150 shows the typical longlasting penetration enhancement.

The choice of a formulation depends on the crop and the pests to be controlled, and also on local practices in the different countries [46]. The OD 150 formulation has the highest risk of plant incompatibility; the adjuvanted SC 100 shows an excellent crop safety while maintaining the level of biological efficacy. In general, the SC 240 formulation is used in countries where the farmers prefer to add their own tank-mix adjuvants.

In the case of foliar-applied herbicides with high water-solubility, the formulation type has no influence on penetration as the active ingredient is always completely dissolved in the spray solution. However, such herbicides often need specific adjuvants in order to improve their cuticular penetration, and consequently other factors may define the selection process of formulation type: such factors include the stability of the active ingredient and the possibility of including complex mixtures of adjuvants into the recipe. In particular, weak acid herbicides show a significantly improved leaf uptake in the presence of ammonium sulfate, and the penetration is further – often synergistically – enhanced by a combination of ammonium salt and surfactant or MSO. Additionally, the leaf uptake of these herbicides is more dependent on humidity and adjuvants prolonging the spray droplet drying time [47].

These examples demonstrate the strong influence of formulation on the bioavailability of agrochemicals. Whilst laboratory tests can provide valuable information on penetration behavior, the final recommendation/selection of a formulation must be evaluated initially in the greenhouse, followed by a broad testing under field conditions in all relevant crops and regions of the world.

9.5 Conclusions and Outlook

The selection of the best-suited formulation type depends on a variety of factors, among which the most important are the physical chemical properties of the active ingredient, its biological profile, and the cost of formulation ingredients and production. Once a formulation type has been selected, the development of the formulation involves an optimization process with two main goals: (i) a formulation recipe must be developed that ensures a (physically) stable and easy-to-handle concentrate; and (ii) the best-suited adjuvants must be identified in order to optimize the bioavailability and, ultimately, the biological performance in the field. Both of these processes are heavily dependent on the physico-chemical properties of the agrochemical and the adjuvants to be included into the formulation. Unfortunately, the most important factors for an optimized biological activity, such as spray deposit properties, cuticle penetration kinetics, and adjuvant effects on leaf uptake, are poorly understood or are extremely complex, with no simple linear relationship to any single physico-chemical parameter of the plant protectant. Furthermore, it has often been found extremely difficult to combine the preferred adjuvants into the selected formulation type, which makes the development of a new formulation difficult and often requires innovative steps. Consequently, more systematic and fundamental investigations are required to characterize and further improve the current knowledge on the MoA of adjuvants, and the impact of physico-chemical parameters on the bioavailability of agrochemicals.

References

1. Seaman, D. (1990) *Pestic. Sci.*, **29**, 437–449.
2. Rodham, D.K. (2000) *Curr. Opin. Colloid Interface Sci.*, **5**, 280–287.
3. Mulqueen, P. (2003) *Adv. Colloid Interface Sci.*, **106**, 83–107.
4. Elsik, C., Tann, S., and Kirby, A. (2005) in *Environmental Fate and Safety Management of Agrochemicals*, ACS Symposium Series, Chapter 26 (eds J. Clark *et al.*), American Chemical Society, Washington, DC, pp. 297–308.
5. Knowles, A. (2006) *Outlook Pestic. Manag.*, **17**, 99–102.
6. Green, J.M. and Beastman, G.B. (2007) *Crop Prot.*, **26**, 320–327.
7. Knowles, A. (2008) *Environmentalist*, **28**, 35–44.
8. Jönsson, B., Lindman, B., Holmberg, K., and Kronberg, B. (1998) *Surfactants and Polymers in Aqueous Solution*, John Wiley & Sons, Inc.
9. Binks, B.P. (1998) *Modern Aspects of Emulsion Science*, The Royal Society of Chemistry.
10. Boh, B., Sajovic, I., and Voda, K. (2003) Microcapsules patents and products, in *Microspheres, Microcapsules and Liposomes* (eds. R. Arshady and B. Boh), Citus Books, London.
11. Beestman, G.B. (2003) in *Chemistry of Crop Protection: Progress and Prospects in Science and Regulation* (eds G. Voss and G. Ramos), Wiley-VCH Verlag GmbH, p. 272.
12. Kwade, A. and Schwedes, J. (2007) in *Handbook of Powder Technology*, Particle Breakage, Vol. 12, Chapter 6 (eds A.D. Salman, M. Ghadiri, and M.J. Hounslow), Elsevier, pp. 251–382.
13. Kwade, A. (1999) *Powder Technol.*, **105** (1-3), 14–20.
14. Vermeer, A.W.P. and Koopal, L.K. (1998) *Langmuir*, **14** (15), 4210–4216.
15. Cohen Stuart, M.A., Scheutjens, J.M.H.M., and Fleer, G.J. (1980) *J. Polym. Sci., Polym. Phys. Ed.*, **18**, 559.
16. Koopal, L.K. (1981) *J. Colloid Interface Sci.*, **83**, 116.
17. Tadros, T.F. (1997) *Solid/Liquid Dispersions*, Academic Press.
18. Myers, D. (1999) *Surfaces, Interfaces, and Colloids, Principles and Applications*, Wiley-VCH Verlag GmbH.
19. Sato, T. and Ruch, R. (1980) *Stabilization of Colloidal Dispersions by Polymer Adsorption*, Marcel Dekker Inc.
20. Olenik, B. (2012) Polymorphism and the organic solid state: Influence on the optimization of agrochemicals, in *Modern Methods in Crop Protection Research*, (eds P. Jeschke, W. Krämer, and U. Schirmer), Wiley-VCH Verlag GmbH, Weinheim.
21. Brittain, H.G. (1999) *Polymorphism in Pharmaceutical Solids*, Marcel Dekker Inc.
22. Vermeer, A.W.P. and Baur, P. (2007) *Pflanzenschutz-Nachr.*, **60** (1), 7–26.
23. Baur, P., Arnold, R., Giessler, S., and Vermeer, A.W.P. (2007) *Pflanzenschutz-Nachr.*, **60** (1), 27–42.
24. Hazen, J.L. (2000) *Weed Technol.*, **14**, 773–784.
25. Butler Ellis, M.C., Tuck, C.R., and Miller, P.C.H. (1997) *Crop Prot.*, **16**, 41–50.
26. Hilz, E., Vermeer, A.W.P., Leermakers, F.A.M., and Cohen Stuart, M.A. (2012) *Aspects Appl. Biol.*, **114**, 71–78.
27. Baur, P. and Pontzen, R. (2007) Basic features of plant surface wettability and deposit formation and the impact of adjuvants in *Proceedings of the 8th International Symposium on Adjuvants for Agrochemicals, ISAA2007, August 6–9, Columbus, Ohio, USA* (ed. R.E. Gaskin) (available as a CD).
28. Rentschler, I. (1971) *Planta*, **96**, 119–135.

29. McCall, P.J. (1988) *Weed Sci.*, **36**, 424–435.
30. Faers, M.A. and Pontzen, R. (2008) *Pest Manag. Sci.*, **64**, 820–833.
31. Stevens, P.J., Baker, E.A., and Anderson, N.H. (1988) *Pestic. Sci.*, **24**, 31–53.
32. Holloway, P.J. (1993) *Pestic. Sci.*, **37**, 203–232.
33. Riederer, M. and Schönherr, J. (1985) *Ecotoxicol. Environ. Safety*, **9**, 196–208.
34. Jeffree, C.E. (1996) in *The Plant Cuticle - An Integrated Functional Approach* (ed. G. Kerstiens), BIOS Scientific Publisher, Oxford, pp. 33–82.
35. Bauer, H. and Schönherr, J. (1992) *Pestic. Sci.*, **35**, 1–11.
36. Kirkwood, R.C. (1999) *Pestic. Sci.*, **55**, 69–77.
37. Schreiber, L. (2005) *Ann. Bot.*, **95**, 1069–1073.
38. Schönherr, J. (2006) *J. Exp. Bot.*, **57**, 2471–2491.
39. Fernández, V. and Eichert, T. (2009) *Crit. Rev. Plant Sci.*, **28**, 36–68.
40. Schönherr, J. and Riederer, M. (1986) *Plant Cell Environ.*, **9**, 459–466.
41. Schönherr, J. and Baur, P. (1996) in *The Plant Cuticle - An Integrated Functional Approach* (ed. G. Kerstiens), BIOS Scientific Publisher, Oxford, pp. 134–155.
42. Stock, D. and Holloway, P.J. (1993) *Pestic. Sci.*, **38**, 165–177.
43. Wang, C.J. and Liu, Z.Q. (2007) *Pestic. Biochem. Physiol.*, **87**, 1–8.
44. Baur, P., Grayson, B.T., and Schönherr, J. (1997) *Pestic. Sci.*, **51**, 131–152.
45. Baur, P., Marzouk, H., and Schönherr, J. (2000) *Res. Adv. Agric. Food Chem.*, **1**, 49–58.
46. Kühnhold, J., Klueken, A.M., de Maeyer, L., van Waetermeulen, X., Brück, E., and Elbert, A. (2008) *Bayer CropSci. J.*, **61**, 279–306.
47. Ramsey, R.J.L., Stephenson, G.R., and Hall, J.C. (2005) *Pestic. Biochem. Physiol.*, **82**, 162–175.

10
Polymorphism and the Organic Solid State: Influence on the Optimization of Agrochemicals

Britta Olenik and Gerhard Thielking

10.1
Introduction

Polymorphs are different crystalline formations of an active ingredient with an identical chemical formula. Their physico-chemical properties, which show variations in terms of melting behavior, hardness, dissolution rate (among others) may influence the bioavailability of a polymorph, and can have a strong impact on the quality and efficacy of agrochemicals.

The production of modern agrochemicals must cope with many requirements. For example, the products must fulfill specifications which define the impurity profile of the active ingredient, while the physico-chemical properties of the batches must be consistently the same. In earlier times, problems such as changes in melting behavior or crystalline habit, as well as difficulties with filtration, drying, or agglomeration, were overcome during the development process, as and when they occurred. However, an increased awareness of the solid state of agrochemicals and its physico-chemical prospects led to changes in industrial workflows and the implementation of different types of study on solids. Based on the results obtained, a better control over the crystallization process of the active ingredients was achieved and the optimization of formulation recipes became much easier.

In this chapter, a survey is provided of the theoretical basics of polymorphism and its analytical characterization. The thermodynamic approach will provide information regarding monotropism and enantiotropism, while the different rules for the stabilities of polymorphs – which form the basic principles of polymorphism screening in modern crop protection – will be described, as will the analytical techniques used to characterize the different polymorphs of agrochemicals. The patentability of polymorphs will also be reviewed, based on one agrochemical example.

Modern Methods in Crop Protection Research, First Edition.
Edited by Peter Jeschke, Wolfgang Krämer, Ulrich Schirmer, and Matthias Witschel.

10.2 Theoretical Principles of Polymorphism

10.2.1 The Solid State

Whilst a first examination of a crystalline organic material might not consider such a material to be amazing, to delve more deeply into its solid state should open up a much wider field of investigation. Depending on the intermolecular interactions, there are many possibilities for creating a crystalline lattice. One crystal may be a polymorph, a hydrate or a solvate, a cocrystal or a salt; moreover, beneath these the amorphous state is equally important as all of these crystalline phases, and offers many possibilities on the basis of an absent lattice (Figure 10.1).

Differences in the crystalline lattices of active ingredients such as agrochemicals are responsible for the various physico-chemical properties that influence important characteristics such as solubility [1], thermodynamic and mechanical stabilities, melting points, hygroscopicity, density, and crystalline habits. Consequently, the choice of an optimal crystalline form is necessary to impart an early influence on process and formulation development, and to create usable agrochemicals. The solid form must fulfill numerous requirements such as good solubility and optimal stability, but to combine these requirements is not a trivial process. For example, whilst the amorphous phase would be the most soluble solid form, its tendency to recrystallize could destroy the formulation and make the product unusable. Hydrates and solvates can lose their water or solvent by recrystallization during milling processes, or they can be generated by exactly the same process without warning. In both cases, agglomeration would occur such that the quality of the formulated product would be changed for the worse. During recent years, however, the implementation of solid-state screenings in modern crop protection has led to a much better understanding of these effects.

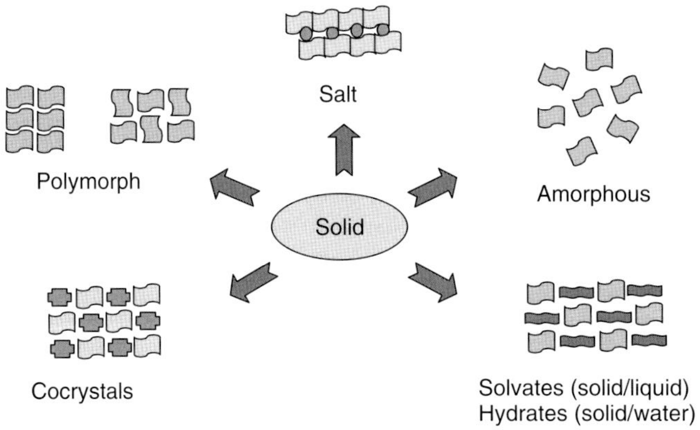

Figure 10.1 The solid state and its phases.

10.2.2 Definition of Polymorphism

Although the term "*polymorphism*" was first proposed by Mitscherlich in 1821 [2], a *polymorph* was first defined by W. C. McCrone in 1965 as "*... a solid crystalline phase of a given compound resulting from the possibility of at least two crystalline arrangements of the molecules of that compound in the solid state*" [3]. Polymorphs are chemically identical but differ in their crystalline lattices, and therefore in their physico-chemical properties.

Some years later, in 1997, Grunenberg presented a statistical evaluation of the polymorphism screening of pharmaceutical active ingredients at Bayer AG [4]. In this case, Grunenberg found that approximately 80% of the organic substances existed in more than one crystalline form – which indicated that polymorphism was not the exception but rather the rule! The importance of polymorphism remained underestimated, however, and it took a remarkably long time – and many bad experiences – before polymorphism screening was established within the industrial workflow. Initially, these types of investigation were implemented into the industrial workflows of dyestuffs and pharmaceutical production, with crop protection [5] followed several years later.

10.2.3 Thermodynamics

In modern crop protection it is necessary to impart an early influence on the crystallization process of a polymorphic system. The quality and usability of formulation types such as suspension concentrates (SCs) depends on the behavior of the solid active ingredient. The recrystallization of a metastable form can affect the physico-chemical properties of solid-based formulations such as SCs and water-dispersible granules (WGs), and can render the whole formulation recipe useless.

The exertion of an influence on a polymorphic crystalline system is based on the thermodynamics of crystallization [6, 7], as the thermodynamic stabilities are dependent on the energetic relationship of the different polymorphs, and also on the temperature.

10.2.3.1 Monotropism and Enantiotropism

The transformation of polymorphs was first investigated by Lehmann in1877 [8], who distinguished two types of transition state – the reversible enantiotropic and the irreversible monotropic conversion. Enantiotropically related systems have a thermodynamic phase transition and a transition point where the enantiotropically related forms are in equilibrium. The transition temperature can be measured by using differential scanning calorimetry (DSC; see Section 10.3.1) [9]. The thermodynamic aspects of the different crystalline phases were elaborated by Buerger and Bloom [10], who presented their energy/temperature diagrams in 1951 [11], and by Burger and Ramberger in 1979 [12, 13].

The determining factor for the existence of polymorphs is the free energy (G) of the crystalline lattice, which is described by the Gibbs–Helmholtz-equation (10.1), where H is the enthalpy, T is the temperature, and S describes the entropy of a given system:

$$\Delta G = \Delta H - T\Delta S \tag{10.1}$$

This equation shows the temperature dependence of the crystallization of different polymorphs for a given substance. The system wishes to minimize its free energy (G) by crystallization; however, the stable polymorph is the crystalline form with the lowest free energy (G) and this should therefore be preferred. Yet, the crystallization process depends also on the kinetics, as is most impressively demonstrated by the carbon system of diamond and graphite. As the metastable form, diamond should be transformed into graphite, but a high activation barrier influences the rate of transformation, which becomes very slow; hence, diamonds can be "a girl's best friend" for quite a long time!

The influences of thermodynamics versus kinetics were described by Bernstein *et al.* in 1999 [14]. Generally, the system wishes to escape supersaturation as quickly as possible, and therefore accepts the formation of a crystalline structure that is not (necessarily) the ideal one. Transformation into the thermodynamically most stable form can occur later, this being referred to as *Ostwald ripening* [15].

10.2.3.2 Energy Temperature Diagrams and the Rules

The thermodynamic evaluation of a given system is most easily summarized by an energy/temperature diagram (Figures 10.2 and 10.3) [16]. This *semi*-schematic interpretation of the Gibbs–Helmholtz-equation (10.1) allows the presentation of complex polymorphic systems in one diagram. The diagram includes the G- and H-isobars of the liquid phase (the melt) and of the different crystalline forms. At absolute zero, the stable polymorph must have the lowest free energy; the free

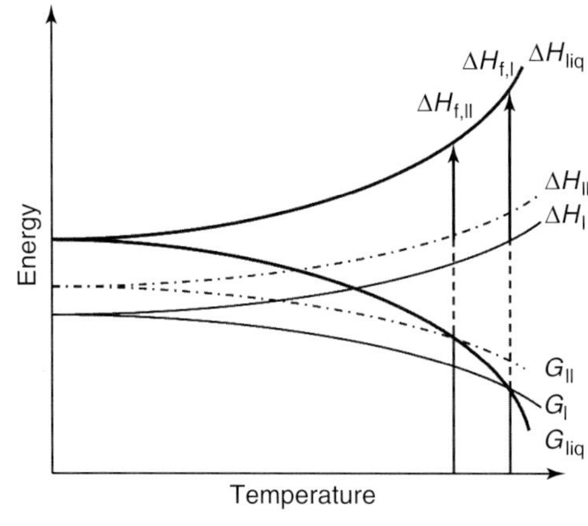

Figure 10.2 Energy/temperature diagram of a monotropically related system of two crystalline forms I and II [17].

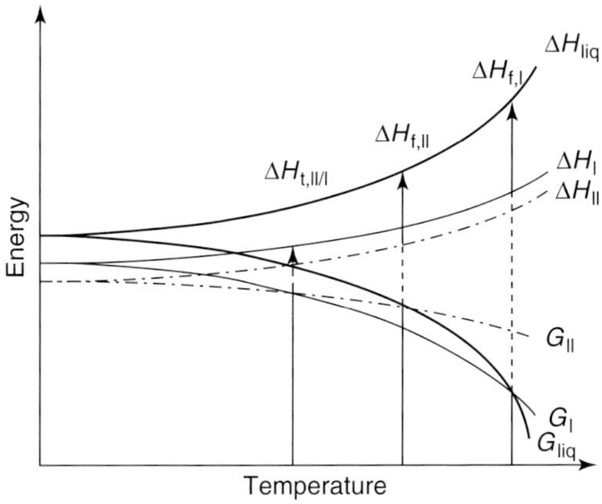

Figure 10.3 Energy/temperature diagram of a monotropically related system of two crystalline forms I and II [17].

energy and enthalpy of a given crystalline form must be equal (T in Equation 10.1 is zero), and the metastable crystalline form must have the higher enthalpy.

For monotropic systems (Figure 10.2), the energy/temperature diagram does not show any intersection of the G-isobars, and the thermodynamic stability of the higher-melting polymorph is not temperature-dependent.

The G-isobars of an enantiotropically related system (Figure 10.3) pass through an intersection – the transition point – where both crystalline forms are in equilibrium. Transformations from the lower melting form II to the higher melting form I are exothermic at temperatures below the transition point, and endothermic at temperatures above it.

For the assessment of thermodynamic stabilities and relationships of different polymorphs of a given system, Burger and Ramberger [12, 13] introduced several "rules" which are very useful for the evaluation of thermoanalytical and spectroscopic data:

- The heat-of-transition rule (HTR) states that an endothermic transition only takes place at temperatures above the thermodynamic transition point in an enantiotropically related system. Exothermic phase conversions occur at temperatures below the thermodynamic transition point, and are therefore preferred in monotropically related systems.
- The entropy-of-fusion rule (EFR) is highly applicable because it can be easily verified by DSC data. It indicates that, when the higher-melting form has the lower entropy of fusion value, then the system should be enantiotropically related; otherwise, the system will be monotropic.
- The heat-of-fusion rule (HFR) correlates to the EFR, and was first mentioned by Burger and Ramberger [12, 13]. When the higher-melting form shows the higher enthalpy of fusion value, the system must be monotropically related, but enantiotropically related when it has the lower enthalpy of fusion. This rapidly

becomes clear when examining the energy/temperature diagrams (see the bold arrows crossing the H-isobars in Figures 10.2 and 10.3).

- The enthalpy-of-sublimation rule (ESR) can be described as follows: The enthalpy of sublimation of an existing polymorph is the direct sum of its enthalpy of fusion and its enthalpy of vaporization. The last-named is identical for all polymorphs of a given compound. Therefore, the higher-melting form of an enantiotropically related system has the lower enthalpy of sublimation.
- The heat-of-capacity rule (HCR) states that the higher-melting form of a given enantiotropic system shows the higher heat capacity at given temperature.
- The density rule (DR) is not based on thermoanalytical data. The crystalline lattice of the stable polymorph of a given compound will have the strongest intermolecular interactions and the highest density at absolute zero. Because temperature changes do not strongly influence the density of a solid form, it is permissible to assume that the higher-melting form of a monotropically related system will have the higher density. In enantiotropic systems, this situation is reversed.

All of these rules are very helpful when evaluating thermoanalytical and spectroscopic data derived from polymorphism screening in modern crop protection.

10.2.4
Kinetics of Crystallization: Nucleation

Nucleation and its thermodynamic and kinetic aspects have been extensively investigated by Davey and Garside [18] and Beckmann [19]. Nucleation serves as the beginning of every crystallization process, when a few molecules (about 10^{20}) come together to form intermolecular interactions and find a crystalline packing. The decrease in ΔG from creating the critical nucleus competes against the decrease of ΔG from re-dissolving. The outer molecules of the nuclei are unable to form all the intermolecular interactions of which they are capable, and therefore returning to solution is, for them, a valid alternative. The ratio of inner to outer molecules influences the solubility of the critical nucleus, and this is the reason why larger crystals are less soluble than their smaller counterparts.

The nucleation depends on the degree of supersaturation, the concentration of crystals in suspension, the hydrodynamic interactions between crystals and solution, and also on the stirring speed and the power input during crystallization.

Without supersaturation, the nucleation process does not take place, and knowledge of the solubility curve is therefore fundamental for each crystallization process. The schematic solubility curve (Figure 10.4) includes supersaturated, unsaturated, and metastable zones. Starting from the composition X_i, an increase in concentration (arrows 1) or a decrease in temperature (arrows 2) would lead to supersaturation and nucleation.

In concentrations below the solubility curve, the new nuclei would redissolve because the system is undersaturated. In the supersaturated area, spontaneous nucleation occurs leading to the metastable zone where the solution is labile and crystal growth takes place without the formation of new nuclei. This is the most

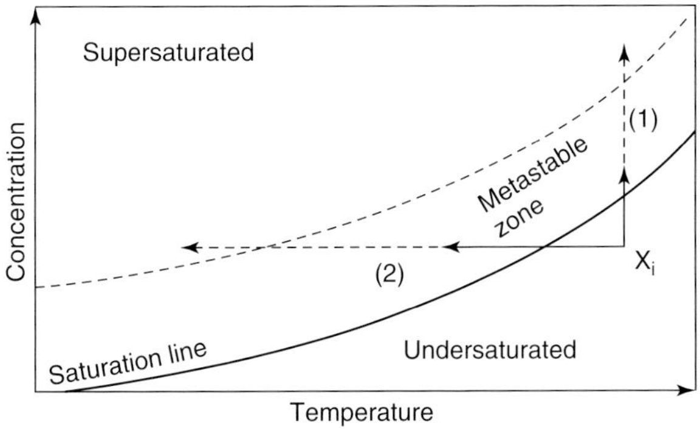

Figure 10.4 Schematic solubility curve including supersaturated, unsaturated, and metastable zones [18].

important area for seeded crystallization processes, where larger crystals can be obtained.

The influence of stirring speed is very important in terms of the secondary nucleation. Existing crystals come into contact with tank walls, the stirring blades, or with each other, and start recrystallization on the phase periphery; this is referred to as *contact or collision nucleation*. The stirring speed and the power input may also have a major influence on solid–solid transformation and Ostwald ripening effects.

For industrial processes, such as in modern crop protection, it is most important to detect the starting point of the nucleation. Some physical properties will change during germ formation, the first of these being the temperature change due to a reduction in the free energy as a result of crystallization. This in turn, leads to an increase in temperature which is easy to detect. Because of the density change between liquid and solid phase, the latter will have a higher density (if it is not the water/ice system); a nucleation process should cause a reduction of the reaction volume. Turbidity is very easily measured by using inline techniques, and this will change dramatically upon the formation of crystalline nuclei. In addition, the generation of a solid phase in the crystallizer will have an influence on the concentration of the solution by decreasing the ratio of the solute.

10.3 Analytical Characterization of Polymorphs

Based on the ordered packing of the molecules in the crystalline lattices of different polymorphs, changes in the physico-chemical properties of these solid phases may occur. The polymorphs will have different intermolecular interactions that will influence the melting point and the enthalpy of fusion values. A change in the crystalline packing can form a new morphology and another space group. The asymmetric unit will be different, which in turn has an influence on the functional groups of the molecules in the crystalline lattice, and also their vibrational states.

Although many techniques are available for the analytical characterization of different polymorphs [20], only a brief overview of the most common systems used in modern crop protection will be provided here. Furthermore, techniques such as solid-state nuclear magnetic resonance (NMR), near-infrared (NIR), and terahertz spectroscopy, or combined techniques such as Raman- or infrared (IR) microscopy, are perfect tools for the analysis of crystalline forms. Berstein [21], Brittain [22], Byrn *et al.* [23], and Hilfiker [24] have each provided comprehensive surveys on polymorphism.

10.3.1 Differential Thermal Analysis and Differential Scanning Calorimetry

For polymorphism screening and studies of active ingredients, it is essential to quantify the different melting processes of the crystalline forms. Knowledge of the melting temperature of the active ingredient, as well as the enthalpy of fusion values, is necessary to evaluate the thermodynamic stabilities. Based on HTR and the HFR, the creation of energy/temperature diagrams is possible via these analytical data (see Section 10.2.3.2).

The endothermic and exothermic effects can be measured using either differential thermal analysis (DTA) or DSC [25–27]. In both techniques the sample, together with an inactive reference, are heated up at a defined rate, whereupon endothermic or exothermic processes in the sample cause differences in the temperature profile compared to the reference. In DTA, these temperature differences are detected as a function of temperature (this system works with temperature sensors only). In contrast, DSC is based on another technique (Figure 10.5), whereby the temperature differences between the sample (P) and the reference (R) are compensated by a second heating element. The differences in heat flow between the oven and the second heating element as a function of temperature or time are detected. The main advantage of DSC is the direct measurement of energy exchange ΔH, but this is accompanied by a higher noise, and means that the analysis of very small or slow effects can became more difficult and imprecise.

The interpretation of thermograms can be very complex because many different endothermic and exothermic processes can be identified that are running sequentially and/or simultaneously. For example, the melting of a sample, as well as the boiling and sublimation processes, are endothermic effects, whereas recrystallization and decomposition are exothermic (Figure 10.6). Hence, it is essential to combine different thermoanalytical techniques such as hot stage microscopy or thermogravimetry with DSC [29].

Typically, DSC experiments are influenced by the physical properties of the sample in use. Amorphous parts of, and impurities within, the sample can decrease the values of melting temperature and the enthalpy of fusion. The particle size also has an influence on the heat capacity of the sample, and homogeneous samples with small particle sizes are best-suited to DSC analysis. Unfortunately, the grinding of a sample is not reasonable because this activity can cause the transformation of metastable polymorphs.

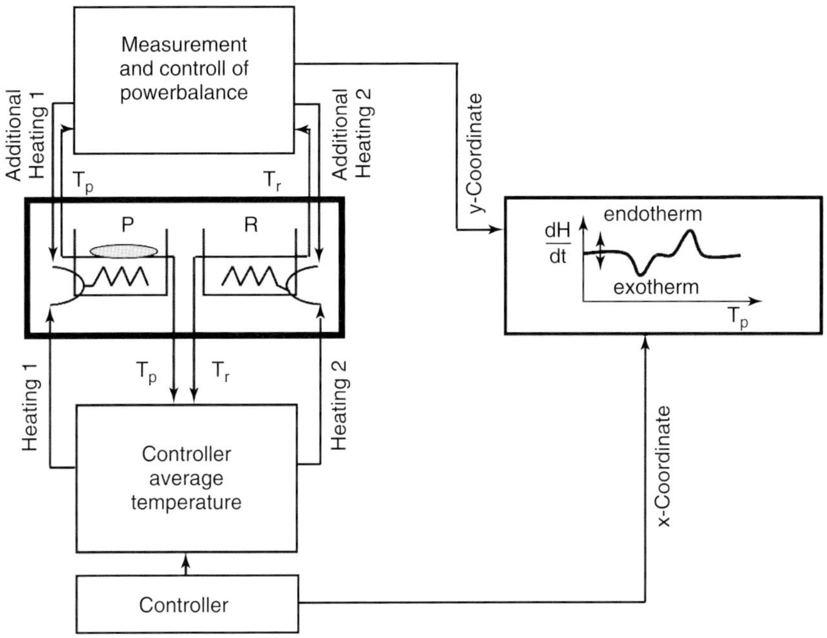

Figure 10.5 Schematic configuration of a differential scanning calorimetry (DSC) system [28], where P describes the sample, R represents the empty reference crucible, and T_p and T_r specify the temperatures of the sample and the reference, respectively.

In addition to these physico-chemical aspects, the measurement conditions must be carefully chosen. For example, the material of the crucible can cause catalytic decomposition, while the nature of the closure may help to detect solvates or hydrates. By using hermetically closed high-pressure crucibles, however, it is possible to inhibit any boiling effects, which is quite useful for safety measurements. In order to detect water or solvent in the sample it is advisable to use pierced cover plates, which are also the best choice for melting point (mp) analyses. Purity analyses should be carried out in hermetically closed aluminum crucibles.

For polymorphism investigations in modern crop protection, it is very important also to consider the rate of heating. The analysis of melting points and purities should be made with low rates of heating, but this may lead to thermal stressing of the sample and may facilitate transition into the more stable polymorph at the given temperature. Operational experience has shown that cyclic DSC experiments with different rates of heating can be helpful in obtaining thermoanalytical data for the potential polymorphs. The differences in rates of heating may have an influence on the nucleation of the sample, and may also cause recrystallization of the potential polymorphs. However, with such rapid experiments it may even be possible to capture all of the thermoanalytical values of a metastable crystalline form.

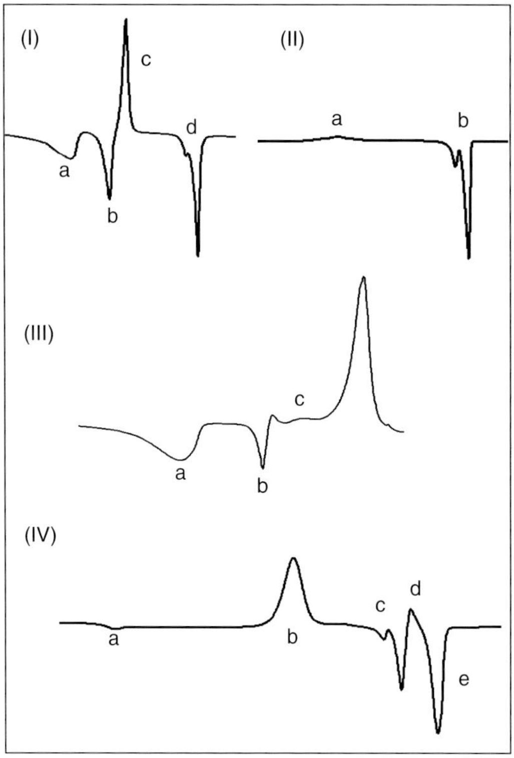

Figure 10.6 Four examples of differential scanning calorimetry (DSC) curves. Curve I describes the loss of absorbed solvent (a), the loss of crystalline solvent (b), followed by recrystallization of a solvent-free form (c), and the two melting processes of two polymorphs (d). Curve II illustrates an exothermic transformation (a) and the melting processes of two crystalline forms (b). Curve III shows the loss of absorbed solvent (a) and the melting process (b), which passes on to the exothermic decomposition (c). Curve IV includes multiple effects starting with a glass transition (a), followed by recrystallization of the amorphous phase (b), and multiple melting effects (c). After recrystallization of the melt (d), a new melting process of the high-melting form (e) is observed.

10.3.2 Thermogravimetry

An alternative thermoanalytical technique used in modern crop protection is that of thermogravimetric analysis (TGA) [27], whereby the detection of weight loss versus temperature or time enables the solvates and hydrates to be characterized. In this case, a small amount of sample is placed on a microbalance and heated up at a defined rate. The resultant thermogram can capture effects such as the loss of solvents or water, any type of decomposition of the sample, oxidation of the sample (via an increase in weight), and the presence of inorganic impurities, as indicated by the deposition of residues at high temperatures.

For the detection of hydrates and solvates (pseudopolymorphs), the temperature dependence of the weight loss may be significant, with adsorbed solvent or water leaving the substance at or near its boiling temperature. If the solvent or water forms part of the crystalline lattice, however, then more energy will be required to break the intermolecular interactions between substance and solvent/water. Mostly, this goes hand-in-hand with a loss of weight (and a step in the thermogram) at a higher temperature than the boiling point. Exceptions to this are non-stoichiometric solvates such as isomorphic desolvates [30].

The combination of TGA with spectroscopic techniques or X-ray powder diffraction enables the characterization of pseudopolymorphs within a short period of time.

10.3.3 Hot-Stage Microscopy

Hot-stage microscopy (HSM) was first developed by Lehmann in 1877 [31], and later improved by A. Kofler and L. Kofler [32, 33]. The use of HSM, and its considerable advantages for polymorphism studies, were later reported in impressive fashion by Kuhnert-Brandstätter [34, 35].

Thermomicroscopy represents one of the best techniques to complete thermoanalytical investigations in modern crop protection since, in combination with DSC, it permits characterization of the different endothermic and exothermic effects by optical analysis. The differentiation of endothermic effects such as melting, transformation, or sublimation for an unknown compound can be quite difficult if only the thermograms are examined; however, the additional use of HSM allows all of these effects to be observed *in situ*.

One major advantage of the HSM technique is the ability to use very small sample quantities (only a few crystals; <1 mg). Likewise, cyclic melting and recrystallization experiments can be conducted without extensive sample preparation, using only a few crystals, and may lead to the identification of different polymorphs at the first attempt. The conversion of polymorphs with essentially the same melting point can also be achieved, and the different effects separated simply by altering the rate of heating.

The DSC heating cycles of fluopicolide [36] (Figure 10.7), an agrochemical fungicide marketed as an SC formulation, that is active against oomycetes (notably *Phytophthora infestans* and *Plasmopara viticola*) and is targeted for use in tomatoes and potatoes, are shown in Figure 10.8. The cyclic DSC measurements showed three endothermic effects at quite close temperatures, and these were also clearly characterized by using HSM. At 140 °C, the melting point of spherolites (form III) was seen, overlaid by recrystallization into plates (form II). The endothermic effect at 151 °C could be assigned to a melting process of a third dendritic polymorph (form I). Finally, the existence of all three polymorphs was confirmed with IR spectroscopy (Figure 10.8).

One of the most interesting phenomena in the investigation of polymorphism is that the microscopy images of the different crystalline habits appear to provide

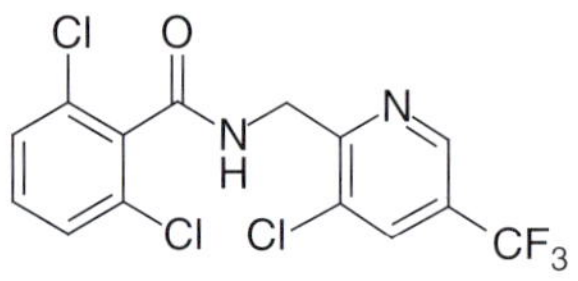

Figure 10.7 Fluopicolide (2005, Infinito®, Bayer CropScience).

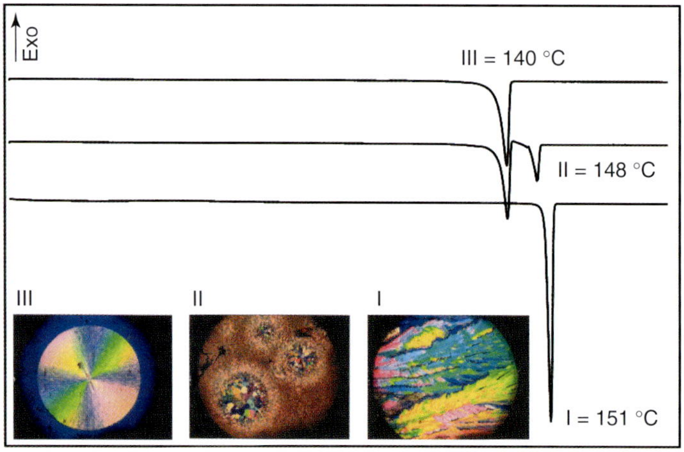

Figure 10.8 Thermograms of three DSC heating cycles of fluopicolide [36]. The three endothermic effects were separated using hot-stage microscopy.

the only convincing evidence that different crystalline forms of a given compound actually exist. However, the differences in habit are not necessarily polymorphs; rather, depending on the crystallization conditions, a crystalline lattice may show differences in growth that might lead to a change in appearance, while the lattice remains the same. As an example, the choice of solvents (with different polarities) in crystallization processes can influence the preference of crystal growth areas. In addition, intense stirring during the production or formulation of agrochemicals can distort the crystalline habit, such that the crystals become more spherical. As noted above, in all of these cases it is necessary to use more than one analytical technique in modern crop protection.

The dehydration of a hydrate is easily detected by using TGA, but may also be observed using HSM. Without further sample preparation, the loss of water or of a solvent during sample heating can be recognized by a small movement of the crystals at a given temperature (mostly above the boiling point of the given solvent); in this situation the crystals appear to "hop" or "dance." However, by mixing the sample with a viscous liquid (in which the substance should not be soluble) such as nujol, the loss of solvent at a given temperature will cause the generation of small bubbles which are very easily observed.

In some cases, the existence of concomitant polymorphs can be difficult to detect. Indeed, mixtures of polymorphs may demonstrate only one broad endothermic

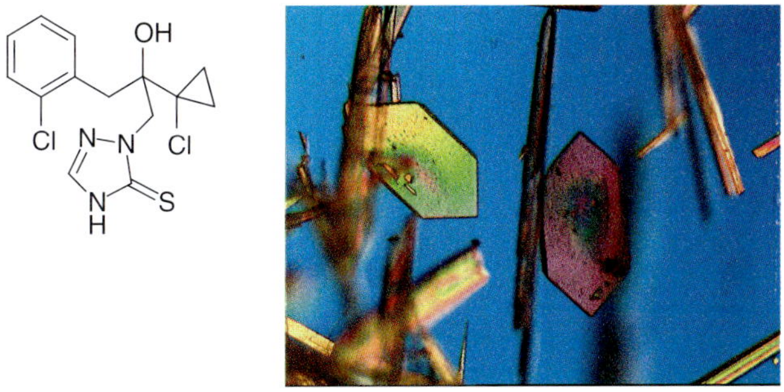

Figure 10.9 Concomitant polymorphs of prothioconazole (2004, Proline®, Bayer CropScience) in plates and needle-like habit.

effect in DSC measurements when their melting points are close together and overlaid in the thermoanalytical experiment. The spectroscopic analysis would lead to mixed spectra, which are quite difficult to interpret without knowledge of the spectroscopic data of the pure polymorphs. In some of these cases, however, microscopy may help to clarify the analytical results.

The concomitant existence of polymorph I and II of prothioconazole [37, 38], a broad-spectrum systemic fungicide with protectant, curative, and eradicative activities, suitable for foliar application and for seed treatment, is shown in Figure 10.9.

Among the varying techniques developed for polymorphism studies in modern crop protection, the crystalline habit of different polymorphs, their interconversion, the melting behavior of these forms, and their recrystallization, are normally explored by using thermomicroscopy.

10.3.4 IR and Raman Spectroscopies

The qualitative and quantitative [39] analysis of different polymorphs in technical active ingredients, as well as in formulated agrochemicals, is often conducted using spectroscopic techniques such as Raman and IR spectroscopies [40, 41]. Both techniques are based on the interaction of electromagnetic radiation with matter, as expressed in the Einstein equation (10.2):

$$E = h \cdot \upsilon \tag{10.2}$$

One photon collides with a particle in ground state; by absorbing a radiation quantum, the particle is promoted to the excited state vibrational mode. Based on the classical mechanics and quantum mechanics, the energy states of vibrations are well described by the model of an anharmonic oscillator (Figure 10.10).

This model explains the decrease in the distances between the energy levels with increasing energy and the existence of dissociation energy by increasing distance between the vibrating masses [42]. If the energy of the radiation quantum corresponds to the energy difference between the ground and excited states, the

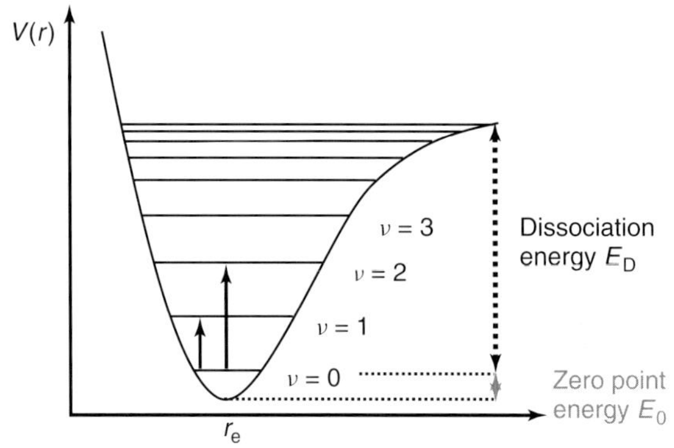

Figure 10.10 Model of the anharmonic oscillator [42].

absorption will lead to an excited vibration mode. The return to the ground state will be followed by a radiation-free loss of energy through transformation into kinetic energy [43].

In mid-IR spectroscopy, the absorption of electromagnetic radiation energy by molecular vibrations must correspond to a change in the dipole moment. The Raman spectra are produced by an inelastic scattering of monochromatic radiation [44]. While most of the monochromatic radiation will be elastically scattered in all directions (Rayleigh scattering), a small part of it will undergo a change in frequency because of its interaction with the substance. Stokes–Raman scattering describes the promotion from ground to a higher excited state of vibrational mode, and the return to the first excited state. This inelastic scattering leads to a reduction of frequency, whereas the anti-Stokes–Raman scattering shows an increase in frequency as a consequence of transition from the first excited state of the vibrational mode to the ground state via a higher excited state. The anti-Stokes bands are less intensive because the first excited state is not as well occupied as the ground state, and therefore this transition is less probable (Figure 10.11).

Typical Raman spectra show the Stokes bands. Because the basis of the Raman effect is the change in polarizability during interaction with the electromagnetic radiation, it is a perfect complement to mid-IR spectroscopy, or vice versa. This can be illustrated by two examples: The analysis of inter- and intramolecular hydrogen bonding by IR spectroscopy is well known and often used, whereas Raman spectroscopy does not identify any OH bands. As the vibrations of the water molecule do not change its polarizability, Raman spectroscopy represents a perfect tool for both qualitative and quantitative analyses of the water-based formulation types (e.g., SC) that are used in modern crop protection. Both of these advantages may be quite useful for the detection of polymorphs.

The Raman spectra of polymorphs I and II of sulcotrione (Figure 10.12) [45, 46] are shown in Figure 10.13. Sulcotrione is a broad-spectrum herbicide for use in monocotyledonous crops, that was originally launched as an SC formulation in 2001. However, problems with agglomeration of Mikado SC 300 led to a polymorphism screening being conducted on the active ingredient. Subsequently,

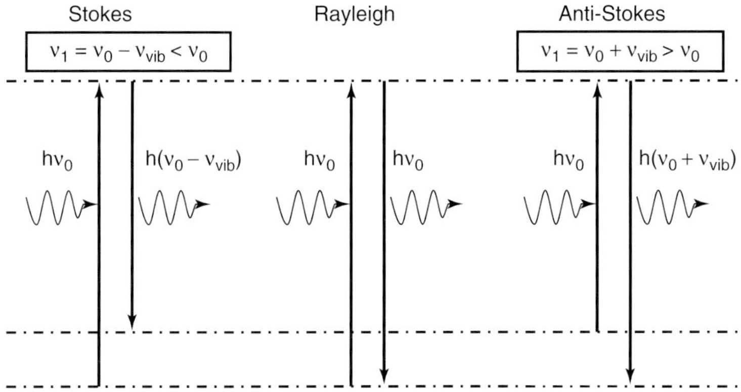

Figure 10.11 Vibrational energy level transitions in Raman spectroscopy, Stokes, anti-Stokes, and Rayleigh scattering [42].

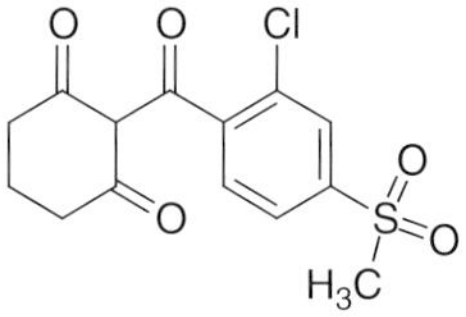

Figure 10.12 Sulcotrione (1990, Mikado®, Stauffer Chemicals Company).

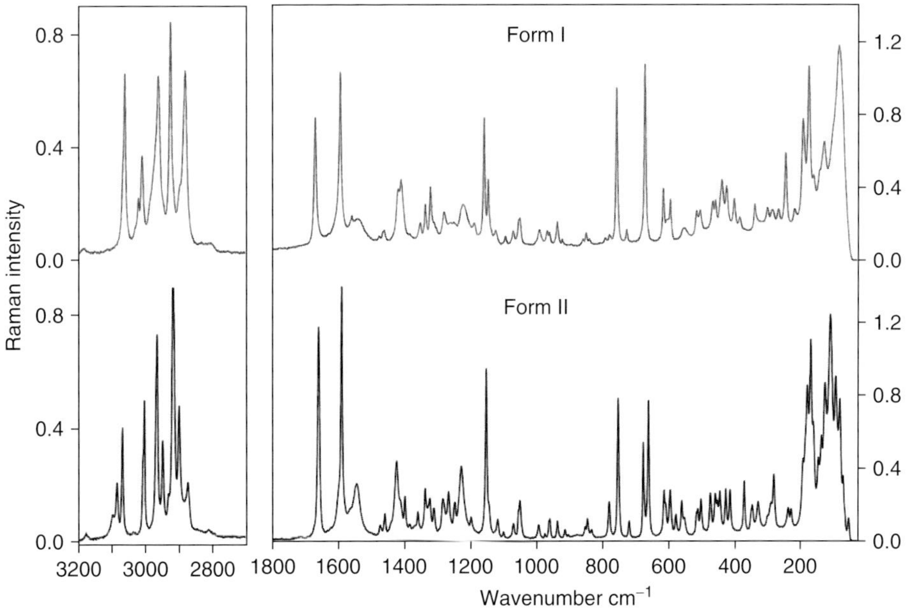

Figure 10.13 Raman spectra of polymorphs I and II of sulcotrione in the technical active ingredient.

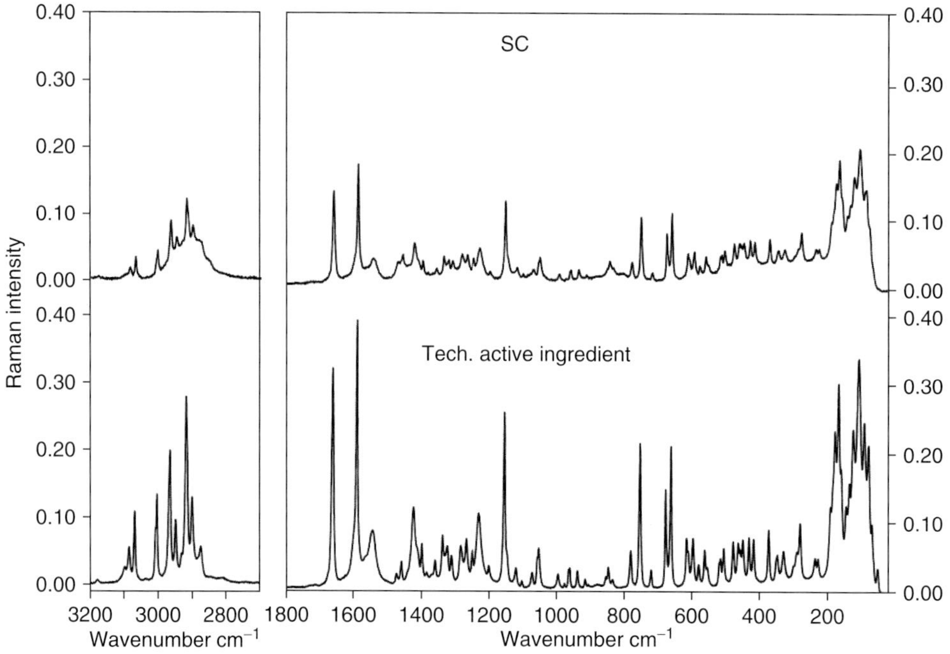

Figure 10.14 Comparison of the Raman spectra of sulcotrione form II in technical active ingredient and in Mikado SC 300 formulation.

sulcotrione was found to exist in two enantiotropically related polymorphic forms in the formulation, and recrystallization of the metastable crystalline form caused agglomeration problems. Thus, it became necessary to develop methods for the quantitative analysis of the two forms not only in the technical active ingredient but also in the SC formulation.

As shown in Figure 10.13, the Raman spectra were perfectly suited to for characterization of the crystalline forms. A comparison of the spectra of the stable polymorph II in the technical active ingredient and the SC formulation is shown in Figure 10.14.

For the spectroscopic analysis of different polymorphs of a given organic compound, it is very important to consider sample preparation. Often, the identification of substances using IR spectroscopy is performed in nujol or potassium bromide; however, the mechanical stressing of the sample that occurs during the milling process required for these analyses can influence the polymorphic systems. For example, the energetic input from the milling may initiate a transformation of a metastable crystalline form into a more stable polymorph, and the analysis of a sample treated in this way would therefore be invalid. In order to overcome these effects, the use of an attenuated total reflection unit (ATR) is advisable, where the sample is placed with a spatula onto a flat diamond head, without any further preparation. The quality of this analysis is influenced by the particle size and reflection effects.

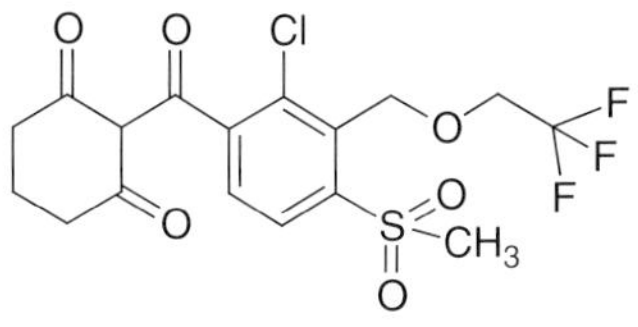

Figure 10.15 Tembotrione (2009, Laudis® Bayer CropScience).

Raman measurements can be performed in glass vessels (small test tubes), without any sample preparation. Although the sample handling is totally non-destructive, the measurement itself can be damaging, with the laser intensity perhaps leading to transformation or, in the worst case, decomposition of the sample.

The possible transformation of a metastable crystalline form may also influence the development of quantitative methods. For calibration purposes, mixtures of the analytes would normally be generated and analyzed, but the mixing process of a stable and a metastable polymorph of a given substance may cause a shift in concentration because of solid–solid transformations of the metastable form. Thus, it is advisable to start the method evaluation with the use of calculated mixed spectra on the basis of original and well-characterized samples of the polymorphs. The associated chemometry and partial least squares (PLS) regression are supported by the software, and will be processed automatically.

The importance of Raman spectroscopy is demonstrated for two polymorphs of tembotrione (Figure 10.15) [47–49], a broad-spectrum herbicide with a 4-hydroxyphenyl-pyruvate-dioxygenase (4-HPPD) mode of action, used especially in maize and corn. Neither polymorph I nor polymorph II of this active ingredient could be clearly characterized by thermal analysis, because there was almost no difference in their melting points. However, the use of Raman spectroscopy enabled qualitative and quantitative analyses of these two crystalline forms to be conducted. The Raman spectra of forms I and II of tembotrione are shown in Figure 10.16. Only the spectroscopic analysis of both crystalline forms provided clear data and allowed a characterization of the two polymorphs. This is one example of the need to combine different analytical methods in polymorphism investigations.

Raman and IR spectroscopy represent fundamental techniques for the qualitative characterization and quantitative determination of polymorphs in technical active ingredients, as well as in formulations. Moreover, as a tool for quality control, they should form part of an industrial routine in modern crop protection.

10.3.5
X-Ray Analysis

Besides the qualitative and quantitative analysis of crystalline forms of active ingredients, it may be helpful to investigate the crystalline lattice on the atomic level. The three-dimensional (3-D) assembly of the atoms in the lattice provides

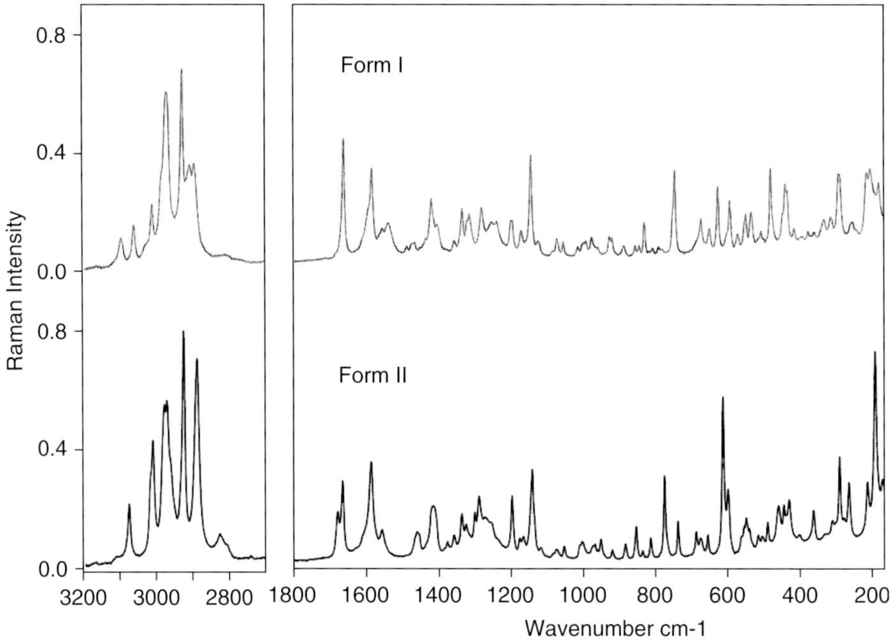

Figure 10.16 Comparison of the Raman spectra of polymorphs I and II of tembotrione.

information on the conformation, the bond length, and angle, intermolecular interactions such as hydrogen-bonding patterns [50], stoichiometric composition, and the density of the polymorph [51].

The normal distance between atoms in a molecule – the bond length – is about 0.15 nm. This means that the wavelength of visible light, which is between 400 and 700 nm, is much too long; thus, simple observation is not possible and another form of radiation is needed to analyze these very small objects. X-ray radiation has a wavelength of 50–230 pm, and shows interaction with the crystalline lattice [52].

In X-ray structure analysis, the diffraction of the X-irradiation by the crystalline lattice is utilized to localize the atoms. The diffraction by the electrons of the lattice atoms leads to interferences without any change in wavelength of the radiation. By analysis of the location and the intensities of the diffraction maxima – the reflexes – it is possible to draw conclusions on the arrangement of the scattering points (i.e., atoms) in the crystalline lattice. Whereas, the location of the reflexes defines the translation lattice, the assembly of the atoms is derived from an evaluation of the intensities of the diffraction maxima [53].

Based on the assumption of Laue, that crystals interact with X-ray radiation like a diffraction grating, a diffraction pattern can be generated that describes the system of parallel lattice planes suitable for interference [54]. Therefore, an amplification of the interferences is needed which occurs only at a defined wave angle of the radiation. The correlation between the wavelength λ, the distance of the lattice planes, the lattice constant d, and the wave angle of the radiation θ, is summarized

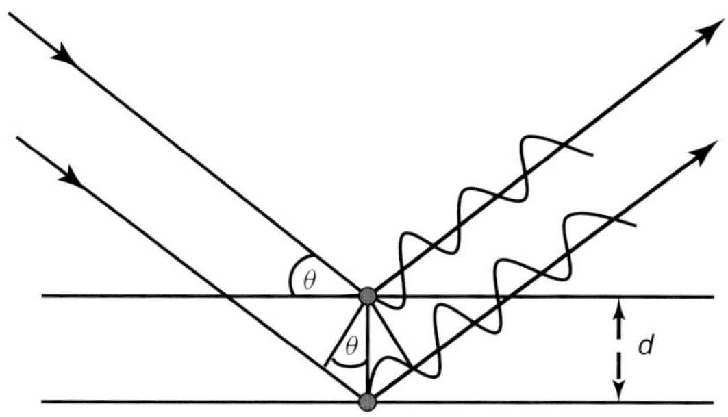

Figure 10.17 Diffraction on two atoms of the parallel lattice planes. Amplification of the interferences occurs at a wave angle of θ [53].

in the Bragg equation (10.3), which describes the reflex condition.

$$n\lambda = 2d \times \sin\theta \qquad (10.3)$$

The physical process of amplification and cancellation/interference behind the Bragg equation (10.3) is illustrated, in schematic fashion, in Figure 10.17.

Based on the symmetry of the crystalline assembly, the lattice can be described by one building block – the unit cell – and appropriate symmetry operations, which describe the repeating 3-D orientation of the unit cell in a crystalline composition. The unit cell itself is characterized by the translation vectors, which form the skeletal structure for the translation lattice.

X-ray radiation is generated by focusing an electron beam onto a very pure metal anode in a total vacuum. The electrons are partly retarded by the electrical field of the metal ions, and their kinetic energy transformed into radiation referred to as "*Bremsstrahlung*." The characteristic X-ray radiation results from the removal of one electron from one electron shell (e.g., the K shell) by ionization of the metal atoms. This state is unstable, but becomes stabilized by an electron transfer from the next highest electron shell (in this example, the L shell). Because of the energetic differences of both shells this transfer generates well-defined radiation of a certain wavelength. In order to obtain perfect monochromatic light (because of influences of the spin and the orbital angular moment a radiation doublet can occur), filter technology is used [53].

The quality and size of the crystal is rather important for good data generation. Lattice defects or twinning may complicate the measurement, or even make it impractical.

In addition to the single-crystal structure analysis, a second technique is available which is based on the same physical interactions of X-ray radiation and the crystalline latticc. This so-called *X-ray powder diffraction* functions by the same principle, but employs a different sample preparation [54], with a crystalline powder (which is built up by randomly orientated crystallites) being used to generate a powder diffraction pattern. The plot of the X-ray intensities against the diffraction angle, 2θ, is termed the *diffractogram*; this is characteristic for different polymorphs, but does not yield 3-D information, as does the single-crystal structure analysis.

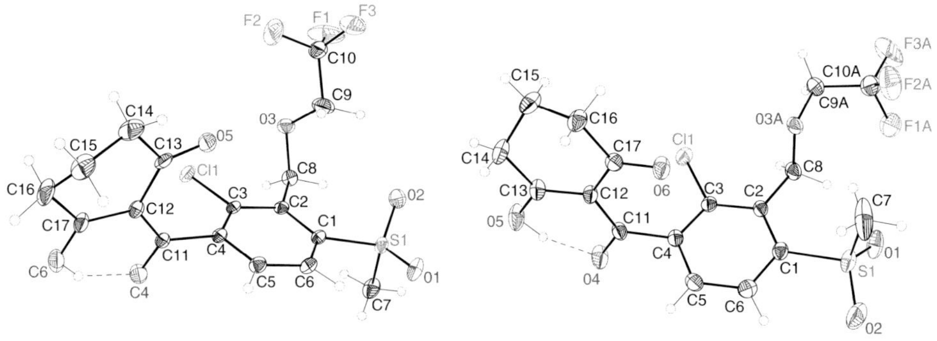

Figure 10.18 Single crystal X-ray analysis of tembotrione crystal forms I and II. (Analysis performed by Dr J. Benet-Buchholz.)

Table 10.1 Crystal data of tembotrione forms I and II.[a]

Parameter	Polymorph I	Polymorph II
Crystal system	Orthorhombic	Monoclinic
Space group	$Pna2_1$	P2(1)/n
Volume (Å)	1788.91	1814.21
Density ($mg\,m^{-3}$)	1.637	1.614

[a]X-ray single crystal structure analysis by Dr J. Benet-Buchholz.

The main benefit of the powder diffraction method is that sample preparation is almost unnecessary.

The results of the single-crystal X-ray analysis of tembotrione forms I and II are shown in Figure 10.18 (see also Section 10.3.4, Raman spectra of this active ingredient). This is an example of the ability of a given molecule to form different crystalline lattices. The differences in the conformation of the molecules in the crystalline lattices of forms I and II of this active ingredient produce different unit cells for these polymorphs (Table 10.1).

10.4
Patentability of Polymorphs

The patentability of different polymorphs of a given substance is well known within the pharmaceutical industry, and in recent years has become increasingly important for modern crop protection. The variation in properties of the crystalline phases has an effect on the bioavailability, stability, and manufacturability of a solid form of an agrochemical. Clearly, the rising number of patents related to the polymorphism of crop protection compounds underlines the increased awareness of the opportunities that the solid state may offer.

A patent covering a hitherto unknown stable crystal modification is of great importance, as the value of the protection obtained is similar to that for a patent on the active ingredient itself, at least as long as formulations containing that substance in the solid state are involved. To file a patent application, a polymorph must to fulfill the requirements of patent law:

- The crystalline form must be new. The novelty of the crystalline phase should be proved by data inquiry, which can be quite difficult because any type of physical property declaration in any form of literature/data is relevant.
- The inventive step is the second requirement. Until now, it has not been possible to reliably predict the number and the crystalline structure of polymorphs by examining the chemical structure. This depends on the high number of degrees of freedom and the temperature dependency of the thermodynamic stabilities of polymorphs. Therefore, the finding of a new crystalline form will be inventive.
- The necessity of an industrial application describes the industrial benefit that an invention must have. This is always given because the choice of the optimal crystalline form has an advantage with regards to the quality of a product such as an agrochemical, the practicability of the production process, or it may influence the efficiency via the bioavailability in insects, plants, or fungi cells.

One example is fipronil (Figure 10.19), an insecticide with contact, stomach, and systemic actions, which has moderate activity against aphids and green leafhoppers. Fipronil is marketed for foliar and soil applications, as well as for seed treatments in different solid-based formulation types.

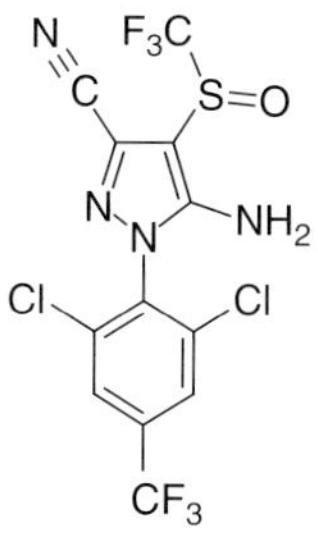

Figure 10.19 Fipronil (1993, Regent®, Rhone-Poulenc).

In 2008, BASF submitted four polymorphism patent applications for different crystalline forms and the amorphous state. A summary of the patent data is represented in Table 10.2.

The different properties of the various crystalline forms do have an industrial benefit:

- Form I [55] and form V [56] are described as having a better stability for production, transportation, and storage.
- Form II has a needle-like habit that is claimed to be better for filtration [57].
- Form IV has a high solubility and a high dissolution rate, both of which have a positive influence on bioavailability [58].

Table 10.2 Crystal data of fipronil forms I, II, IV, and V.

Parameter	Polymorph I	Polymorph II	Polymorph IV	Polymorph V
Crystal system	Monoclinic	Monoclinic	Triclinic	Triclinic
Space group	C2/c	P2(1)c	P-1	P-1
Density (g cm^{-3})	1.81	0.94	1.64	1.73

The increasing awareness of the crystalline state is reflected in the importance of polymorphism patents, which may enhance the risk of litigations. Some illustrious cases from the pharmaceutical area, such as ranitidine hydrochloride or the cefadroxil process, impressively demonstrate the potential for controversies over polymorphic forms (see Ref. [21]).

10.5 Summary and Outlook

The solid state – and especially the polymorphism of agrochemicals – is of utmost importance. Knowledge of the thermodynamic stabilities and physico-chemical properties of all potential solid forms support the design of agrochemical products. Solid formulation types are very sensitive to recrystallization and phase changes. In addition to polymorphic forms, other solid states such as co-crystals, salts, or pseudopolymorphs enable a direct influence to be applied to important properties that include melting behaviors, dissolution rates, substance release and, finally, bioavailability.

As modern crop protection involves the creation of high-tech products with regards to the needs of the customer, the design of modern, solid-state agrochemical active ingredients is essential in order to support product quality.

Acknowledgments

The authors would like to thank Dr. Graham Holmwood for his careful and valuable revision of this manuscript and his support in finding the "right words."

References

1. Kuhnert-Brandstätter, M. and Martinek, A. (1965) *Mikrochim. Acta*, **5–6**, 909–919.
2. Mitscherlich, E. (1821) *Ann. Chim. Phys.*, **19**, 350.
3. McCrone, W.C. (1965) in *Physics and Chemistry of the Organic Solid State* (eds D. Fox, M.M. Labes, and A. Weissberger), John Wiley & Sons, Inc., New York, pp. 726–767.
4. Grunenberg, A. (1997) *Pharm. Unserer Zeit*, **26**, 224–231.
5. Burger, A. and van den Boom, C. (2000) *Mikrochim. Acta*, **135**, 63–69.

6. Burger, A. (1982) *Pharm. Int.*, **3**, 158–163.
7. Burger, A. (1982) *Acta Pharm. Tech.*, **28**, 1–19.
8. Lehmann, O. (1877) *Z. Kristallogr.*, **1**, 97ff.
9. Burger, A. (1982) *Pharm. Unserer Zeit*, **6**, 177–189.
10. Buerger, M.J. and Bloom, M.C. (1937) *Z. Kristallogr.*, **96**, 182–200.
11. Buerger, M.J. (1951) *Phase Transformations in Solids*, John Wiley & Sons, Inc., New York, pp. 183–211.
12. Burger, A. and Ramberger, R. (1979) *Mikrochim. Acta*, **II**, 259–271.
13. Burger, A. and Ramberger, R. (1979) *Mikrochim. Acta*, **II**, 273–316.
14. Bernstein, J., Davey, R.J., and Henck, J.O. (1999) *Angew. Chem. Int. Ed.*, **38**, 3440–3461.
15. Ostwald, W. (1897) *Z. Phys. Chem.*, **22**, 289–330.
16. Henck, J.O. and Kuhnert-Brandstätter, M. (1999) *J. Pharm. Sci.*, **88**, 103–108.
17. Grunenberg, A., Henck, J.O., and Siesler, H.W. (1996) *Int.J. Pharm.*, **129**, 147–158.
18. Davey, R. and Garside, J. (2000) *From Molecules to Crystallizers – An Introduction to Crystallization*, Oxford University Press.
19. Beckmann, W. (2004) *Kristallisation in der industriellen Praxis*, Wiley-VCH Verlag GmbH, Weinheim, pp. 63–98.
20. Haleblian, J. and McCrone, W. (1969) *J. Pharm. Sci.*, **58**, 911–929.
21. Bernstein, J. (2002) *Polymorphism in Molecular Crystals*, Clarendon Press, Oxford.
22. Brittain, H.G. and Swarbrick, J.(eds) (2007) Polymorphism in pharmaceutical solids, in *Drugs and the Pharmaceutical Science*, Informa Healthcare, New York.
23. Byrn, S.R., Pfeiffer, R.R., and Stowell, J.G. (1999) *Solid State Chemistry of Drugs*, SSCI Inc., West Lafayette.
24. Hilfiker, R. (2006) *Polymorphism in the Pharmaceutical Industry*, Wiley-VCH Verlag GmbH, Weinheim.
25. Giron, D. (1995) *Thermochim. Acta*, **248**, 1–59.
26. Kuhnert-Brandstätter, M. (1996) *Pharmazie*, **51**, 443–457.
27. Hemmiger, W.F. and Cammenga, H.K. (1989) *Methoden Der Thermischen Analyse*, Springer, Berlin, Heidelberg.
28. Höhne, G., Hemminger, W., and Flammersheim, H.J. (1996) *Differential Scanning Calorimetry*, Springer, Berlin, Heidelberg.
29. Kuhnert-Brandstätter, M. (1975) *Pharm. Unserer Zeit*, **5**, 131–137.
30. Stephenson, G.A., Groleau, E.G., Kleeman, R.L., Xu, W., and Rigsbee, D.R. (1998) *J. Pharm. Science*, **87** (5), 536–542.
31. Lehmann, O. (1877) *Z. Kristallogr.*, **1**, 43–48, 97–131.
32. Kofler, L. and Kofler, A. (1954) *Thermomikromethoden zur Kennzeichnung Organischer Stoffe und Stoffgemische*, Verlag Chemie, Weinheim.
33. Kofler, A. (1940) *Z. Phys. Chem.*, **187**, 201–210.
34. Kuhnert-Brandstätter, M. (1971) Thermomicroscopy in the analysis of pharmaceuticals, in *International Series of Monographs in Analytical Chemistry*, vol. 45, Pergamon, Oxford.
35. Kuhnert-Brandstätter, M. and Aepkers, M. (1961) *Mikrochim. Acta*, **16**, 189–197.
36. Toquin, V., Barja, F., Sirven, C., Gamet, S., Latorse, M.P., Zundel, J.L., Schmitt, F., and Beffa, R. (2006) *Pflanzenschutz-Nachr. Bayer*, **59**, 171–184.
37. Seidel, E., Vermeer, R., Hasenack, K., and Olenik, B. (2004) WO 2004/008860, Bayer CropScience.
38. Jautelat, M., Eble, H.L., Benet-Buchholz, J., and Etzel, W. (2004) *Pflanzenschutz-Nachr. Bayer*, **57**, 145–162.
39. Stephenson, G.A., Forbes, R.A., and Reutzel-Edens, S.M. (2001) *Adv. Drug Delivery Rev.*, **48**, 67–90.
40. Bugay, D.E. (2001) *Adv. Drug Delivery Rev.*, **48**, 43–65.
41. Brittain, H.G. (1997) *J. Pharm. Sci.*, **86** (4), 405–412.
42. Schrader, B. (1995) *Infrared and Raman Spectroscopy*, Wiley-VCH Verlag GmbH, Weinheim, pp. 7–34.

43. Beckmann, W. and Schmidt, J. (2000) *Struktur-Und Stoffanalytik Mit Spektroskopischen Methoden*, Teubner, Stuttgart.
44. Lin-Vien, D., Colthup, N.B., Fateley, W.G., and Grasselli, J.G. (1991) *The Handbook of Infrared and Raman Characteristic Frequencies of Organic Molecules*, Academic Press, San Diego.
45. Eble, A., Seidel, E., Olenik, B., Benet-Buchholz, J., and Hinz, M.H. (2003) EP 1314724, Bayer CropScience.
46. Eble, A., Sirges, W., Schwiedop, U., and Heyn, A. (2003) WO 03/099409, Bayer CropScience.
47. Hupe, E., Gewehr, M., Erk, P., Saxell, H.E., Griesser, U., and Tischler, M. (2008) WO 2008/110621, BASF.
48. Olenik, B., van Almsick, A., Hinz, M.H., Patel, S., Sixl, F., Thielking, G., and Dworacek, S. (2009) WO 2009/027004, Bayer CropScience.
49. van Almsick, A., Benet-Buchholz, J., Olenik, B., and Willms, L. (2009) *Bayer CropSci. J.*, **62**, 5–16.
50. Etter, M.C. (1990) *Acc. Chem. Res.*, **23**, 120–126.
51. Massa, W. (1996) *Kristallstrukturbestimmung*, Teubner, Stuttgart.
52. Steed, J.W. and Atwood, J.L. (2009) *Supramolecular Chemistry*, John Wiley & Sons, Ltd, Chichester, pp. 55–56, 474–476.
53. Bennet, D.W. (2010) *Understanding Single-Crystal X-Ray Crystallography*, Wiley-VCH Verlag GmbH, Weinheim.
54. Allmann, R. (2003) *Röntgenpulverdiffraktometrie*, Springer, Berlin, Heidelberg.
55. Saxell, H.E., Erk, P., Taranta, C., Kröhl, T., Cox, G., Desiraju, G.R., Banerjee, R., Bhatt, P.M., Sukopp, M., Scherer, S., and Ojala, A. (2008) WO 2008/055881, BASF.
56. Saxell, H.E. (2008) WO 2008/055883, BASF.
57. Saxell, H.E., Erk, P., Taranta, C., Kröhl, T., Cox, G., Desiraju, G.R., Banerjee, R., and Bhatt, P.M. (2008) WO 2008/055882, BASF.
58. Saxell, H.E., Erk, P., Taranta, C., Kröhl, T., Cox, G., Desiraju, G.R., Banerjee, R., and Bhatt, P.M. (2008) WO 2008/055884, BASF.

11
The Determination of Abraham Descriptors and Their Application to Crop Protection Research

Eric D. Clarke and Laura J. Mallon

11.1
Introduction

The seminal studies of Briggs and Bromilow and their coworkers on the uptake of pesticides into plants, and their subsequent movement in the xylem and phloem, has long highlighted the importance of physical properties in the assessment of bioavailability profiles [1–5]. Properties relating to lipophilicity, acid–base dissociation, water solubility, and volatility remain the primary compound inputs to a diverse range of soil/root and foliar uptake models [6–17]. Such models, which continue to be developed, re-evaluated, and validated, have been comprehensively reviewed in recent times [18–25].

The subject of this chapter goes beyond the conventional assessment of physical properties [26–28], their application in Lipinski style rules [29–32] and direct use in uptake and movement models to explore a different way of profiling the bioavailability of agrochemicals based on a small number of experimentally defined and chemically intuitive molecular descriptors [33].

Properties more generally termed as mobility or transport, which relate to the passive movement of a compound from one phase to another, can be described by as few as five molecular descriptors which encode chemical information relating to the ability of a compound to interact with a given phase via dispersion forces, dipolarity/polarizability, hydrogen bond donor and hydrogen bond acceptor potential, and size [34–36].

These molecular descriptors, as selected and defined by Abraham, have been used extensively in drug research to describe physical properties and mobility-related processes such as cell permeability, intestinal adsorption, and blood–brain distribution via a linear free energy relationship (LFER) [37, 38]. Critically, the resulting LFERs have the potential to predict and provide chemical insight to processes dependent upon the equilibrium transfer or the rate of transfer of a compound (solute) between gas–liquid, liquid–liquid, liquid–solid, and more complex biological phases and compartments [35].

A pragmatic example-based approach has been taken to show how Abraham descriptors can be determined for agrochemicals from readily measured physical

Modern Methods in Crop Protection Research, First Edition.
Edited by Peter Jeschke, Wolfgang Krämer, Ulrich Schirmer, and Matthias Witschel.

properties and chromatographic data. In addition, examples are given of how these descriptors can be used to set up LFERs of relevance to agrochemical research and environmental fate.

11.2
Definition of Abraham Descriptors

The background to the key research led by Prof. Michael Abraham to define solute descriptors in terms of physically relevant solvation models, that are chemically interpretable and of general applicability, has been detailed in two highly recommended reviews [35, 36].

Of particular relevance is the cavity theory of solvation, in which the transfer of a solute from the gas phase into a solvent, or between two different solvent phases, is considered as a three-step process:

1) A cavity of suitable size to accommodate the solute is created in the solvent. This first step involves the energetically unfavorable breaking of solvent–solvent interactions.
2) The solvent molecules are then reorganized into their equilibrium position around the cavity.
3) Finally, the solute is inserted into the reorganized cavity and energetically favorable solute–solvent interactions are set up to facilitate the processes of solvation.

The cavity and solute–solvent interaction terms can be represented by an LFER and the Abraham method is based on two LFERs relating a given solute property (SP) to the solute descriptors (E, S, A, B, V, L) through equation coefficients (e, s, a, b, v, l) defining the property.

$$\mathrm{SP} = \mathrm{c} + \mathrm{e\,E} + \mathrm{s\,S} + \mathrm{a\,A} + \mathrm{b\,B} + \mathrm{v\,V} \quad (11.1)$$

$$\mathrm{SP} = \mathrm{c} + \mathrm{e\,E} + \mathrm{s\,S} + \mathrm{a\,A} + \mathrm{b\,B} + \mathrm{l\,L} \quad (11.2)$$

Equation 11.1 is used to define the process of transferring a compound between condensed phases such as in water–solvent partitioning (log P) and reversed-phase high-performance liquid chromatography (RP-HPLC). Equation 11.2 covers the transfer of a compound from gas to a condensed phase in processes such as air–water partitioning, log K_W, and gas–liquid chromatography (GLC).

While the terms E, S, A, B, V, and L define or describe the compound of interest, there is a significant difference in the meaning and interpretation of the coefficients e, s, a, b, v, and l in Equations 11.1 and 11.2 For Equation 11.1, the coefficients represent the difference in properties of the two phases taking part in the partitioning process; however, with Equation 11.2, as there are no solute interactions within the gas phase to consider, the coefficients describe the properties of the single solvent phase. Note that by definition, as shown below, the

L descriptor is a more relevant cavity term for Equation 11.2, which leads to better LFER correlations for gas–solvent systems than V.

The solute descriptors are defined as follows:

- **E**: excess molar refraction arising from dispersion force interactions due to the polarizability of π and n-electrons relative to an alkane of equivalent molar volume, V – expressed as $(cm^3\ mol^{-1})/10$.
- **S**: dipolarity/polarizability due to solute–solvent interactions between dipoles and induced dipoles.
- **A**: overall hydrogen bond acidity defining the strength and number of H-bonds formed by solute donor groups with lone pairs of solvent acceptor groups.
- **B**: overall hydrogen bond basicity defining the strength and number of H-bonds formed by lone pairs of solute acceptor groups with solvent donor groups.
- **V**: McGowan characteristic volume – expressed as $(cm^3\ mol^{-1})/100$.
- **L**: preferred size descriptor for gas-condensed phase systems defined as the log of the gas-hexadecane partition coefficient at 298 K.

11.3 Determination of Abraham Descriptors: General Approach

Procedures for determining experimental descriptors have been comprehensively detailed in the 2004 review by Abraham *et al.* [35]. The descriptors V and E can be calculated from structure, leaving the descriptors A, B, S, and L to be determined from experimental data. L and S were originally obtained from GLC measurements, while A and B were initially derived from measurements of 1 : 1 donor : acceptor complexation constants (K) in tetrachloromethane [34].

Now that reliable LFERs exist for a large number of solute properties, it is common practice to use measurements of at least four (and often many more) distinctly different solute properties to derive the A, B, S, and L descriptors via the Excel "Solver" add-in tool. The Solver method simply produces a "trial-and-error" fit to solute properties from their relevant LFER equations by minimizing their sum of squares to give the best fit to the unknown or "floating" terms, which in this case are the descriptors A, B, S, and L. Note that, due to the way in which their scales have been set up, values for the hydrogen-bonding descriptors A and B cannot be less than zero, and this constraint is applied in Excel Solver ($A \geq 0, B \geq 0$) [34]. While the Solver inputs for the E and V descriptors are usually fixed as the values calculated from structures, it should be noted that they are scaled by a factor of 1/10 and 1/100 respectively to make all descriptors of similar numeric order so as not to skew the fitting process.

The quality of fit is judged in terms of the overall sum of the squared errors (SSEs) for which, based on practical experience, $\leq$0.1 is assigned as reliable, $\geq$0.1 to $\leq$0.2 as acceptable, $\geq$0.2 to $\leq$0.5 as approximate, and $\geq$0.5 as unacceptable. While the resulting experimental descriptors are most often statistically assessed as reliable or acceptable, they also need to be reviewed to ensure that they make chemical sense. For example, an A value of 0.5 arising from Solver analysis for a compound which clearly does not have a hydrogen bond donor group is chemically unacceptable. This can usually be rectified by fixing the A value as zero and re-fitting, often without serious loss in statistical quality. A more generally useful approach to reviewing chemical credibility is to determine descriptors for two, preferably more, compounds within a chemical class to look for consistency and trends in descriptor profiles.

11.3.1 V and E Descriptors

Abraham selected V as the McGowan characteristic volume as it can be readily calculated for a molecule by simple summation of atom constants and subtraction of a fixed value of 6.56 $cm^3\ mol^{-1}$ for each bond, be they single, double, or triple [39].

$$V = (\Sigma \text{ atom contribution } - (6.56 \times B))/100$$

The number of bonds can be readily obtained from the algorithm [34]:

$$B = N - 1 + R$$

where

B = number of bonds, with all bonds (single, double, triple) counting as one.
N = total number of atoms.
R = total number of ring structures.

Commonly used atom contributions, in $cm^3\ mol^{-1}$, are listed in Table 11.1.

Excess molar refraction, E, chosen as a measure of polarizability arising from dispersion force interactions, is simply the molar refraction of the molecule minus the molar refraction of an alkane of equivalent McGowan volume.

Molar refraction, MR_X, is defined as:

$$MR_X = [(n^2 - 1) / (n^2 + 2)]\ V$$

where n is the refractive index of the molecule calculated as a liquid at 20 °C using ACD/ChemSketch [40].

As MR_X for the equivalent alkane can be obtained from the correlation [36]

$$MR_{X\,(alkane)} = 2.83195V_{(alkane)} - 0.52553$$

it follows that

$$E = MR_X - 2.83195V + 0.52553$$

Table 11.1 Atom contributions for calculation of *V*.

C = 16.35	N = 14.39	O = 12.43	S = 22.91	H = 8.71
P = 24.87	B = 18.32	Si = 26.83	Se = 27.81	Sn = 39.35
F = 10.48	Cl = 20.95	Br = 26.21	I = 34.53	

Both V and E can also be reliably calculated using structural fragment contribution methods [35, 36].

11.3.2
A, B, and S Descriptors

Abraham and coworkers have shown that as few as four specifically selected measured organic/water partition coefficients, namely for octanol, cyclohexane, toluene, and chloroform, expressed as log P_S, will allow the determination of the A, B, and S descriptors for drugs [41]. Enomoto, in her PhD studies (UCL/Syngenta), selected almost the same set of organic/water partition systems – that is, for octanol, hexane, toluene, and dichloromethane (DCM) – to determine descriptors for agrochemicals [42, 43].

These sets of organic/water partition coefficients are quite similar in nature to the "critical quartet" of octanol (amphiprotic), alkane (inert), chloroform (proton donor), and propylene glycol dipelargonate (PGDP; proton acceptor) proposed by Taylor and coworkers for quantitative structure–activity relationship (QSAR) studies involving membrane permeation [44, 45]. Seiler used octanol and cyclohexane to define the term $\Delta \log P$ as the difference in their organic/water partition coefficients [46], to in effect give an octanol/alkane partition coefficient as an indicator of hydrogen-bonding potential relevant to membrane permeability [47]. The studies of Hansch and coworkers, to firmly establish the octanol/water partition coefficient as a model for lipophilicity, are well known [48–51]. Perhaps less well known are the earlier investigations of Collander to set up the first LFERs for organic/water systems using, for example, butanol/water, ether/water, and olive oil/water partition coefficients [26, 52].

11.3.3
A, B, S, and L Descriptors

The "descriptor quartet" approach outlined in Section 11.3.2 has been further developed to allow the L descriptor to be simultaneously obtained with the A, B, and S descriptors [35]. The same limited number of measured organic/water partition coefficients can be used, but with an expanded number of LSER equations accessible via the relationships implicit to the Abraham method, as outlined in Figure 11.1 [35].

For example, if there are four measured organic/water partition coefficients (log P_S), this then leads to a further four organic/air equilibrium constants (log K_S),

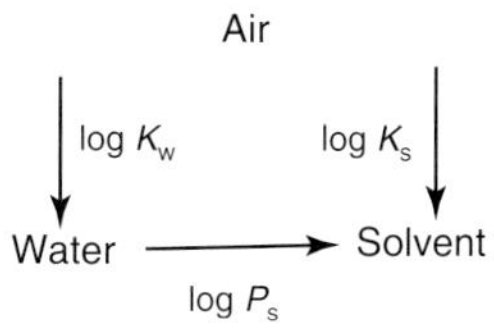

$$\log K_s = \log P_s + \log K_w$$

Figure 11.1 Relationship between log P_S, log K_S, and log K_W.

which with two equations for log K_W in both V (log P_S) and L (log K_S) gives a total of 10 equations to set up in Excel Solver to obtain the best overall fits for the A, B, S, and L descriptors – and additionally for the important parameter log K_W. If a measured value for the water/air partition coefficient K_W, is known it can be used directly, but more usually it is an undefined input to Excel Solver whose value is allowed to float and be optimized through best fits to its equations in V and L.

11.3.4 LFER Equations for Use in Determining Descriptors

Table 11.2 provides the LFER equations for all solvents mentioned in Section 11.3.2, to allow a ready comparison of their organic/water (log P_S) and organic/air (log K_S) coefficients, and to make them readily available to the reader for the determination of the A, B, S, and L descriptors. The use of these solvents is not prescriptive, and LFERs for a limited number of additional solvents for which data often appear in the literature are included. However, it is important to ensure diversity of solvent type; for example, simply using log P_S values for a set of alcohols or a set of alkanes, which would have similar LFER equation coefficients within each solvent class, would not yield reliable or chemically sensible descriptors.

Log P_S values for compounds can either be directly measured, for example, by the shake flask method, or calculated indirectly from their measured solubility in water and measured solubility in organic solvent, subject to certain constraints [35].

$$P = \text{Solubility in solvent } (S_S) \text{ / Solubility in water } (S_W)$$

$$\text{Giving } \log P_S = \log S_S - \log S_W$$

Note that solubility must be in the same units (i.e., mol l^{-1} or ppm) for solvent and water to give the dimensionless P value.

For some solvents – notably, octanol – that have some degree of miscibility with water, direct, partitioning measurements result in a different LFER equation for log P_S, designated "wet," than the LFER equation for log P_S obtained indirectly from the measurement of solubility in organic solvent and water designated "dry." This also applies to log K_S depending on the source of log P_S in the relationship shown in Figure 11.1. It is important to be aware that, under these circumstances, the hydrogen bond basicity scale may not be completely universal [35]. It is well documented that for some specific chemical types or classes, an alternative B descriptor known as B^0, may be required to obtain a good fit from Excel Solver for log $P_{octanol}$ obtained from the commonly used direct "wet" system [35]. However, this is not the case for log $P_{octanol}$ values obtained from the much less frequently used indirect "dry" system [35]. For many solvents, miscibility with water is not a practicable issue and only one LFER equation applies, which for simplicity is designated as "dry" in Table 11.2 – that is, is independent of the method used to obtain log P_S. There are of course solvents such as propan-1-ol and acetonitrile for which organic/water partition coefficients cannot be directly measured due to their

Table 11.2 LFER coefficients for log P_S (organic/water) and log K_S (organic/air) systems.

Solvent	LFER Type	c	e	s	a	b	v	e
Octan-1-ol	log *P* "Wet"	0.088	0.562	−1.054	0.034	−3.460	3.814	0.000
Octan-1-ol	log *P* "Dry"	−0.034	0.489	−1.044	0.024	−4.235	4.218	0.000
Octan-1-ol	log *K* "Wet"	−0.222	0.088	0.701	3.478	1.477	0.000	0.851
Octan-1-ol	log *K* "Dry"	−0.147	−0.214	0.561	3.507	0.749	0.000	0.943
Butan-1-ol	log *P* "Wet"	0.376	0.434	−0.718	−0.097	−2.350	2.682	0.000
Butan-1-ol	log *P* "Dry"	0.152	0.438	−1.177	0.096	−3.919	4.122	0.000
Butan-1-ol	log *K* "Wet"	−0.095	0.262	1.396	3.405	2.565	0.000	0.523
Butan-1-ol	log *K* "Dry"	−0.039	−0.276	0.539	3.781	0.995	0.000	0.934
Propan-1-ol	log *P* "Dry"	0.148	0.436	−1.098	0.389	−3.893	4.036	0.000
Propan-1-ol	log *K* "Dry"	−0.028	−0.185	0.648	4.022	1.043	0.000	0.869
Cyclohexane	log *P* "Dry"	0.159	0.784	−1.678	−3.740	−4.929	4.557	0.000
Cyclohexane	log *K* "Dry"	0.163	−0.110	0.000	0.000	0.000	0.000	1.013
Hexane	log *P* "Dry"	0.361	0.579	−1.723	−3.599	−4.764	4.344	0.000
Hexane	log *K* "Dry"	0.292	−0.169	0.000	0.000	0.000	0.000	0.979
Hexadecane	log *P* "Dry"	0.087	0.667	−1.617	−3.587	−4.869	4.433	0.000
Hexadecane	log *K* "Dry"	0.000	0.000	0.000	0.000	0.000	0.000	1.000
Chloroform	log *P* "Dry"	0.191	0.105	−0.403	−3.112	−3.514	4.395	0.000
Chloroform	log *K* "Dry"	0.157	−0.560	1.259	0.374	1.333	0.000	0.976
DCM[a]	log *P* "Dry"	0.319	0.102	−0.187	−3.058	−4.090	4.324	0.000
DCM[a]	log *K* "Dry"	0.192	−0.572	1.492	0.46	0.847	0.000	0.965
1,2-DCE[b]	log *P* "Dry"	0.183	0.294	−0.134	−2.801	−4.291	4.180	0.000
1,2-DCE[b]	log *K* "Dry"	0.017	−0.337	1.600	0.774	0.637	0.000	0.921
Toluene	log *P* "Dry"	0.143	0.527	−0.720	−3.010	−4.824	4.545	0.000
Toluene	log *K* "Dry"	0.121	−0.222	0.938	0.467	0.099	0.000	1.012
Ethyl acetate	log *P* "Wet"	0.441	0.591	−0.699	−0.325	−4.261	3.666	0.000
Ethyl acetate	log *K* "Wet"	0.130	0.031	1.202	3.199	0.463	0.000	0.828
PGDP[c]	log *P* "Wet"	0.256	0.501	−0.828	−1.022	−4.640	4.033	0.000
Diethyl ether	log *P* "Wet"	0.248	0.561	−1.016	−0.226	−4.553	4.075	0.000
Diethyl ether	log *P* "Dry"	0.330	0.401	−0.814	−0.457	−4.959	4.320	0.000
Diethyl ether	log *K* "Wet"	0.206	−0.169	0.873	3.402	0.000	0.000	0.882
Diethyl ether	log *K* "Dry"	0.288	−0.347	0.775	2.985	0.000	0.000	0.973
Acetone	log *P* "Dry"	0.313	0.312	−0.121	−0.608	−4.753	3.942	0.000
Acetone	log *K* "Dry"	0.127	−0.387	1.733	3.060	0.000	0.000	0.866
Acetonitrile	log *P* "Dry"	0.413	0.077	0.326	−1.566	−4.391	3.364	0.000
Acetonitrile	log *K* "Dry"	−0.007	−0.595	2.461	2.085	0.418	0.000	0.738
Butanone	log *P* "Dry"	0.246	0.256	−0.080	−0.767	−4.855	4.148	0.000
Butanone	log *K* "Dry"	0.112	−0.474	1.671	2.878	0.000	0.000	0.916
Olive oil	log *P* "Dry"	−0.035	0.574	−0.798	−1.422	−4.984	4.210	0.000
Olive oil	log *K* "Dry"	−0.159	−0.277	0.904	1.695	−0.090	0.000	0.876
Water/Air[d]	log *P* n/a	−0.994	0.577	2.549	3.813	4.841	−0.869	0.000
Water/Air[d]	log *K* n/a	−1.271	0.822	2.743	3.904	4.814	0.000	−0.213

[a]DCM = Dichloromethane;
[b]1,2-DCE = 1,2-dichloroethane;
[c]PGDP = Propylene glycol dipelargonate;
[d]Water/Air = water/air partition system.

Table 11.3 LFER coefficients for chromatographic hydrophobicity index (CHI) systems.

CHI system[a]	c	e	s	a	b	v	r^2
Luna C18, AcN	40.89	5.80	−17.80	−22.26	−62.34	67.067	0.95
Develosil CN, AcN	−0.38	9.51	−12.03	−1.34	−44.99	51.12	0.87
PLRP, AcN	50.10	12.30	−14.24	−31.01	−55.40	49.81	0.95
Luna C18, MeOH	44.77	5.09	−12.97	−7.00	−42.92	51.68	0.89
Develosil CN, MeOH	−5.98	9.61	−10.95	2.99	−44.02	54.15	0.90
Luna C18, TFE	50.99	4.46	−14.59	−19.00	−25.32	39.64	0.94
Perfluorooctyl silica FO, TFE	46.21	−2.62	−5.37	−20.94	−17.73	29.57	0.86

[a]details of the RP-HPLC systems for CHI are given in refs [35, 53, 54] and [66]; all columns 4.6 x 50 mm (5 μm); AcN: acetonitrile; MeOH = methanol; TFE = trifluoroethanol; C18 = Luna C18(2) column (Phenomenex); CN = CN–UG column (Nomura Chemical Company); PLRP = PLRP–S–100 polystyrene/divinyl benzene column (Polymer Laboratories); FO = Perfluorooctyl silica column (ES Industries).

complete miscibility with water. This results in virtual or hypothetical log P_S values which can only be obtained from their solubility in "dry" solvents and water.

Organic/water partition coefficients are not the only options for analysis by Excel Solver to provide experimental descriptors. Abraham and coworkers have shown that the retention parameter, log *k*, for seven diverse gradient RP-HPLC systems when standardized as chromatographic hydrophobicity index (CHI) values, can yield reliable descriptors [53, 54].

Table 11.3 lists the LFER equations in CHI for the seven RP-HPLC systems based on an analysis of 80 organic and pharmaceutical compounds, as reported by Zissimos *et al.* [53].

In principle, LFER equations can be drawn from Tables 11.2 and 11.3 to give a combined partitioning and chromatographic systems input to Excel Solver. However, as already shown for the V and E descriptors, appropriate scaling is required prior to Solver analysis. In the case of the LFERs for the RP-HPLC CHI systems, the CHI values (~0 to 100) and their equation coefficients have been arbitrarily divided by 30, whereas Abraham chose to divide by 20 [35]. The actual scaling figure used is not important as long as the inputs to Excel Solver are of similar numeric order.

11.3.5 Prediction of Abraham Descriptors

Abraham descriptors can be predicted using the Absolv program [55]. This is a group contribution method originally based on a structural fragment analysis of measured descriptors for approximately 4000 diverse chemicals, compiled by Prof. Abraham at University College, London [38, 56]. The current version of Absolv is based on a more comprehensive analysis of descriptors for about 6000 compounds [55, 57]. Absolv is used routinely to calculate the V and E descriptors, after which the Absolv predictions for the A, B, S, and L descriptors are taken as the initial inputs to Excel Solver to facilitate the fitting process. A detailed account of the application of Absolv to profile agrochemicals has recently been published [33].

11.4 Determination of Abraham Descriptors: Physical Properties

The need for high-quality measured physical property and chromatographic data cannot be overemphasized [33]. Many of the methods used to measure properties in support of agrochemical research activities from lead generation through to the selection of development candidates can be used directly, or modified to generate data of the standard required for LFER analysis. The quality of data required in terms of error limits can broadly be considered to be the same as for regulatory purposes. For active ingredients used in crop protection products, their physical properties are in the public domain, although they are not always easy to find. Fortunately, there are several electronic data compilations which include the Medchem Database [58], The e-Pesticide Manual MacBean [59], the EPI Suite [60], and the Pesticide Properties Database [61], as well as hard-copy sources such as The Pesticide Manual Tomlin [62], and the MacKay *et al. Handbook of Physical-Chemical Properties* [63]. While indispensable, there remain a number of typographic errors and questionable values in these data sources which become apparent on data collation for analysis and as major anomalies in Excel Solver best fits compared to database values [33].

For research compounds, properties must be measured "in house" or outsourced to experienced contractors. Mobility-related properties routinely measured by physical chemists at Syngenta in support of research are shown in Figure 11.2. The

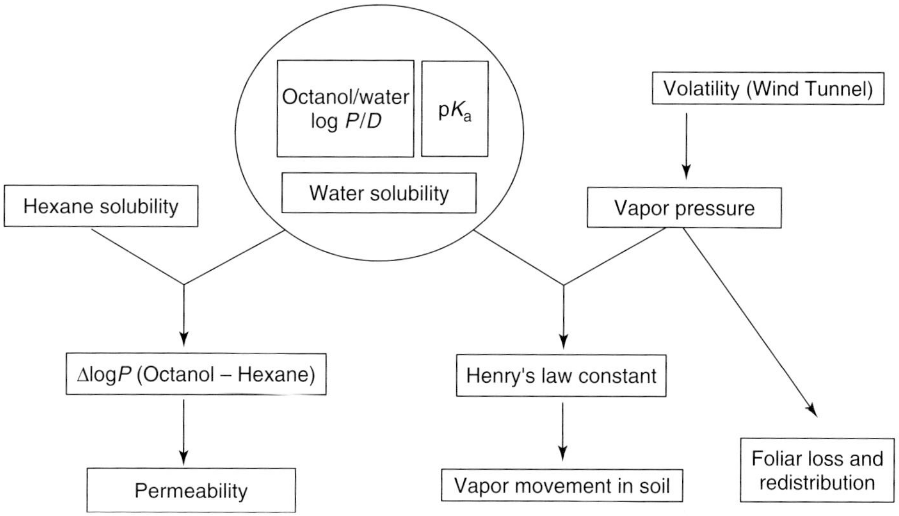

Figure 11.2 Typical mobility-related properties measured for research compounds.

methods used to determine some of these properties are briefly outlined here to indicate which are appropriate for LFER analysis.

The simple "Wind Tunnel" approach for assessing the relative volatility of compounds as deposits on glass has been described [64]. However, while fit for its original purpose, significant additional calibration with "in test" standards and some form of normalization for the variable nature of the compound deposits on glass would be required to generate data suitable for credible LFER analysis.

The CHI is routinely measured in up to three gradient HPLC systems at up to three values of pH, to provide an early assessment of lipophilicity, acid–base dissociation, and stability in support of lead generation activities [54]. While the CHI data as a chromatographic parameter are used to determine descriptors [35, 53], log $P_{octanol}$ values derived from CHI data [65, 66] have not proved to be sufficiently reliable across the wide range of chemical types required. In fact, it was found that only about 60% of compounds had a CHI-derived log $P_{octanol}$ value within 0.5 log units of conventionally measured log $P_{octanol}$ for a set of 800 structurally diverse compounds taken from the Syngenta physical properties database. More generally, error limits of up to ±0.5 log unit have been cited for the indirect determination of octanol/water partition coefficients using retention data from isocratic RP-HPLC systems [67].

The determination of Abraham descriptors is mainly based on the routine measurement of pK_a, water solubility, log $P_{octanol}$, and log P_{hexane}, to which typically are added two further measurements of log $P_{toluene}$ and log P_{DCM}. The organic/water partition coefficients arise either indirectly from solubility measurements for hexane, toluene, DCM, and water buffered at a pH to give the neutral form of compounds that are acids and bases, or directly by the "shake-flask" method following established guidelines [68, 69], with due regard to solvent specific practical issues [44]. While the conventional "shake-flask" method [68] is used to directly measure low octanol/water log P values in the range of $\sim+1$ to -2, the preferred method for direct measurements of octanol/water partition (log P) or pH-dependent distribution (log D) coefficients in the range of $\sim+0.5$ to $+4.5$ utilizes octanol-coated RP-HPLC columns [70].

Currently, HiChrom RPB (C8/C18) hybrid columns (1–5 cm) are physically coated with octanol by recirculating an aqueous mobile phase buffered between pH 2 and 9 (usually pH 7), pre-saturated with HPLC-grade octanol at room temperature. The log P/D value of a compound is obtained by measuring its retention time and comparing this with the retention time of a set of standards of known log P, as shown in Figure 11.3. With experience, log $P_{octanol}$ values higher than 4.5 can be measured using this method, the highest independently validated measurement being 5.6 for the azo dye Sudan I [71]. More often, the Generator Column method is used for reliable measurements of log $P_{octanol}$ in the range of ~4.5 to 8 [72, 73].

There are many structure-based prediction methods available for log $P_{octanol}$, and for "broad brush" Lipinski-type analyses of log $P_{octanol}$ profiles based on thousands of compounds the choice of prediction method used is unlikely to be important [29, 31, 33]. However, as judged against the many thousands of log $P_{octanol}$ measurements for research compounds in the Syngenta physical properties

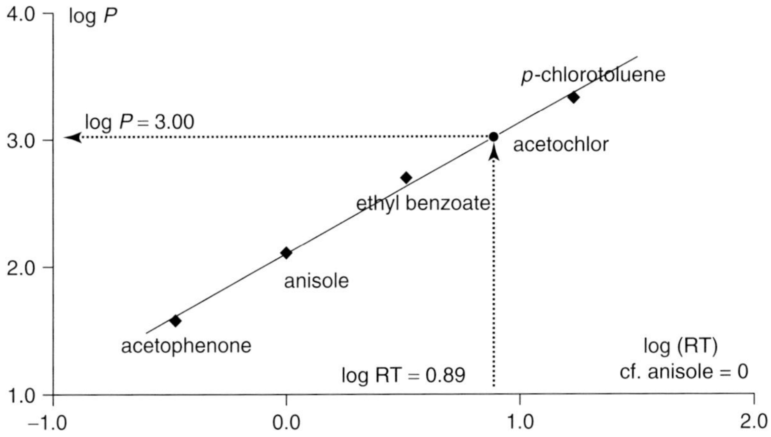

Figure 11.3 Direct determination of log $P_{octanol}$ from retention time (RT) data using a calibrated octanol-coated HPLC column.

database, a single prediction method suitable for "fine-detail" QSAR studies, and which consistently provides reliable predictions across diverse chemical classes, has yet to be identified. The results of in-house studies further suggest that it is unlikely that any "global" prediction method developed from published data sets could lead to correlations with measured log $P_{octanol}$ values with an $r^2 > 0.7$ and mean absolute errors of <0.5 when applied to research data sets. Whilst undoubtedly extremely useful in agrochemical research, and capable of performing at a level comparable to some indirect RP-HPLC approaches, there remain too many uncertainties in methods used for log $P_{octanol}$ predictions for their recommendation as a primary data input for the determination of descriptors.

11.5 Determination of Abraham Descriptors: Examples

Examples of descriptor determination are given for herbicides, insecticides, and fungicides based on the Excel Solver analysis approach which utilizes the "log K_S = log P_S + log K_W" relationship, as outlined in Section 11.3. For continuity and brevity, organic/water partition coefficients (log P_S) directly measured or indirectly calculated from measured solubility values have in the main been taken from the Syngenta physical properties database. The use of published data has been indicated in relevant examples.

In most of the examples, Solver simultaneously gives best fits to 10 LFER equations derived from four measured log P_S values, that is, four in log P_S using LFERs in V, four in log K_S using LFERs in L, and two in log K_W for which there are LFERs in V and L. When the alternative B^0 descriptor has needed to be assigned for octanol log P_S, a refit was done to its LFER to obtain a B^0 value as that which fitted to the measured value, then the octanol log K_S LFER was removed, and the remaining eight equations refitted to give final values for the A, B, S, and L descriptors, and log K_W. There are instances where a chromatographic CHI value has been used in place of the log P_S for DCM, which results in nine equations for analysis. In all examples, Solver best fits compared to measured log P_S or log P_S and CHI values are given, which are representative of the complete "log K_S = log P_S + log K_W" Solver output.

The examples for diuron and atrazine/simazine include comparison with descriptors obtained from extensive literature studies. It is important to note that descriptor profiles assessed as being reliable are not necessarily absolute, but are subject to subtle differences reflecting the number and type of measured data inputs to Excel Solver. Those descriptor profiles shown here supersede some of those previously published [33, 43].

11.5.1
Herbicides: Diuron (1)

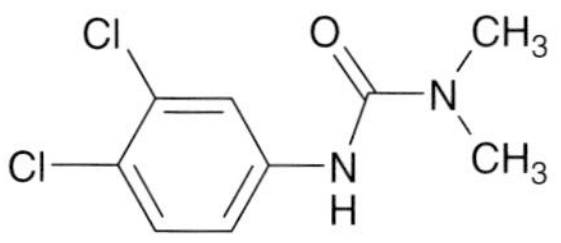

1

In 2000, Green *et al.* carried out an extensive study of the diphenylurea herbicide diuron and six other analogs that varied in their phenyl ring substituents [74]. The solubility ($mol\,l^{-1}$) of diuron was measured in 19 organic solvents and divided by a published water solubility value of 42 ppm (converted to $mol\,l^{-1}$) to give dimensionless organic/water partition coefficients *P*. Likewise, a published vapor pressure value of 9.2×10^{-6} Pa was initially used to derive a further set of 19 dimensionless organic/air equilibrium constants, *K*. The report authors queried the reliability of the vapor pressure value that they had selected, and fully described the process used to calculate a preferred value of 4.2×10^{-6} Pa from LFER analysis, which is actually quite close to the initial value used. Ultimately, a total of 38 equations in log P_S and log K_S were solved to give the descriptors designated as Ref. [74] in Table 11.4.

The V-value used was readily confirmed as 1.5992 (i.e., 159.92/100), by calculating from structure (see Section 11.3.1). The same value is also predicted by the Absolv

Table 11.4 Experimental descriptors for diuron.

Solver	S	A	B	E	V	L	log K_w
Diuron (Ref. [74])	1.60	0.57	0.70	1.28	1.599	8.06	–
Diuron (4 log *P*)	1.71	0.54	0.70	1.28	1.599	8.12	8.23
Diuron (4 log *P*, E 1.58)	1.78	0.52	0.73	1.58	1.599	8.55	8.63

Table 11.5 Solver analysis: log P_S inputs for diuron and final LFER fits.

		log $P_{octanol}$	log P_{hexane}	log $P_{toluene}$	log P_{DCM}
Diuron	Measured	2.68	−0.21	1.78	2.54
SSE = 0.047	Solver fit (E 1.58)	2.68	−0.21	1.86	2.44
SSE = 0.032	Solver fit (E 1.28)	2.68	−0.21	1.82	2.49.

SSE = Overall Sum of the Squared Errors

program [55], as expected for this relatively simple additive atom fragment-based descriptor. The report cited a calculated E-value of 1.28, which is the same as the Absolv prediction, but 0.3 lower than a value of 1.58 calculated using a refractive index of 1.606 (see Section 11.3.1).

Table 11.4 includes the descriptors obtained using the current Excel Solver approach based on the four log P_S measurements listed in Table 11.5, comparing both E values of 1.28 and 1.58 as fixed inputs. Excellent Solver fits to measured values were obtained, as illustrated for the log P_S data (Table 11.5).

There is clearly good agreement in the descriptor profiles obtained from the very extensive analysis by Green *et al.* [74], and the more practical analysis based on four log P_S values, irrespective of which E-value is used. The preferred descriptor set (E = 1.28) was used to provide reliable predictions of RP-HPLC-derived CHI values for three systems (see Table 11.3), namely C18, AcN (acetonitrile; predicted 69.1, measured 72.5), PLRP, AcN (predicted 65.2, measured 68.4), and DCN, AcN (predicted 40.5, measured 46.1). This demonstrated the potential for interchange of log P_S and CHI data in Excel Solver analysis and the general applicability of descriptor profiles for the prediction of unrelated properties, save that they can be represented by credible LFERs.

11.5.2
Herbicides: Simazine (2) and Atrazine (3)

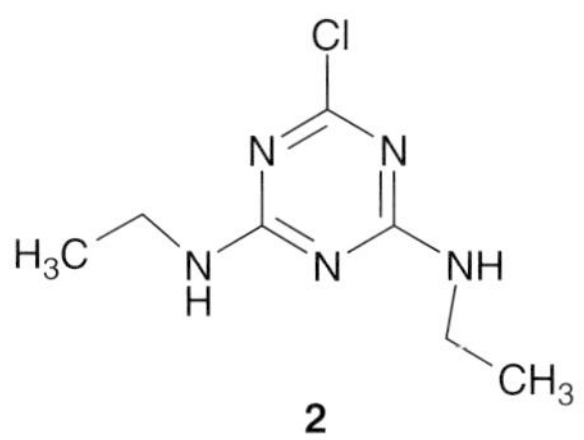

2

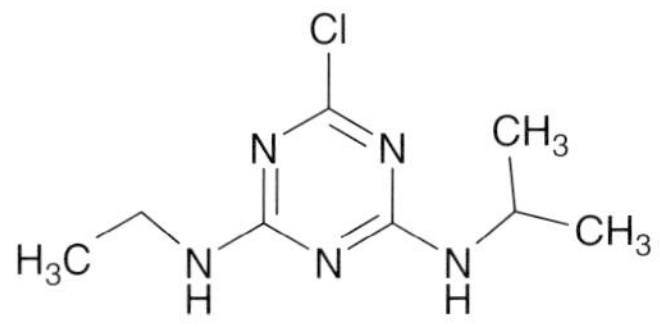

3

In 2007, Abraham *et al.* determined descriptors for 19 substituted 1,3,5-triazines in order to predict their water/air partition coefficients, log K_W, which are closely related to Henry's law constants [75]. A combination of organic/water partition

Table 11.6 Experimental descriptors for atrazine and simazine.

Solver	S	A	B	B^0	E	V	L	log K_w
Atrazine	1.29	0.17	1.01	0.88	1.22	1.620	7.78	7.10
Simazine	1.32	0.18	0.98	0.84	1.25	1.479	7.32	7.24

From Ref. [75].

Table 11.7 Solver analysis: log P_S inputs for simazine and initial LFER fit.

		log $P_{octanol}$	log P_{hexane}	log $P_{toluene}$	log P_{DCM}
Simazine	Measured	2.14	−0.29	1.33	2.33
	Solver fit	1.65	−0.08	1.30	2.01

coefficients (log P_S) and micellar electrokinetic chromatography (MEKC) retention factors were used as the primary measured data inputs to Excel Solver. In total, 18 equations were analyzed for each compound, and the descriptors obtained for atrazine and its close analog simazine are shown in Table 11.6.

The V- and E-values used as fixed inputs to Excel Solver are essentially the same as calculated by the current Absolv program [55]. However, as was the case for diuron, the E-value calculated for atrazine of 1.52 based on its refractive index value of 1.605 is 0.3 higher.

The evidence for assignment of the B^0 descriptor is exemplified for simazine in Table 11.7, using the usual four organic/water partition coefficients (log P_S) as representative of the overall 18 LSER equations analyzed. A difference of ~0.5 between the measured and calculated log $P_{octanol}$ values appeared to justify the specific use of a B^0 value for this "wet" partition coefficient. However, there were also significant – albeit lower – differences in measured and calculated values for log P_{DCM} (~0.3) and log P_{hexane} (~0.2), which as "dry" partition coefficients cannot require the B^0 descriptor. It is possible that these somewhat high errors in log P may reflect a relative bias toward the MEKC data in the overall analysis.

The descriptors for atrazine and simazine were re-determined using the current Excel Solver approach. Taking V and E from Absolv as fixed values gave the Solver best fit log P results shown in Table 11.8. In this case, Solver provided good fits with no need to resort to a B^0 descriptor. As a further check, the DCM/water partition coefficient (log P_{DCM}) was replaced by the chromatographic retention-related parameter CHI obtained from the C18, AcN) system. Again, the Solver best fit values were in good agreement with measurements (Table 11.9).

The descriptors and log K_W values obtained from both datasets are listed in Table 11.10.

Table 11.8 Solver analysis: log P_S inputs for atrazine and simazine and final LFER fits.

		log $P_{octanol}$	log P_{hexane}	log $P_{toluene}$	log P_{DCM}
Atrazine	Measured	2.58	0.49	1.82	2.32
SSE = 0.038	Solver fit	2.57	0.53	1.83	2.28
Simazine	Measured	2.14	−0.29	1.33	2.33
SSE = 0.093	Solver fit	2.09	−0.33	1.49	2.21

Table 11.9 Solver analysis: log P_S and CHI inputs for atrazine and simazine and final LFER fits.

		log $P_{octanol}$	log P_{hexane}	log $P_{toluene}$	CHI/30
Atrazine	Measured	2.58	0.49	1.82	2.41
SSE = 0.026	Solver fit	2.56	0.51	1.84	2.37
Simazine	Measured	2.14	−0.29	1.33	2.03
SSE = 0.019	Solver fit	2.09	−0.33	1.49	2.07

Table 11.10 Experimental descriptors for atrazine and simazine.

Solver	S	A	B	E	V	L	log K_w
Atrazine (using DCM)	1.12	0.34	0.93	1.22	1.620	7.57	7.03
Atrazine (using CHI)	1.14	0.34	0.93	1.22	1.620	7.62	7.06
Simazine (using DCM)	1.61	0.39	0.77	1.25	1.479	7.54	7.77
Simazine (using CHI)	1.42	0.41	0.82	1.25	1.479	7.38	7.59

The self-consistency of the atrazine descriptor data is excellent and, though more variable, is reasonable for simazine (Table 11.10). Interestingly, the log K_W values were in good agreement with the preferred published values of 7.0 for atrazine and 7.5 for simazine, as cited by Abraham *et al.* in 2007 [75]. Also, the L descriptor value of 7.6 for atrazine agreed quite well with a recently reported gas chromatographic measurement of 7.3 for its air/hexadecane log K [76]. It has been reported that, in LSER equations for RP-HPLC systems such as CHI, thc B^0 descriptor is preferred to B [35, 36]. Given that CHI from the C18, AcN system has been readily interchanged with log P_{DCM} for Solver analysis and the quality of additional CHI/30 data for atrazine from the DCN, AcN (measured 1.30, predicted 1.28), and PLRP, AcN (measured 2.22, predicted 2.26) systems further supports the view from this re-evaluation study that B is the appropriate general descriptor.

11.5.3
Herbicides: Acetochlor (4) and Alachlor (5)

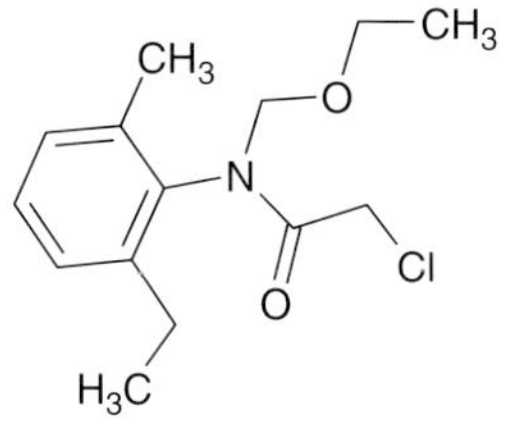

4

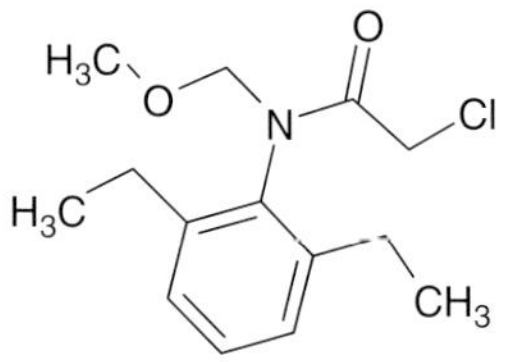

5

Due to the lack of absolute values for organic solubility in physical property databases, and the practical difficulties in measuring shake-flask organic/water partition coefficients for oils such as acetochlor, there is uncertainty as to the reliability of previously reported descriptors for chloroacetanilide herbicides [33, 44]. Also, the apparent anomaly of the currently cited log $P_{octanol}$ value for acetochlor (4.14) compared to alachlor (3.09), given their close structural similarity and identical molecular weights, further prompted a revisit of the descriptor analysis for this class of chemistry [77].

Using the octanol-coated column method (see Section 11.4), the direct log $P_{octanol}$ measurement for acetochlor (3.00) is, as expected, almost the same as for alachlor (3.07), and indeed the same (3.03) as reported in the 11th edition of the *Pesticide Manual*. Tomlin [78]. Shake-flask organic/water partition coefficient measurements (log P_S) were made for alachlor and re-made for acetochlor, with hexane and toluene as the organic phase, but not for DCM as it was not deemed confident that a reliable value could be obtained [44]. In addition, CHI values were measured in the C18, AcN chromatographic system. The measured log P_S and CHI data used for analysis by Excel Solver and their "best fits" are listed in Table 11.11.

The descriptors and log K_w values obtained are shown in Table 11.12.

Both compounds had essentially the same measured organic/water partition coefficients in octanol, hexane, and toluene and, not surprisingly, the same experimental descriptors and predicted dimensionless water/air partition coefficients,

Table 11.11 Solver analysis: log P_S and CHI inputs for acetochlor and alachlor and final LFER fits.

		log $P_{octanol}$	log P_{hexane}	log $P_{toluene}$	CHI (C18/AcN)/30
Acetochlor	Measured	3.03	2.43	3.23	3.09
SSE = 0.177	Solver fit	3.25	2.34	3.23	3.02
Alachlor	Measured	3.06	2.44	3.26	3.25
SSE = 0.171	Solver fit	3.30	2.40	3.26	3.04

Table 11.12 Experimental descriptors for acetochlor and alachlor.

Solver	S	A	B	E	V	L	log K_w
Acetochlor	0.816	0.000	1.376	1.110	2.140	8.874	6.594
Alachlor	0.806	0.000	1.373	1.160	2.140	8.913	6.585

given as log K_W. The L descriptor value of 8.91 for alachlor is in reasonable agreement with a recently reported gas chromatography-based measurement of 8.41 for its air/hexadecane log K_S [76]. The DCM/water partition coefficient predicted from its LSER equation and the Solver experimental descriptors is 3.9 for both compounds, which is in the region reported as being potentially too high for practical measurement of chlorinated solvents by the shake-flask method [44].

11.5.4
Insecticides: Fipronil (6)

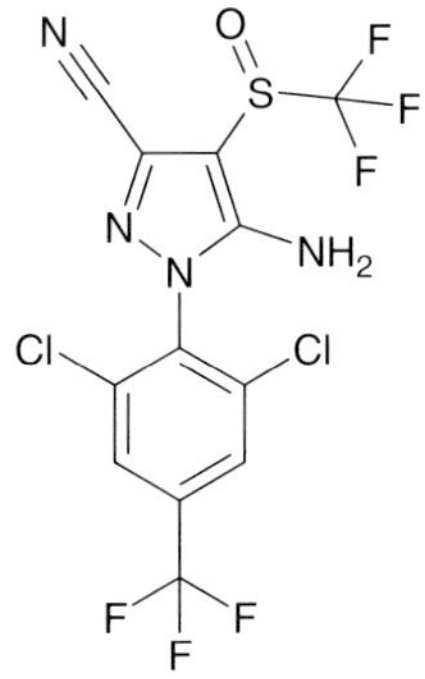

6

The measured data used to determine descriptors for fipronil were in the main taken from Macbean [79]. The organic solubility measurements for hexane (28 ppm), toluene (3000 ppm), and DCM (22 300 ppm) were used as cited, but the authors' own measurement of water solubility (1.4 ppm) was used preferably to calculate log P for these solvents rather than the (albeit close) value of 1.9 ppm cited. The four log P_S values used and the excellent Solver fits obtained are listed in Table 11.13.

The reliable descriptor profile obtained and log K_W value are shown in Table 11.14.

A nuclear magnetic resonance (NMR) method was also used to independently determine the hydrogen bond acidity descriptor A for fipronil. The full methodology is described in the report made in 2006 by Abraham *et al.* [80]. The value for A is calculated using the relationship:

$$A = 0.0066 - 0.128\text{IS} + 0.133\Delta\delta$$

Table 11.13 Solver analysis: log P_S inputs for fipronil and final LFER fits.

		log $P_{octanol}$	log P_{hexane}	log $P_{toluene}$	log P_{DCM}
Fipronil	Measured	4.00	1.30	3.33	4.20
SSE = 0.073	Solver fit	3.93	1.29	3.46	4.11

Table 11.14 Experimental descriptors for fipronil.

Solver	S	A	B	E	V	L	log K_w
Fipronil	1.87	0.39	1.13	1.97	2.254	11.49	9.94

where the "indicator variable" IS, is zero, except for thiols (IS = 1) and $\Delta\delta$ is the difference in chemical shift of the relevant protons in d_6-DMSO and $CDCl_3$.

For fipronil, the shift of interest is in exocyclic N–H protons for which $\Delta\delta$ (d_6-DMSO–$CDCl_3$) is 3.17, giving an A value of 0.43, which is in good agreement with the Solver value of 0.39.

11.5.5
Insecticides: Imidacloprid (7)

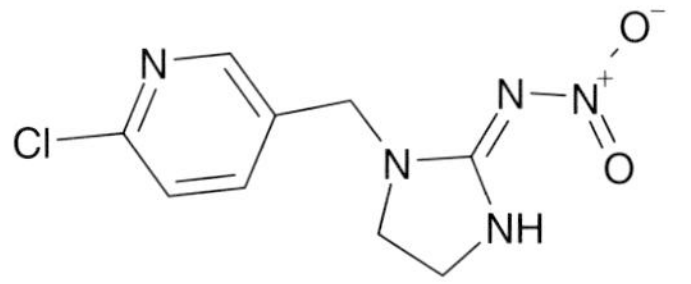

7

For imidacloprid, the data used to determine descriptors were taken from the Pesticide Properties Database (PPDB) [61] and the present authors' own measurements. The PPDB cited a log $P_{octanol}$ value of 0.57, together with a toluene solubility measurement of 690 ppm and a water solubility measurement of 610 ppm, to give a calculated toluene log P_S of 0.053. The four log P_S values used and the excellent Solver fit obtained are listed in Table 11.15. The resulting descriptors and log K_W value are given in Table 11.16.

However, on applying the NMR method only a small chemical shift was observed in the cyclic N–H proton, to give a $\Delta\delta$ (d_6-DMSO–$CDCl_3$) of 0.79, which resulted in an A-value of 0.11, significantly lower than determined by Excel Solver analysis. Repeating the Solver analysis with A fixed as 0.11 gives the results shown in Table 11.17. The very large discrepancy in the measured and fitted log $P_{octanol}$ data

Table 11.15 Solver analysis: log P_S inputs for imidacloprid and initial LFER fits.

		log $P_{octanol}$	log P_{hexane}	log $P_{toluene}$	log P_{DCM}
Imidacloprid	Measured	0.57	−3.64	0.053	1.81
SSE = 0.067	Solver fit	0.64	−3.72	0.14	1.75

Table 11.16 Initial experimental descriptors for imidacloprid.

	S	A	B	E	V	L	log K_w
Solver	3.54	0.55	0.90	1.67	1.683	10.59	13.86

Table 11.17 Solver analysis: log P_S inputs for imidacloprid and final LFER fits.

		log $P_{octanol}$	log P_{hexane}	log $P_{toluene}$	log P_{DCM}
Imidacloprid	Measured	0.57	−3.64	0.05	1.81
SSE = 0.444	Solver fit (B)	−2.23	−3.22	0.09	1.82
SSE = 0.042	Solver fit (B^0)	0.57	−3.62	−0.01	1.86

Table 11.18 Final experimental descriptors for imidacloprid.

	S	A	B	B^0	E	V	L	log K_w
Solver (B)	3.27	0.11	1.22	–	1.67	1.683	10.12	13.10
Solver (B^0)	3.57	0.11	1.20	0.90	1.67	1.683	10.45	13.86

is both unusual and unacceptable. If taken at face value as a case for assigning the alternative B^0 descriptor using the process outlined in Section 11.5, then the modified analysis yields what appeared to be reliable data with a 10-fold lower SSE.

The resulting overall descriptor profiles and log K_W values shown in Table 11.18 are quite similar, which would be expected given that the B^0 value only applies to the log P_S for octanol in this set of log P_S data.

The measured CHI value from the C18, AcN RP-HPLC system gave a poor prediction when using what appeared to be reliable original descriptors (Table 11.16) – that is, with A as 0.55, CHI measured is 42.5 measured and 31.7 predicted. An even worse CHI prediction of 22.7 was made using the similar Solver (B) descriptors with A fixed as 0.11 (Table 11.18). However, when the Solver B^0 descriptor was used in place of B with A fixed as 0.11, a surprisingly good prediction of 41.3 was obtained. Given the preference for use of B^0 over B in LFERs for RP-HPLC systems, this result would seem to support the use of the Solver (B^0) descriptor profile for imidacloprid given in Table 11.18.

11.5.6
Insecticides: Chlorantraniliprole (8)

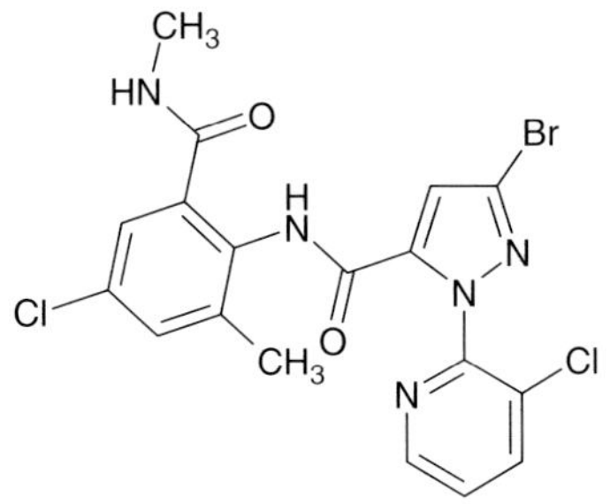

8

The measured log P_S data used for chlorantraniliprole and the Solver fits obtained are given in Table 11.19.

In this case, the overall Solver fit was assessed as acceptable rather than reliable, consistent with the CHI/30 (C18, AcN) RP-HPLC data shown in Table 11.19, calculated post Solver analysis from the resulting experimental descriptors (Table 11.20). The actual CHI values were 79.4 measured and 64.2 predicted.

Table 11.19 Solver analysis: log P_S inputs for chlorantraniliprole and final LFER fits.

		log $P_{octanol}$	log P_{hexane}	log $P_{toluene}$	log P_{DCM}	CHI/30
	Measured	2.78	−1.52	1.84	2.57	2.65
SSE = 0.16	Solver fit	2.77	−1.39	1.62	2.73	2.14

Table 11.20 Final experimental descriptors for chlorantraniliprole.

	S	A	B	E	V	L	log K_W
Solver	2.72	0.38	2.18	3.44	2.925	16.52	17.45

11.5.7
Insecticides: Thiamethoxam (9)

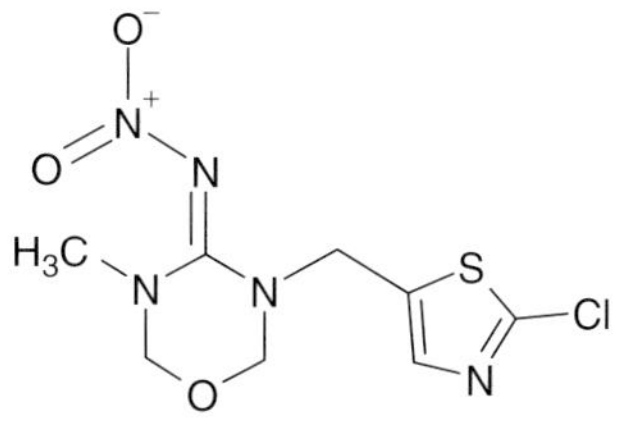

9

The log P_S values and Excel Solver fits for thiamethoxam shown in Table 11.21 initially looked acceptable. However, in this example the resulting descriptor profile did not make chemical sense (Table 11.22).

An A-value of 0.39 is obviously incorrect, as thiamethoxam does not have a hydrogen bond donor, which reassuringly was confirmed as a test of the NMR method [80]. On correctly setting A to zero for thiamethoxam, it was evident from the discrepancy of 0.8 in the measured and fitted log P_S values for octanol that the alternative hydrogen bond basicity B^0 value needed to be assigned, as outlined in Section 11.5 (Table 11.23).

An overall acceptable fit to the measured data was obtained with an A value of zero and a B^0 value of 1.25, as shown in Table 11.23, with the final descriptor set shown in Table 11.24.

The availability of a CHI value from the C18, AcN chromatographic system again proved useful in supporting the assignment of a B^0 value. An excellent CHI prediction of 32.1 was obtained compared to the measured CHI of 33.6 using the B^0 value of 1.25. When the B-value of 1.56 was used a very poor CHI prediction of 11.7 was obtained.

Table 11.21 Solver analysis: log P_S inputs for thiamethoxam and initial LFER fits.

		log $P_{octanol}$	log P_{hexane}	log $P_{toluene}$	log P_{DCM}
	Measured	−0.13	−4.33	−0.78	1.20
SSE = 0.183	Solver fit	−0.19	−4.41	−0.63	1.21

Table 11.22 Preliminary experimental descriptors for thiamethoxam.

	S	A	B	E	V	L	log K_W
Solver	3.55	0.39	1.28	1.76	1.808	11.28	15.25

Table 11.23 Solver analysis: log P_S inputs for thiamethoxam and LFER fits with A set to zero.

		log $P_{octanol}$	log P_{hexane}	log $P_{toluene}$	log P_{DCM}
	Measured	−0.13	−4.33	−0.78	1.20
SSE = 0.440	Solver fit (B)	−0.95	−3.96	−0.68	1.27
SSE = 0.235	Solver fit (B^0)	−0.13	−4.34	−0.80	1.28

Table 11.24 Final experimental descriptors for thiamethoxam.

	S	A	B	B^0	E	V	L	log K_W
Solver (B)	3.31	0.00	1.57	–	1.76	1.808	10.94	14.45
Solver (B^0)	3.57	0.00	1.56	1.25	1.76	1.808	11.21	15.10

11.5.8
Fungicides: Azoxystrobin (10)

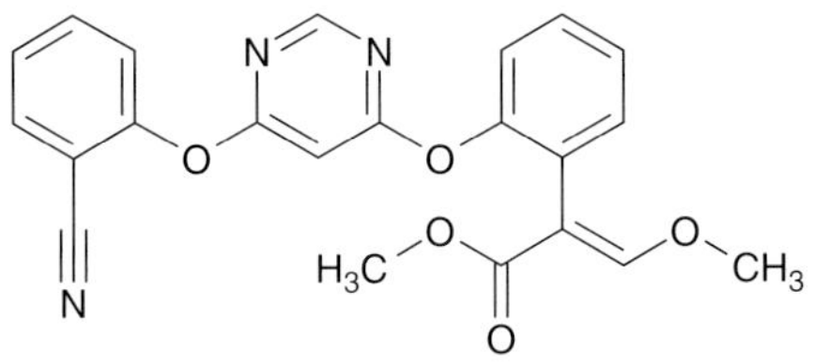

10

The measured log P_S data used for azoxystrobin and the Solver fits obtained are shown in Table 11.25. In this example, a log P_S for AcN was used in place of DCM as the latter was expected to be very high and unreliable.

Azoxystrobin with a discrepancy of 0.7 between the measured and fitted log P_S values for octanol was a relatively uncomplicated example requiring the assignment

Table 11.25 Solver analysis: log P_S inputs for azoxystrobin and LFER fits.

		log $P_{octanol}$	log P_{hexane}	log $P_{toluene}$	log P AcN
	Measured	2.50	1.11	3.96	3.60
SSE = 0.429	Solver fit (B)	3.24	0.80	3.90	3.55
SSE = 0.072	Solver fit (B^0)	2.50	1.14	4.04	3.47

Table 11.26 Final experimental descriptors for azoxystrobin.

	S	A	B	B^0	E	V	L	log K_W
Solver (B^0)	2.75	0.00	1.78	2.06	2.32	2.917	15.00	13.53

of the alternative hydrogen bond basicity B^0 descriptor, as outlined in Section 11.5 (Table 11.25). Accordingly, a reliable descriptor profile was obtained as shown in Table 11.26.

Once again, holding back the CHI value from the C18, AcN chromatographic system from the Solver analysis provided a useful check on the B^0 value. The measured CHI of 87.1 compared well with the prediction of 89.9 with the B^0 value of 2.06. It is also worth noting that the predicted log P_S value for DCM was ~6, well beyond the practical limit for a reliable measurement for chlorinated solvents [44].

11.5.9
Plant Growth Regulator: Paclobutrazol (11)

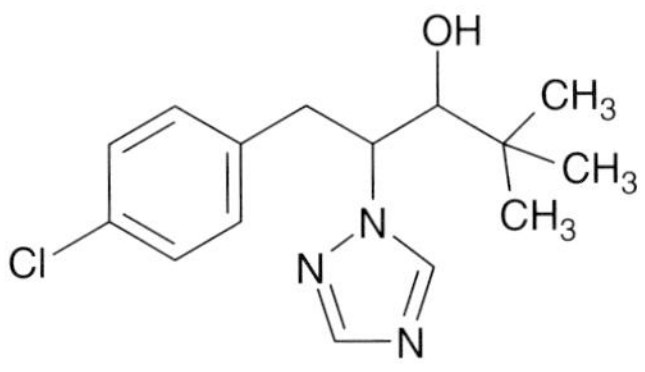

11

The measured log P_S data used for paclobutrazol and the Solver fits obtained are given in Table 11.27. This is a straightforward example with the resulting reliable descriptors shown in Table 11.28.

Table 11.27 Solver analysis: log P_S inputs for paclobutrazol and LFER fits.

		log $P_{octanol}$	log P_{hexane}	log $P_{toluene}$	log P_{DCM}
	Measured	3.14	0.96	2.53	3.39
SSE = 0.061	Solver fit	3.10	1.01	2.60	3.30

Table 11.28 Final experimental descriptors for paclobutrazol.

	S	A	B	E	V	L	log K_W
Solver	1.39	0.21	1.46	1.53	2.270	10.45	9.39

11.6 Application of Abraham Descriptors: Descriptor Profiles

The application of descriptor profiles to agrochemical research can initially be considered in a manner similar to the "Lipinski"-style analysis of physical properties, such as octanol/water partition coefficients (log $P_{octanol}$) and molecular parameters based on simple counts of molecular features such as hydrogen bond donors and acceptors. Hence, in 2009 Clarke reported simple bioavailability guidelines for agrochemicals in terms of the descriptors A ($\leq$1), B ($\leq$3), and V ($\leq$3) [33], building on the 2003 report of Clarke and Delaney [31]. The Lipinski approach [29], which was closely followed by Tice in his 2001 report on agrochemicals [30], and the approach taken by Clarke are of course quite similar. Predictions and counts, or predictions alone were made for large numbers of compounds representing either drugs with oral bioavailability or agrochemicals utilized as herbicides, fungicides, or insecticides, to provide what ultimately proved to be relatively simple and readily understood guidelines to these classifications. There are parallels in the attempts of drug and crop protection scientists to extend this type of analysis to provide insights, for example, to lead-like profiles, though at least with respect to agrochemicals such studies – while useful – have arguably not proved to be as incisive as would be liked in terms of chemical design [33].

The process of determining experimental Abraham descriptors can in itself be enlightening, as exemplified in Section 11.5. The in-principle simple, but at times frustratingly difficult, act of reliable data collation has shown on the one hand the breadth of values that agrochemicals can have with respect to organic/water partition coefficients (log P_S) across the four solvents used for descriptor determination but, on the other hand, the similarity that can occur for compounds with, for example, a common application mode and route of uptake. Hence, while there are similarities in these "descriptor quartet" log P_S profiles for pre-emergence herbicides such as diuron and atrazine, the overall log P_S profile is different for acetochlor, possibly reflecting a shift in balance between uptake by the roots and the emerging shoots. Such observations are of course completely in line with the principle of the "critical quartet" of solvents promoted within drug research to more fully understand bioavailability [45].

The current utilization of the "log K_S = log P_S + log K_W" approach to descriptor determination adds further potential for profiling, not just in providing reliable organic/air (log K_S) equilibrium constants but also a reliable assessment of the important water/air parameter log K_W which is a dimensionless form of Henry's Law Constant [74, 75]. Models relating to the uptake and movement of agrochemicals already make use of log K_W and, more recently, the octanol/air log K_S parameter [17]. These parameters can either arise directly from the descriptor determination process or in cases where experimental descriptors are already known via their respective LSER equations (see Table 11.2).

Given the differences in "descriptor quartet" log P_S profiles, it is a reasonable expectation that the five-component experimental descriptor profiles obtained from them could lead to a greater insight and subtlety in the present understanding of the uptake and movement properties of agrochemicals. In order to acquire this knowledge, LFERs are used to relate the experimental descriptors for a given compound to relevant physical and physiological end points.

11.7
Application of Abraham Descriptors: LFER Analysis

The requirements for generating credible LFERs through multiple linear regression analysis (MLRA) have been reviewed in detail by Vitha and Carr, in 2006 [36]. With respect to the datasets used, two aspects are of critical importance. The first aspect relates to the number of compounds and the diversity of chemistry for which reliable, or at least acceptable, Abraham descriptors are available. The second aspect relates to the quality of the end-point data and the numeric range that it covers. For a five-component MLRA the minimum requirement is arguably 20 compounds, which should ideally cover variably substituted aliphatic aromatic and heterocyclic structures. The end-point data should span at least a 10-fold numeric range and have well defined and acceptable error limits. While tempting, published data sets which have low compound numbers and limited structural diversity or limited numeric range and high error limits for end-point values should be avoided as a basis for LFER analysis. With respect to compound sets and selection, it is useful to check for and to avoid a high colinearity between the E and V, S and V, and E and S experimental descriptors. The following examples of LFER analysis of data sets relevant to crop protection research serve to illustrate the points raised.

11.7.1
LFERs for RP-HPLC Systems

LFER equations for isocratic RP-HPLC systems utilizing different C18 stationary phases with a range of methanol/water or AcN/water ratio mobile phases vary in their equation coefficients. However, when the coefficients are normalized to v, the resulting e/v, s/v, a/v, and b/v coefficient ratios are essentially constant [81]. Consequently, while these systems essentially encode the same chemical information, their actual use in descriptor analysis requires a careful selection of the appropriate LFER for the C18 column and mobile phase used [82], or the calibration of a personal laboratory system with perhaps 25–30 compounds with known descriptors to set up the system specific LFER. While it is evident that the measure of lipophilicity from such C18 isocratic RP-HPLC systems is not the same as for octanol/water partition coefficients, Lombardo and coworkers have shown that the use of octanol as a mobile phase additive (typically 0.25% by volume) can lead to near-equivalent LFERs for neutral unionized compounds [54, 83].

Table 11.29 LFER equations for the CHI (C18, AcN) system compared to log $P_{octanol}$.

	c	e	s	a	b	v	*n*	r^2	sd
Ref. [53]	1.363	0.193	−0.593	−0.742	−2.078	2.236	80	0.95	0.19
Normalized/v	0.61	0.09	−0.26	−0.33	−0.93	1.00	–	–	–
This work	1.125	0.206	−0.405	−0.627	−1.945	2.118	39	0.93	0.23
Normalized/v	0.53	0.10	−0.19	−0.30	−0.92	1.00	–	–	–
log $P_{octanol}$	0.088	0.562	−1.054	0.034	−3.460	3.841	613	0.99	0.12
Normalized/v	0.023	0.15	−0.28	0.01	−0.91	1.00	–	–	–

Valko and coworkers overcame the C18 system variability issue by calibrating the compound retention output from an AcN/water gradient RP-HPLC against a set of standards to give CHI as a universal lipophilicity parameter [54, 65]. Thus, CHI values have been measured in the C18, AcN/water system for 39 compounds covering registered agrochemicals and research leads for which experimental descriptors have been determined; subsequently, an LFER equation was set up for comparison with the published LFER based on 80 pharmaceutical compounds [53]. In Table 11.29, the two LFER equations, given in their Excel Solver-ready "1/30" form, are similar and essentially have the same coefficients when normalized against v. This is to be expected given that the end point data is measured in the same way, and taking into account the wide structural diversity of both (albeit different) sets of compounds.

Further comparison with the LFER for log $P_{octanol}$ clearly indicates that the major difference from chromatographic hydrophobicity index (C18, AcN) relates to hydrogen bond acidity, which can lead to lower than expected "CHI/log *P*" values. Consequently, the correlation equation for the prediction of log $P_{octanol}$ from CHI values (C18, AcN system) is reported to be improved by the inclusion of a hydrogen bond donor (HBD) count term or the Abraham A descriptor [54, 84]. However, it should be noted that care must be taken to avoid errors due to tautomerism in structure-based HBD counts and in Absolv predictions of the A descriptor [33]. With regard to the latter chemical class, corrections can be made to A-value predictions using a limited number of experimental A-values.

In 2002, Donovan and Pescatore reported the details of an alternative RP-HPLC approach to directly determine log $P_{octanol}$ based on a octadecyl-poly(vinyl alcohol) (ODP) stationary phase and methanol/water gradient mobile phase (ODP, MeOH) [82]. Here, an LFER has been generated for their "HPLC log *P*" values based on the available experimental descriptors for 51 of the 120 diverse general organic, pharmaceutical, and agrochemical compounds cited. The equation coefficients are compared to the established log $P_{octanol}$ equation in Table 11.30.

On the basis of this analysis, the LFER for the ODP, MeOH system has an overall balance of descriptor coefficients which are reasonably similar to log $P_{octanol}$, thus supporting the potential use of "HPLC log *P*" values in structure–activity and structure–property studies, as suggested by Donovan and Pescatore [82].

Table 11.30 LFER equation for the "HPLC log *P*" (ODP, MeOH) system compared to log $P_{octanol}$.

	c	e	s	a	b	v	*n*	r^2	sd
"HPLC log *P*"	0.035	1.001	−0.812	0.236	−3.340	3.180	51	0.85	0.50
Normalized/v	0.011	0.32	−0.26	0.07	−1.05	1.00	–	–	–
log $P_{octanol}$	0.088	0.562	−1.054	0.034	−3.460	3.841	613	0.99	0.12
Normalized/v	0.023	0.15	−0.28	0.01	−0.91	1.00	–	–	–

11.7.2
LFERs for Soil Sorption Coefficient (K_{OC})

Soil binding data have been published for several hundred organic compounds which has been normalized for organic carbon content and expressed as soil sorption coefficient, K_{OC} values [85]. In 1999, Poole and Poole generated LFERs for soil/water (K_{OC}) and soil/air (K_{OCA}) distribution coefficients using 138 and 69 compounds respectively [86]. In 2004, Abraham, Clarke, and Enomoto presented a LFER for K_{OC} [33] based on 209 compounds taken from the dataset of Tao *et al.* [85]. The LFER equations arising from these studies are shown in Table 11.31.

The Poole and Poole LFER for log K_{OC} used in Table 11.31 is "equation 8" from Ref. [86] – that is, the initial equation prior to further refinement for outliers. These authors also defined their LFER using the B^0 rather than the B term; however, it is assumed that in reality B^0 applied to only a small number of compounds in the dataset and that in fact most hydrogen bond basicity descriptors were B. In any event, the two equations for log K_{OC} are effectively the same, again illustrating the reliability of LFER analysis when applied to different datasets meeting the compound and end-point criteria outlined in Section 11.7. Comparison of the log K_{OC} LFERs with the equation for log $P_{octanol}$ indicates that the main difference in

Table 11.31 LFER equations for log K_{OCA} and log K_{OC} compared to log $P_{octanol}$.

	c	e	s	a	b	v	l	*n*	r^2	sd
log K_{OCA} (Ref. [85])	−0.46	0.65	2.40	3.39	2.57	n/a	0.36	69	0.99	0.24
log K_{OC} (Ref. [85])	0.55	0.95	−0.39	−0.39	−1.51	1.76	n/a	138	0.94	0.39
Normalized/v	0.31	0.54	−0.22	−0.22	−0.86	1.00	–	–	–	–
log K_{OC} (Ref. [33])	0.39	0.88	−0.34	−0.36	−1.98	2.01	n/a	209	0.92	0.38
Normalized/v	0.19	0.44	−0.17	−0.18	−0.99	1.00	–	–	–	–
log $P_{octanol}$	0.088	0.562	−1.054	0.034	−3.460	3.841	n/a	613	0.99	0.12
Normalized/v	0.023	0.15	−0.28	0.01	−0.91	1.00	–	–	–	–

n/a = not applicable.

their correlation arises from sensitivity to hydrogen bond donors, A, and dispersion force interactions, E.

11.7.3 LFERs for Partitioning into Plant Cuticles

This example, which is based solely on the 2000 report of Platt and Abraham [87], serves to exemplify the approach of combining suitable end-point data from two studies using isolated cuticles from the same plant species to give a good-sized dataset. Consequently, the LFER equations detailed in Table 11.32 for air/plant cuticle (log K_{MXa}) and plant cuticle/water (log K_{MXw}) partition coefficients were obtained using cuticles from tomato fruit for a total set of 62 compounds.

Table 11.32 LFER equations for log K_{MXa} and log K_{MXw} compared to log $P_{octanol}$.

	c	e	s	a	b	v	l	*n*	r^2	sd
log K_{MXa}	−0.617	0.082	1.282	3.120	0.820	n/a	0.860	62	0.99	0.23
log K_{MXw}	−0.415	0.596	−0.413	−0.508	−4.096	3.908	n/a	62	0.98	0.24
Normalized/v	−0.106	0.15	−0.11	−0.13	−1.05	1.00	n/a	–	–	–
log $P_{octanol}$	0.088	0.562	−1.054	0.034	−3.460	3.841	n/a	613	0.99	0.12
Normalized/v	0.023	0.15	−0.28	0.01	−0.91	1.00	n/a	–	–	–

n/a = not applicable.

On the basis of the normalized coefficients, there is clearly quite a close relationship between plant cuticle/water and octanol/water partition coefficients, the main differences in their correlation in this case being sensitivity to hydrogen bond donors and polarizability. The potential for plant species dependency of this relationship is currently unknown due to a lack of datasets for analysis.

11.7.4 LFERs for Root Concentration Factor (RCF)

Here, the data reported by Briggs *et al.* [1] and Shone and Woods [88] have been combined to produce the minimum acceptable data set of 20 compounds for root concentration factor (RCF) for barley from water.

As can be seen from the data in Table 11.33, a credible LFER was obtained for log RCF in terms of its R^2 and SD. In this case, there is clearly no relationship between the normalized coefficients for log RCF and log $P_{octanol}$. This is in agreement with the published biphasic plot of log $P_{octanol}$ versus log RCF [1], which shows that there is not a simple relationship between these two parameters across the log $P_{octanol}$ range, from about −0.5 to 4.5. A previously reported LFER for RCF [33] could be considered misleading, as it is based on a compilation of mean RCF values with standard deviations of similar order to the mean values used which results from the use of data from diverse plant species in hydroponic and soil test systems described

Table 11.33 LFER equation for log RCF compared to log $P_{octanol}$.

	c	e	s	a	b	v	*n*	r^2	sd
log RCF	−0.285	1.226	0.360	−0.996	−0.696	−0.163	20	0.90	0.18
Normalized/v	1.75	−7.52	−2.21	6.11	4.27	1.00	–	–	–
log $P_{octanol}$	0.088	0.562	−1.054	0.034	−3.460	3.841	613	0.99	0.12
Normalized/v	0.023	0.15	−0.28	0.01	−0.91	1.00	–	–	–

by Polder *et al.* [89]. Also, it relates only to compounds in the ~2 to 4.5 log $P_{octanol}$ range, for which a tentative linear correlation appeared to exist with log RCF.

11.7.5
LFER for Transpiration Stream Concentration Factor

Attempts have been made to generate LFERs for transpiration stream concentration factor. However, despite the fact that over 100 measurements from diverse plant species and test systems have been described (as compiled by Dettenmaier *et al.* in 2009), none of these meets the requirements for credible LFER analysis [90]. Consequently, while compound sets were chosen with good reason for individual studies (e.g., on the basis of log $P_{octanol}$), they do not meet the requirements for either number or chemical diversity. For example, whilst 21 descriptor profiles of the 25-compound set of Dettenmaier *et al.* [90] were available, they spanned a narrow range of A and B and had a high colinearity of E and V ($R^2 \sim 0.7$). None of these attempts at pooling data sets resulted in a robust analysis, which is why none is shown here. It should be noted that, in 2007, Carvalho *et al.* provided two reports giving LFER equations for plant concentration factor [91] and transpiration stream concentration factor [92]; however, as each of the studies included only 10 compounds, they failed to meet the generally excepted criteria for reliable LFER analysis.

11.8
Application of Abraham Descriptors: Generality of Approach

In this chapter, the LFER equations for processes relevant to agrochemical research have been briefly outlined, such as the prediction of chromatographic parameters related to lipophilicity, soil sorption coefficients, partitioning into plant cuticles, and RCF. The LFER equation generated for CHI using agrochemicals is the same as that reported previously for drugs, and shows the generality of the experimental descriptor based LFER approach. A new LFER equation is also presented for log RCF which does not show a simple relationship with log $P_{octanol}$, and supersedes a previously described tentative LFER. It is also shown that, despite the existence of a significant amount of data, it has not been possible to generate a reliable LFER for transpiration stream concentration factor. Currently, more than 6000 experimental descriptors are available for organic compounds, but only about 100 are for crop protection compounds. Nonetheless, sufficient examples of experimental descriptor determination have been provided in this chapter to encourage the determination of many more.

Abraham descriptors and LFER analysis are widely applied across the scientific disciplines. Many LFERs exist for physiological and toxicological systems, as recently summarized by Clarke [33]. Recent investigations by Abraham and coworkers have addressed issues such as the prediction of solubility in organic solvents [93], solubility in mixed solvents [94], diffusion coefficients [95] the partitioning of ionized compounds [96], descriptors for oximes [97], and descriptors for pyridines and pyridine *N*-oxides [98].

Ultimately, the time spent generating experimental Abraham descriptors may be well worthwhile, as it can lead to insights to diverse physical and physiological processes through LFER analysis beyond that achieved by conventional physical properties alone.

Acknowledgments

The authors are very grateful to Prof. Michael Abraham for his longstanding support and advice, and also to many of Prof. Abraham's former PhD and post-doctoral students whose studies have been cited in this chapter. They are also indebted to past and current members of Physical Chemistry at Syngenta, both for the quality of their measurements and their willingness to go beyond conventional physical chemistry into the world of experimental descriptors.

References

1. Briggs, G.G., Bromilow, R.H., and Evans, A.A. (1982) *Pestic. Sci.*, **13**, 495–504.
2. Briggs, G.G., Bromilow, R.H., Evans, A.A., and Williams, M. (1983) *Pestic. Sci.*, **14**, 492–500.
3. Bromilow, R.H., Rigitano, R.L.O., Briggs, G.G., and Chamberlain, K. (1987) *Pestic. Sci.*, **19**, 85–99.
4. Briggs, G.G., Rigitano, R.L.O., and Bromilow, R.H. (1987) *Pestic. Sci.*, **19**, 101–112.
5. Rigitano, R.L.O., Bromilow, R.H., Briggs, G.G., and Chamberlain, K. (1987) *Pestic. Sci.*, **19**, 113–133.
6. Ryan, J.A., Bell, R.M., Davidson, J.M., and O'Conner, G.A. (1988) *Chemosphere*, **17**, 2299–2323.
7. Kleier, D.A. (1988) *Plant Physiol.*, **86**, 803–810.
8. Grayson, B.T. and Kleier, D.A. (1990) *Pestic. Sci.*, **30**, 67–79.
9. Kleier, D.A., Grayson, B.T., and Hsu, F.C. (1998) *Pestic. Outlook*, **9**, 26–30.
10. Satchivi, N.M., Stoller, E.W., Wax, L.M., and Briskin, D.P. (2000) *Pestic. Biochem. Physiol.*, **68**, 67–84.
11. Satchivi, N.M., Stoller, E.W., Wax, L.M., and Briskin, D.P. (2006) *Pestic. Biochem. Physiol.*, **84**, 83–97.
12. Hung, H. and Mackay, D. (1997) *Chemosphere*, **35**, 959–977.
13. Chiou, C.T., Sheng, G.Y., and Manes, M. (2001) *Environ. Sci. Technol.*, **35**, 1437–1444.

14. Trapp, S. and Matthies, M. (1995) *Environ. Sci. Technol.*, **29**, 2333–2338.
15. Trapp, S. (2000) *Pest Manag. Sci.*, **56**, 767–778.
16. Trapp, S. (2004) *Environ. Sci. Pollut. Res.*, **11**, 33–39.
17. Collins, C.D. and Finnegan, E. (2010) *Environ. Sci. Technol.*, **44**, 998–1003.
18. Bromilow, R.H. and Chamberlain, K. (1995) in *Plant Contamination: Modelling and Simulation of Organic Chemical Processes*, Chapter 3 (eds S. Trapp and J.C. McFarlane), Lewis Publishers, Boca Raton, FL, pp. 37–68.
19. Trapp, S. (1995) in *Plant Contamination: Modelling and Simulation of Organic Chemical Processes* (eds S. Trapp and J.C. McFarlane), Lewis Publishers, Boca Raton, FL, pp. 107–151.
20. Collins, C.D. and Fryer, M.E. (2003) *Environ. Sci. Technol.*, **37**, 1617–1624.
21. Collins, C.D., Martin, I., and Fryer, M.E. (2006) Evaluation of models for predicting plant uptake of chemicals from soil. Science Reports SC050021/SR, Environment Agency, Bristol.
22. Katayama, A., Bhula, R., Burns, G.R., Carazo, E., Felsot, A., Hamilton, D., Harris, C., Kim, Y.H., Kleter, G., Koedel, W., Linders, J., Peijnenburg, J.G.M.W., Sabljic, A., Stephenson, R.G., Racke, D.K., Rubin, B., Tanaka, K., Unsworth, J., and Wauchope, R.D. (2010) *Rev. Environ. Contam. Toxicol.*, **203**, 1–86.
23. Trapp, S. and Legind, C.N. (2011) in *Dealing with Contaminated Sites: From Theory Towards Practical Application*, Chapter 9 (ed. F.A. Swartjes), Springer, New York, pp. 369–408.
24. Collins, C.D., Martin, I., and Doucette, W. (2011) *Plant Ecophysiol.*, **8**, 3–16.
25. Doucette, W., Dettenmaier, E., Bugbee, B., and Mackay, D. (2011) in *Handbook of Chemical Mass Transport in the Environment*, Chapter 14 (eds L.J. Thibodeaux and D. Mackay), CRC Press, Boca Raton, FL, pp. 389–411.
26. Briggs, G.G. (1981) *J. Agric. Food Chem.*, **29**, 1050–1059.
27. Briggs, G.G. (1990) *Philos. Trans. R. Soc. Lond. B*, **329**, 375–382.
28. Donovan, S.F. (2007) in *Synthesis and Chemistry of Agrochemicals VII*, ACS Symposium Series 948, Chapter 2 (eds J.W. Lyga, G. Theodoridis), American Chemical Society, Washington, DC, pp. 7–22.
29. Lipinski, C.A., Lombardo, F., Dominy, B.W., and Feeney, P.J. (1997) *Adv. Drug Delivery Rev.*, **23**, 3–25.
30. Tice, C.M. (2001) *Pest Manag. Sci.*, **57**, 3–16.
31. Clarke, E.D. and Delaney, J.S. (2003) *Chimia*, **57**, 731–734.
32. Delaney, J.S., Clarke, E.D., Hughes, D., and Rice, M. (2006) *Drug Discovery Today*, **17**, 839–845.
33. Clarke, E.D. (2009) *Bioorg. Med. Chem.*, **17**, 4153–4159.
34. Abraham, M.H. (1993) *Chem. Soc. Rev.*, **22**, 73–83.
35. Abraham, M.H., Ibrahim, A., and Zissimos, A.M. (2004) *J. Chromatogr. A*, **1037**, 29–47.
36. Vitha, M. and Carr, P.W. (2006) *J. Chromatogr. A*, **1126**, 143–194.
37. Abraham, M.H., Chadha, H.S., Martins, F., Mitchell, R.C., Bradbury, M.W., and Gratton, J.A. (1999) *Pestic. Sci.*, **55**, 78–88.
38. Abraham, M.H., Ibrahim, A., Zissimos, A.M., Zhao, Y.H., Comer, J., and Reynolds, D.P. (2002) *Drug Discovery Today*, **7**, 1056–1063.
39. Abraham, M.H. and McGowan, J.C. (1987) *Chromatographia*, **23**, 243–246.
40. ACD/PhysChem Suite, version 12, ChemSketch Properties Module. For details and ACD/ChemSketch Freeware version, see: *www.ACD/Labs.com*.
41. Zissimos, A.M., Abraham, M.H., Barker, M.C., Box, K.J., and Tam, K.Y. (2002) *J. Chem. Soc., Perkin Trans. 2*, 470–477.
42. Enomoto, K. (2003) Application of linear solvation energy relationships to the prediction of important physico-chemical properties of agrochemicals. PhD Thesis, University College London.
43. Enomoto, K., Clarke, E.D., and Abraham, M.H. (2003) in *Designing Drugs and Crop Protectants: Processes, Problems and Solutions* (eds M. Ford, D. Livingstone, J. Dearden, and H. Vande Waterbeemd), Blackwell, Oxford, pp. 20–22.
44. Leahy, D.E., Morris, J.J., Taylor, P.J., and Wait, A.R. (1992) *J. Chem. Soc., Perkin Trans. 2*, 723–731.

45. Leahy, D.E., Morris, J.J., Taylor, P.J., and Wait, A.R. (1992) *J. Chem. Soc., Perkin Trans. 2*, 705–722.
46. Seiler, P. (1974) *Eur. J. Med. Chem.*, **9**, 473–479.
47. Toulmin, A., Wood, J.M., and Kenny, P.W. (2008) *J. Med. Chem.*, **51**, 3720–3730.
48. Fujita, T., Iwasa, J., and Hansch, C. (1964) *J. Am. Chem. Soc.*, **86**, 5175–5180.
49. Leo, A. and Hansch, C. (1971) *J. Org. Chem.*, **36**, 1539–1544.
50. Hansch, C. and Klein, T. (1986) *Acc. Chem. Res.*, **19**, 392–400.
51. Hansch, C. (1993) *Acc. Chem. Res.*, **26**, 147–153.
52. Collander, R. (1951) *Acta Chem. Scand.*, **5**, 774–780.
53. Zissimos, A.M., Abraham, M.H., Du, C.M., Valko, K., Bevan, C., Reynolds, D., Wood, J., and Tam, K.Y. (2002) *J. Chem. Soc., Perkin Trans. 2*, 2001–2010.
54. Valko, K. (2004) *J. Chromatogr. A*, **1037**, 299–310.
55. ACD/ADME, version 5, Absolv module. For details, see: *www.ACD/Labs.com*.
56. Platts, J.A., Butina, D., Abraham, M.H., and Hersey, A. (1999) *J. Chem. Inf. Comput. Sci.*, **39**, 835–845.
57. Japertas, P., Sazonovas, A., Clarke, E.D., and Delaney, J.S. (2007) Abstracts of Papers, 234th National Meeting of the American Chemical Society, Boston, MA, 19–23 August 2007, poster pdf available from the authors.
58. Medchem Database (2011) Daylight Chemical Information Systems Inc., Laguna Niguel, CA, *http://www.daylight.com* (accessed 2 November 2011).
59. MacBean, C.(ed.) (2011) *The e-Pesticide Manual Version 5.1*, British Crop Protection Council, Alton.
60. US Environmental Protection Agency (EPA) & Syracuse Research Corporation (SRC) (2011) The Estimations Programs Interface (EPI) Suite Version 4.1 *http://www.epa.gov/opptintr/exposure/docs/episuite.htm* (accessed 2 November 2011).
61. PPDB (2009) The Pesticide Properties Database (PPDB) Developed by the Agriculture and Environment Research Unit (AERU), University of Hertfordshire, funded by UK national sources and the EU-funded FOOTPRINT project (FP6-SSP-022704), *http://www.herts.ac.uk/aeru/footprint* (accessed 2 November 2011).
62. Tomlin, C.D.S.(ed.) (2009) *The Pesticide Manual*, 15th edn, British Crop Protection Council, Alton.
63. MacKay, D., Shiu, W.Y., and Ma, K.C. (1997) *Illustrated Handbook of Physical-Chemical Properties and Environmental Fate for Organic Chemicals*, Pesticide Chemicals, Vol. 5, Lewis Publishers, CRC Press, Boca Raton.
64. Smith, S.C., Clarke, E.D., Ridley, S.M., Bartlett, D., Greenhow, D.T., Glithro, H., Klong, A.Y., Mitchell, G., and Mullier, G. (2005) *Pest Manag. Sci.*, **61**, 16–24.
65. Valko, K., Bevan, C., and Reynolds, D. (1997) *Anal. Chem.*, **69**, 2022–2029.
66. Valko, K., Abraham, M.H., Du, C.M., Bevan, C., and Reynolds, D.P. (2001) *Curr. Med. Chem.*, **8**, 1137–1146.
67. OECD (2004) Test No. 117: partition coefficient (n-octanol/water), HPLC method, *OECD Guidelines for the Testing of Chemicals, Section 1: Physical-Chemical Properties*, OECD Publishing.
68. OECD (1995) Test No. 107: partition coefficient (n-octanol/water): shake flask method, *OECD Guidelines for the Testing of Chemicals, Section 1: Physical-Chemical Properties*, OECD Publishing.
69. Hansch, C., Leo, A., and Hoekman, D.H. (1995) *Exploring QSAR: Fundamentals and Applications in Chemistry and Biology*, Chapter 4, American Chemical Society, Washington, DC.
70. Mitchell, G., Clarke, E.D., Ridley, S.M., Greenhow, D.T., Killen, K.J., Vohra, S.K., and Wardman, P. (1995) *Pestic. Sci.*, **44**, 49–58.
71. Abraham, M.H., Amin, M., and Zissimos, A.M. (2002) *Phys. Chem. Chem. Phys.*, **4**, 5748–5752.
72. EPA (1996) OPPTS 830.7560: partition coefficient (n-Octanol/Water), generator column method, *Product Properties Test Guidelines, Prevention, Pesticides and Toxic Services (7101)*, United States Environmental Protection Agency.

73. Woodburn, K.B., Doucette, W.J., and Andren, A.W. (1984) *Environ. Sci. Technol.*, **18**, 457–459.
74. Green, C.E., Abraham, M.H., Acree, W.E. Jr, De Fina, K.M., and Sharp, T.L. (2000) *Pest Manag. Sci.*, **56**, 1043–1053.
75. Abraham, M.H., Enomoto, K., Clarke, E.D., Roses, M., Rafols, C., and Fuguet, E. (2007) *J. Environ. Monit.*, **9**, 234–239.
76. Bronner, G., Fenner, K., and Goss, K.U. (2010) *Fluid Phase Equilib.*, **299**, 207–215.
77. Tomlin, C.D.S. (ed.) (2009) *The Pesticide Manual*, 15th edn, British Crop Protection Council, Alton (Entry 6 - Acetochlor and entry 16 - Alachlor).
78. Tomlin, C.D.S. (ed.) (1997) *The Pesticide Manual*, 11th edn, British Crop Protection Council, Farnham (Entry 7 - Acetochlor).
79. Macbean, C. (ed.) (2011) *The e-Pesticide Manual Version 5.1*, British Crop Protection Council, Alton (Entry 377 - Fipronil).
80. Abraham, M.H., Abraham, R.J., Byrne, J., and Griffiths, L. (2006) *J. Org. Chem.*, **71**, 3389–3394.
81. Abraham, M.H., Roses, M., Poole, C.F., and Poole, S.K. (1997) *J. Phys. Org. Chem.*, **10**, 358–368.
82. Donovan, S.F. and Pescatore, M.C. (2002) *J. Chromatogr. A*, **952**, 47–61.
83. Lombardo, F., Shalaeva, M.Y., Tupper, K.A., Gao, F., and Abraham, M.H. (2000) *J. Med. Chem.*, **43**, 2922–2928.
84. Valko, K., Du, C.M., Bevan, C., Reynolds, D.P., and Abraham, M.H. (2001) *Curr. Med. Chem.*, **8**, 1137–1146.
85. Tao, S., Piao, H., Dawson, R., Lu, X., and Hu, H. (1999) *Environ. Sci. Technol.*, **33**, 2719–2725.
86. Poole, S.K. and Poole, C.F. (1999) *J. Chromatogr. A*, **845**, 381–400.
87. Platts, J.A. and Abraham, M.H. (2000) *Environ. Sci. Technol.*, **34**, 318–323.
88. Shone, M.G.T. and Woods, A.V. (1974) *J. Exp. Bot.*, **25**, 390–400.
89. Polder, M.D., Hulzebos, E.M., and Jager, D.T. (1995) *Environ. Toxicol. Chem.*, **14**, 1615–1623.
90. Dettenmaier, E.M., Doucette, W.J., and Bugbee, B. (2009) *Environ. Sci. Technol*, **43**, 324–329.
91. de Carvalho, R.F., Bromilow, R.H., and Greenwood, R. (2007) *Pest Manag. Sci.*, **63**, 789–797.
92. de Carvalho, R.F., Bromilow, R.H., and Greenwood, R. (2007) *Pest Manag. Sci.*, **63**, 798–802.
93. Abraham, M.H., Smith, R.E., Luchtefeld, R., Boorem, A.A., Luo, R., and Acree, W.E. Jr (2010) *J. Pharm. Sci.*, **99**, 1500–1515.
94. Abraham, M.H. and Acree, W.E. Jr (2011) *J. Solut. Chem.*, **40**, 1279–1290.
95. Hills, E.E., Abraham, M.H., Hersey, A., and Bevan, C.D. (2011) *Fluid Phase Equilib.*, **303**, 45–55.
96. Abraham, M.H. and Acree, W.E. Jr (2010) *J. Org. Chem.*, **75**, 1006–1015.
97. Abraham, M.H., Gil-Lostes, J., Enrique, C.M., Cain, W.S., Poole, C.F., Atapattu, S., Sanka, N., Abraham, R.J., and Leonard, P. (2009) *New J. Chem.*, **33**, 76–81.
98. Abraham, M.H., Honcharova, L., Rocco, S.A., Acree, W.E. Jr, and De Fina, K.M. (2011) *New J. Chem.*, **35**, 930–936.

Part IV
Modern Methods for Risk Assessment

Modern Methods in Crop Protection Research, First Edition.
Edited by Peter Jeschke, Wolfgang Krämer, Ulrich Schirmer, and Matthias Witschel.

12
Ecological Modeling in Pesticide Risk Assessment: Chances and Challenges

Walter Schmitt

12.1
Introduction

An important step in the regulatory process for the approval of plant protection products is that of ecological risk assessment. Such an assessment, which must be performed for all uses of the product for which approval is intended, consists mainly of two steps. The first step includes the determination or estimation of environmental concentrations of the active substances contained in the product, and also of their environmental metabolites. In the second step, these concentrations serve as an exposure estimate and as such form the basis of an actual risk assessment where they are related to the ecotoxicological endpoints of the substances under consideration. These assessments must be performed for all relevant environmental compartments, including soil and surface water, and also for the biological taxa present in these compartments.

In general, ecological risk assessments are performed by comparing exposure values in various compartments [e.g., predicted environmental concentrations (PECs) or estimated exposure concentrations (EECs)], with effect values generated in ecotoxicological studies on indicator species [e.g., lethal concentration (LC) or effect concentration (EC)].

One expression of this risk assessment – the toxicity to exposure ratio (TER) – is the quotient between the relevant toxicological endpoint and the maximum concentration of the considered substance as it is estimated to occur in the respective environmental compartment. Typically, additional assessment factors are applied which usually depend not only on different regulations but also on the type of ecotoxicological endpoint and the taxon which is being considered. The reason for such inclusions is to account for any uncertainty and inter-individual and inter-species variabilities of toxicological susceptibility.

The traditional risk assessment as described above deals implicitly with individuals of the considered species, since the typical ecotoxicological endpoints (as determined in standard toxicological tests) provide information concerning the probability that an individual will be affected by a certain exposure. The underlying

Modern Methods in Crop Protection Research, First Edition.
Edited by Peter Jeschke, Wolfgang Krämer, Ulrich Schirmer, and Matthias Witschel.

assumption of the approach is that a population, or even an ecosystem, will be sufficiently protected if each individual in such a system is protected.

Within the past one to two decades, more realistic – though far more complex – higher-tier approaches on the exposure and effect sides have been developed, and are needed to demonstrate the safety of plant protection products because, for various reasons, the very conservative lower tiers tend to fail. Particularly in Europe, more detailed estimates of environmental exposure – as well as increased standards for the assessment process itself – have been introduced, and this process is continuing on a permanent basis.

In higher-tier risk assessments – which today are increasingly gaining in importance – two very difficult-to-tackle issues frequently arise. These are to: (i) account for temporally and spatially varying exposures, as introduced by more realistic exposure assessments; and (ii) to assess effects at the population level for considering the respective protections. The use of computer simulation techniques has been frequently discussed as a potential solution to both cases.

Today, environmental exposures caused by the use of plant protection products are in most cases estimated by numerical simulations based on realistic scenarios. These simulations include various kinetic effects that are relevant to the respective environmental compartments, and result in temporal variations in environmental concentrations. Particularly in the case of surface water concentrations, very short-lasting exposures must occasionally be considered for risk assessments. Yet, such peak-like exposure patterns stand in contrast to the exposures used in many ecotoxicological studies, which typically are conducted under static or semi-static conditions in order to guarantee a defined exposure. Discrepancies between temporal exposure patterns in ecotoxicological studies and reality also occur in the case of terrestrial vertebrates, where acute toxicity tests are usually performed using gavage administration such that the total daily dose of a test substance can be administered within a very short time. Of course, in Nature the exposure may occur via a contaminated feed which is taken up over longer time periods or in several batches, and which often leads to significantly lower internal concentrations than would occur in tests with a uniform daily dose.

Apart from temporally varying exposures, the spatial variation of exposures may also be derived from environmental exposure simulations, which must then be considered together with toxicity endpoints derived from tests with a homogeneously applied substance. In theory, it is possible to develop higher-tier ecotoxicological studies with conditions adapted to the exposure patterns that will be considered in the risk assessment. In practical terms this is rarely an option, however, because the tests become very sophisticated and often can no longer be managed from a technical standpoint. Moreover, exposure patterns are specific to the use of a product, and very different usage scenarios must often be considered during a risk assessment. Clearly, as the testing of these different scenarios in an experimental fashion is essentially impossible, simulations based on appropriate computer models that allow extrapolations to be made between exposure patterns may offer a realistic solution.

An even greater challenge than considering realistic exposure patterns in risk assessments would be to assess the impact of using plant protection products on whole populations or even communities of populations. Experimentally, the development of a population or its recovery from an adverse effect can only be investigated in special cases. If at all possible, respective studies should be conducted under field conditions which represent a realistic worst-case scenario, in order to extrapolate the results to comparable uses. Unfortunately, such outdoor studies are very labor-intensive and can be afforded only on seldom occasions. Moreover, even if a study were successful the results are generally burdened by the fact that they may depend on specific environmental conditions, and cannot easily be generalized. Yet, to perform studies under all possible relevant conditions is virtually impossible. Apart from this downside, ecotoxicological studies are usually performed only with surrogate species because, for both practical and ethical reasons, it is not possible to conduct toxicology tests with all species of interest. Moreover, species which are of the greatest interest for a risk assessment often are endangered and thus not available for experiments. For all of these reasons, it is generally assumed that population models represent a suitable option. Indeed, such models may serve as the only means by which the informative value of experimental studies in risk assessments for plant protection products can be extended toward increased realism, especially in cases where lower-tier approaches might be overprotective.

In this chapter, an overview will first be presented of the potential uses of ecological models in regulatory risk assessments. This will be followed by a description of the various modeling approaches that are available, after which the problems of applying simulation models to risk assessments when approving plant protection products will be discussed.

12.2 Ecological Models in the Regulatory Environment

The European directives and regulations on environmental risk assessments for plant protection products and other chemicals include several options for refinements of risk assessments. Hommen *et al.* [1] evaluated respective guidances in the field of fresh water risk assessment, and identified five refinement areas where ecological effect models could significantly contribute to the risk assessment:

1) Extrapolation of effects between different exposure profiles.
2) Extrapolation of recovery processes.
3) Extrapolation of organism-level effects to the population level.
4) Analysis and prediction of indirect effects.
5) Prediction of bioaccumulation within food chains.

Although this evaluation was restricted to fresh water risk assessment, it is clear that the same areas of application would also be valid in other types of risk assessment. Likewise, as the refinement options are not restricted to the European approval processes for plant protection products, very similar (even identical) considerations hold also for other areas, particularly North America.

Additional background information on the benefits that ecological models can contribute to the above-mentioned areas, and which types of model are best-suited for application, are provided in the following subsections. An excellent, and more extensive, overview of the available models, together with a discussion of their suitability for application to the five refinement areas, is has been provided by Galic *et al.* [2].

12.2.1
Consideration of Realistic Exposure Patterns

In the past, environmental exposure is been estimated by using simplistic models that generally allow estimates to be made of the maximum environmental concentrations for different, single-exposure pathways. An example of this is the concentration in surface water bodies at the edge of a treated field, for which the upper limit can be calculated by assuming an overspray situation and simply distributing the amount per unit area that strikes the water surface into the entire volume below. The concentrations derived in this way can easily be compared to toxicological endpoints derived from standard tests. Assuming that the species being tested is the most sensitive one in the respective environmental compartment, or by applying an assessment factor for considering the potential higher sensitivity of another, untested species will allow a very simple risk assessment to be made. In fact, even refinements in several tiers might be possible without the approach becoming over-sophisticated. For the above-mentioned case, it is possible to regard the different distances between a sprayed field and a water body by considering only the spray drift with reduced amounts instead of the overspray situation.

During the past two decades, however, increasingly sophisticated environmental exposure models have been introduced in the regulatory process for the registration of plant protection products. In Europe, this process started with the implementation of the Forum for the Co-ordination of Pesticide Fate Models and their Use (FOCUS), which released a first guidance on leaching models in the European Union (EU) registration [3]. By using this as a starting point, a complete set of guidelines was developed and implemented in the EU directive for the registration of plant protection products [4] that covers exposure modeling for all relevant environmental compartments [5–7]. Several mechanistic simulation models were introduced, which allow a much more realistic prediction of environmental concentrations than do the simplistic models. Moreover, it is a characteristic of the FOCUS reports that they aim to consider the variability of environmental conditions over the whole of Europe by defining several different scenarios for which the environmental exposure is to be assessed. This has, however, two implications to the ecological risk assessment process, particularly with regards to aquatic risk assessments:

- The simulation models used for predicting aquatic exposure now integrate different exposure pathways which results, due to their dynamic nature, in temporal exposure patterns instead of single concentration values.
- As mentioned above, results from several different scenarios – that is, climate and soil conditions – must be considered for the assessment.

The issue imposed by these facts is an incompatibility between the calculated exposure patterns and the exposure given in standard and even higher-tier toxicology tests. Due to the plurality of calculated concentration patterns, it is also not possible to adapt the test respectively and to determine the effects under comparable exposure. Deriving single characteristic concentrations (e.g., maximum concentration) from the predicted patterns and using such values for the risk assessment solves the problem only in first tier, because it often leads to over-conservative estimates and does not allow an adequate risk assessment. In this situation, ecological models that include toxicodynamic and/or toxicokinetic submodels [8, 9] are needed to combine standard test results with realistically predicted environmental concentrations for the risk assessment.

By allowing an extrapolation between the different exposure patterns and considering realistic environmental exposures, toxicokinetic/toxicodynamic (TK/TD) models represent a suitable means to cover the first two refinement areas in the list above. However, recovery with these models can only be demonstrated if an effect is transient, and will occur only after a short-term exposure is stopped.

12.2.2 Extrapolation to Population Level: The Link to Protection Goals

In current regulatory documents, the ecological protection goals for pesticide risk assessments are usually defined quite ambiguously [10]. In most cases it is, however, clear that the entity to be protected is at least a population. A clear exception to this is given for vertebrates, where usually no lethal effects to individuals are tolerated. For sublethal effects, however, also in this case the goal is to protect the population in general. The concretization (definition and practical implementation) of protection goals is presently the subject of intensive discussions among the scientific and regulatory communities. In the EU, a scientific opinion of the European Food and Safety Agency (EFSA) [11] provides several criteria which need to be considered when protection goals are defined, as there are the magnitude, the duration of observable effects which can be tolerated, and the ecological entity that should be considered. Here an important development is, that in many cases it is now clearly stated that certain, reversible effects on population level are acceptable. This statement contrasts a trend towards a no-effect policy, as has been observed in the past.

For risk assessments according to protection goals such as those described above, the issue arises that toxicological endpoints are in most cases only available on individual level, and it is not easily possible to directly derive population effects from them. In some cases, higher-tier ecotoxicological tests – using model ecosystems such as aquatic mesocosm systems – allow the investigation of pesticide effects on whole populations, even though this may involve a quite small confined system. In some cases, "real" field studies may be possible in which the effect of pesticide applications on different species of a whole ecological community can be studied under more or less undisturbed conditions. Such higher tier studies are, however, very elaborate and costly, with the effort increasing in line with their complexity. Ultimately, it cannot be excluded that the outcome of such studies may be influenced by the specific environmental conditions under which they have been performed. However, to test this point explicitly would be virtually impossible because the number of tests required to include all possible conditions would be much too large.

Simulations of the impact of pesticides on population development can address a good deal of the limitations of higher-tier studies. Provided that a suitable and validated population model for the species to be assessed is available, it can be used to calculate effects on a population level based on toxicological information obtained from lower-tier studies, and potentially even from laboratory-based studies. Usually, it will also be possible to include submodels which describe the influence of environmental conditions on population dynamics and/or toxicokinetic/toxicodynamic models, and which will enable the consideration of dynamically varying exposure. Simulation models with these capabilities can then be used to perform many types of extrapolation and to allow the investigation of various environmental and exposure scenarios with much less effort than can be achieved experimentally. An important option here is the possibility to extend population models to meta-populations, and to consider in addition the spatial variation not only of the population density but also of the environmental conditions and exposure.

Due to their interesting opportunities, ecological models are becoming increasingly recognized as options for the risk assessment of pesticides in the regulatory environment, and have as such been mentioned in recent guidance documents [12]. The use of such simulation models in risk assessments allows not only estimations to be made of the relevance of effects on an individual level to whole populations, but also investigations of the recovery potential of populations. In contrast to the TK/TD-models discussed above, recovery mediated by recolonization can be considered in meta-population models.

12.2.3
Extrapolation to Organization Levels above Populations

Several questions in risk assessments address biological organizations on an even higher level than the population of a single species. One effect to be considered in a risk assessment is the propagation of exposure from one trophic level to the next and, in consequence, the accumulation of substances with a high persistence in species of higher trophic levels. This is especially of interest in environmental risk assessments when considering the indirect poisoning of wild live animals, as well as in human risk assessment for estimating exposure to humans as the species on the highest trophic level. Refined assessments in this field are classically made using food-web models [13, 14] that originally were developed to investigate the accumulation of persistent environmental pollutants in different trophic levels of an ecosystem. Such models can, in principle, be regarded as toxicokinetic models of whole food-webs.

Recently, environmental protection goals have tended to be defined on the more abstract level of ecosystem services [15]. Such ecosystem services are defined from an anthropocentric point of view, and describe various services that ecosystems provide to their human inhabitants. As such services can seldom be attributed to a single species in the ecosystem, it becomes even more ambiguous as to how the effects of using plant protection products experimentally on such a level can be assessed. Whilst simulations might offer an alternative solution to the problem, models are required that go far beyond the description of single-species populations. Instead, multispecies models are needed that include the consideration of interaction between the different populations, and which include competition for resources as in food-web models and also other types of interactions. Such models, albeit with a reduced number of species, can also serve to assess any indirect effects that might occur if a species of concern were not affected directly itself by a plant protection product, but might instead be influenced by impacts on the population of a different species – for example, a prey species – which itself is of less concern for the protection goal.

12.3
An Overview of Model Approaches

Different modeling approaches are needed for application to the different refinement areas in the environmental risk assessments discussed in the previous section. With the increasing level of biological organization that must be addressed, the complexity of the models required will also be increased. A schematic representation of the eco(toxico)logical modeling approaches that are suitable for these different purposes is shown in Figure 12.1.

The extrapolation of effects between different exposure profiles requests a TK/TD model. Such models may also be applicable for the modeling of recovery processes, if such recovery occurs simply due to transient exposure and reversible effects. TK/TD models usually are applied on an organism (individual) level. Of course, the extrapolation of organism-level effects to a population level requires a population model, but this might be based on an individual TK/TD-model. For species with high mobility, recolonization may be an important factor for the recovery of

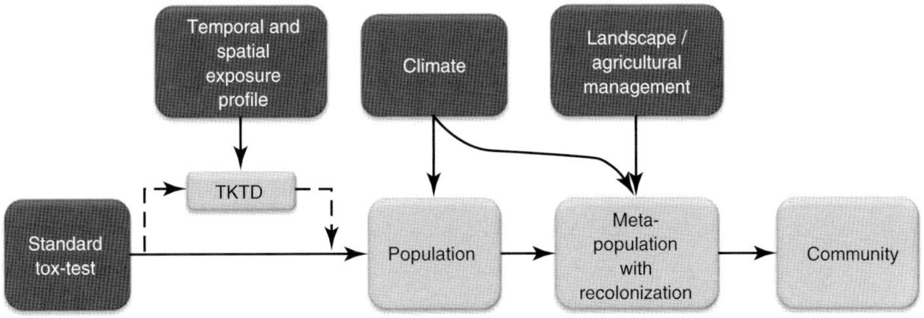

Figure 12.1 Systematic of ecotoxicological models with increasing levels of biological organization to which the models extrapolate from standard test results. The dark gray boxes represent non-biological information that may be involved in the respective models. TKTD: Toxicokinetic/toxicodynamic.

affected populations. Such processes can only be considered in models that take into account spatial landscape information and which potentially can deal with meta-populations. For the consideration of indirect effects and the prediction of bioaccumulation, models are requested that describe the dynamics of different species in parallel considering their interactions; these are food-web models and ecosystem models.

Some short descriptions of the principles of the various modeling approaches are provided in the following sections, together with some overviews of their uses for risk assessment.

12.3.1
Toxicokinetic Models

Toxicokinetic models for use in ecological risk assessments are have been described for very different species. Respectively, there is a large variance in the complexity and the use of these models. The primary use of toxicokinetic models is the translation of external exposure into internal concentrations [16]. In their least complex form, such models consist of one unstructured body compartment, and contain mathematical terms that describe the uptake and elimination of the considered substance mathematically as first-order kinetics:

$$\frac{dC_{\text{int}}}{dt} = k_{\text{uptake}} C_{\text{ext}} - k_{\text{elimination}} C_{\text{int}} \tag{12.1}$$

The solutions of such first-order differential equations are exponential functions. Important characteristics of these are the typical time constants, such as half-life times, which are given by the rate constants k. The elimination half-life can, for example, be calculated as:

$$DT_{50}^{\text{elim}} = \frac{\ln(2)}{k_{\text{elimination}}} \tag{12.2}$$

The toxicokinetics of a compound is only of relevance for its effect, if the typical time constant of any of the toxicokinetic processes such as uptake, distribution or elimination is longer than the time constants of the other processes involved in the action of the substance, in particular, the rates with which exposure changes. In such cases, toxicokinetics may determine the temporal course of an effect. It can be assumed that the toxicokinetic time constants are correlated with the size of an organism. Surface-to-volume ratios and body mass-specific metabolic rates [17] decreasing with increasing body size generally lead to characteristic time constants that increase with body size [18]. Thus, it can be expected that toxicokinetics does usually not have to be considered for very small organisms, while it will play a significant role in the risk assessments for larger ones. In consequence, TK models are usually not applied to protozoae (e.g., algae), but for aquatic invertebrates TK/TD models have been described by various authors [19, 20]. In the earlier attempts at defining TK/TD models, the term in the model equations dedicated to toxicokinetics had more the function of a general delay term that dissolves a lag between changes in exposure and respective reactions of the effect. Recently, a new model was developed that unifies the different previous approaches and also more clearly identifies the role of toxicokinetics [9]. Additional support has been provided from an experimental aspect, with Ashauer *et al.* having investigated the toxicokinetics of different pesticides in *Gammarus pulex* [21]. These authors showed that the typical time constants for uptake and depuration in fact are in the range of days and thus comparable to the duration of peak exposures as typically occur for pesticides in flowing surface water bodies close to treated fields. Therefore, toxicokinetics clearly has an impact on the effect of pesticides on aquatic organisms under environmental conditions [22]. Moreover, variation of the TK parameters between different species [23] can, at least in part, explain differences in sensitivity for different species [22] (see Box 1).

Box 1: Toxicokinetic/Toxicodynamic (TK/TD) Models

The impact of toxicokinetics on the evolution of toxic effects is well explained in a schematic taken from Ref. [22] (see Figure B1a). For the toxic action of a chemical, the concentration within the organism at the target site is relevant. Depending on the properties of the compound, the considered species and the type of exposure, this internal concentration may deviate from the external concentration in both magnitude and temporal pattern of variation. Thus, in many cases a TK/TD model relating the physiological response of the organism to the exposure to a toxicant will lead to unrealistic results, if the external exposure is considered instead of the relevant internal exposure.

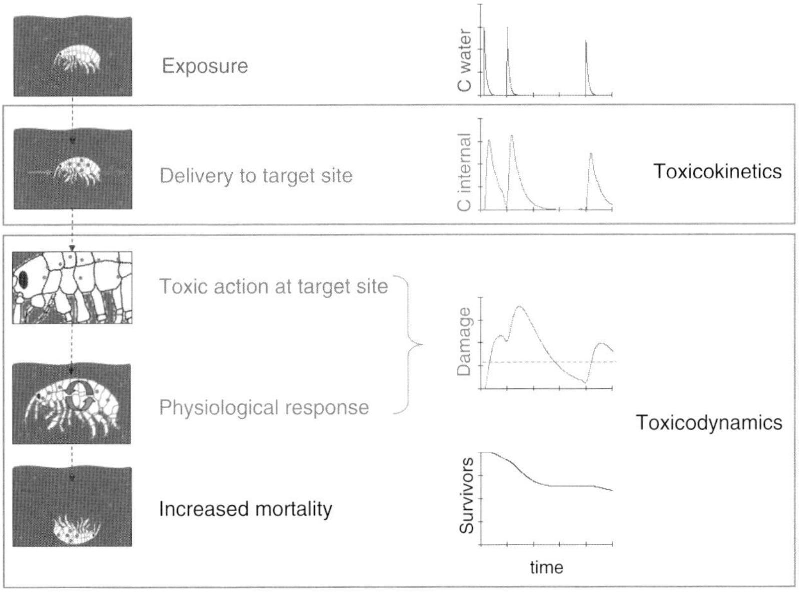

Figure B1a Reproduced from Ref. [22].

An important note at this point is that, in TK/TD models, the EC relations used to describe the physiological response must be based on the same internal concentration that is considered for the TK/TD approach. However, when EC relationships from standard tests are considered this is usually not the case, because they are normally based on experimentally more easily accessible external concentrations. In such cases, the concentration scale must first to be transformed into an internal concentration scale, using the toxicokinetic model applied to the situation in the standard test. The resulting transformed EC relation can then be used in the TK/TD model.

In risk assessments for vertebrates, toxicokinetic models can play a role when the exposure in the natural case which is assessed clearly deviates from the situation in toxicological studies. This is primarily the case when in the study the application of a test substance was by gavage (as typically in acute toxicology studies). Under environmental conditions, the primary route of exposure with pesticides for terrestrial vertebrates is by the consumption of contaminated food or water. This uptake is does not usually occur as one brief event per day, but rather as several events distributed over different time periods. Depending on the toxicokinetic properties of the chemical, this may lead to significantly lower internal exposures as compared to a gavage application, despite the daily dose being the same. Moreover, in many cases avoidance effects occur which allow an animal to cease feeding on the contaminated food and thus to further reduce intracorporal exposure. For the insecticide pirimicarb, a simple toxicokinetic model has been applied in the risk assessment that takes both temporally distributed uptake and avoidance into account [24].

Simple TK models allow modeling of the mean total internal concentration or the concentration in blood only. If the dose in target tissues is explicitly considered, then a physiologically based toxicokinetic (PBTK) model must be used [8, 25] in which the different organs and physiological processes of an organism are explicitly considered, so as to allow the simulation of tissue concentrations. A further advantage of this modeling system compared to other methods that are not based on physiological processes and properties is the high suitability for extrapolation to untested scenarios. As most of the parameters of these models can be determined independently, they are predetermined for extrapolations between different species and exposure routes simply by choosing the respective parameterization. This feature is particularly interesting in the field of environmental risk assessment, where very often different scenarios must be considered in a single assessment. Today, PBTK modeling is intensively discussed and increasingly accepted as a method for human risk assessment [26–28]. In contrast, in ecological risk assessment the use of PBTK models has so far been limited, with no reports having been made on any specific applications in this area.

12.3.2 Population Models

12.3.2.1 Differential Equation Models

Depending on the aim of an environmental risk assessment, ecological models of very different types and complexity might be helpful. A common feature of such computer models is the mathematical description of the temporal evolution of a population of one or several species. Moreover, for use in risk assessments for pesticides, the inclusion of some form of TD model is obligatory, as this can relate the exposure to a toxicant to an effect of the chemical to the simulated biota. In some cases it may be reasonable to base the TD model on internal rather than an external exposure (see previous section), although a toxicokinetic submodel must then also be included. Even in the least-complex models it is possible to consider the influence of environmental conditions such as temperature and nutrition on population growth. Hence, it is possible to use these models with different environmental scenarios to consider, for example, different geographical regions in a risk assessment.

The simplest approach for describing the development of a population over time is the so-called "exponential growth model," where the change in number of individuals or the total biomass of a population per time is proportional to the respective present state, with the population growth rate as a proportionality factor. If the growth rate is constant, this leads to an unlimited exponential growth with time. Unfortunately, such a behavior is unrealistic, and in reality the growth rate is influenced by various factors that include temperature (T), nutrient supply (Nut), and population density (N). Such varying growth rates may be considered when the model is mathematically formulated as a differential equation. In particular, density-dependent factors lead to an upper limit of the population size, the capacity limit, and result in S-shaped population growth curves, much like the logistic function. Usually, it is sensible also to consider a second rate in the model that reflects the limited lifespan of the individuals in a population, and is therefore referred to as the *mortality rate*. The general equation for an exponential growth model is then:

$$\frac{dN}{dt} = \left(K_{\text{growth}}\left(T, Nut, N, \ldots\right) - k_{\text{death}}\right) N \tag{12.3}$$

If the rates in Equation (12.3) are constant over time, the equations has a simple solution given by the exponential function:

$$N\left(t\right) = N_{o}\exp\left[\left(k_{\text{growth}}\left(T, Nut, N \ldots\right) - k_{\text{death}}\right) t\right] \tag{12.4}$$

In case that the rates are time-dependent – either directly or indirectly – due to a temporal variability of the influencing factors, the equation in most cases must be solved numerically because an analytical solution is not available.

Differential equation models are generally adequate for use in problems where a differentiation of individuals or subpopulations, such as different live stages, is not necessary or is even impossible. A typical example is the growth of populations of cellular organisms such as bacteria or algae. Different algal models of this type have been reported that were developed primarily to investigate the influence of different environmental factors on the population growth rate [29, 30]. In general, these models were developed to investigate and optimize the production of biomass or the remediation of nutrient from eutrophicated water bodies under different conditions. It is, however, possible simply to use the same model for pesticide risk assessments by introducing the toxicant concentration as an additional factor that alters the growth rate. Similar modeling approaches were also applied to simulate the population growth of the aquatic macrophyte species *Lemna* [31], which usually also shows a simple exponential growth behavior.

Generally, for ecological risk assessments differential equation models are applicable in cases where it is sufficient to regard the population as a whole, without any need to differentiate between subpopulations of life stages. In the case of risk assessments for pesticides, one important prerequisite is that the substance affects all individuals equally, independent of gender and/or age, or that such differences can at least be ignored. Apart from the above-mentioned cases, differential equation models have also been used for assessing the effect of pesticides on the populations of aquatic [32, 33] and terrestrial [34] arthropods. Furthermore, some examples of differential equation models describing fish populations have been reported which mainly deal with the effects of endocrine disruptors on populations [35, 36].

12.3.2.2 **Matrix Models**

Whilst differential equation models usually consider populations as a whole, it is in many cases desirable to subdivide the population – for example, into different live stages – in order to achieve a realistic description either of the population dynamics or of the effects of a toxicant. A rather simple mathematical approach to describe the dynamics of age or life-stage structured populations is referred to as matrix models [37]. This type of model employs the specific characteristics of matrix mathematics to describe the development of a population on the basis of transition probabilities between different stages/ages, instead of working with rates (as is the case for differential equation models).

The simplest form of matrix model – the Leslie matrix [38] – is age-structured and based on probabilities for transition to the next age-class and/or to age-class zero – that is, newborns. The former are equivalent to survival rates at a time step, while the latter are fecundity factors in a respective differential equation model. Thus, a Leslie matrix could also be transformed into a differential equation model. The sum of fecundity factors is then equivalent to the population growth rate and summing up (1 – survival rates) leads to the mortality rate. The main advantage of matrix models is that the probability factors may differ from age-class to age-class, and thus allow the consideration of age- or life stage-specific toxicodynamics. This ability is lost when using simple differential equation models, as the rate constants are equal for the whole population. One downside of matrix models is that they provide the state of the population only at fixed predefined time steps, whereas differential equations can be solved at any time step and provide such results on a more or less continuous time scale.

A simple example of a Leslie matrix for an age-structured model is as follows:

$$\begin{pmatrix} f_0 & f_1 & f_2 & \cdots & f_{k-1} & f_k \\ S_0 & 0 & 0 & \cdots & 0 & 0 \\ 0 & S_1 & 0 & \cdots & 0 & 0 \\ 0 & 0 & S_2 & \cdots & 0 & 0 \\ \vdots & \vdots & \vdots & \ddots & \vdots & \vdots \\ 0 & 0 & 0 & \cdots & S_{k-1} & 0 \end{pmatrix} \tag{12.5}$$

This matrix is suitable for calculating the population dynamics of a species with individuals living k years. The matrix elements s_i are survival rates from year $(i-1)$ to year i. Respectively, f_i are the fecundities of individuals of age i.

If $N(t-1)$ is the population vector of length k containing the number of individuals of a population in any age-class at time step $(t-1)$, then the state of this population at time step t can be calculated with the equation:

$$N(t) = L\ N(t-1) \tag{12.6}$$

Consequently, it is:

$$N(t) = L^t\ N(0) \tag{12.7}$$

Leslie matrices can be converted into life-stage structured matrices by assigning several age classes to a life-stage, according to the life-history of the species to be modeled. The survival probabilities are then split into a probability for individuals to survive and stay in a life-stage, or to survive and transit to the next life stage. The former are placed on the diagonal of the matrix, while the latter replace the S-values in Equation 12.5 on the subdiagonal.

The approach of matrix models can be adapted to many different cases, and allows extensive investigations to be made of the population dynamics (see Ref. [37] and references therein). The generalized form of the Leslie matrix is then termed the *projection matrix*. A notable option for pesticide risk assessment is the extension to spatially explicit matrix models [39, 40] which can also be used to describe meta-populations [41]. Thus, matrix models allow a high degree of flexibility. Although matrix models were for a long time used only to investigate purely ecological questions, they can simply be adopted for use in ecotoxicological risk assessments [42]. This can be achieved by considering a suitable toxicodynamic equation which relates the matrix elements – that is, the transition probabilities – to the concentration of a toxicant.

Several examples of the application of matrix models to ecotoxicological risk assessments have been reported. For example, a case study of the effects of methiocarb on *Chironomus riparius* is available in Ref. [42], while studies investigating the effects of pesticides on populations of aquatic invertebrates have been described by Billoir *et al.* [43], Chandler *et al.* [44], and Raimondo and McKenney [45]. In addition to aquatic invertebrates, applications to terrestrial arthropods [46, 47], fish [48], and birds [49] have also been made. These examples provide a selection which demonstrate not only the wide variety of species investigated but also the broad span of complexity. Whilst many models have been used simply to investigate the influence of a static exposure on a local population, others have been much more complex and allowed the temporal variation of toxicant concentrations to be studied [45, 50], as well as the spatial distribution of the population [48].

Taking into account exposures that vary realistically in time is, however, a challenging prospect with matrix models [51] due to their discrete time steps and the fact that the projection matrix is always a $N \times N$ matrix for N time steps. This limits the resolution in time that can be handled, for computational reasons. A further disadvantage of the matrix model is its inability to consider the inter-individual variability of properties in a population (see Box 2).

12.3.2.3 **Individual-Based Models**

The least limitations in flexibility when considering different influences on the development of populations are provided in individual-based population models (IBMs) [53]. In these approaches, each individual of the population is described separately as a single object with individual properties. Each of these individuals undergoes a whole life cycle and interacts with other individuals and the environment, which in turn allows a very detailed and realistic description of a population. Of particular interest – and particularly for use in pesticide risk assessments – is the possibility of making IBMs spatially explicit. This means that the virtual individuals in the model "live" in an environment with spatially varying properties. Depending on their interaction with the environment, this may lead to a spatial variability of population density. Moreover, it allows spatial differences in exposure to a pesticide to be considered, an example being that of treated and untreated regions. The potential for spatially explicit models, and arguments for their use in risk assessments, has been discussed by Wickwire [54].

The main downside of the flexibility provided by IBM is their complexity, and this is mainly reflected in the large number of parameters and rules for which sufficient species-specific information is required in order to achieve a realistic description of the behavior of a population. Information is needed not only regarding the values of such parameters but also of their variability. Some of the parameters – and their distribution – can be determined by observations, and this is mainly the case for life cycle parameters as spans of life stages or the number of offspring. Other parameters, however, cannot easily be measured, notably parameters of spatially explicit models which describe the behavior of individuals. Typical examples are characteristic moving distances or the attractiveness of habitats or other individuals within the population, which determine the behavior of an individual. Despite these issues, the details of well-validated models are available, and these have proved to show a high predictability. Examples include the fish models of Railsback *et al.* [55, 56] and various models of arthropod (ground beetle, linyphiid spider) and vertebrate (hare, field vole, deer, partridge, and skylark) populations which have been developed with the complex ALMaSS-framework [57]. The latter examples are based on very detailed rule sets for the behavior of individuals and complex realistic landscapes. In addition, they contain realistically described interactions between the animals and the landscape in terms of food provision and consumption. Models of small mammal populations with less-detailed, species-specific life histories and behavior rules and more schematic landscapes have also been shown to lead to good predictions of the observed characteristics of population development [58, 59].

Individual-based population models have also been developed for aquatic arthropods. Although, for these species, the rules of behavior are much less complex than for vertebrates, it emerged that the individual-based approach has clear advantages over alternative modeling approaches. In the case of an individual-based *Daphnia magna* population (IDamP) model [21], the methodology allowed the explicit consideration of both size- and life stage-dependent properties. The stochastic attribution of varying property values to the individuals in a population led to realistic variations in population dynamics. The model also proved capable of predicting the effects of a toxicant on population level, based on toxicity data on lethal and sublethal effects derived from a standard test with *D. magna* [60]. A different approach was developed to model particularly the effects of pesticides on the *Assellus aquaticus* population [52, 61]. In this case, use was made of the individual-based methodology to describe the dynamics of spatially distributed meta-populations, taking into consideration the impact of temporally and spatially varying pesticide exposures in different water bodies (see also Box 12.2).

Box 2: Spatially Explicit Individual-Based Population Model

Recently, van den Brink *et al.* [52] described a spatially explicit individual-based population model for the aquatic invertebrate *Asellus aquaticus*, which neatly demonstrated the use of this model in the risk assessment of insecticides. The ecosystems simulated water bodies (e.g., pond, stream, and ditch) that corresponded to those which were considered in regulatory surface water exposure assessments in Europe. In these water bodies, which had typical dimensions of few hundred meters, individuals of *A. aquaticus* were simulated according to the life-cycle and behavior (see Figure B2a). Their mortality was influenced by exposure to the pesticide, the spatial and temporal variations of which were considered to have resulted from exposure simulations according to European requirements for the assessment of plant protection products.

Figure B2b shows the typical curves of the development of population size of the simulated species, without any influence of toxicant and with different levels of exposure to an insecticide that increased the mortality. Whereas, in Figure B2b the temporal development of the total number of individuals in the whole water body is shown, Figure B2c illustrates the spatial distribution of the population in a ditch of 600 m length. Starting on January 1st, the population initially developed

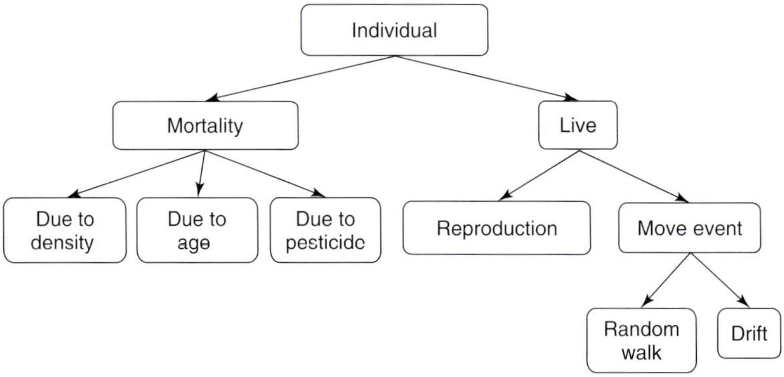

Figure B2a Reproduced from Ref. [52].

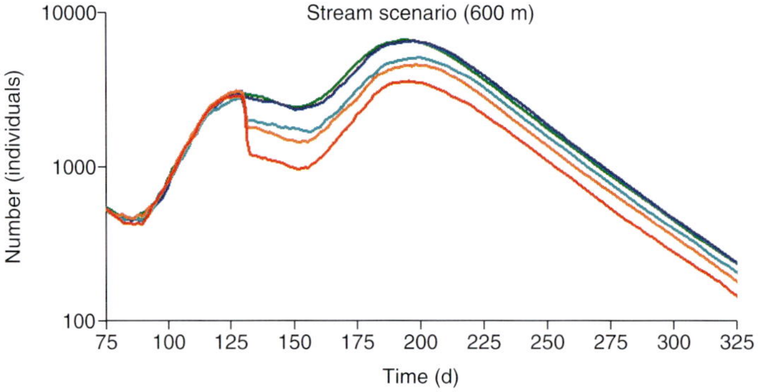

Figure B2b Reproduced from Ref. [52].

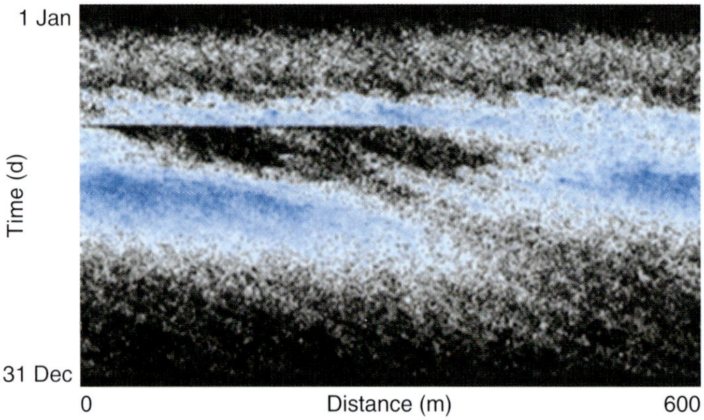

Figure B2c Reproduced from Ref. [52].

homogeneously over the whole water body (the blue points depict the individuals of the population). When the pesticide was applied and suddenly entered the ditch by spray drift, the number of individuals quickly decreased at the point of application due to the insecticide's rapid lethal action. The respective "empty" spot in Figure B2c then moved downstream with the flow of the water in the ditch, but gradually was re-filled as the *Asellus* population recovered.

12.3.3 Ecosystem or Food-Web Models

Ecosystem and food-web models are generalized predator–prey models and thus are, in principle, derived from the well-known Lotka–Volterra equations. Generally, these models consist of systems of coupled differential equations in which the state variables are the biomass or numbers of individuals of different species. Usually, the species belong to different trophic levels, and the coupling between the single differential equations is given by the different predator–prey relationships and/or competition between the different species. In principle, such models are thus extended versions of the differential equation population models discussed above.

While food-web models originally were developed to investigate the principles of ecosystem dynamics, they have in the meantime also become accepted as tools for environmental planning and risk assessment. The inclusion of toxic effects also allows an assessment of the influence of environmental contaminants on whole ecosystems, which in turn permits the investigation not only of the direct effects of toxicants on sensitive species but also an assessment of indirect effects on tolerant species caused by an alteration of the function of sensitive species in the food-web. An example of a food-web model for the assessment of pesticide effects on an aquatic community of species on different trophic levels was described by Traas *et al.* [62] (see Box 3), who investigated the effect of chlorpyrifos on aquatic microcosms in simulations and compared the results to observations made experimentally.

Box 3: Food-Web Model

The food-web model, which included different planktonic, plant, arthropod, and mollusk species, described the respective predator–prey interactions (Figure B3a). In addition, the fate of chlorpyrifos in the water sediment system was simulated in a submodel.

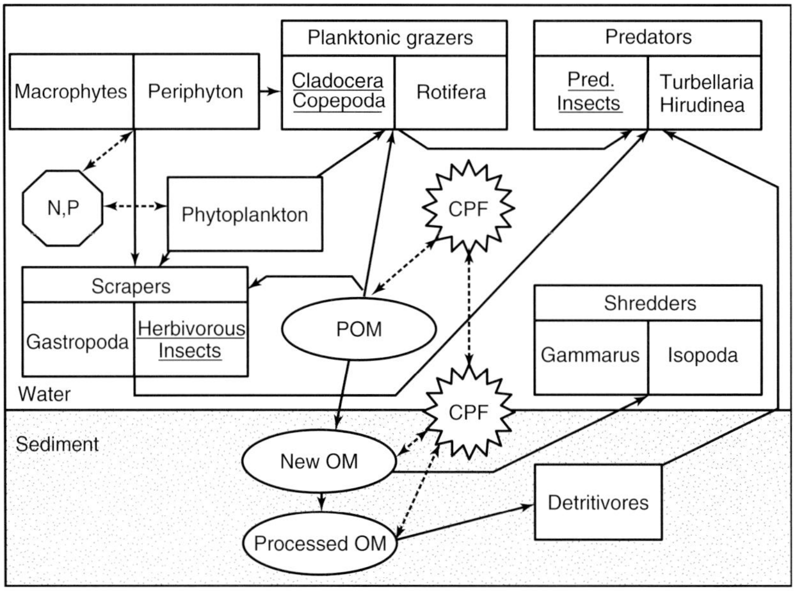

Figure B3a Reproduced from Ref. [62].

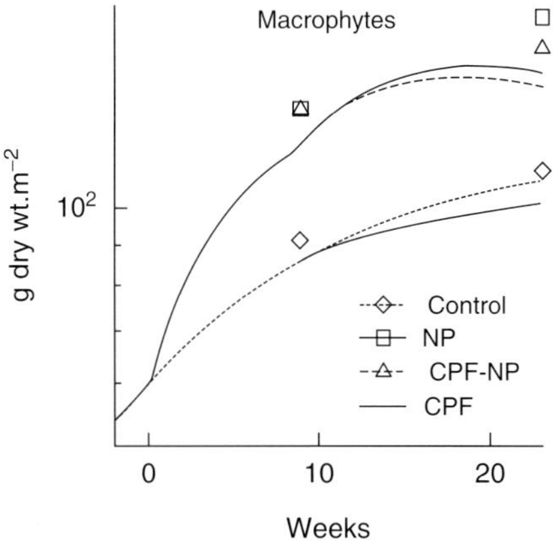

Figure B3b Reproduced from Ref. [62].

Model simulations were compared to observations made in a microcosm study in which the effect of nutrients and chlorpyrifos on the biomass of different aquatic species was investigated. The model could be used to predict some observed features, including the positive influence of nutrients on the macrophyte biomass (Figure B3b). Other characteristics were well reflected by the model in terms of their tendency, but quantitatively the prediction differed significantly from the experimental findings. Notably, this was the case for the indirect effect of chlorpyrifos on the mollusk population, whereby exposure to chlorpyrifos led to a significant increase in the mollusk biomass, mainly because the arthropod population competing with the mollusks for peryphyton mass as feed, was reduced by the insecticide. Although this effect was also observed in the simulation, it occurred to a much smaller degree however (Figure B3c).

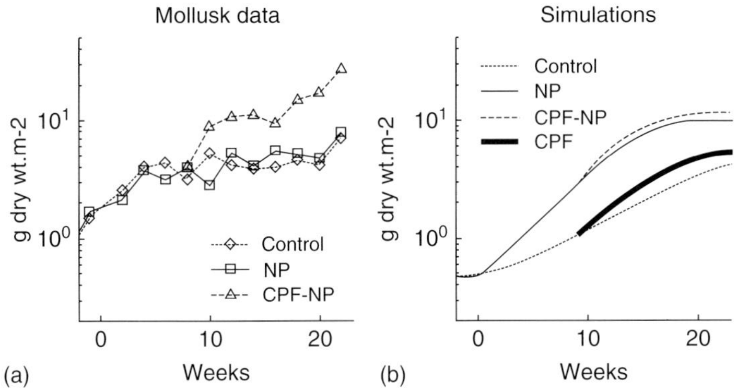

Figure B3c Reproduced from Ref. [62].

For aquatic ecosystems, several generic modeling tools are currently available which can be adapted to answer specific questions by suitable parameterization. One such system is AQUATOX, as developed by the US Environmental Protection Agency ([63]; *http://water.epa.gov/scitech/datait/models/aquatox/index.cfm*). AQUATOX contains a model that describes algae and aquatic plants as producer species, and invertebrates and fish as higher trophic levels in the community. This food-web model is combined with a model describing the fate of toxicants introduced into the water body. Another example of a generic aquatic ecosystem model is CASM [64, 65]. Both of these models have been used for various case studies in which the influence of different contaminants on whole ecosystems has been investigated. For example, Sourisseau used AQUATOX to investigate the effect of the insecticide deltamethrin on ecosystems [66, 67], while CASM was used for a case study on the risk posed by the herbicide diquat-dibromide to a generic lake in Florida [65]. A further reported application of CASM involved the risk assessment of various chemicals, including an insecticide and two herbicides, to a Japanese lake [68]; very recently, CASM was also applied to a risk assessment for atrazine in US Midwestern streams [69].

An even more comprehensive (albeit commercial) generic model for simulating ecosystems is RAMAS (*http://www.ramas.com/software.htm*). This is a rather complete modeling environment than a single model, as it consists of several tools that allow not only the simulation of ecosystem networks but also the consideration of geographic information. Unlike AQUATOX and CASM, however, RAMAS is not restricted to aquatic systems.

12.4 Regulatory Challenges

Despite the obvious potential of ecological models in environmental risk assessment, the methodology is at present only rarely employed in regulatory risk assessments for pesticides. While the reasons for this are manifold, a major factor is that in the past the regulatory framework and its requirements were mainly adapted to the available experimental methods, with virtually all risk assessments being performed on lower tiers and the need for more sophisticated methods being low. For a long time, the situation differed from that employed in the exposure assessment for pesticides, although with the introduction of the EU directive on the protection of groundwater [70] a threshold value for groundwater concentrations of potentially dangerous substances of $0.1\,\mu g\,l^{-1}$ was introduced that was subsequently applied to pesticides [4]. The experimental proof of compliance with this threshold, however, proved to be difficult, since direct measurements were complicated and could provide answers only after several years of a substance being used. Although, subsequently, lysimeters were used as experimental model systems, they proved to be very labor-intensive and needed to be run for at least two years before any results could be obtained. Perhaps the major drawback was that a single experimental test could cover only one climate–soil combination, and even if the experimental set-up was designed to reflect a worst-case situation this could not ultimately be proven. Consequently, in order to conduct a risk assessment for a larger eco-climatic region, such as a country or even a whole continent, a large number of tests would be required to reflect the different conditions.

It was during the 1990s that the use of groundwater-leaching simulations with mechanistic hydrological models for predicting groundwater concentrations was fostered. These models allowed an easy investigation of the dependence of groundwater concentrations on environmental conditions, and thus were better suited to a risk assessment that would be valid for larger regions than would experimental methods. A breakthrough in the simulation methodology was achieved when a harmonized use of these models was reported by the FOCUS working group on groundwater scenarios [7]. With the FOCUS report available as guidance, environmental concentrations in groundwater predicted by simulation models became accepted by regulatory authorities in Europe and, as a consequence, they form today the basis of all risk assessments for new pesticide registrations. Moreover, following the introduction of groundwater modeling, corresponding frameworks were developed for the risk assessment of other environmental compartments such as surface water, soil, and air [5–7, 71].

As noted above, the situation in ecological risk assessment is today in many respects comparable to the situation that existed for exposure assessment 20 years ago. Today, new requirements for the regulations are becoming increasingly difficult to fulfill with experimental approaches and affordable efforts. However, the application of eco(toxico)logical models to regulatory risk assessments for plant protection products is clearly more complex than for exposure modeling. Whereas, in order to predict environmental concentrations the approaches can largely be standardized, this is not easily achieved in case models for ecological risk assessments. The main reason for this is the much wider variety of questions to be answered, due primarily to the diversity of species that must be considered (at least potentially) and for which respective species-specific models are required. Moreover, there will be generic differences in modeling approaches for aquatic, terrestrial, or soil species, and last – but not least – different types of toxic effect might require different modeling approaches. Nevertheless, it is conceivable that such diversity could be significantly reduced with the availability of a manageable number of framework-like applications that possessed sufficient flexibility to be adapted to different, albeit related, applications. For example, a similar modeling approach could be used to simulate populations of different species if it were to be implemented in a modular fashion, with species-specific modules. Unfortunately, however, no such solutions are yet in sight, due mainly to the fact that most presently available models have been developed in academic environments and were intended for use in specific scientific issues, without any need for either generalization or standardization.

Although a wide variety of ecological models has been reported, and many are – at least in principle – well suited to use in pesticide risk assessment [2, 72], virtually all were originally not developed for this purpose, as noted above. Although, from a scientific point of view, they could most likely be applied for actual risk assessments, regulatory uses have more stringent requirements for the documentation and validation of a model than do scientific uses. This is increasingly the case, since there is at present very little experience with ecological models within the regulatory environment. Thus, results obtained with scarcely documented and validated models are only minimally assessable for their reliability. This situation has been clearly identified in the scientific and regulatory communities, and guidance for the appropriate development, documentation and validation of models is currently under development [73, 74]. The implementation and consideration of such guidance will simplify the assessment of the models themselves, although with more technical guidance in place it will remain to be seen how these models can be used for regulatory risk assessments. Typical questions include, for example, which scenarios will be simulated, which spatial range will be considered so as to allow conclusions to be drawn on population development, how and where in the process is uncertainty to be considered, and finally how should the quantitative results of a simulation be considered in the risk assessment process? The latter question is aimed at the significance of models in comparison to experimental higher-tier options. So, might models be used to derive new higher-tier endpoints that then could be used in the assessment process corresponding to experimental endpoints? Clearly, further discussions among the regulatory and scientific communities is of paramount importance if these problems are to be resolved in the near future.

At present, a lack of guidance appears to be the main obstacle to the more intense use of models, and an acceptance of the results obtained in ecological risk assessments for plant protection products. For those applicants, the missing of guidance relates to a high degree of uncertainty if ultimately an intended modeling approach is to be accepted or not, and if the effort to be invested for its development will pay dividends. On the other hand, from the viewpoint of the regulatory agencies, the assessment of new approaches without useful guidance relates to an increased effort which often cannot be made. Thus, respective approaches may be refused either generically in respective guidelines [75], or in the reaction of concrete applications for approvals.

The best progress could certainly be made if, similar to the FOCUS framework for exposure modeling, standard scenarios could be developed and implemented for different cases of risk assessment, in combination with guidance as to which model would best be used. Such a scheme would make modeling results more comparable and thus much easier to assess. Likewise, a respective approach would make the acceptability of risk assessments based on simulations much more calculable for applicants, and would therefore enhance the use of models in regulatory risk assessments.

Apart from the scarce regulation of model uses, one further factor limits the wider spread of modeling approaches in regulatory risk assessment, namely the limited experience available with this technology. There is clearly a high demand for further training in the development, use and assessment of these models and their applications, and this issue has been identified and efforts made to improve the situation. In this respect, the Society of Environmental Toxicology and Chemistry (SETAC), under whose roof an advisory group was formed in the European branch that is specifically dedicated to modeling in risk assessment [76], has been highly active. Indeed, SETAC has already organized workshops on modeling-related topics, and more are planned for the future. In addition, a project funded by the EU has two main goals: (i) to train young scientists in eco(toxico)logical modeling; and (ii) to foster communication between academia, industry, and regulatory authorities [77].

In conclusion, it is clear that environmental effect modeling has great potential in regulatory risk assessments for plant protection products although, due to its complexity and lack of experience in its use, it has not yet been fully accepted. Nonetheless, most of the obstacles involved have been identified, such that during the past few years many activities have been initiated aimed at improving the situation. In comparison with other fields, where modeling and simulation techniques have found their ways into regulatory assessment processes (e.g., environmental fate modeling and pharmacokinetic/pharmacodynamics modeling), it is surely just a matter of time until these models become a standard part of the "toolbox" of methods for environmental risk assessment. Often, the fact that various aspects are continually and increasingly introduced for regulatory assessments makes it difficult – and perhaps even impossible – to resolve problems in an experimental fashion. However, there is clearly an ever-increasing need to seek alternative methods that will lead to significant developments in risk assessment.

References

1. Hommen, U., Baveco, J.M., Galic, N., and van den Brink, P.J. (2010) *Integr. Environ. Assess. Manage.*, **6** (3), 325–337.
2. Galic, N., Hommen, U., Baveco, H., and van den Brink, P.J. (2010) *Integr. Environ. Assess. Manage.*, **6** (3), 338–360.
3. Boesten, J. *et al.* (2005) Leaching Models and EU Registration, EC document DOC 4952/VI/95, *http://ec.europa.eu/food/plant/protection/evaluation/guidance/gw_en.pdf* (accessed 10 August 2011).
4. European Union (1991) Council Directive 91/414/EEC of 15 July 1991 Concerning the Placing of Plant Protection Products on the Market.
5. FOCUS (2001) FOCUS Surface Water Scenarios in the EU Evaluation Process Under 91/414/EEC. Report of the FOCUS Working Group on Surface Water Scenarios, EC Document Reference SANCO/4802/2001-rev.2, 245pp.
6. FOCUS (2006) Pesticides in Air: Considerations for Exposure Assessment. Report prepared by the FOCUS Working Group On Pesticides in Air (FOCUS Air Group), SANCO/10553/2006 Rev 2 June 2008.
7. FOCUS (2011) Generic Guidance for Tier 1 FOCUS Ground Water Assessments (Version 2.0), *http://focus.jrc.ec.europa.eu/gw/docs/Generic_guidance_FOCUS_GW_V2.pdf* (accessed 10 August 2011).
8. Krishnan, K. and Peyret, T. (2009) Physiologically based toxicokinetic (PBTK) modeling in ecotoxicology, in *Ecotoxicology Modeling, Emerging Topics in Ecotoxicology: Principles, Approaches and Perspectives* (ed. J. Devillers), Springer Science+Business Media, LLC, New York.
9. Jager, T., Albert, C., Preuss, T.G., and Ashauer, R. (2011) *Environ. Sci. Technol.*, **45**, 2529–2540.
10. Brock, T.C.M., Arts, G.H.P., Maltby, L., and van den Brink, P.J. (2006) *Integr. Environ. Assess. Manage.*, **2** (4), e20–e46.
11. European Food and Safety Agency (2010) *EFSA J.*, **8** (10), 1821.
12. European Food and Safety Agency (2009) *EFSA J.*, **7** (12), 1438.
13. Mackay, D. and Fraser, A. (2000) *Environ. Pollut.*, **110** (3), 375–391.
14. Pimm, S.L. (2002) *Food Webs*, University of Chicago Press, Chicago.
15. Nienstedt, K.M., Brock, T.C., van Wensem, J., Montforts, M., Hart, A., Aagaard, A., Alix, A., Boesten, J., Bopp, S.K., Brown, C., Capri, E., Forbes, V., Köpp, H., Liess, M., Luttik, R., Maltby, L., Sousa, J.P., Streissl, F., and Hardy, A.R. (2011) *Sci. Total Environ.*, **415** (1), 31–38.
16. Escher, B.I. and Hermens, J.L.M. (2004) *Environ. Sci. Technol.*, **38**, 455A–462A.
17. White, C.R. and Seymour, R.S. (2005) *J. Exp. Biol.*, **208**, 1611–1619.
18. Hendriks, A.J., van der Linde, A., Cornelissen, G., and Sijm, D.T.H.M. (2001) *Environ. Toxicol. Chem.*, **20** (7), 1399–1420.
19. Ashauer, R. and Brown, C.D. (2008) *Environ. Toxicol. Chem.*, **27** (8), 1817–1821.
20. Preuss, T.G., Hammers-Wirtz, M., Hommen, U., Rubach, M.N., and Ratte, H.T. (2008) *Ecol. Model.*, **220**, 310–329.
21. Ashauer, R., Caravatti, I., Hinterneister, A., and Escher, B.I. (2010) *Environ. Toxicol. Chem.*, **29** (7), 1625–1636.
22. Ashauer, R. and Escher, B.I. (2010) *J. Environ. Monit.*, **12** (11), 2056–2061.
23. Rubach, M.N., Ashauer, R., Maund, S.J., Baird, D.J., and van den Brink, P.J. (2010) *Environ. Toxicol. Chem.*, **29** (10), 2225–2234.
24. European Food and Safety Agency (2005) *EFSA J.*, **240**, 1–21.
25. Reddy, M., Yang, R.S., Andersen, M.E., and Clewell, H.J. III (2005) *Physiologically Based Pharmacokinetic Modeling: Science and Applications*, Wiley-Interscience, Hoboken, NJ.
26. U.S. Environmental Protection Agency (EPA) (2006) *Approaches for the Application of Physiologically Based Pharmacokinetic (PBPK) Models and Supporting Data in Risk Assessment*, National Center for Environmental Assessment, Washington, DC, EPA/600/R-05/043F. Available from: National Technical Information Service, Springfield, VA and online at: *http://cfpub.epa.*

gov/ncea/cfm/recordisplay.cfm?deid=157668 (accessed 18 April 2012).

27. Chiu, W.A. *et al.* (2007) *J. Appl. Toxicol.*, 27, 218–237.
28. World Health Organization (WHO) (2010) Harmonization Project Document No. 9: Characterization and Application of Physiologically Based Pharmacokinetic Models in Risk Assessment.
29. Goldman, C. and Carpenter, E.J. (1974) *Limnol. Oceanogr.*, **19**, 756.
30. Thornton, A., Weinhart, T., Bokhove, O., Zhang, B., van der Sar, D.M., Kumar, K., Pisarenco, M., Rudnaya, M., Savcenco, V., Rademacher, J., Zijlstra, J., Szabelska, A., Zyprych, J., van der Schans, M., Timperio, V., and Veerman, F. (2010) Modeling and Optimization of Algae Growth, Internal Report University of Twente, *http://doc.utwente.nl/73345/* (accessed 28 August 2011).
31. Driever, S.M., van Nes, E.H., and Roijackers, R.M.M. (2005) *Aquat. Bot.*, **81**, 245–251.
32. Barnthouse, L.W. (2004) *Environ. Toxicol. Chem.*, **23**, 500–508.
33. Péry, A.R., Babut, M.P., Mons, R., and Garric, J. (2006) *Environ. Toxicol. Chem.*, **25** (1), 144–148.
34. Adams, B.M., Banks, H.T., Banks, J.E., and Stark, J.D. (2005) *Math. Biosci.*, **196**, 39–64.
35. Brown, A.R., Riddle, A.M., Cunningham, N.L., Kedwards, T.J., Shillabeer, N., and Hutchinson, T.H. (2003) *Hum. Ecol. Risk Assess.*, **9** (10), 761–788.
36. Gurney, W.S.C. (2006) *Environ. Health Perspect.*, **114** (S-1), 122–126.
37. Caswell, H. (2001) *Matrix Population Models: Construction, Analysis and Interpretation*, 2nd edn, Sinauer Associates, Sunderland, MA.
38. Leslie, P.H. (1945) *Biometrica*, **33**, 183–212.
39. Lebreton, J.D. and Gonzales-Davila, G. (1993) *J. Biol. Syst.*, **1**, 389–423.
40. Hunter, C. and Caswell, H. (2005) *Ecol. Model.*, **188**, 15–21.
41. Hanski, I. and Gilpin, M.E. (1997) *Metapopulation Dynamics: Ecology, Genetics and Evolution*, Academic Press, San Diego, CA.
42. Charles, S., Billoir, E., Lopes, C., and Chaumot, A. (2009) Matrix population models as relevant modeling tools in ecotoxicology, in *Ecotoxicology Modeling, Emerging Topics in Ecotoxicology: Principles, Approaches and Perspectives* (ed. J. Devillers), Springer Science+Business Media, LLC, New York.
43. Billoir, E., Pery, A.R.R., and Charles, S. (2007) *Ecol. Model.*, **203**, 204–214.
44. Chandler, G.T., Cary, T.L., Bejarano, A.C., Pender, J., and Ferry, J.L. (2004) *Environ. Sci. Technol.*, **38**, 6407–6414.
45. Raimondo, S. and McKenney, C.L. (2005) *Integr. Comp. Biol.*, **45** (1), 151–157.
46. Stark, J.D. and Bamfo, S. (2002) *Proceedings of the 1st International Symposium on Biological Control of Arthropods, Honolulu, Hawaii, 14–18 January 2002*, United States Department of Agriculture, Forest Service, Washington, DC, pp. 314–317.
47. Stark, J.D. and Banks, J.E. (2004) *Biol. Control*, **29**, 392–398.
48. Chaumot, A., Charles, S., Flammarion, P., Garric, J., and Auger, P. (2002) *Ecol. Appl.*, **12**, 1771–1784.
49. Gervais, J.A., Hunter, C.M., and Anthony, R.G. (2006) *Ecol. Appl.*, **16**, 666–677.
50. Raimondo, S. and McKenney, C.L.J. (2005) *Environ. Toxicol. Chem.*, **24**, 564–572.
51. Banks, E., Dick, L.K., Banks, H.T., and Stark, J.D. (2008) *Ecol. Model.*, **210** (1-2), 155–160.
52. van den Brink, P.J., Verboom, J., Baveco, J.M., and Heimbach, F. (2007) *Environ. Toxicol. Chem*, **26**, 2226–2236.
53. Grimm, V. and Railsback, S.F. (2005) *Individual-Based Modeling and Ecology*, Princeton University Press, Princeton, NJ.
54. Wickwire, T., Johnson, M.S., Hope, B.K., and Greenberg, M.S. (2011) *Integr. Environ. Assess. Manage.*, **7** (2), 158–168.
55. Railsback, S.F. (2001) *Nat. Resour. Model.*, **14**, 465–474.
56. Railsback, S.F., Harvey, B.C., Lamberson, R.R., Lee, D.E., Claasen, N.J., and Yoshihara, S. (2002) *Nat. Resour. Model.*, **15**, 83–110.

57. Topping, C.J., Hansen, T.S., Jensen, T.S., Jepsen, J.U., Nikolajsen, F., and Odderskær, P. (2003) *Ecol. Model.*, **167** (1-2), 65–82.
58. Wang, M. and Grimm, V. (2007) *Ecol. Model.*, **205**, 397–409.
59. Wang, M. and Grimm, V. (2010) *Environ. Toxicol. Chem.*, **29** (6), 1292–1300.
60. Preuss, T.G., Hammers-Wirtz, M., and Ratte, H.T. (2010) *J. Environ. Monit.*, **12**, 2070–2079.
61. van den Brink, P.J. (2011) MASTEP – An Individual Based Model to Predict Recovery of Aquatic Invertebrates Following Pesticide Stress, *http://www.mastep.wur.nl/Download/MASTEP.pdf* (accessed 10 August 2011).
62. Traas, T.P., Janse, J.H., van den Brink, P.J., Brock, T.C.M., and Aldenberg, T. (2003) *Environ. Toxicol. Chem.*, **23** (2), 521–529.
63. Park, R.A., Clough, J.S., and Wellman, M.C. (2008) *Ecol. Model.*, **213**, 1–15.
64. Bartell, S.M., Lefebvre, G., Kaminski, G., Carreaqu, M., and Campbell, K.R. (1999) *Ecol. Model.*, **124**, 43–67.
65. Bartell, S.M., Campbell, K.R., Lovelock, C.M., Nair, S.K., and Shaw, J.L. (2000) *Environ. Toxicol. Chem.*, **19**, 1441–1453.
66. Sourisseau, S., Basseres, A., Perie, F., and Caquet, T. (2008) *Water Res.*, **42**, 1167–1181.
67. Sourisseau, S., Basseres, A., Perie, F., and Caquet, T. (2008) Stochastic simulation of aquatic ecosystem models for the ecological risk assessment of chemicals. SETAC Europe Conference, Warsaw, Poland.
68. Naito, W., Miyamoto, K., Nakanishi, J., Masunaga, S., and Bartell, S.M. (2002) *Water Res.*, **36** (1), 1–14.
69. Brain, R., Hendley, P., Bartell, S., Nair, S., Hosmer, A., and Wall, S. (2011) Use of the comprehensive aquatic system model for atrazine ($CASM_{ATZ}$) to estimate potential changes in primary producer community structure based on intensive stream monitoring chemographs. SETAC Europe Conference, 15–19 May, Milan, Italy.
70. European Union (1980) Council Directive 80/68/EEC of 17 December 1979 on the Protection of Groundwater Against Pollution Caused by Certain Dangerous Substances as Amended by Council Directive 91/692/EEC (Further Amended by Council Regulation 1882/2003/EC).
71. FOCUS (1997) Soil Persistence Models and EU Registration. The final report of the work of the Soil Modelling Work group of FOCUS.
72. Schmolke, A., Thorbek, P., Chapman, P., and Grimm, V. (2010) *Environ. Toxicol. Chem.*, **29** (4), 1006–1012.
73. Grimm, V., Berger, U., Bastiansen, F., Eliassen, S., Ginot, V., Giske, J., Goss-Custard, J., Grand, T., Heinz, S., Huse, G., Huth, A., Jepsen, J.U., Jørgensen, C., Mooij, W.M., Müller, B., Pe'er, G., Piou, C., Railsback, S.F., Robbins, A.M., Robbins, M.M., Rossmanith, E., Rüger, N., Strand, E., Souissi, S., Stillman, R.A., Vabø, R., Visser, U., and DeAngelis, D.L. (2006) *Ecol. Model.*, **198**, 115–126.
74. Schmolke, A., Thorbek, P., DeAngelis, D.L., and Grimm, V. (2010) *Trends Ecol. Evol.*, **25** (8), 479–486.
75. Swedish Chemicals Agency (2011) Guidance Document on Work-Sharing in the Northern Zone in the Registration of Plant Protection Products, *http://www.kemi.se/upload/Bekampningsmedel/Vaxtskyddsmedel/Northern%20Zone%20work-sharing%20guidance%20July%202011.pdf* (accessed 30 August 2011).
76. Preuss, T.G., Hommen, U., Alix, A., Ashauer, R., van den Brink, P., Chapman, P., Ducrot, V., Forbes, V., Grimm, V., Schäfer, D., Streissl, F., and Thorbek, P. (2009) *Environ. Sci. Pollut. Res.*, **16**, 250–252.
77. Grimm, V., Ashauer, R., Forbes, V., Hommen, U., Preuss, T.G., Schmidt, A., van den Brink, P.J., Wogram, J., and Thorbek, P. (2009) *Environ. Sci. Pollut. Res.*, **16**, 614–617.

13
The Use of Metabolomics *In Vivo* for the Development of Agrochemical Products

Hennicke G. Kamp, Doerthe Ahlbory-Dieker, Eric Fabian, Michael Herold, Gerhard Krennrich, Edgar Leibold, Ralf Looser, Werner Mellert, Alexandre Prokoudine, Volker Strauss Tilmann Walk, Jan Wiemer, and Bennard van Ravenzwaay

13.1
Introduction to Metabolomics

Metabolite profiling describes the analysis of endogenous low-molecular-weight compounds such as carbohydrates, amino acids, lipids, and organic acids, that are the products of biochemical pathways [1]. Although metabolite profiling has a long history of application in the plant sciences [2], it has within the past few years also become a more-often used technology in toxicology studies, to elucidate changes in biochemical pathways following the administration of test compounds [3–5].

In the context of toxicology research, the analysis is performed routinely by using blood or urine and applying mainly two different technologies: (i) profiling by nuclear magnetic resonance (NMR) spectroscopy [1, 6]; and (ii) profiling via chromatography coupled with mass spectrometry (MS) detection systems [4, 7–9]. For a more detailed comparison of the two techniques, see the reviews of van Ravenzwaay *et al.* [5] and Gomase *et al.* [10].

The use of sensitive liquid chromatography-mass spectrometry (LC-MS) and gas chromatography-mass spectrometry (GC-MS) techniques offers the possibility to detect a broad range of metabolites, and thus increases the chance of identifying relevant biomarkers or patterns of change. In addition, the advantage of using small samples of blood or urine for analysis enables the refinement of animal testing, since sample taking is less invasive, does not lead to the death of the test animal and, in parallel, makes possible time-course sampling. But above all, metabolomics has been demonstrated as being statistically more powerful for detecting effects compared to other "omics" technologies [11, 12].

It has been reported that the levels of endogenous metabolites can be altered as a result of toxicological responses. For example, metabolite changes have been described in the context of liver and/or kidney toxicity and other even more specific effects [4, 5, 13–17]. For the purpose of establishing a large metabolite profiling data base for chemicals, agricultural chemicals, and pharmaceuticals, BASF has established a specific and highly standardized 28-day testing procedure that includes

Modern Methods in Crop Protection Research, First Edition.
Edited by Peter Jeschke, Wolfgang Krämer, Ulrich Schirmer, and Matthias Witschel.

sampling the blood and urine of rats at several time points [4]. Extensive studies have been conducted to determine the robustness and reproducibility of this technology (H.G. Kamp *et al.*, unpublished results).

13.2
MetaMap®Tox Data Base

13.2.1
Methods

The methods applied for the studies performed to establish BASF's metabolomics data base are described in detail in Refs [5, 18, 19]. Briefly, the metabolome studies were performed as follows.

13.2.1.1 Animal Treatment and Maintenance Conditions

Wistar rats were housed singly in standard cages (floor area 800 cm^2) and maintained in an air-conditioned room at a temperature of 20–24 °C, a relative humidity of 30–70%, and a 12 h light/12 h dark cycle. Both, diet and drinking water were available *ad libitum* (except before blood sampling) and were assayed regularly for chemical contaminants and the presence of microorganisms. For these animal experiments, the animals were aged 42 or 70 days at the start of the study, which were conducted according to the German Animal Welfare legislation. The laboratory was also certified by the Association for Assessment and Accreditation of Laboratory Animal Care (AAALAC).

For each dose group, five rats of each sex were used. The doses were chosen based either on a BASF internal study or on published data, and reflected the 28-day maximum-tolerated dose (MTD) for these animals. The compounds were administered either via the feed or by gavage. When building the data base MetaMap®Tox, more than 500 reference compounds with known toxicity were used, of which approximately 20% could be regarded as agrochemicals, approximately 40% as chemicals, and the remainder (ca. 40%) as pharmaceutical active ingredients. A few further reference compounds comprises, for example, vitamins or nutritionals.

All animals were checked daily for any clinically abnormal signs and mortalities. Their food consumption was determined on study days 6, 13, 20, and 27, while their body weights were determined before the start of the administration period (in order to randomize the animals) and again on study days 0, 3, 6, 13, 20, and 27.

At the end of the treatment period, the animals were sacrificed by decapitation under isoflurane anesthesia.

13.2.1.2 Blood Sampling and Metabolite Profiling

For MS-based metabolite profiling analysis, blood samples were taken (using potassium-EDTA as anticoagulant) from the retro-orbital sinus in all rats under isoflurane anesthesia, after a fasting period of 16–20 h, on study days 7, 14, and 28. The plasma was separated and extracted using a proprietary method, with three types of analysis being applied to all samples: GC-MS and LC-MS/MS were used for broad profiling, as described by van Ravenzwaay *et al.* [4], while solid-phase extraction-liquid chromatography-mass spectrometry/mass spectrometry (SPE-LC-MS/MS) was applied for the determination of catecholamine and steroid hormone levels. After protein precipitation using acetonitrile, the polar and nonpolar fractions were separated for both GC-MS and LC-MS/MS analyses. For the GC-MS analysis, the nonpolar and polar fractions were further derivatized before analysis [20]. For LC-MS analysis, both fractions were reconstituted in appropriate solvent mixtures, and HPLC was performed by gradient elution on reversed-phase separation columns. An MS detection technology was applied which allowed target and high sensitivity multiple reaction monitoring (MRM) profiling in parallel to a full screen analysis.

For GC-MS and LC-MS/MS profiling, data were normalized to the median of the reference samples, which were derived from a pool formed from aliquots of all samples to account for any inter- and intra-instrumental variation. Steroid hormones, catecholamines, and their metabolites were measured using online SPE-LC-MS/MS [21]. Absolute quantification was performed by means of stable isotope-labeled standards.

The methods applied resulted in 290 plasma analytes for semi-quantitative analysis, 234 of which were chemically identified and 56 were unknown.

13.3 Evaluation of Metabolome Data

13.3.1 Data Processing

13.3.1.1 Metabolite Profiling

The data were analyzed by univariate and multivariate statistical methods. The sex and day-stratified heteroscedastic *t*-test ("Welch test") was applied to log-transformed quantitative and semi-quantitative metabolite data to compare the treated groups with the respective controls. *p*-values, *t*-values, and ratios of corresponding group medians were collected as metabolic profiles and fed into a database (MetaMap®Tox).

13.3.1.2 Metabolome Patterns

It has been shown that, by using reference compounds with known, similar toxicity, common sets of metabolite changes can be identified, which are specific to the particular toxicity of interest. The metabolic profile of a novel class of herbicide, which inhibits 4-hydroxyphenylpyruvate dioxygenase (HPPD) enzyme activity, was reported by van Ravenzwaay *et al.* [4]. Other examples of mode of action (MoA)-specific metabolome patterns were described by Strauss *et al.* [19] for hemolytic anemia, and by van Ravenzwaay *et al.* [5] for peroxisomal proliferation.

The development of such patterns has been described elsewhere [22]. Briefly, metabolite patterns correlating to specific toxicological modes of action are based on at least three different chemicals from the MetaMap®Tox data base, which share a common toxicological MoA (reference compounds). After identification of the significantly changed metabolites and a consistency check for changes over time and dose–response relationship through an expert panel, the pattern is validated against the data base. A pattern should correctly identify at least one further reference compound sharing the same MoA, which has not been used to establish the pattern. Furthermore, reference compounds in MetaMap®Tox which do not share this particular toxicity should not be identified.

In total, more than 110 specific metabolite patterns describing more than 40 different toxicological modes of action have been developed based on the MetaMap®Tox data base, covering a broad range of target organs and effects, such as liver (enzyme induction, peroxisome proliferation, liver toxicity), nervous system (dopamine agonism/antagonism, noradrenaline agonism, acetylcholinesterase inhibition, nicotinic receptor agonist), kidney (tubular toxicity, organic anion transporter inhibition), adrenals (corticosteroid synthesis inhibition), testes (impaired spermatogenesis), ovaries (estrogenic receptor modulation), endocrine modulation (aromatase inhibition, anti-androgenic effect, estrogenic effects), thyroid (direct: hormone synthesis inhibition, indirect: increased metabolism), blood (porphyrin synthesis inhibition, aplastic anemia, hemolytic anemia, platelet aggregation inhibition), and bone (osteoblast inhibition, mineralization).

The test compound-induced changes can be compared against these patterns by using proprietary algorithms. The result of this comparison is a "similarity score," which provides evidence for a respective toxicological potency. In a second evaluation step, the particular metabolite changes can be assessed and interpreted by an expert panel, which guarantees the exclusion of false-positive or false-negative findings. If, for example, high similarity scores for the test compound-induced changes to patterns for liver toxicity are identified, and if these are considered by the expert panel as being biologically meaningful, the test compound should be regarded as being hepatotoxic.

13.3.1.3 **Whole-Profile Comparison**

Means are calculated of all metabolite values measured in blood samples taken at study days 7, 14, and 28 of the control, low- and high-dose groups for each sex separately. Using the "Welch test," t-values are calculated comparing the means of the dose groups with controls. With the whole-profile comparison, the profiles (t-values) of an unknown compound are compared with each profile of the compounds in the MetaMap®Tox database. These comparisons are quantified by calculating either the parametric Pearson product moment correlation coefficient or the nonparametric Spearman rank correlation coefficient. The similarity between the profiles of the unknown compound and the compounds of the MetaMap®Tox database is ranked by the resulting correlation coefficients. A discrepancy between the Pearson and Spearman coefficient is often due to some extreme changes in metabolite levels that strongly affect the Pearson – but not the Spearman – correlation coefficient, the latter being more reliable in this case. A coefficient of about >0.500 is estimated as a good correlation when regarding the distribution of the correlation coefficients calculated with all profiles in the MetaMap®Tox database.

13.4 Use of Metabolome Data for Development of Agrochemicals

13.4.1 General Applicability

The data base MetaMap®Tox is based on 28-day systemic toxicity studies, and comprises plasma metabolome data for sampling time points 7, 14, and 28 days. Consequently, the data base can be used as an additional tool for any short-term systemic toxicity study, either a 7- or 14-day rangefinder/screening study or a regulatory OECD 407 study. Especially when applied to the latter type of study, careful consideration must be given as to whether multiple blood sampling is appropriate, or whether a single sampling towards the end of the study is preferred.

The inclusion of the metabolome analysis into an OECD 407 study, on the other hand, provides the integrative view on clinical, physiological, and also histopathological changes. The combination of these data provides a more holistic view on the effects induced by a certain test compound, thus enabling decisions to be made on further mechanistic testing instead of continuing to perform studies simply because they are demanded by the regulations. The mechanistic information obtained by metabolome analysis will help to target toxicological research. Moreover, early information on the possible mechanisms will help in the conduct of further mechanistic studies at an earlier stage, which will in turn assist in the risk assessment of compounds (see Section 13.5.2). Finally, the metabolome analysis provides valuable data to investigate the toxicological differences of sets of compounds. In this way, it enables the selection of the best candidates for development from a regulatory perspective on a more solid basis than by applying only the classical parameters (clinical and histopathology).

Even when used in stand-alone short-term studies, MetaMap®Tox provides information that is sufficient for improved decision-making during compound development. In the context of such screening/range finder studies, the results of metabolome analysis can significantly influence the nature of subsequent studies, and their design (e.g., additional end-points to be assessed).

13.4.2 Case Studies

13.4.2.1 Liver Enzyme Induction

The induction of hepatic metabolizing enzymes is a common finding associated with functional and histopathological changes in subchronic and chronic rodent studies. Moreover, such changes have also been related to increased incidences of liver tumors. Indirectly, enzyme induction has also been associated with the increased excretion of thyroid hormones, which concomitantly could lead to thyroid tumors induced by the dysregulation in thyroidal cell proliferation. However, those effects in rodents are considered irrelevant for humans for two reasons:

- The phase II metabolism leading to increased thyroid hormone excretion is different between both species, and is not severely affected by liver enzyme induction in humans.
- The pool of thyroid hormones in humans is much higher than in rats, which may avoid severe changes in homeostasis in humans.

Using reference compounds such as pentobarbital sodium, Aroclor 1254, pentachlorobenzene, ethyl-benzene, and vinclozoline, metabolites were grouped that showed the same changes in samples from animals treated with these different reference compounds for liver enzyme induction. The pattern based on these compounds with liver enzyme-inducing properties can be readily found following metabolite profiling. In Table 13.1, examples are given for metabolites which are commonly changed by liver enzyme inducers, for male and females, respectively [5].

In Table 13.1, also the changes observed for a test compound, which had been developed as an agrochemical, are listed. For this compound, a 28-day study in rats was conducted according to the OECD 407 test guideline. Blood samples were taken towards the end of the study and plasma was used for metabolome analysis and comparison against the MetaMap®Tox data base. Based on the similarity with the changes induced by reference compounds known to be liver enzyme inducers, the changes shown provide evidence that the test compound also elicits effects on the liver enzymes.

Figure 13.1 shows a display of the pattern ranking for the above-mentioned test compound. In the 28-day study, clear matches were observed – besides others – for liver enzyme induction, liver toxicity, and indirect effects on the thyroid gland.

In addition, by using whole-profile comparison, information concerning a possible liver toxic activity of the compound was obtained. Those reference compounds which correlated best with the above-mentioned test compound in MetaMap®Tox are listed in Tables 13.2 and 13.3. All reference compounds within the top 20 correlations (sorted by Pearson correlation coefficient) in females are known as liver enzyme inducers. For male animals, 18 of the top 20 correlations (sorted by Pearson correlation coefficient) are also known to be liver enzyme inducers.

Taken together, these results provide strong evidence that the test compound would have the potential to induce liver enzymes, and therewith to cause subsequent toxicological effects such as hepatocellular hypertrophy, hepatotoxicity and, potentially, the induction of hepatic tumors. Furthermore, strong evidence for inducing secondary effects in the thyroid gland was obtained.

In the OECD 407 study, increased liver weights were observed without any histopathological correlation. Furthermore, there was no clear evidence for liver enzyme induction from clinical pathology. Given such a data set without the metabolome analysis, the most likely procedure would have been to proceed with

Table 13.1 Metabolite pattern in rats for a BASF development compound compared to a set of known liver enzyme inducers as reference compounds.

Metabolite	BASF test compound				Reference compounds for liver enzyme induction			
	Direction of regulation	Strength of regulation (× fold)	Direction of regulation	Strength of regulation (× fold)	Direction of regulation	Strength of regulation (× fold)	Direction of regulation	Strength of regulation (× fold)
	Males		Females		Males		Females	
Arachidonic acid	Up	1.87–1.99	Up	1.70–2.63	Up	1.16–2.20	Up	0.91–3.18
Behenic acid	Up	1.57–1.99	Up	1.60–2.28	Up	1.19–2.60	Up	1.17–4.30
Cholesterol	Up	1.69–2.81	Up	1.89–3.60	Up	1.16–3.29	Up	0.99–4.54
Eicosanoic acid	–	–	Up	1.25–2.43	–	–	Up	0.83–3.19
Glycerol	–	–	Up	2.74–4.99	–	–	Up	1.01–8.03
Lignoceric acid	Up	2.08–2.47	Up	1.44–2.40	Up	1.35–2.93	Up	1.00–5.01
Linoleic acid	Up	1.94–2.40	Up	2.58–3.69	Up	1.08–2.97	Up	1.02–5.23
Palmitic acid	Up	1.87–2.34	Up	2.15–3.03	Up	1.11–1.34	Up	1.05–3.53
Stearic acid	Up	1.89–2.27	Up	1.23–2.64	Up	1.21–2.59	Up	0.87–3.03

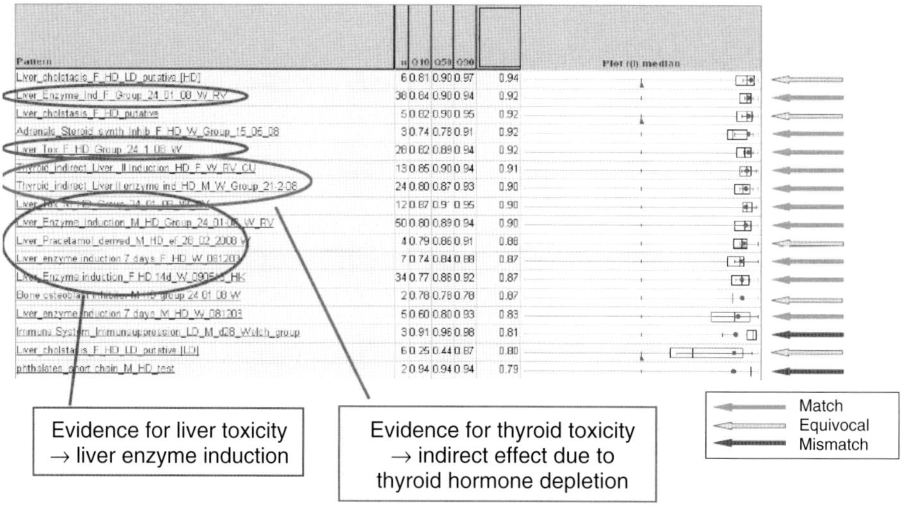

Figure 13.1 Graph showing a ranking list of specific metabolite patterns compared to the test compound. Ranks are based on the Pearson correlation coefficient of *t*-values of group medians for each metabolite change in the respective pattern. In the plot, the reference interval comprises 10 and 90% of the correlation factors calculated with reference compounds (black box). Comparison of the test compound with reference compounds is represented by the red dot (median of the correlation factors) and the red ticks (10 and 90%). The arrows display the expert judgment on matches to toxicity patterns. The color coding is described in the figure.

classical regulatory testing (i.e., a 90-day study followed by a chronic/carcinogenicity study), without being aware of the potential problems that lay ahead. In fact, the outcome of the carcinogenicity study with the BASF developmental compound was an increased incidence in liver and thyroid tumors. Without the availability of the metabolome data, mechanistic studies in the liver and thyroid gland would most likely have been initiated following the detection of tumors in these organs.

By making use of the results of the metabolome analysis, targeted mechanistic studies could already be conducted in parallel to the chronic/carcinogenicity study, and the data were available at the time of the histopathological analysis of the cancer study. By employing this approach, it was possible to limit the studies to those that confirmed the indicated MoA (thus reducing the number of animals required and other resources to the absolute minimum), such that approximately six to nine months in development time were saved. The net result was that the compound could be marketed one cultural season earlier than would have been the case had the classical approach been followed.

13.4.2.2 **Liver Cancer**

The liver is one of the main target organs in toxicological studies, since this organ plays a central role in the metabolism of endogenous and exogenous substances. In regulatory animal studies, artificially high dose levels of test substances are administered. Due to its function as a central metabolizing organ, and its exposure to these very high quantities of test substance, the liver is susceptible to toxicity from toxic agents or their metabolites, and through different modes of action such toxic insults can ultimately lead to the formation of tumors. It is commonly agreed that liver tumor formation is the result of a multi-step (initiation, promotion, progression) process:

- The first step in tumor formation (initiation) is characterized by the induction of mutations in somatic cells through exposure to genotoxic compounds, although it might also result from pre-existing genetic conditions.
- The second phase (promotion) involves different processes, which ultimately result in a proliferative stimulus and an environment allowing for clonal growth of the initiated cells.
- Progression again involves genetic changes, either induced from external sources (genotoxic agents) or through genetic instability of the (preneoplastic) tumorigenic lesion [23].

Within the 28-day treatment period in the studies used to set up MetaMap®Tox, it is not possible to cover all steps of liver tumor formation as described above. Consequently, it is also not possible to identify metabolome changes which could be specific and predictive for tumorigenic tissue in the liver. However, van Ravenzwaay

Table 13.2 Top 20 reference compounds ranking with the test compound (female animals).

Treatment	Pearson *R*	Pearson Rank
Oxcarbazepine HD	0.80	1
Test compound low-dose	0.73	2
Liver enzyme inducer (test compound)	0.73	3
Liver enzyme inducer (test compound)	0.72	4
Carvedilol HD	0.71	5
Liver enzyme inducer (test compound)	0.71	6
Beta-naphthoflavone (MOA73) HD	0.70	7
Beta-ionone HD	0.70	8
Toxaphene HD	0.69	9
Phenytoin HD	0.68	10
Liver enzyme inducer (test compound)	0.68	11
Liver enzyme inducer (test compound)	0.68	12
Cyproterone acetate LD	0.68	13
Pentachlorobenzene HD	0.68	14
Liver enzyme inducer (test compound)	0.68	15
Vinclozolin HD	0.67	16
Liver enzyme inducer (test compound)	0.66	17
Alpha-methylstyrene HD	0.66	18
Dimethylformamide HD	0.66	19
Dimethylformamide LD	0.66	20

HD, high-dose; LD, low-dose.

Table 13.3 Top 20 reference compounds ranking with the test compound (male animals).

Treatment	Pearson *R*	Pearson Rank
Test compound low-dose	0.79	1
Beta-ionone HD	0.69	2
Liver enzyme inducer (test compound)	0.68	3
Butylated hydroxytoluene HD	0.67	4
Liver enzyme inducer (test compound)	0.66	5
Liver enzyme inducer (test compound)	0.66	6
Liver enzyme inducer (test compound)	0.64	7
Tetrahydrofuran LD	0.64	8
Liver enzyme inducer (test compound)	0.64	9
Liver enzyme inducer (test compound)	0.63	10
Liver enzyme inducer (test compound)	0.62	11
Liver enzyme inducer (test compound)	0.62	12
Liver enzyme inducer (test compound)	0.62	13
Phenytoin HD	0.61	14
Liver enzyme inducer (test compound)	0.61	15
Liver enzyme inducer (test compound)	0.61	16
Test compound	0.61	17
Liver enzyme inducer (test compound)	0.60	18
Tetrahydrofuran HD	0.60	19
Dazomet HD	0.60	20

HD, high-dose; LD, low-dose.

et al. described the use of metabolomics to identify toxicological modes of action, which are potentially involved in tumor promotion in the liver and that finally could result in liver tumorigenesis [5]. Besides liver enzyme induction, these include liver toxicity resulting in hepatocyte loss and a subsequent regenerative cell proliferation stimulus (hepatic necrosis, steatosis, cholestasis) and receptor-mediated stimulus of cell proliferation (e.g., peroxisome proliferator-activated alpha receptor (PPAR), aryl hydrocarbon (Ah) receptor).

13.4.3
Chemical Categories

The formation of chemical categories (often referred to as *grouping*) and subsequent read across from data-rich chemicals belonging to this category, is probably the most efficient way of providing the required safety information during compound development and for regulatory purposes, while maintaining the amount of animal testing to an absolute minimum. However, the prerequisite is a transparency of the grouping process and the quality of the groups/categories formed. Initially, grouping was thought to be possible based on chemical structure, for example, via quantitative structure–activity relationship (QSAR) models alone. However, chemical similarity does not always indicate a similarity of the toxicological profile. Many examples have been published that describe where small changes in structure have resulted in strong differences with regards to biological/toxicological activity, and consequently only a few QSAR models with satisfactory predictive capacities are available. Nonetheless, in order to overcome this limitation, biological data could be used in the grouping process. Such biological data could be obtained from different sources, including *in vitro* studies or limited animal studies. This form of "read-across" has been used successfully in the US/OECD High Production Volume program, and might become an important tool for reducing the extent of animal testing under REACH (Registration, Evaluation, Authorisation and restriction of Chemicals). For agrochemical products, such methods could be applied for the assessment of derivatives or relevant metabolites. Under the REACH legislation, some guidance is provided with respect to the formation of chemical categories and subsequent read-across:

> "Substances whose physico-chemical, toxicological, and ecotoxicological properties are likely to be similar or follow a regular pattern as a result of structural similarity may be considered as a group, or "category" of substances. Application of the group concept requires that physico-chemical properties, human health effects, and environmental effects or environmental fate may be predicted from data for reference substance(s) within the group by interpolation to other substances in the group (read-across approach). This avoids the need to test every substance for every endpoint."

Recently, it has been shown that metabolomics used in short-term toxicity studies in rats could also be used to form groups/chemical categories [22, 24].

13.5 Discussion

Both, LC-MS/MS- and GC-MS-based metabolomics approaches have been used to construct a data base containing the rat plasma metabolite profiles of more than 500 reference compounds. By using this technique, 290 plasma metabolites with molecular weights below approximately 1500 Da could be reliably detected and quantified. For these metabolites, changes are calculated relative to untreated control animals, whereby for the untreated animals a separation of plasma metabolite profiles between male and female rats has been noted. Based on the above-mentioned reference compounds, more than 110 specific sets of metabolite changes (patterns) have been established for more than 40 different toxicological modes of action.

13.5.1 Challenges and Chances Concerning the Use of Metabolite Profiling in Toxicology

An extensive analysis of the challenges relating to the use of metabolomics in toxicology was originally provided by van Ravenzwaay *et al.* [5] in "The use of metabolomics in cancer research," a chapter in *An Omics Perspective of Cancer* (ed. W.C.S. Cho), although new opportunities have also been identified. The challenges discussed by van Ravenzwaay *et al.* [5] remain basically valid and are briefly discussed in the following paragraph (for more detail, see the original publication).

Metabolomics can be discussed as being complementary to other "omics" techniques such as genomics, transcriptomics, and proteomics, though it might have certain advantages compared to other "omics" technologies [1, 5, 6, 11, 25]. The chain of events after a toxic insult, starting with an interaction of the test substance with the components of cells and tissues (e.g., receptors, membrane activity), followed by changes in gene expression and protein levels, results in an altered physiology, as displayed by changes in the reaction products of biological pathways. In this context, metabolomics may provide insight into the later events in the cascade, as described above. However, it is likely that metabolomics – in combination with classical toxicological investigations and even in combination with other "omics" technologies – will be able to provide a maximum understanding of the toxicological effects caused by a test substance.

The early recognition of toxic effects caused by test substances is of major importance in the development of substances with a desired activity, such as agricultural chemicals and pharmaceuticals. Besides other technologies, such as *in vitro* tests, metabolite profiling in combination with the comparison against appropriate data bases can assist decision-making processes during substance development, and thus focus the development on candidates that show not only a desirable target activity but also a noncritical toxicity profile.

The challenges and the opportunities of using a combination of data from genomics, proteomics, and metabolite profiling for the toxicity assessment of test substances have been demonstrated by Craig *et al.* [26]. Approaches taken in the past have been limited to the search for correlations between specific genes and defined metabolite changes whereas, for the future, comprehensive multivariate statistical analysis seems to be the route to acquire significant knowledge from such integrated approaches [6, 27]. Nevertheless, research is ongoing to combine toxicological data from different resources in order to achieve a more comprehensive assessment of toxicological modes of action (this is referred to as *systems toxicology*). Such knowledge could also help to avoid the conduct of unnecessary studies within the development and assessment of the regulatory profile of an agrochemical. It is known that only a handful of studies from the more than 40 that are required from a regulatory standpoint, are used to define risk assessment and risk management measures. A deeper understanding of the toxicological modes of action – thus excluding any remaining possible targets of toxicity – would focus toxicological testing on exactly those studies which are needed for defining possible risks. These targeted studies would still be needed for risk assessment, due to the fact that, at present, metabolomics data would not suffice the quantitative aspects of risk assessment – that is, the definition of no observable adverse effect levels (NOAELs). In fact, according to the European Centre for Ecotoxicology and Toxicology of Chemicals (ECETOC), the findings from omics technologies would have to be combined with other observable changes at both the microscopic and macroscopic level in order to define a NOAEL. Furthermore, the changes observed with "omics" technologies would have to be correlated with pathways related to an adverse effect [28].

Another aspect to be considered here is that metabolomics is a highly sensitive technique by which changes in the physiological status of an organism can be detected. Consequently, the test system (in the case of MetaMap®Tox, the corresponding rat study including blood sampling and plasma preparation) as well as the analytical procedures must be strictly controlled [18, 19].

13.5.2
Applicability of the MetaMap®Tox Data Base

As metabolic profiling can be performed in most of the classical systemic toxicity studies, without causing any major interference on the study itself (200 μl of plasma is sufficient for an analysis), such profiling can be performed in any of these studies (e.g., OECD guideline 407 or 408). In addition, the MetaMap®Tox data base can be used to generate the specific metabolite profiles of test substances or test substances classes, and to identify the toxicological modes of action or target organs. By employing tools such as whole-metabolome correlation analysis and comparison against specific toxicity patterns, direct conclusions can be drawn on the potential toxicity of test compounds. Furthermore, the use of these tools could assist in the identification of groups of chemicals that show the same effects in repeated-dose toxicity studies. Hence, a grouping approach based on the similarity of biological activity is possible, thereby providing an enhanced scientific basis for read-across approaches – that is, from a QSAR to a quantitative biological activity relationship (QBAR). Another option is to assist in the identification of those compounds from a given set that demonstrate the most preferable regulatory profile, thereby supporting the decision-making processes during early compound development.

13.6
Concluding Remarks

In the recent past, metabolomics has demonstrated its potential to yield additional valuable toxicological information as compared to classical parameters, such as histopathology and clinical pathology alone. In order to make maximum use of the metabolomics data acquired from rat studies, a reference data base (MetaMap®Tox) has been set up which contains the metabolite profiles of more than 500 data-rich compounds. These data can be used to generate the specific metabolite profile of test substances or test substances classes, and also to identify the toxicological effects or target organs. Within the context of agrochemical development, metabolomics could also provide a deeper insight into the toxicological modes of action of a compound, thus enabling an earlier and more targeted testing to elucidate any potential toxicological hazards. In addition, as metabolomics can be applied during early repeated dose toxicity testing, it could also assist in the decision-making processes within compound classes at an early stage. In this way, development resources could be directed toward those structures that have a lower potential for inducing adverse effects – and thus a better regulatory profile – at an earlier point in time than would be possible via classical *in vivo* testing strategies.

References

1. Lindon, J.C., Holmes, E., Bollard, M.E., Stanley, E.G., and Nicholson, J.K. (2004) *Biomarkers*, **9**, 1–31.
2. Trethewey, R.N., Krotzky, A.J., and Willmitzer, L. (1999) Metabolic profiling: a Rosetta stone for genomics? *Curr. Opin. Biotechnol.*, **2**, 83–85.
3. Lindon, J.C., Holmes, E., and Nicholson, J.K. (2004) *Prog. Nucl. Magn. Reson. Spectrosc.*, **45**, 109–143.
4. van Ravenzwaay, B., Coelho-Palermo Cunha, G., Leibold, E., Looser, R., Mellert, W., Prokoudine, A., Walk, T., and Wiemer, J. (2007) *Toxicol. Lett.*, **172**, 21–28.
5. van Ravenzwaay, B., Coelho-Palermo Cunha, G., Fabian, E., Herold, M., Kamp, H., Krennrich, G., Krotzky, A., Leibold, E., Looser, R., Mellert, W., Prokoudine, A., Strauss, V., Trethewey, R., Walk, T., and Wiemer, J. (2010) in *An Omics Perspective of Cancer* (ed. W.C.S. Cho), Springer Science+Business Media B.V., pp. 141–166.
6. Griffin, J.L. and Bollard, M.E. (2004) *Curr. Drug Metab.*, **5**, 389–398.
7. Looser, R., Krotzky, A.J., and Trethewey, R.N. (2005) in *Metabolome Analyses – Strategies for Systems Biology* (eds S. Vaidyanathan, G.G. Harrigan, and R. Goodacre), Springer, New York, pp. 103–118.
8. Weckwerth, W. and Morgenthal, K. (2005) *Drug Discovery Today*, **10**, 1551–1558.
9. Wilson, I.D., Plumb, R., Granger, J., Major, H., Williams, R., and Lenz, E.M. (2005) *J. Chromatogr. B*, **817**, 67–76.
10. Gomase, V.S., Changbhale, S.S., Patil, S.A., and Kale, K.V. (2008) *Curr. Drug Metab.*, **9** (1), 89–98.
11. Ankley, G., Daston, G.P., Degitz, S.J., Denslow, N.D., Hoke, R.A., Kennedy, S.W., Miracle, A.L., Perkins, E.J., Snape, J., Tillitt, D.E., Tyler, C.R., and Versteeg, D. (2006) *Environ. Sci. Technol.*, **40** (13), 4055–4065.
12. Robertson, D.G. (2005) *Toxicol. Sci.*, **85** (2), 809–822.
13. Aoki Konya, Y., Takagaki, T., Umemura, K., Sogame, Y., Katsumata, T., and Komuro, S. (2011) *Rapid Commun. Mass Spectrom.*, **25** (13), 1847–1852.
14. Boudonk, K.J., Rose, D.J., Karoly, E.D., Lee, D.P., Lawton, K.A., and Lapinskas, P.J. (2009) *Bioanalysis*, **1** (9), 1645–1663.
15. Clark, J. and Haselden, J.N. (2008) *Toxicol. Pathol.*, **36**, 140–147.
16. Holmes, E., Nicholson, J.K., and Trante, G. (2001) *Chem. Res. Toxicol.*, **14**, 182–191.
17. Suter, L., Schroeder, S., Meyer, K., Gautier, J.C., Amberg, A., Wendt, M., Gmuender, H., Mally, A., Boitier, E., Ellinger-Ziegelbauer, H., Matheis, K., and Pfannkuch, F. (2011) *Toxicol. Appl. Pharmacol.*, **252** (2), 73–84.
18. Mellert, W., Kapp, M., Strauss, V., Wiemer, J., Kamp, H., Walk, T., Looser, R., Prokoudine, A., Fabian, E., Krennrich, G., Herold, M., and van Ravenzwaay, B. (2011) *Toxicol. Lett.*, **207** (2), 173–181.
19. Strauss, V., Wiemer, J., Leibold, E., Kamp, H., Walk, T., Mellert, W., Looser, R., Prokoudine, A., Fabian, E., Krennrich, G., Herold, M., and van Ravenzwaay, B. (2009) *Toxicol. Lett.*, **191**, 88–85.
20. Roessner, U., Wagner, C., Kopka, J., Trethewey, R.N., and Willmitzer, L. (2000) *Plant J.*, **23** (1), 131–142.

21. Yamada, H., Yamahara, A., Yasuda, S., Abe, M., Oguri, K., Fukushima, S., and Ikeda-Wada, S. (2002) *J. Anal. Toxicol.*, **26**, 17–22.
22. van Ravenzwaay, B., Herold, M., Kamp, H., Kapp, M.D., Fabian, E., Looser, R., Krennrich, G., Mellert, W., Prokoudine, A., Strauss, V., Walk, T., and Wiemer, J. (2012) *Mutat. Res.*, 26 January [Epub ahead of print]. PMID: 22305969.
23. Pitot, H.C. (1986) *Fundamentals of Oncology*, 3rd edn, Marcel Decker Inc., New York.
24. ECETOC (2010) Omics in (Eco)toxicology: case studies and risk assessment 22–23 February 2010, Malaga, ECETOC Workshop Report No. 19, ECETOC, Brussels.
25. Bilello, J.A. (2005) *Curr. Mol. Med.*, **5**, 39–52.
26. Craig, A., Sidaway, J., Holmes, E., Orton, T., Jackson, D., Rowlinson, R., Nickson, J., Tonge, R., Wilson, I., and Nicholson, J. (2006) *J. Proteome Res.*, **5** (7), 1586–1601.
27. Thomas, C.E. and Ganji, G. (2006) *Curr. Opin. Drug Discovery Dev.*, **9** (1), 92–100.
28. ECETOC (2008) Workshop on the Application of Omics Technologies in Toxicology and Ecotoxicology: Case Studies and Risk Assessment 6–7 December 2007, Malaga, Workshop Report No. 11, ECETOC, Brussels.

14
Safety Evaluation of New Pesticide Active Ingredients: Enquiry-Led Approach to Data Generation[1)]

Paul Parsons

14.1
Background

Over many years, the approach to the toxicological evaluation of new pesticide active ingredients has been to conduct a sequential series of studies aimed at meeting all of the various data requirements worldwide. Toxicologists have therefore traditionally been faced with the task of completing a "shopping list" of studies in order to meet global legislative requirements. The data requirements across different regions of the world are not fully harmonized, and this has resulted in a combined set of global data requirements that are larger than the requirements in any one given region. This has in turn resulted in the generation of a vast amount of toxicology data, much of which does not directly address the needs of risk assessors. While there have been many advances in the various scientific disciplines that underpin toxicology, these have not yet resulted in any significant changes in the toxicology testing paradigm.

Over the past two decades, the number of required studies has grown larger, as have the number of end-points covered by individual studies. During this period considerable effort has been devoted to trying to minimize the number of animals used in toxicology studies, and to modify the design of these studies in order to minimize the impact on the test animals. Thus, a tension exists between the wish to have more toxicological data in order to make more informed risk management decisions and, at the same time, the desire not to increase the number of animals used to evaluate safety. The challenge of striking a balance between these two conflicting objectives was evident during the development of the REACH (Registration, Evaluation, and Authorization and Restriction of Chemicals EC/1907/[2)]) regulations in

1) The views expressed in this chapter are those of the author and do not represent the views of Syngenta, Ltd.

2) Regulation (EC) No. 1907/2006 of the European Parliament and of the Council of 18 December 2006 Concerning the Registration, Evaluation, Authorization and Restriction of Chemicals (REACH), Establishing a European Chemicals Agency, Amending Directive 1999/45/EC and Repealing Council Regulation (EEC) No 793/93 and Commission Regulation (EC) No 1488/94 as well as Council Directive 76/769/EEC and Commission Directives 91/155/EEC, 93/67/EEC, 93/105/EC and 2000/21/EC.

Modern Methods in Crop Protection Research, First Edition.
Edited by Peter Jeschke, Wolfgang Krämer, Ulrich Schirmer, and Matthias Witschel.

Europe. Under REACH, emphasis is placed on considering opportunities to use quantitative structure–activity relationships (QSARs), read-across, study waivers based on lack of human exposure potential and data sharing; all of which are intended to reduce the number of animals used in toxicity testing. Nevertheless, the fact remains that many chemicals will be brought under the scope of the REACH regulations, and many of these will require some level of *in vivo* toxicity testing that may not have been conducted had REACH not been implemented. This illustrates the challenge facing risk assessors who require at least some minimum level of safety information in order to inform risk management decisions.

As mentioned above, there have been many advances in the different scientific disciplines that underpin toxicology, and some of these have challenged basic underlying principles such as dose–response. For example, it has long been assumed that a threshold does not exist for genotoxic carcinogenicity, and this assumption forms the basis of the United States Environmental Protection Agency (US EPA) linear extrapolation approach to cancer risk assessment [1]. However, it is now generally accepted that thresholds may exist for genotoxic carcinogens, as certain levels of DNA damage may be tolerated due to repair processes [2–4]. Another example of a challenge to the traditional dogma of dose–response relationships is the concept of hormesis. It has been proposed that some chemicals may exhibit a biphasic dose–response relationship in which a chemical exerts opposite effects dependent on the dose; this results in a so-called "U-shaped" dose–response curve [5].

A detailed consideration of the potential impact of emerging methods and technologies was published by the USA National Academy of Sciences National Research Council (NASNRC) in 2007, in a US EPA commissioned paper that was entitled "Toxicity Testing in the 21st Century: A Vision and a Strategy" [6]. This review focused heavily on the advances in molecular biology and biotechnology, and how these may be used in conjunction with computational toxicology and bioinformatics to make toxicity testing more relevant to low-level human exposures, to reduce the need for *in vivo* tests in animal models, and to make the whole process of hazard data generation quicker and less expensive. This led to the development of the US EPA's "Strategic Plan for Evaluating the Toxicity of Chemicals" and to the initiation of the US EPA's ToxCast™ program in 2007 [7]. The latter was intended to develop ways to predict the potential for toxicity in humans and to develop a cost-effective approach for prioritizing the thousands of chemicals that may require toxicity testing.

There has also been a recognition of the need to advance the science of risk assessment to bring greater focus on the assessment of the options for managing risk, rather than risk assessment itself being the end-goal. The evaluation of some high-volume commercially important chemicals (e.g., formaldehyde, trichloroethylene), undertaken in Europe and the USA, took many years to conclude. This was due primarily to difficulties in reaching a consensus on the interpretation of the available toxicology data and their translation into end-points for human health risk assessment and agreed positions on hazard classification. Consequently, there has been a call for improvements in uncertainty and variability analysis within risk assessments, as well as a greater consideration to planning, scoping, and problem formulation at the outset of the risk assessment process [8].

Aside from the focus on generating single-chemical hazard data, there is also now a greater recognition of the challenges posed to risk managers by the need to consider the potential impact arising from exposure to multiple chemicals and non-chemical agents (stressors) that may modulate an individual's ability to respond to such exposures [8–10]. This may occur through a single exposure to a defined mixture of chemicals, or through co-exposure to several chemicals through different pathways. The complexity of the dynamics of multiple chemical exposures presents significant challenges to risk assessors, both in terms of obtaining relevant hazard data and estimating the likelihood of co-exposure.

From the above, it is clear that there are both political and scientific drivers to change the existing toxicology testing paradigm, with the objective of generating data of greater relevance to low-level human exposures than is currently provided by the conduct of high-dose *in vivo* animal studies. The goal shared by most of the stakeholders is to move away from an over-reliance on an ever-increasing suite of complex animal studies of questionable relevance to humans. Ultimately, this may result in a total replacement of the existing *in vivo* experimental models by *in vitro* and *in silico* methodologies that allow for an integrated systems-based approach to evaluating the potential for perturbation of underlying homeostatic pathways. The transition to a new testing paradigm is likely to be a continuously evolving process, as confidence is gradually gained in the predictive ability of *in vitro* methods. In the meantime, it will be necessary to continue to evaluate the potential for chemicals and chemical mixtures to cause adverse effects on human health through the conduct of *in vivo* toxicity studies. However, even within the existing testing paradigm, there are opportunities to adopt a more focused approach to data generation rather than working systematically through lists of required studies.

In this chapter, consideration is given to an enquiry-lead approach towards data generation, where each study that is conducted has a clearly defined purpose within the context of human health risk assessment and risk management. The aim of such an approach is to seek to minimize the generation of toxicity data that have little or no relevance to the risk management of the chemical concerned. This is considered both from a qualitative perspective (what studies are needed and what end-points need to be measured?) and a quantitative perspective (on the basis of anticipated levels of human exposure, what dose ranges should be administered to experimental animals?). Another important perspective is that *in vivo* toxicity studies should not be the only source of information for making risk management decisions. Today, many tools are available that can be employed in order to assimilate the knowledge required to evaluate human safety, such as exposure modeling, structure–activity relationships (SARs), *in vitro* models to evaluate phenotypic and gene expression changes, computational toxicology, bioinformatics, and systems biology.

14.2 What Is the Purpose of Mammalian Toxicity Studies?

The extensive toxicological database generated for most pesticides is for the purpose of gaining a positive approval (i.e., a decision taken by a regulatory authority that a given pesticide product is safe to use under certain specified conditions). The decision as to whether or not an approval is acceptable from a human safety perspective depends mainly on the outcome of human health risk assessments, although there has recently been a shift towards a more hazard-based approach. For the development of a new pesticide active ingredient, it is necessary to generate a toxicological database that will meet the legislative requirements of the key regions in which the product is to be registered. Traditionally, this has meant addressing the various toxicity data requirements that exist in all the different regions of the world [10–12].[3)]

It is often said that the purpose of toxicity studies is to define the intrinsic toxicological properties of a chemical in order to provide reassurance regarding human safety, or to identify compounds that are viewed as unsafe. The judgment as to whether or not a product is safe should be made primarily on the basis of human health risk assessment, although increasingly the observation of certain toxicological hazards has also been used as a basis for regulation, irrespective of any consideration of the relevance of the dose levels causing toxicity in animals to the level of human exposure. Thus, the two main outputs from the mammalian toxicity database are reference doses for use in human health risk assessment and hazard classification. Acute toxicity and genotoxicity studies are conducted primarily for the purpose of hazard identification and contribute little to the selection of reference doses, whereas repeat dose studies and studies to address carcinogenicity and reproductive toxicity (fertility and development) are used primarily to derive reference doses, and also frequently contribute to the hazard classification of a substance.

While the focus of the pesticide toxicity data requirements is on the active ingredient, there has been an increasing need to evaluate environmental degradates ("metabolites"), particularly in relation to European guidance concerning the potential for soil metabolites to be present in groundwater [13].

Most of the mammalian toxicity test guidelines are designed in a manner that maximizes the ability of the study to elucidate the hazardous properties of a chemical. For example, acute inhalation toxicity studies require that the mass median aerodynamic diameter (MMAD) of the test material should be $\leq 4\ \mu m$, even though, in most cases, real-life exposures are to a form of material that has a much larger particle size distribution. Most test guidelines require the toxicologist to select dose levels with the intention of pushing the biological system in question to its limits, but stopping short of dosing at levels that would cause unnecessary suffering of the experimental animals. Consequently, the current testing paradigm frequently results in the generation of high-dose toxicity data that have questionable relevance to human health, especially when extrapolating to a prediction of effects

3) The guidelines related to the study reports for the registration application of pesticide Appendix to Director General Notification, No. 12-Nousan-8147, 24 November 2000, Agricultural Production Bureau, Ministry of Agriculture, Forestry and Fisheries of Japan.

that may occur after low-level human exposure. Dealing with this uncertainty is a challenge that toxicologists and risk assessors have had to address for many years.

A *hazard* can be defined as the potential for something to cause harm. When that something is a chemical, it is often said that the hazard reflects the intrinsic properties of the chemical. Toxicologists frequently describe the effects observed in toxicity studies as representing the intrinsic hazardous properties of the test substance. Consequently, a chemical may be described as being "carcinogenic" or as being a "reproductive toxicant," with such terms being used to convey its intrinsic hazardous properties. Traditionally, *risk* has been defined as a function of hazard and exposure with hazard identification (what adverse effects does a chemical cause in experimental animals?) and hazard characterization (what is the dose response for the adverse effects observed in experimental animals?) representing a starting point in the human health risk assessment process (Figure 14.1). However it can be argued that, rather than starting the risk assessment process with the generation of mammalian toxicity data, the starting point should be an evaluation of the exposure potential for the chemical concerned. Having developed a understanding of the magnitude and frequency of exposure in relation to the proposed uses of the chemical, risk assessors and toxicologists would then be in a position to define the minimum data needs to evaluate the potential risks to human health and would have an insight in to the relevant dose-range top evaluate.

When considering the definition of the word "intrinsic" ("belonging to a thing by its very nature"; "of or relating to the essential nature of a thing; inherent"), it is clear that this term does not apply to the hazardous properties of a chemical as defined by observations made in mammalian toxicity studies. This is because the intrinsic properties of a chemical are something that belong to its very nature and, importantly, are independent of external circumstances. Such properties may be described as "essential" or "inherent," and include items such as molecular mass, melting point, charge, density, pK_a, and volatility. These properties are independent of how much of the material is present, and also independent of

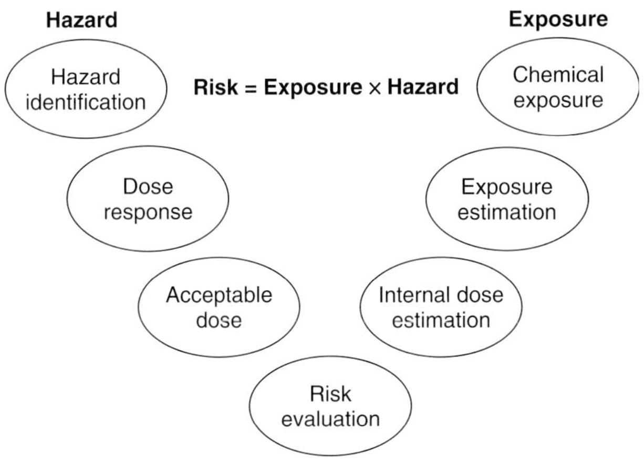

Figure 14.1 Traditional risk assessment paradigm.

the form of the material, as they depend mainly on chemical composition and structure. Thus, when talking about the hazardous properties of a chemical in terms of the responses observed in toxicity studies, what is actually being described are its extrinsic properties under defined experimental conditions. A property that is not essential or inherent is termed an *extrinsic property*. For example, mass is an intrinsic property of any physical object, whereas weight is an extrinsic property that varies depending on the strength of the gravitational field in which the respective object is placed.

These extrinsic properties will vary depending on the nature of the environment into which the substance is placed, and also the tools and techniques that are employed to observe the interaction between that substance and the experimental environment. The misconception that a chemical may possess intrinsic toxicological properties is widely engrained within the various regulatory frameworks for chemical regulation. For example, one of the stated aims of REACH is " ... to improve the protection of human health and the environment through the better and earlier identification of the intrinsic properties of chemical substances" (*http://ec.europa.eu/environment/chemicals/reach/reach_intro.htm*).

The empirical observations made during the conduct of toxicity studies – be they in-life observations, observations made following analysis of biological samples collected in-life or at necropsy – do not define the intrinsic properties of a chemical. Rather, these observations are a function of the experimental model and design, the nature of the interaction between test substance, and the various biological systems present within the experimental model and the sensitivity and specificity of the different methodologies employed to make observations throughout the experiment. The findings of a toxicity study should therefore be considered as a defined set of observations made under specified experimental conditions, and should not be considered to represent the intrinsic properties of the test article. The fundamental point being that, in contrast to the intrinsic properties of a chemical which are constant and do not change, the observations made in toxicity studies will vary from study to study, even when the design of separate studies is identical and they are conducted in the same laboratory. This variability is one of the key challenges for risk assessors and risk managers to address when faced with making judgments on human safety based on observations made in toxicity studies that have been conducted in experimental animals.

It is clear that a chemical's toxicological potential is described by the various observations that are made in different experimental models, and there are many variables which can influence the outcome of individual studies (Figure 14.2). Some of these variables are intentionally controlled (e.g., species, dose, vehicle, strain, number of dose groups), while others are more difficult to control (e.g., composition of diet and water, handling of the animals during dosing, impact of bedding and housing conditions, processing, storage, and analysis of biological tissues). All, however, may have an impact on the outcome of a study. Of critical importance is the judgment as to whether or not the observations made in toxicity studies would be reproduced in humans under defined conditions of exposure. It is the interaction between the conditions of exposure (frequency, pattern, and

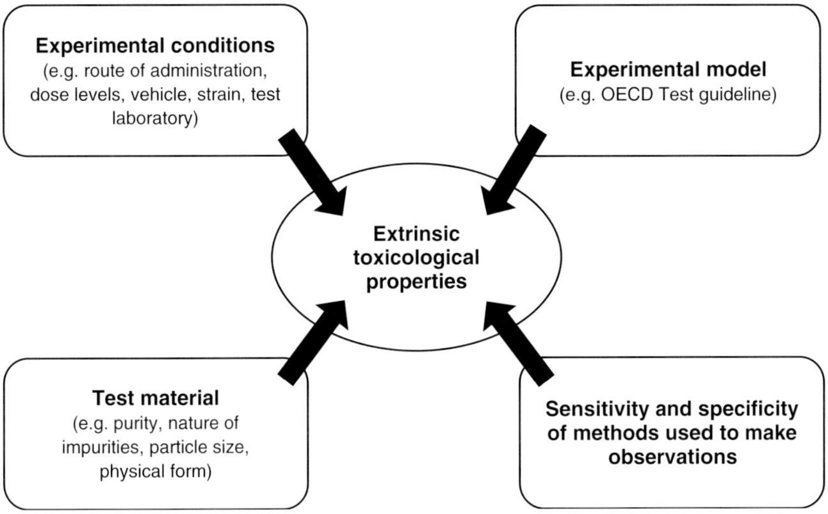

Figure 14.2 Factors influencing the extrinsic toxicological properties of a chemical.

magnitude) and the characteristics of the biological system which is exposed that will define the likelihood of observing an adverse health effect, and not the intrinsic hazardous properties of a chemical. For this reason, it is extremely difficult to make judgments on the human safety of a chemical based solely on the consideration of hazard, as large margins of safety may exist for chemicals causing toxicity only at very high dose levels (see Section 14.4.1).

The purpose of toxicity studies is, therefore, to provide a description of the observed toxic responses under the conditions within which a chemical has been evaluated. Typically, toxicity studies use three dose levels:

- The low dose is selected on the basis that it will produce no adverse toxicological effects (no observed adverse effect level; NOAEL).
- The highest dose is selected to cause observable adverse toxic effects.
- The mid-dose is selected to cause some intermediate level of toxicity in order to help describe the dose–response for the observed effects.

Collectively, this information may be used to describe the potency of a chemical, although as there are often only two dose levels causing observable effects, the dose–response is usually poorly described. Typically, the NOAELs are used as a basis for the derivation of reference doses for human health risk assessment. In addition, the qualitative nature of the response and, in some cases, the potency are evaluated against specific criteria in order to assign a hazard classification. However, it should be clear from what is outlined above that hazard classification does not represent a description of the intrinsic hazardous properties of a chemical. Rather, it is a framework within which chemicals can be categorized based on the observations made under the various experimental conditions used in toxicity studies. Hazard classification should therefore be viewed as something that occurs as a consequence of generating mammalian toxicity data rather than the reason for generating the data in the first place. Indeed, the same toxicological database can result in different hazard classification outcomes depending on the specific classification criteria applied in different regions of the world. Therefore, it follows that the primary reason for conducting mammalian toxicity studies on pesticides is to provide the necessary data to undertake the human health risk assessments that provide a basis for either allowing or refusing the approval of a product.

To reiterate, chemicals do not possess intrinsic toxicological properties. Rather, their toxicity is defined by the experimental models in which they are tested, the design and conditions under which the tests are conducted, the nature of the interaction between the test substance and the various biological systems present within the experimental model, and the sensitivity and specificity of the various methodologies employed to make observations throughout the experiment.

14.3 Addressing the Knowledge Needs of Risk Assessors

It is beyond the scope of this chapter to provide a detailed commentary on the numerous methods and approaches to human health risk assessment. However, given that mammalian toxicity studies are conducted primarily for the purpose of human health risk assessment, those studies that are undertaken – and their design– should be heavily influenced by the information needs of the risk assessors. These needs will vary from one region to another, depending on the risk assessment methodologies employed by different regulatory authorities. The different knowledge needs of risk assessors must be taken into account when developing toxicology study programs, and also in the design of individual toxicity studies. Ultimately, the regulators will need to be satisfied that their protection goals can be achieved, and that the chemical concerned can be used with reasonable certainty of causing no harm (e.g., the US Food Quality Protection Act 1996; [14]). However, it is clear that the initial judgments on human safety and acceptability of risk will be made by the companies who develop the chemicals. If a company is not convinced that a particular product can be used safely, then it will not be developed for commercialization.

The approach that has typically been taken is to produce a consolidated list of all required toxicity studies based on the various data requirements worldwide (as per Table 14.1), and then to systematically conduct all of the studies on the list according to relevant international test guidelines. This approach can be viewed as reassuring from the perspective of generating a globally compliant database. However, the approach is intensive in terms of animal use, a long time is required to complete the full package, it is expensive, and a large proportion of the data obtained will not be used to derive end-points for risk assessment. Consequently, several initiatives have considered alternative testing paradigms (e.g., the tiered testing approach proposed by ILSI-HESI; [15]) that follow an approach which is more focused on the data needs of risk assessors. There has also been a move towards using more integrated study designs, with the aim of obtaining more information from any

Table 14.1 Summary of mammalian toxicology studies typically conducted for the registration of a new pesticide active ingredient.

Toxicity study	OECD test guideline	Toxicity study	OECD test guideline
Acute oral	425	Dog maximum tolerated dose	None
Acute dermal	402	28-day dog	None
Acute inhalation	403	90-day dog	409
Skin irritation	404	1-year dog	452
Eye irritation	405	Preliminary rat developmental toxicity	None
Skin sensitization	429	Rat developmental toxicity	414
Bacterial mutation	471	Preliminary rabbit developmental toxicity	None
Mammalian cell gene mutation	476	Rabbit developmental toxicity	414
Mammalian cell clastogenicity	473	Preliminary rat reproductive toxicity	None
In vivo micronucleus	474	Rat reproductive toxicity	416
In vivo unscheduled DNA synthesis	486	Immunotoxicity	US EPA OPPTS 870.8700
28-day rat	407	Preliminary acute neurotoxicity	None
90-day rat	408	Acute neurotoxicity	424
1-year rat	453	28-day preliminary neurotoxicity	None
2-year rat	453	Sub-chronic neurotoxicity	424
28-day mouse	None	Preliminary developmental neurotoxicity	None
90-day mouse	None	Developmental neurotoxicity	426
80-week mouse	453	Toxicokinetics	417

given study, thus reducing any redundant testing. For example, an acute dermal toxicity study does not provide any data for use in human health risk assessment, and also has a limited value for hazard classification, especially if an acute oral toxicity study has already been conducted on that material [16]. Another example of a move towards avoiding the conduct of studies that add little incremental knowledge to the evaluation of a chemical has been the decision by the European Commission and US EPA to no longer require the conduct of a one-year study in dogs, as this is no longer considered scientifically justifiable [17]. The modification of standard toxicity study designs in order to obtain data across a broader range of end-points is another approach to modifying the traditional testing paradigm, in a manner that reduces the overall number of animals required in a testing program. For example, it is possible to incorporate multiple end-points such as neurotoxicity, immunotoxicity, toxicokinetics, and mode of action (MoA) studies into a standard OECD 408 repeat dose study in the rat [18].

Clearly, toxicologists should avoid conducting studies that will have little or no impact on the regulation of a chemical. However, the legitimate concern is that by not conducting such studies, they could jeopardize the ability to successfully register the compound concerned as the database may be viewed as incomplete. Thus, greater flexibility is required in the generation of the rationale behind the scope of a particular toxicology data package in order to allow the toxicologist to develop a scientifically robust database that meets the needs of risk assessors and risk managers. Such flexibility will demand the ability to conduct studies, the design of which may not be fully compliant with international test guidelines, as well as the ability to develop toxicity data packages that may not conform rigidly to the required studies listed in the various data requirements. The aim should be to provide the best available information in order to make a sound scientific judgment as to whether or not the substance in question can be used with reasonable certainty of no harm. This type of flexibility is evident within the safety testing philosophy for pharmaceuticals, where specific guidance is given on the need for certain studies, based primarily on the projected clinical use and predicted pattern of human exposure. For example, carcinogenicity studies are recognized to be time-consuming and resource-intensive, and should only be performed when human exposure warrants the need for information from life-time studies in animals in order to assess a material's carcinogenic potential (see International Conference on Harmonization; ICH [19]).

Given the option to develop data-generation strategies that are more focused on the needs of risk assessors and risk managers, the design and conduct of toxicology studies can be focused on addressing specific questions of relevance to the regulation of a chemical. Consequently, risk management decisions will be made in the knowledge that only relevant toxicity data are generated, thereby avoiding the conduct of unnecessary studies that will have no impact on regulation. Within the context of an individual toxicity end-point – for example, carcinogenicity – the toxicologist must consider what information is needed in order to evaluate whether or not the product concerned could pose a carcinogenic risk to humans under the proposed conditions of use. The key output of such an assessment is an expert judgment on the carcinogenic risk posed to humans. However, until such time as the outcome of the rodent carcinogenicity studies is known, toxicologists must assimilate all available information in order to predict the risk of carcinogenicity for humans. These initial predictions will be based mainly on SARs and expert judgment, and will utilize the ever-increasing knowledge database relating to the underlying biological mechanisms of human carcinogenicity. Over time, however, as more data are developed for a substance under evaluation, the initial predictions can be refined, and the process of data development may be visualized as a "knowledge pyramid" whereby there is an incremental increase in the available knowledge with time. During this time, a variety of tools may be employed to generate the information required to evaluate any carcinogenic risk to humans (Figure 14.3).

For new pesticide active ingredients that are to be registered worldwide, the current toxicity testing paradigm requires the conduct of life-time carcinogenicity studies in rodents. In the past, there has been some debate as to the utility of

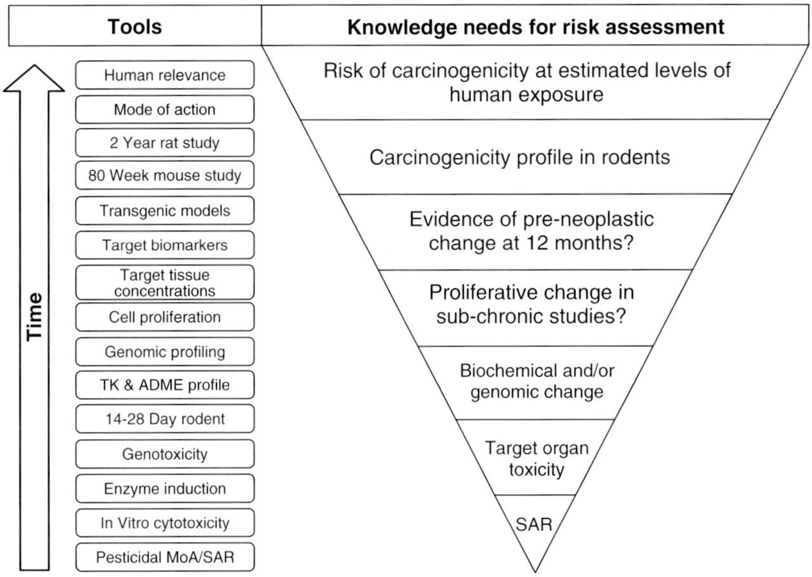

Figure 14.3 "Knowledge pyramid" for carcinogenicity.

conducting a mouse carcinogenicity study in addition to a rat study [20, 21]; however, both assays are still performed routinely, based on the understanding that they are required for a global registration. It is likely that, as data requirements are updated, there will be a greater flexibility with regard to the use of alternatives to long-term *in vivo* bioassays.

Both, the current scientific understanding of the underlying biological mechanisms of carcinogenicity, and the tools available to elucidate such mechanisms, have made major advances during recent years, particularly with regards to the use of genomic technologies for evaluating changes in gene expression. The mapping of the human genome, together with focused studies of the genomic changes that underlie key human diseases such as cancer, will lead to new techniques and approaches for screening chemicals, the aim being to determine their potential for perturbing key homeostatic pathways. For example, by combining high-throughput mutation detection techniques with knowledge of the human genome sequence, it should become possible to identify somatically acquired sequence variants/mutations, and hence to identify genes that are critical in the development of human cancers (see the Wellcome Trust Cancer Genome Project [22]).

The ability to use bioinformatics to map and relate multiple changes in homeostatic pathways, and to anchor these to phenotypically observed adverse outcomes in traditional toxicology studies, provides toxicologists with new insights into the sequence of changes that may precede the observation of an adverse effect. In future, this should allow toxicologists to describe detailed modes of action for not only carcinogenic but also non-carcinogenic end-points. Yet, this should not be viewed as an ability to elucidate "pathways of toxicity," as arguably such pathways do not exist and what is being observed is in fact a perturbation of the underlying homeostatic and physiological mechanisms which, ultimately, leads to an adverse health outcome. Nonetheless, it is hoped that by acquiring a greater understanding of the relationships between key biological changes in homeostatic pathways observed early in the sequence of events leading to a carcinogenic outcome, it will no longer be necessary to conduct life-time rodent carcinogenicity studies. Indeed, evidence from toxicogenomic studies has suggested that the dose response for molecular changes in key pathways and the traditional toxicity end-point are very similar [23].

From the perspective of the risk assessor, an expert scientific judgment on the carcinogenic potential of a substance is a key factor in understanding the risks posed by long-term exposure to a compound. As illustrated in Figure 14.3, toxicologists have a variety of tools at their disposal in order to evaluate the carcinogenic potential of a substance. Whereas, the key questions faced by risk assessors have remained fairly constant over time, the tools available to toxicologists are constantly changing. Yet, the pace of change – particularly with respect to the development of harmonized international test guidelines and revised data requirements – can be very slow, taking many years. It is clear that the shared goal in toxicology is to move to a testing paradigm that provides information that is directly relevant to human responses, but at realistic human exposure levels. Ideally, this would no longer involve *in vivo* testing in experimental animals but, at the same time, would provide risk assessors with scientifically robust information for them to be able to evaluate any risks to human health. Therefore, in future the aspiration is that the left-hand side of Figure 14.3 will depict a set of tools for data generation ranging from *in silico* predictive models and short-term *in vitro* studies to more involved methodologies for integrating diverse sets of complex data using systems biology approaches [24].

14.4 Opportunities for Generating Data of Direct Relevance to Human Health Risk Assessment within the Existing Testing Paradigm

14.4.1 Dose Selection for Carcinogenicity Studies

Continuing with the example of the end-point of carcinogenicity, risk assessors are concerned with the question of whether the chemical has the potential to cause carcinogenicity at dose levels to which humans may be exposed. In contrast, the current application of the test guidelines for evaluating carcinogenicity (OECD 453 [25]; OPPTS 870:4200 and 4300 [26, 27]) is more focused on the question of whether the chemical has any potential to cause carcinogenicity at the highest dose which can be tolerated by the test animals over the study duration, regardless of the level of human exposure. The disconnect between these two approaches presents a challenge to risk assessors and risk managers when considering the relevance of high-dose threshold-based tumors observed in animal studies to low-dose exposures in humans. Furthermore, although the primary purpose of rodent carcinogenicity studies is to generate data for human health risk assessment, these studies are also used to make a judgment on cancer hazard classification. The test guidelines address both considerations from a study design perspective, although this can create conflicting objectives with regard to dose level selection (i.e. low dose risks versus high dose hazards).

For compounds which exhibit low mammalian toxicity, the requirement to conduct carcinogenicity studies at a limit dose (1000 mg kg^{-1} day^{-1}; roughly equivalent to a 20 000 ppm dietary inclusion level) is somewhat paradoxical, in that there is a very real risk of causing tumors due to an overload of physiological and homeostatic processes. Thus, lower-toxicity pesticides, which are judged to have high margins of safety in human health risk assessments, may be at an increased risk of causing tumors in rodent carcinogenicity studies because they can be administered at very high dose levels. As outlined above, the utility of highlighting such high-dose hazards through classification and labeling is questionable, as this does not reflect an intrinsic property of the chemical, but rather is a consequence of the design of the study in which it has been tested. The combination of a toxicology testing paradigm that maximizes the ability to detect hazard with a classification and labeling system that takes no account of toxicological potency means that it can be very difficult to convey an objective view on the risks posed to human health for a classified product.

A clear opportunity for refinement of the current testing paradigm exists through the use of doses that represent multiples of human exposure in long-term carcinogenicity studies conducted with pesticides. Although the magnitude of estimated human exposure may be a point of debate, a wealth of information is available concerning the residue levels of pesticides in raw agricultural commodities and in processed foodstuffs. It is therefore possible to make sound scientific judgments as to the approximate order of magnitude of predicted human exposure within the context of dose levels that may potentially be used in carcinogenicity studies. By way of illustration, Table 14.2 presents the highest estimated daily chronic intake for a range of agrochemicals currently registered in Europe. The values used for these estimates are supervised trials median residue (STMR) data. Although often still considered as conservative estimates of chronic human exposure, the STMRs provide a more realistic indication of exposure potential than maximum residue levels (MRLs). The compounds were selected on the basis that chronic consumer dietary risk assessments were available that had been refined through the use of STMR data, and a range of compounds was chosen to represent different target pests and a range of crop uses. The intention of Table 14.2 is to provide an illustration of a typical range of chronic exposure values, to enable these to be compared to the limit dose for carcinogenicity testing of 1000 mg kg^{-1} day^{-1}, as specified in OECD 453 [25].

It can be seen from the examples in Table 14.2 that estimated chronic daily human exposure, as defined by the population giving the worst-case (highest) value, ranges from 0.00097 to 0.01047 mg kg^{-1} day^{-1}. Thus, the limit dose for carcinogenicity studies is approximately 95 000-fold higher than the highest estimated chronic daily

Table 14.2 Typical estimates of chronic consumer exposure.

Compound	Indication	Worst-case subpopulation	Highest estimated human intake ($mg\,kg^{-1}\,day^{-1}$)	Ratio of carcinogenicity study limit dose to highest chronic exposure estimate
A	Fungicide	UK infant	0.00121	826 446
B	Fungicide	WHO Cluster B	0.00295	338 983
C	Fungicide	WHO Cluster B	0.01029	97 181
D	Fungicide	German child	0.01047	95 510
E	Insecticide	French infant	0.00097	1 030 927
F	Insecticide	Dutch child	0.00208	480 769
G	Insecticide	German child	0.00473	211 416
H	Insecticide	Irish adult	0.00356	280 898
I	Herbicide	UK toddler	0.00510	196 078

human exposure. Hence, when considering a hypothetical (but not unrealistic) scenario of a two-year rat carcinogenicity study conducted with a low-toxicity pesticide at doses of 10, 100, and 1000 $mg\,kg^{-1}\,day^{-1}$ (broadly equivalent to dietary inclusion levels of 200, 2000, and 20 000 ppm), how relevant would a non-genotoxic carcinogenic response observed at the top dose be to the human exposure scenario? If it is assumed that the top dose caused liver tumors with no established MoA, the mid-dose caused non-neoplastic findings in the liver which are considered toxicologically adverse, and the low dose was a no observed effect level (NOEL), the substance would, in all likelihood, be considered to pose a carcinogenic "hazard" to humans and would be assigned a cancer hazard classification on this basis, despite the fact that there is a very large margin of safety between estimated human exposure and the carcinogenic dose (Figure 14.4). This reflects the fact that most of the existing criteria for the hazard classification of substances do not consider carcinogenic potency. It could be argued that the toxicologist should be obliged to elucidate the MoA for these high-dose rat liver tumors, and to establish its relevance to humans. However, given the fact that human exposure is estimated to be tens of thousands of times lower than the carcinogenic dose, the utility of generating further data at dose levels that have no relevance to human exposure must be questioned, especially if this were to involve additional *in vivo* animal testing.

A more appropriate basis for dose selection would be to use dose levels that are multiples of the highest estimated human exposure, such as 10-, 100-, and 1000-fold. Based on the illustration in Table 14.2, it would be proposed to conduct the hypothetical carcinogenicity study at doses of 0.1, 1.0, and 10 $mg\,kg^{-1}\,day^{-1}$ (equivalent to dietary inclusion levels of 2, 20, and 200 ppm, assuming that 20 ppm equates to 1 $mg\,kg^{-1}\,day^{-1}$).

As a greater understanding of human biology and the relationship between genomic and phenotypically observed changes in physiological and homeostatic

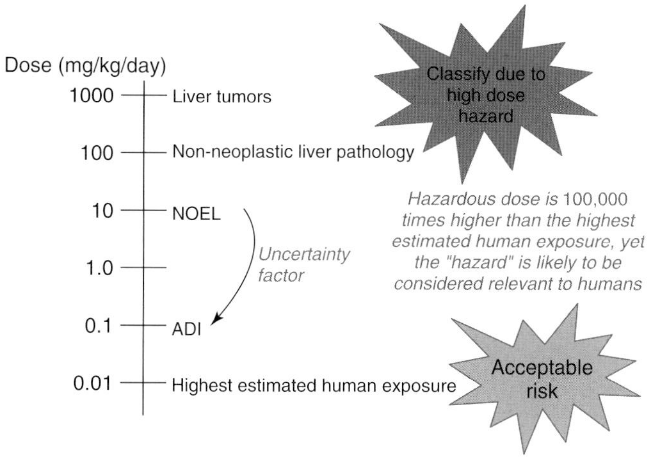

Figure 14.4 Illustration of hazard classification of low-risk substance.

pathways of regulation leading to adverse health outcomes is developed, it is to be expected that the need for long-term high-dose *in vivo* animal studies will diminish. It may then be possible to establish reference doses for human health risk assessment based on systemic concentrations of toxicant that have been shown not to perturb any key homeostatic pathways. In order to achieve this, however, it will be essential to be able differentiate genomic and biochemical changes that occur as part of the normal physiological process of handling exogenous agents (e.g., consuming a meal) from those that represent an irreversible change that is part of a known human disease process.

14.4.2
Integrating Toxicokinetics into Toxicity Study Designs

The purpose of toxicity studies is to provide a description of the observed toxic responses under the conditions within which a chemical has been evaluated, in order to make a judgment on the likelihood of such effects occurring in exposed humans. The guidelines for these studies dictate that a certain level of toxicity should be demonstrated through attempts to maximize systemic exposure. *Toxicokinetics* can be used to describe the pattern and magnitude of systemic exposure that occurs during the conduct of toxicity studies and allows an understanding of the exposure–response relationships to be established, which in-turn can be used to compare and contrast different outcomes across species. This enables points of departure to be described at a systemic level, and may also allow the identification of a perturbation in homeostatic and physiological mechanisms due to underlying toxicity. Whilst aiding the interpretation of the outcome of toxicity studies is valuable in itself, toxicokinetics can also be helpful for setting the dose levels for any subsequent studies.

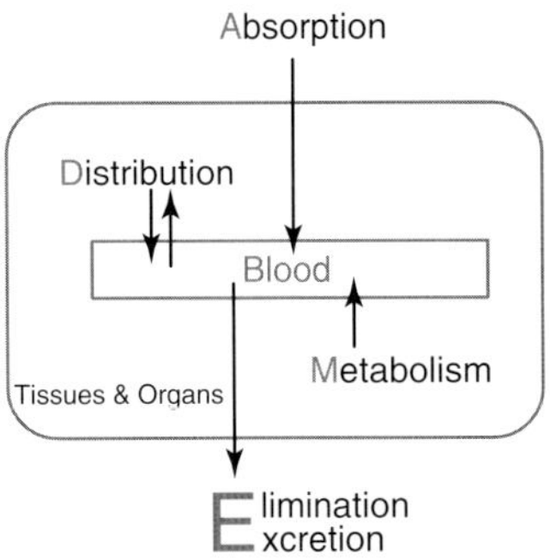

Figure 14.5 Illustration of the ADME process.

Systemic exposure to an active ingredient is determined by monitoring how the concentration in blood/plasma changes with time. Following its oral administration, the systemic exposure of a chemical is determined by the absorption, distribution, metabolism, and excretion (ADME) of the compound within the animal. The ADME process provides a description of how the chemical enters the body, is distributed via the blood throughout the body, and is then metabolized and excreted (Figure 14.5).

In addition to the use of toxicokinetics as part of *in vivo* studies, it is also possible to use *in vitro* approaches to define certain pharmacokinetic parameters such as hepatic clearance, plasma protein binding and the ability to be transported across cells derived from the gastrointestinal tract (i.e., predicted oral absorption). Data from *in vitro* studies can be incorporated into the overall evaluation of a chemical in three main ways:

- The use of cross-species (including human) *in vitro* metabolism studies to aid in the selection of toxicity species most relevant to humans.
- In the absence of *in vivo* toxicokinetic data, *in vitro* approaches and modeling can be used to predict systemic exposure in experimental animals.
- All available *in vivo* kinetic data can be incorporated with *in vitro* data, and can be used to model systemic exposure under a scenario or at dose levels that were not tested.

Taken together with a knowledge of the physico-chemical properties of a substance, these *in-vitro* data complement the pharmacokinetic and mass balance data obtained from regulatory ADME studies (OECD 417, [28]), allowing the metabolic fate and disposition of a substance to be fully described.

A key challenge when considering the incorporation of toxicokinetics into toxicity study designs is to ensure that the purpose of the study with regards to toxicological evaluation is not compromised, and to avoid the use of excessive animal numbers. However, modern analytical methods allow a great deal to be achieved with small blood volumes such that, with careful design and consideration of sampling time-points in relation to the evaluation of key toxicity end-points (e.g., functional observation battery in 28- and 90-day rodent studies), it is possible to obtain meaningful toxicokinetic data without affecting the toxicity component of the study [18, 29–32].

Another key decision point is to identify exactly which components should be analyzed in the collected samples. The initial focus may involve measurements of the parent compound in whole blood samples, although if the substance under evaluation is known to be rapidly metabolized then other analyses – including early metabolite profile assessments – may be more appropriate. Where repeat dosing studies are carried out, the opportunity to obtain information on the effect of time and dose on potential induction and accumulation can be obtained by determining the systemic exposure after the first dose, and again after several doses when it is likely that steady-state conditions have been achieved. At steady-state, the full extent of any induction and/or accumulation resulting from the dosing regimen employed can be characterized. The time to reach steady-state is determined by the half-life of the compound (effectively after at least three times the half-life), with any assessment of whether steady-state is likely to have been achieved being based on data obtained from previously conducted pharmacokinetic studies. If such information is not available, however, the estimate of systemic exposure obtained on study completion can be assessed against what would be expected through modeling and simulation based on the first dose.

Toxicokinetics may be of particular value when considering dose selection for carcinogenicity studies. Many of the underlying processes that define the toxicokinetic profile of a chemical (i.e., absorption, distribution, metabolic activation or deactivation, and excretion) can become saturated at high dose levels, such that the qualitative and/or quantitative nature of exposure in an animal model may not be representative of that in humans exposed to considerably lower dose levels. Of particular importance here is the need to understand whether the kinetics are linear or nonlinear within the tolerable dose-range. Departures from linearity may be indicative of the saturation of one or more of the underlying kinetic processes, and may represent a perturbation of homeostasis of the underlying physiology (Draft guidance within OECD 116; [32]). While it may be possible to further increase systemic exposure by using higher dose levels, the relevance of data generated in the range of perturbed kinetics is questionable. In cases where the saturation of systemic exposure can be demonstrated, such that any further increase in the administered dose has little or no further impact on the systemic exposure, then the toxicokinetics should be considered as the primary dose-limiting factor that influences selection of the top dose, even in the absence of any observable toxicity [33, 34].

14.5 Enquiry-Led Data Generation Strategies

Within the existing toxicology testing paradigm, opportunities exist for data development strategies that focus on the key questions that must be addressed by risk assessors in order to make decisions on the ability of a product to be used safely. While there is still the reality of the need to meet formal data requirements with studies that conform to international test guidelines, the design and sequence of toxicity studies can be influenced significantly by maintaining a focus on the specific purpose of each study, and the use to which the data will be put. Moreover, by careful use of integrated study designs, there is opportunity for the refinement of experimental procedures and a reduction in overall animal numbers. This approach should minimize or avoid the conduct of studies that have no impact on the regulation of a chemical.

During the product development life cycle, the initial emphasis is on the selection of a suitable candidate molecule from many other similar molecules, in order that this can be taken forward to full development. During the initial phases of development, when compounds are evaluated for their potential to become candidates for full development, the molecules under consideration will be relatively data-poor. Consequently, in order to make judgments on human safety it will be necessary to use predictive models and expert judgment. However, even at these early stages there will be key questions to address, and a variety of tools and approaches are available for this purpose. It is important to make the best use of the existing information for the substance itself, and of any information that may exist for structurally similar molecules or different structures with similar pesticidal modes of action. The incremental assimilation of knowledge in response to questions focused on the needs of risk assessors provides the opportunity to design a focused program of toxicity studies using integrated study designs, with the aim of providing a scientifically robust data package to evaluate if there is reasonable certainty of no harm. The challenge with such an approach is to recognize when there is sufficient information to make a reliable risk management decision, such that no further toxicity testing is required. In many respects, this challenge could be met more readily if there was a more cooperative approach towards data development between regulators and industry. The REACH regulation goes some way towards this in requiring registrants to submit testing proposals for higher-tier studies before they are initiated. A more open dialogue between data developers and data reviewers during the data development process may facilitate a greater acceptance of alternative and innovative approaches. Rather than industry submitting a full new data package that the regulators have not previously seen, opportunities to discuss data-development strategies before key studies are initiated should be encouraged.

The key questions to address pertaining to the ability of a substance to be used with reasonable certainty of no harm will be similar throughout the development life cycle. However, as more knowledge is assimilated for a specific substance, the focus of the questions will shift from one of prediction to one of data evaluation and expert judgment. For example, given only a structure and a view on the probable pesticidal MoA, it is likely that only very tentative predictions can be made about the carcinogenic potential of the substance. However, with knowledge of the predicted pattern and magnitude of exposure, gentoxicity profile, *in vitro* gene expression data, and some short-term toxicity data in rodents, a more informed judgment will be possible. The following example of an enquiry-led approach to data generation therefore considers examples of key questions at different stages in the development process and the tools that may be available to address them.

As mentioned earlier, the currently available tools for a full toxicology development program depend predominantly on *in vivo* toxicity studies in experimental animals. However, as alternative approaches to evaluating key toxicity end-points are developed and accepted, there will be a significant shift towards the use of *in vitro* and *in silico* methodologies. Together with the use of systems biology-based approaches and a rapidly growing knowledge of human disease processes, it be possible to predict the downstream consequences of the genomic, biochemical, and physiological changes observed *in vitro* in terms of their potential to adversely impact homeostatic pathways *in vivo*. It should also needs to be recognized that data-generation strategies should be heavily influenced by exposure potential. By understanding the pattern and magnitude of predicted human exposure, the appropriate duration of dosing and relevant dose levels can be used for the toxicity evaluation.

14.5.1
Key Questions to Consider While Identifying Lead Molecules

At this early stage of development, it is likely that a series of candidate molecules will be under evaluation. The available knowledge for the prediction of adverse health outcomes in humans will be limited, and the focus will be on SARs and a judgment on the likelihood of the pesticidal MoA being expressed in mammals (Table 14.3).

While it may be difficult to predict the likelihood of specific outcomes at this stage in the development of a compound, it is possible to evaluate a number of key factors that may contribute to significant toxicity findings. Broadly speaking, there are three ways in which a pesticide active ingredient may cause toxicity (Table 14.4):

- Toxicity may occur as a consequence of the pesticidal MoA being expressed in mammals; this is referred to as *Type 1 toxicity*. Examples of chemicals causing Type 1 toxicity are the organophosphorus and pyrethroid insecticides.
- The chemical properties of the molecule or biological properties other than its pesticidal MoA may cause toxicity; this is known as *Type 2 toxicity*. Many compounds will cause Type 2 toxicities, but these may be very difficult to predict unless the compound concerned has a close structural similarity to compounds that have been more fully evaluated.
- Finally, toxicity may occur as a consequence of physiological overload due to the use of very high dose levels; this is referred to as *Type 3 toxicity*. It is not possible to predict Type 3 toxicities based on knowledge of the structure of the molecule, or its pesticidal MoA. An example of Type 3 toxicity would be the observation of tumors only at the highest dose in carcinogenicity studies conducted at dose levels producing excessive levels of toxicity.

Table 14.3 Key questions while identifying lead molecules.

Identifying lead molecule	
Question	**Tools and options**
What is the pesticidal MoA?	SAR; *in vitro* assays to elucidate biochemical targets, electrophysiology
What is the potential for the pesticidal MoA to be expressed in mammals, and what would be the consequences?	SAR, knowledge of selectivity between pest target and toxicity testing species and humans; *in vitro* screens against potential mammalian targets; *in vitro* ADME to evaluate the potential for targets to be reached
Does the chemical structure indicate any alerts for potential mammalian toxicity and/or genotoxicity?	SAR; analog database searches; *in vitro* biochemical screens; *in vitro* genotoxicity testing; predictive modeling
What do we know about structurally related compounds or compounds with a similar MoA?	Internal company databases; open literature; regulatory views of related compounds
Can the relationship between structure and predicted toxicity be modulated? Does the potential for toxicity reside in the same moiety as that for efficacy?	Design-out potentially toxic moieties and retain efficacy
Given the intended use profile, what can be predicted about the potential for human exposure?	Obtain early view on key crops, application rate, frequency of application, method of application, formulation types. Undertake initial evaluation of exposure potential (magnitude and frequency) and consider this is dose selection
What can be predicted about metabolism in different mammalian toxicity testing species, livestock (hen and goat) and humans? Which mammalian toxicity testing species is the most relevant to humans?	SAR; comparative *in vitro* metabolism studies in rat, dog, mouse, rabbit, goat, hen, and human
What can be predicted about kinetics in the various species used to evaluate mammalian toxicity compared to humans?	*In vitro* evaluation of metabolism; hepatic clearance; blood protein binding; potential for oral absorption

14.5.2
Key Questions to Consider When Selecting Candidates for Full Development

As new active ingredients progress through the development life cycle, lead compounds are identified that are candidates for full development. This is often referred to as the *optimization phase* of a project. A key objective at this stage is

Table 14.4 Possible types of toxicity for pesticide active ingredients.

Toxicity type	Cause of toxicity	Can this be predicted from mode of action	Can this be predicted by SAR?
1	Expression of the pesticidal MoA in mammals	Yes	Sometimes based on known *in vitro/in vivo* correlations
2	Chemical properties of the molecule or biological activity other than the pesticide MoA	Can be difficult, but possible for known reactive species or for site of contact toxicity	Sometimes (e.g., genotoxicity), but can be difficult for other end-points (e.g., developmental toxicity)
3	Effects seen only at high doses in *in vivo* studies for compounds which otherwise have a low toxicity	No	No

to identify any human health concerns that may have the potential to preclude a successful registration of the compound. From a mammalian toxicity perspective, knowledge generation is more focused on developing data that allow predictions to be made regarding the likely outcome of key regulatory studies (Table 14.5). It is assumed that, during the pre-development phase, there is usually only one molecule under evaluation; however, this may not always be the case and differentiation between compounds may also be an objective.

14.5.3 Key Questions to Consider for a Compound in Full Development

Once a compound has been committed to full development, the over-riding objective is to produce a data package that will comply with global regulatory requirements. While the need to address these data requirements using study designs that conform to international test guidelines may limit flexibility with regard to the tools that can be employed, the possibility of using alternative approaches that provide scientifically robust data to address risks to human health should not be overlooked. The likelihood of new methodologies being accepted will be increased if regulators are consulted and are given opportunity to accept alternative approaches. The tools and options presented in Table 14.6 mainly reflect those that are used within the existing paradigm. As noted earlier in this chapter, as time progresses the expectation is that the key questions for risk assessors can be addressed using approaches other than *in vivo* testing in experimental animals.

14.6 Conclusions

It is clear that the current toxicology testing paradigm needs to be changed from one which is focused heavily on data generation through the conduct of

Table 14.5 Key questions when selecting candidates for full development.

Question	Tools and options
What factors may potentially differentiate the candidate molecules from a human safety perspective?	SAR; potency against key biological targets in mammals; acute toxicity; ADME; toxicokinetic parameters relating to detoxification and activation pathways
Are there any differences in the predicted hazard classification of the candidate molecules?	Any studies or information sources that could further elucidate the potential for differences in hazard classification
Are there any differences in predicted reference doses between the candidate molecules?	Predicted NOAELs across species; views as to which species may be more relevant to humans; relative potency of candidate molecules against key targets
Are there any alerts for genotoxicity?	SAR; *in vitro* genotoxicity screening
What is the predicted hazard classification?	Short-term repeat dose studies; *in vitro* models for key hazard end-points; preliminary studies (e.g., developmental toxicity); genotoxicity; gene expression changes in short-term studies
What are the predicted reference doses?	Acute toxic potency; NOAELs from short-term studies; reference doses for modes of action/structural analogs
Are there any alerts for the potential to cause perturbation to endocrine homeostasis?	SAR; *in vitro* screening; *in vivo* assays for alerting structures
What is the predicted oral absorption?	Predictive modeling based on physico-chemical properties; *in vitro* models of oral absorption; cold analysis of samples from short-term toxicity studies; preliminary ADME studies with radiolabel
What is the predicted dermal absorption?	Predictive modeling based on physico-chemical properties; *in vitro* models of dermal absorption

What is the potential for human exposure? What commodities/routes/uses are the key drivers for exposure?	Key crops; application rate; frequency of application; method of application; formulation types; oral absorption; dermal absorption; predicted reference doses. Undertake initial assessment of exposure and relate this to predicted reference doses
Are there any potential concerns arising from inter-compartmental (soil, water, livestock, crop) differences in metabolism?	Preliminary *in vivo* mammalian metabolism; comparative *in vitro* metabolism; crop metabolism; soil metabolism
Are there any inter-species differences in metabolism/disposition?	Comparative *in vitro* metabolism studies in rat, dog, mouse, and rabbit. Cold analysis of metabolites in plasma samples from toxicity studies. Observed inter-species differences in toxicity
Is there potential for genotoxic impurities in the test material to be used for critical toxicity studies?	Chemical analysis of impurity profile in toxicity testing batches. SAR on impurities. Genotoxicity testing of toxicity testing batches
Is there a need to consider developing a physiologically based pharmacokinetic model for predicting tissue dosimetry in humans?	Evaluate the ability to pass human health risk assessments and consider if there is a need to model target tissue concentrations in humans?

Table 14.6 Key questions for a compound in full development.

Hazard profile	
Question	**Tools and options**
Is the compound carcinogenic in rodents and, if so, what is its carcinogenic potential in humans?	Repeat-dose toxicity studies in rodents; cell proliferation in target tissues; gene expression profiling; genotoxicity testing; transgenic models; MoA studies
Is the compound a specific developmental toxin and, if so, what is the potential for the effects observed in animals to be expressed in humans?	Pre-natal developmental toxicity studies in the rat and rabbit
Does the compound have specific effects on fertility in animals? Are the observed effects relevant to humans?	Reproductive toxicity study (over one or two generations); effects on reproductive organs in short-term toxicity studies; *in vitro* screens against specific biochemical targets
Is the compound mutagenic *in vivo* in germs cells?	*In vitro and in vivo* somatic cell genotoxicity assays
Is the compound more toxic by one route than another? Is there an explanation for the route-specific toxicity?	Comparison of acute oral, acute dermal, and acute inhalation toxicity studies. Extent of dermal absorption. Comparisons of NOAELs from 28-day oral and dermal toxicity studies. Particle size of test material versus that to which humans are exposed
Is the compound neurotoxic?	SAR; relevance of pesticidal mode of action to mammals (particularly insecticides); functional observations in short-term toxicity studies; acute and sub-chronic neurotoxicity studies

Risk profile

Question	Tools and options
How much compound is systemically available after oral dosing? Does first-pass metabolism enhance or reduce toxicity?	Bioavailability after oral versus intravenous dosing. Extent of biliary and urinary excretion of radiolabeled compound after oral dosing
What is the dermal absorption of the active substance? Which formulation should be tested? What dilutions should be used?	*In vitro* and *in vivo* dermal absorption studies with commercial formulations
What are the reference doses for human health risk assessment (AOEL/ARfD/CRfD-ADI)?	Key NOAELs for treatment-related adverse effects of relevance to humans from studies of the relevant duration. Dose–response and dose–spacing across data package. Extent of oral absorption. Demonstration of threshold for carcinogenic end-points (i.e., MOE versus Q*). FQPA safety factor
Does the available knowledge indicate that the risk assessments may require additional uncertainty factors to be applied to ensure reasonable certainty of no harm?	Evidence for increased sensitivity of specific life-stages and/or sub-populations (can the FQPA safety factor be reduced from 10?); steepness of dose–response; severity of effect; extrapolation from short-term to long-term reference dose; absence of no effect level for end-point of concern
Will any metabolites be included in the residue definition? What information is needed to adequately evaluate the toxicity of the metabolites?	Quantitative and qualitative nature of exposure; comparison of crop and livestock metabolism with rat metabolism. Prediction of hazard profile for metabolites not evaluated in the rat; SAR, *in vitro*, and/or *in vivo* testing. Evaluation of risk based on estimated exposure compared to parent ADI, risk assessment based on estimated metabolite reference doses (e.g., threshold of toxicological concern approach), risk assessment based on toxicity data for the metabolite

AOEL: acceptable operator exposure level; ARfD: acute reference dose; CRfD: chronic reference dose; ADI: acceptable daily intake; MoA: mode of action; NOAEL: no-observed-adverse-effect-level; MOE: margin of exposure (non-genotoxic carcinogens versus genotoxic (q1*) carcinogens); FQPA SF; Food Quality Protection Act safety factor.

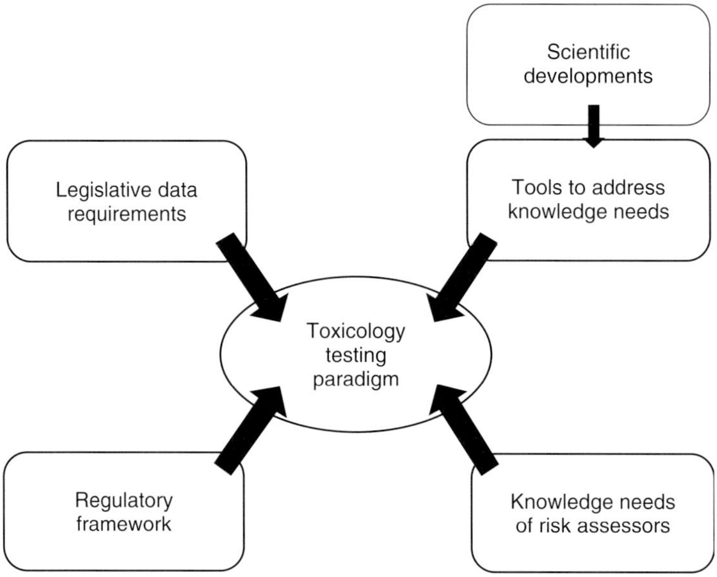

Figure 14.6 Key factors influencing changes in the toxicology testing paradigm.

in vivo studies in mammals, to one that uses *in vitro* and *in silico* models to predict adverse health effects in humans. Whilst the various scientific disciplines that underpin toxicology have developed at a rapid pace during recent years, the ability to translate these scientific advancements into new legislation which defines internationally harmonized data requirements is limited by the diverse political landscape within which such changes need to be made. The fact that this change must be implemented by achieving consensus in a complex multi-stakeholder environment means that such a process can take many years (Figure 14.6).

The agreement of new data requirements is dependent on the regulator's acceptance of new methodologies, and their confidence in the ability of a new testing paradigm to provide a scientifically robust evaluation of the risks posed to human health. There is a danger that the new technologies are seen as useful additions to the exiting paradigm rather than opportunities to replace a flawed system. Nevertheless, it is clear that there will need to be some form of objective evaluation of the ability of the new models to predict adverse health effects in humans compared to the existing methods. What should be avoided above all is an evaluation of the ability of the new models to predict the outcome of existing *in vivo* animal models, as the intention should not be to recreate a testing paradigm with similar flaws.

It is clear that traditional approaches to method validation will not be applicable to the new testing paradigm, and nor would they be desirable given that they are extremely time-consuming and often result in an ambiguous outcome. It is, therefore, likely that new approaches will be required in order to evaluate the ability of an integrated systems biology-based approach to predict adverse health effects

Table 14.7 Working toward a new mammalian toxicity testing paradigm.

Imperative	Key hurdles to overcome
Objective evaluation of the ability of new models to predict adverse health outcomes in humans compared to that of existing models	Ability to integrate diverse data sets with existing knowledge Need to develop new approaches to method "validation" and need to focus on predicting adverse health outcomes for humans, not for experimental animals Need to reduce the number of false-positive outcomes and need to be able to evaluate potential for false negatives Ideally, any new paradigm needs to be applicable to all sectors of chemical regulation
To reduce the number of animals used for *in vivo* testing with the aim of total replacement with *in vitro* and *in silico* approaches	Development of *in vitro* and *in silico* models that gain international acceptance as being fit for a specific scientific purpose Regulator's acceptance of the new approaches as fit-for-purpose for chemical regulation
A more cost-efficient testing paradigm	Need greater harmonization of data requirements across different regions – avoid regional specific data requirements Need to conduct fewer animal-intensive long-term *in vivo* studies in mammals Need less expensive studies – avoid cost being a barrier to testing
A faster timeline to generate the data required to determine if there is reasonable certainty of no harm	Use of short-term studies to predict health outcomes arising from long-term exposures Need to spend less time investigating human relevance of experimental artifacts Need a faster timeline for risk characterization to be completed

in humans. The pharmaceutical industry faces the same challenge in seeking to avoid the failure of development drug candidates based on toxicities observed in clinical trials in humans that were not observed in preclinical toxicity studies in experimental animals. A number of imperatives are required in order to transition from the current testing paradigm to a new one. Moreover, it is important that these imperatives are addressed in concert in a manner that brings together the key stakeholders, so that they can develop a new regulatory framework that will allow chemicals to be approved based on a judgment of reasonable certainty of no harm with minimal or no *in vivo* animal testing (Table 14.7).

In recognizing that the move towards a new testing paradigm will be gradual, it is important to consider the opportunities to modify the existing paradigm in a way that provides more information of direct relevance to risk assessors and risk managers, and also uses fewer animals and more refined experimental procedures. Such opportunities are afforded by the use of integrated toxicity study designs, the use of dose levels that are relevant to predicted human exposure (and, conversely, avoiding the use of dose levels that are tens of thousands time higher than predicted human exposure), and by only conducting toxicity studies that are designed to address specific questions of relevance to the regulation of a chemical. By remaining focused on the key questions that face risk assessors, and also on the rapidly increasing knowledge of human disease processes, it should be possible to develop a new set of tools for predicting adverse health outcomes for defined exposure scenarios in humans.

References

1. US EPA (Environmental Protection Agency) (2005) Guidelines for Carcinogen Risk Assessment, EPA/630/P-03/001B, Risk Assessment Forum, U.S. Environmental Protection Agency, Washington, DC.
2. O'Connor, P.J., Manning, F.C.R., Gordon, A.T., Billett, M.A., Cooper, D.P., Elder, R.H., and Margison, G.P. (2000) *Toxicol. Pathol.*, **28**, 375–381.
3. Hengstler, J.G., Bogdanffy, M.S., Bolt, H.H., and Oesch, F. (2003) *Rev. Pharmacol. Toxicol.*, **43**, 485–520.
4. US FDA (2008) Guidance for Industry Genotoxic and Carcinogenic Impurities in Drug Substances and Products: Recommended Approaches U.S. Department of Health and Human Services, Food and Drug Administration Center for Drug Evaluation and Research (CDER), December 2008 Pharmacology and Toxicology.
5. Fukushima, S., Kinoshita, A., Puatanachokchai, R., Kushida, M., Wanibuchi, H., and Morimura, K. (2005) *Carcinogenesis*, **26** (11), 1835–1845.
6. National Research Council and National Academy of Sciences (2007) Toxicity Testing in the 21st Century: A Vision and a Strategy. Committee on Toxicity Testing and Assessment of Environmental Agents, (The National Academies Press, 5th October, 2007).
7. US EPA Toxcast™ (2007) Testing Program *http://www.epa.gov/ncct/toxcast/*.
8. US EPA National Research Council of the National Academies (2009) *Science and Decisions Advancing Risk Assessment, Committee on Improving Risk Analysis Approaches used by the US EPA, Board on Environmental Studies and Toxicology, Division on Earth and Life Studies, National Research Council of the National Academies*, The National Academies Press, Washington, DC.
9. EU Commission: Regulation (EC) No 396/2005 of the European Parliament and of the Council of 23 February 2005 on Maximum Residue Levels of Pesticides in or on Food and Feed of Plant and Animal Origin and Amending Council Directive 91/414/EEC. Official Journal of the European Union L 70, 16.3.2005, pp. 1–16.
10. EU Commission. Commission Regulation (EU) No 544/2011 of 10 June 2011 Implementing Regulation (EC) No 1107/2009 of the European Parliament and of the Council as Regards the Data Requirements for Active Substances. Official Journal of the European Union L 155, 11.6.2011, pp. 67–126.
11. US EPA Code of Federal Regulations (2011) 40 - Protection of Environment Chapter I - Environmental Protection Agency Sub-chapter e - Pesticide Programs Part 158 - Data Requirements for Pesticides, September 28, 2011.I
12. EU Commission. Directive 98/8/EC of the European Parliament and of the Council of 16 February 1998 Concerning the Placing of Biocidal Products on the Market. Official Journal of the European Union L 123, 1998, pp. 1–63.

13. EU Commission (2003) Guidance Document on the Assessment of the Relevance of Metabolites in Groundwater of Substances Regulated Under Council Directive 91/414/EEC, Sanco/221/2000 – Rev.10-Final, February 25, 2003.
14. US Congress (1996) Food Quality Protection Act (FQPA).
15. Carmichael, N.G., Barton, H.A., Boobis, A.R., Cooper, R.L., Dellarco, V.L., Doerrer, N.G., Fenner-Crisp, P.A., Doe, J.E., Lamb, J.C., and Pastoor, T.P. (2006) *Crit. Rev. Toxicol.*, **36**, 1–7.
16. Creton, S., Dewhurst, I.C., Earl, L.K., Gehen, S.C., Guest, R.L., Hotchkiss, J.A., Indans, I., Woolhiser, M.R., and Billington, R. (2010) *Crit. Rev. Toxicol.*, **40**, 50–58.
17. Kobel, W., Fegert, I., Billington, R., Lewis, R., Bentley, K., Bomann, W., Botham, P., Stahl, B., van Ravenzwaay, B., and Spielmann, H. (2010) *Crit. Rev. Toxicol.*, **40** (1), 1–15.
18. Terry, C., Rasoulpour, R.J., Gollapudi, B., and Billington, R. (2011) *Toxicologist*, **120** (Suppl. 2, PL 858), 184 and (Platform Presentation at Society of Toxicology Annual Meeting 2011).
19. ICH (1995) International Conference on Harmonisation of Technical Requirements for Registration of Pharmaceuticals for Human Use. ICH Harmonised Tripartite Guideline. Guideline on the Need for Carcinogenicity Studies of Pharmaceuticals S1a. Current Step 4 Version. Dated 29 November 1995.
20. Billington, R., Lewis, R., Mehta, J., and Dewhurst, I. (2010) *Crit. Rev. Toxicol.*, **40**, 35–49.
21. Doe, J.E., Boobis, A.R., Blaker, A., Dellarco, V., Doerrer, N.G., Franklin, C., Goddman, J.I., Kronenberg, J.M., Lewis, R., McConnel, E.E., Mercier, T., Moretto, A., Nolan, C., Padilla, S., Phang, W., Solecki, R., Tilbury, L., van Ravenzwaay, B., and Wolf, D.C. (2006) *Crit. Rev. Toxicol.*, **36**, 37–68.
22. Wellcome Trust, Sanger Institute, Cancer Genome Project. Available at: *http://www.sanger.ac.uk/*. Accessed 4th June 2012.
23. Thomas, R.S., Clewell, H.J. III, Allen, B.C., Wesselkamper, S.C., Wang, N.C., Lambert, J.C., Hess-Wilson, J.K., Zhao, Q.J., and Andersen, M.E. (2011) *Toxicol. Sci.*, **120** (1), 194–205.
24. Hartung, T. (2009) *Nature*, **460**, 208–212.
25. OECD (2009) Guideline for the Testing of Chemicals: OECD 453 Combined Chronic Toxicity\Carcinogenicity Studies. Adopted 7 September 2009.
26. EPA (1998) US EPA Health Effects Test Guidelines OPPTS 870.4200 Carcinogenicity. EPA 712–C–98–211, August 1998.
27. EPA (1998) US EPA Health Effects Test Guidelines OPPTS 870.4300 Combined Chronic Toxicity/Carcinogenicity, EPA 712– C– 98–212, August 1998.
28. OECD (2010) Test Guideline 417: Toxicokinetics. OECD Guidelines for the Testing of Chemicals. Organisation for Economic Co-operation and Development, Paris.
29. Saghir, S.A., Marty, M.S., Clark, A.J., Zablotny, C.L., Bus, J.S., Perala, A.W., Yano, B.L., and Neal, B.H. (2009) *Off. J. Soc. Toxicol.*, **108**, S1161.
30. Saghir, S.A., Rick, D.L., Clark, A.J., Staley, J.L., Bartels, M.J., Terry, C., and Billington, R. (2011) *Toxicologist*, **120** (Suppl. 2, PS 2107), 452.
31. Creton, M., Billington, R., Davies, W., Dent, M.P., Hawksworth, G.M., Parry, S., and Travis, K.Z. (2009) *Regul. Toxicol. Pharmacol.*, **55**, 291–299.
32. OECD (2010) Draft Guidance Document No 116 on the Design and Conduct of Chronic Toxicity and Carcinogenicity Studies, Supporting TG 451, 452, 17 December 2010.
33. ECETOC (1996) Practical Concepts for Dose Selection in Chronic Toxicity and Carcinogenicity Studies in Rodents. European Centre for Ecotoxicology and Toxicology of Chemicals, Brussels, Belgium, Monograph No. 25.
34. US EPA (2003) Health Effects Division (HED) Interim Guidance Document (# G2003.02) Rodent Carcinogenicity Studies: Dose Selection and Evaluation, 1 July 2003.

15
Endocrine Disruption: Definition and Screening Aspects in the Light of the European Crop Protection Law

Susanne N. Kolle, Burkhard Flick, Tzutzuy Ramirez, Roland Buesen, Hennicke G. Kamp, and Bennard van Ravenzwaay

15.1
Introduction

Endocrine-disrupting compounds (EDCs) encompass a variety of substance classes, including natural and synthetic hormones, plant constituents, crop protection products, substances used in industry and in consumer products, and other industrial byproducts and pollutants. On the molecular level, there are a number of mechanisms by which EDCs can affect endocrine systems and potentially cause adverse effects. These include binding to hormone receptors so as to exert agonistic or antagonistic effects, binding to transport proteins in the blood (thus altering the amounts of natural hormones present in the circulation), or interfering with metabolic processes in the body so as to affect the synthesis or breakdown rates of the natural hormones. Although it must be noted that the mere potential of endocrine disruption (=hazard) does not necessarily produce an adverse effect, there has recently been a growing scientific concern, in addition to intense public debate and media attention, regarding possible deleterious effects in humans and wildlife that may result from their exposure to substances that have the potential to interfere with the endocrine system.

The endocrine system is a communications system that maintains normal physiological balance across multiple central and peripheral organ systems. It accomplishes this by modulating or regulating the activity of almost every body system in reaction to variations in body temperature, activity levels, stress, and circulating levels of nutrients and hormones required for growth, reproduction, and metabolism. Besides other substances, some crop protection products have long been known to interact with endocrine function (for a review, see Ref. [1]).

Modern Methods in Crop Protection Research, First Edition.
Edited by Peter Jeschke, Wolfgang Krämer, Ulrich Schirmer, and Matthias Witschel.

15.2
Endocrine Disruption: Definitions

Since the 1990s, several definitions of endocrine disruption (ED) have been proposed and developed by national and international governmental agencies. An early definition of EDC dating back to the US Environmental Protection Agency (EPA) workshop in 1995 was (Kavlock definition; [2]):

> "... an exogenous agent that interferes with the production, release, transport, metabolism, binding, action or elimination of natural hormones in the body responsible for the maintenance of homeostasis and the regulation of developmental processes."

Within the scope of the European Union (EU), the ED definition was amended as a result of the Weybridge workshop, by an inclusion of the "adverse effects" terminology as well as the "intact organism," to read (Weybridge definition; [3]):

> "... an exogenous substance that causes adverse health effects in an intact organism, or its progeny, secondary to changes in endocrine function."

The US EPA has adopted the Kavlock definition [2] and has stated ED not to be adverse *per se* [4].

A consensus working definition for ED has been established by the International Programme for Chemical Safety [World Health Organization (WHO) definition [5]]:

> "... an exogenous substance or mixture that alters function(s) of the endocrine system and consequently causes adverse health effects in an intact organism, or its progeny, or (sub) populations."

In summary, the different ED definitions share common themes, while the Weybridge and WHO definitions take into account that many alterations of the endocrine system can be regarded as adaptive and, therefore, do not necessarily endanger the health of an intact organism. In this respect, an interference with normal endocrine homeostasis must result in an adverse effect (e.g., functional impairment or pathological findings) in an intact organism (i.e., *in vivo*), and the detection of an adverse effect *per se* is insufficient to conclude a substance to be an EDC. Furthermore, *in vitro* test methods alone should be considered inappropriate to identify and EDC as they do not involve intact organisms.

15.3
Current Regulatory Situation in the EU

A hazard-based non-authorization of EDCs is foreseen according to the European Plant Protection Product Commission Regulation [6] and Biocidal Products Directive [7]; that is, active substances, safeners, or synergists in products considered to have ED properties that may be of toxicological significance in humans or non-target organisms will not be approved. Further, the Registration, Evaluation, and Authorization and Restriction of Chemicals (REACH) requires the specific evaluation of ED properties before a decision on authorization can be taken [8]. While a risk assessment approach including a thorough scientific assessment of all available data and exposure factors is used by US authorities, and has been proposed by the European Center for Ecotoxicology and Toxicology of Chemicals (ECETOC) [9], the hazard-based cut-off criterion will tentatively determine the use of such substances in Europe. A draft of the measures concerning specific scientific criteria for the determination of endocrine-disrupting properties shall be presented by the Commission by 14 December 2013 [6]. In summary, the EDCs are specifically regulated in the following European legislations:

- Plant Protection Product Commission Regulation 1107/2009 [6]: EDC cannot be approved unless exposure is negligible.
- Biocidal Products Directive 98/8/EC [7]: EDC cannot be approved unless exposure is negligible.
- REACH Commission Regulation 1907/2006 [8]: EDC substance may be included in the list of substances subject to authorization.

While in a risk assessment approach (risk estimation by comparison of hazard-based reference value and exposure) an acceptable risk can be determined, the presence of the property defined as hazard itself presents a cut-off. Nevertheless, there are cases in which the safe use of substances with inherent hazardous properties can be demonstrated, for example, by negligible exposure ("[. . .] the product is used in closed systems or in other conditions excluding contact with humans [. . .]" [6]). Furthermore, after risk assessment using appropriate risk management procedures, the safe use of a potentially hazardous substance can be assured, for example, by limiting exposure and taking appropriate precautionary measures. The market for a particular plant protection product, however, needs to be determined before exposure scenarios (e.g., for the risk assessment) can be performed. Importantly, in the light of the hazard-based marketing ban as laid out in Plant Protection Product Commission Regulation [6], regulatory criteria characterizing a substance as EDC remain to be defined. Therefore, until December 2013, it remains unclear under which conditions a substance would be considered as EDC, potentially causing adverse health effects to humans.

During the past years, a multitude of *in vitro* and *in vivo* assays of various complexities, as well as validation and regulatory acceptance statuses, have been described, and a detailed review and discussion of the methods employed would be beyond the scope of this chapter. In the following sections, the frameworks of the Organisation for Economic Co-operation and Development (OECD) and the US EPA for the testing and assessment of potential EDCs, will be introduced. While OECD and US EPA propose largely the same clearly defined methods, other approaches including the ECETOC approach take into consideration also non-guideline studies of sufficient relevance and reliability to be included in a weight of evidence approach to assess the mode of action of a putative EDC.

15.4
US EPA Endocrine Disruptor Screening Program and OECD Conceptual Framework for the Testing and Assessment of Endocrine-Disrupting Chemicals

In 1996, based on the fact that certain contaminants induced developmental and reproductive problems in wildlife, the US Congress enacted the Food Quality Protection Act. The act directs and gives authority to the US EPA to design a program for the detection of chemical compounds, crop protection products and environmental chemicals that can potentially affect the estrogen (E) hormone system. The EPA, with the recommendations of the Endocrine Disruption Screening and Testing Advisory Committee, designed a two-tiered program (Endocrine Disrupter Screening Program, EDSP [10]), combining *in vitro* and *in vivo* studies (including fish and wildlife) to identify effects on the E, androgen (A), and thyroid (T) hormone systems. The EPA has determined 11 assays to be included in Tier 1, which is designed to identify substances that have the potential to interact with the E, A, and/or T hormone systems. Those compounds that do exhibit the potential to interact to the E, A, or T systems shall be further tested in the Tier 2 screening battery. Tier 2 includes exclusively *in vivo* testing, and is designed to identify specific endocrine effects and establish the dose levels at which the effects occur. This approach is intended to enable the EPA to gather the information needed to identify EDCs, to validate their assays, and to take appropriate regulatory action.

Similarly, the OECD has proposed a Conceptual Framework (CF) for the Testing and Assessment of Endocrine-Disrupting Chemicals consisting of five different levels of *in silico, in vitro*, and *in vivo* tests to provide guidance in assessing mammalian and non-mammalian toxicity [11]. In Level 1, existing data and non-testing information such as physico-chemical properties and *in silico* predictions are taken into account, whereas Levels 2 and 3 include *in vitro* assays (e.g., receptor binding and receptor-mediated transcriptional activation) and *in vivo* assays (e.g., Hershberger or uterotrophic assays) to provide data about selected ED mechanisms. Level 4 of the OECD CF includes studies to assess adversity on mammalian and non-mammalian endocrine end-points (e.g., reproductive screening test). Level 5 assays are intended to provide more comprehensive data on adverse effects on endocrine-relevant end-points (e.g., extended one-generation reproductive toxicity study) [11, 12]. The major difference between the OECD CF and US EPA EDSP approaches is that the OECD CF proposed different methods classified in five different levels with assays that can be conducted at any stage during the hazard assessment process, depending on the perceived need for information. However, the data generated at various levels have a range of differing applications and implications, and must be interpreted accordingly [11]. While the US EPA EDSP recommends the testing of all assays included in its Tier 1 in order to identify potential EDCs, and only after this first assessment, Tier 2 assays will be performed only for those compounds that result positive in Tier 1. *In vitro* test methods considered as Level 2 in the OECD CF are screening assays used for hazard detection as well as the identification of a possible mode of action. A positive *in vitro* test result indicates the possibility of ED effects *in vivo*; however, negative *in vitro* results alone are insufficient to exclude possible ED activity because of their inherent limitations [11].

15.5 ECETOC Approach

As the ECETOC guidance is a science-based approach, the determination of the endocrine-disrupting properties of a substance should be re-evaluated and improved as research in this field progresses and more test methods are validated. The ECETOC approach classifies the available tests into three categories:

- Targeted *in vitro* screens (e.g., estrogen receptor binding, estrogen receptor transcriptional activation (OECD TG 455 [13]), androgen receptor binding, steroidogenesis assay (OECD TG 456 [14]), and recombinant aromatase screens),
- Targeted mechanistic *in vivo* screening assays (e.g., uterotrophic, Hershberger, pubertal male, and pubertal female assays, the 21-day fish screening assay (OECD TG 230, [15]), fish short-term reproduction assay (OECD TG 229, [16]), and amphibian metamorphosis assay (TG OECD 231, [17]).
- Apical (definitive) and supporting *in vivo* assays (e.g., chronic and oncogenicity assays (OECD TGs 451, 452, 453, [18–20]), the mammalian one- and two-generation studies (OECD TGs 415 and 416 [21, 22]), the extended one-generation reproductive toxicity study (OECD TG 443, [23]), the prenatal development study (OECD 414, [24]), subchronic assays (OECD TGs 408 and 409, [25, 26]), and the updated 28-day study (OECD TG 407) [27]).

A holistic evaluation of all data from the separate assays will be required to assess whether a substance should be regarded as an endocrine disrupter according to the Weybridge definition [3]. An objective, systematic, and structured weight-of-evidence evaluation should be conducted. Thereby, toxicological findings critical to the postulated mode of action (MoA), a dose–response relationship, and a temporal association of the key events and the toxicological response has to be described. The strength, consistency, and specificity of effects are determined, and the biological plausibility of the MoAs and effects is proposed to be evaluated. A framework for hypothesis-driven weight of evidence for evaluating data within the US EPA EDSP has been described recently [28].

In 2011, an ECETOC task force reported on the science-based guidance on the assessment of ED. In this guidance the task force proposed that, besides the ED potency, the quality of the adverse effect caused by an EDC should be evaluated scientifically based on the MoA of the observed adverse effect [9]. The proposed approach, which is based on a scientific ECETOC workshop of regulators, academia, and industry held in Barcelona in 2009 [29], has been described in detail by Bars *et al.* [9], and a short overview is provided here. To demonstrate the principle of the guidance, attention was focused on assessing the relevant effects in humans. A corresponding guidance has also been provided for non-target-species [9]. The task force considers the Weybridge definition of ED (see above [3]) to be the appropriate definition for practical use and proposed specific scientific criteria for the determination of endocrine-disrupting properties, that integrates information from both regulatory (eco)toxicity studies and mechanistic/screening studies. The latter should provide an understanding of the MoA underlying the toxicity findings. Therefore, research studies may also be useful (e.g., in covering end-points not considered by standard regulatory tests), although the relevance, reliability and quality of such studies must also be considered (e.g., Klimisch scores [30]). For any case where an ED property is indicated, an assessment of specificity, human relevance and potency is also proposed to discriminate chemicals of high concern from those of lower concern.

The ECETOC approach considers five scenarios (see also Figure 15.1) to guide the evaluation of available mammalian data to determine whether a substance has endocrine properties; this is followed by an assessment of endocrine specificity, human relevance, and potency [9].

Scenarios with no or insufficient evidence of endocrine disruption include:

Scenario A	Absence of adverse effects using end-points relevant for the assessment of ED in apical studies can be taken as definitive evidence of no ED properties
Scenario B	Absence of adverse effects in the apical tests can also be taken as definitive evidence of no ED properties, even if there are positive outcomes from non-apical targeted *in vitro* or *in vivo* screening studies
Scenario D	If, after exhaustive testing using the full battery of validated *in vitro* and *in vivo* targeted end-point studies, there is no evidence to support an ED-mediated mechanism, the adverse effects in the apical studies should not be considered as evidence of ED
Scenario E	In the absence of all other data, negative outcomes in an exhaustive combination of *in vitro* and *in vivo* targeted end-point studies can be taken as evidence of the absence of ED properties

Scenario with endocrine properties in laboratory mammalian species:

Scenario C	When adverse effects on endocrine-relevant end-points in apical or supporting non-apical *in vivo* studies are supported by mechanistic data from *in vitro* and *in vivo* studies – that is, the sequence of the biochemical and cellular events that underlies the adverse effect is described and understood – then conclusive proof of ED can be considered as established

If an endocrine property of a compound has been indicated, the ECETOC guidance recommends that the specificity, human relevance and potency of this effect be considered. An assessment of specificity is required to determine whether the adverse effects observed occur at dose levels lower than other manifestations of toxicity. Human relevance must be considered regarding the endocrine mechanism

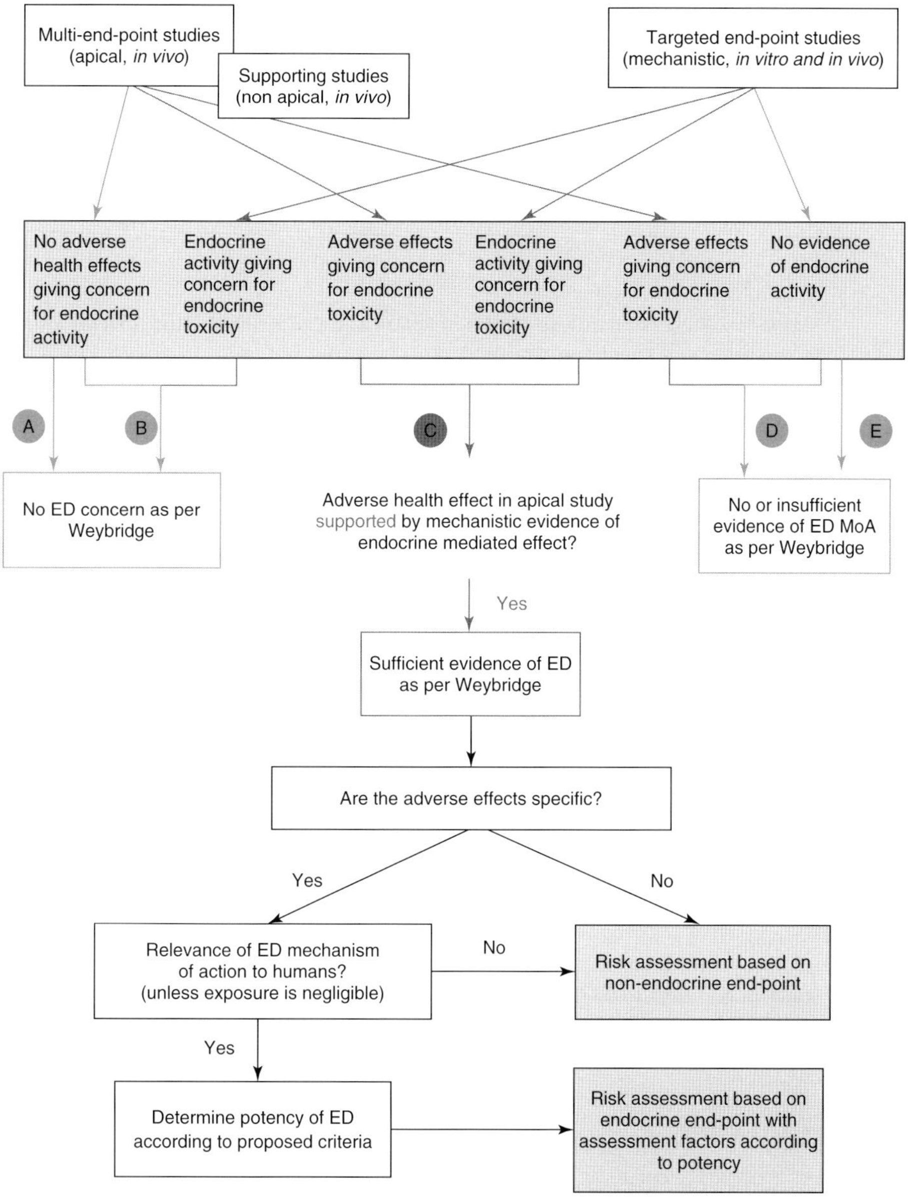

Figure 15.1 Flow chart for the assessment of endocrine-disrupting (ED) properties for human health. Outlined is the decision tree to decide whether a substance should be considered to be an endocrine-disrupting compound (EDC) according to the Weybridge definition, incorporating human relevance. Scenarios A to E are indicated (see text for details). MoA, mode of action. Reproduced with permission from Ref. [9]; © 2011, Elsevier, Amsterdam.

of action and negligibility of exposure. Finally, to discriminate endocrine disrupters of high and low concern, several criteria for potency are proposed which should be considered collectively; these include: (i) the dose level at which an adverse effect on any endocrine end-point occurs; (ii) the duration of exposure that is required for an adverse effect to be induced; (iii) the nature, incidence, and severity of the adverse effect; (iv) the number of species in which an effect is observed; and (v) the occurrence of the effect in one or both sexes.

15.6 Methods to Assess Endocrine Modes of Action and Endocrine-Related Adverse Effects in Screening and Regulatory Contexts

To reduce the number of animals used in testing and, if possible, to replace animal testing completely, the "3R" principle of reduction, refinement, replacement (reducing animal testing, refining test methods, and conducting tests without using animals) has been applied for many years. In addition, *in vitro* and *in vivo* screening tests can be systematically applied that should provide information about specific toxicological mechanisms such as endocrine disruption. Hence, the use of alternative methods as part of the screening strategies helps to set priorities for further toxicological tests, to provide mechanistic information, and also have the potential for application in a regulatory context. Mechanism of action investigations have been proposed to be addressed using the *in vitro*- and *in vivo*-targeted end-point studies such as OECD Level 2–3 assays [11] and US EPA EDSP Tier 1 [10]. Others, including the aforementioned ECETOC approach, suggest the inclusion of assays beyond the classical regulatory tests for the identification of EDCs [9]. A pragmatic and resource-efficient alternative for the detection of sex steroid-related ED mechanisms has been proposed as a combination of two *in vitro* and one *in vivo* assays (Figure 15.2). The animal-free components of the strategy are a yeast-based receptor-mediated transcriptional activation assay and a mammalian cell-based steroidogenesis assay. In addition, a repeated-dose toxicity study (e.g., OECD 407 [27]; including a plasma metabolome analysis) is used to provide valuable information of the putative ED mechanism) [31, 32].

15.6.1 *In-Vitro* Assays

The identification of substances with potential ED effects has increased the activities to develop sensitive and predictive tools to detect EDCs and to determine their putative endocrine MoAs. The use of *in-vitro* assays provides several advantages. For example, they are in accordance with the 3R principle for animal testing, reduction, refinement, and replacement of animal testing. Moreover, they not only reduce costs but can also provide quick answers using minimal quantities of test substance, which represents a major benefit when developing a new compound. Recent advances in technologies, as well as a better understanding of the cellular

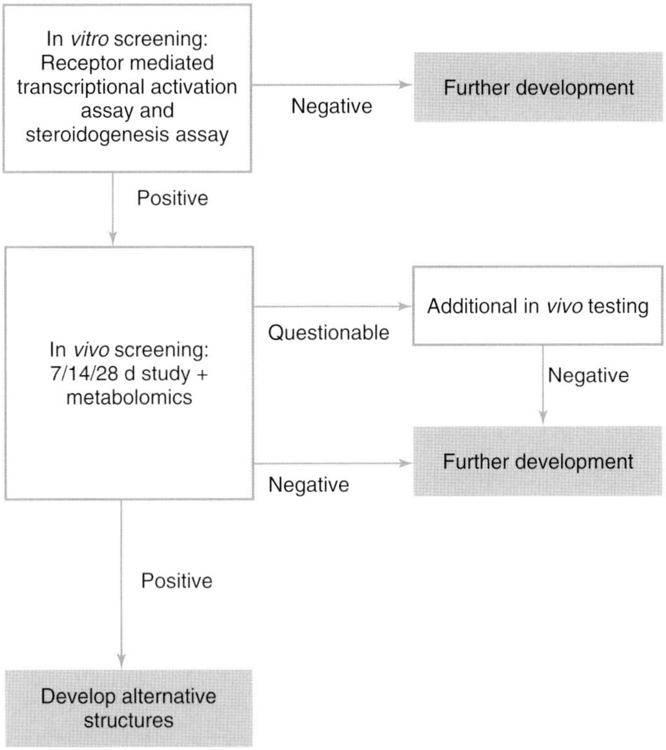

Figure 15.2 Assessment of mammalian endocrine effects using a combination of *in-vitro* and *in-vivo* assays to target toxicity testing during product development. In a first step, alerts for direct or indirect endocrine modes of action are generated in *in-vitro* assays, such as receptor-mediated transcriptional activation assays and a steroidogenesis assay. Endocrine modes of action are then confirmed or disconfirmed in a repeated-dose toxicity study, including plasma metabolome analysis. While questionable outcomes may require additional *in-vivo* testing, additional structure development is initiated if an endocrine mechanism is confirmed; development is continued if no endocrine mechanisms of action are detected *in vivo*. Modified from Ref. [32].

processes of endocrine MoAs, *in-vitro* assays based on nuclear receptor signaling or hormone biosynthesis, have facilitated a better qualitative elucidation of the ED mechanisms. Due to the fact that single methods are usually limited to the specific ED mechanisms that are being assessed, the combination of several *in-vitro* assays into testing "batteries" has been proposed.

Several *in-vitro* tests have been proposed within the US EPA EDSP, OECD CF, and others [9, 32] for the detection of EDCs. These tests have been selected based on their capacity to identify E- and A-mediated effects through different MoAs – that is, receptor binding, transcriptional activation, or impact on hormone biosynthesis. The EDC may have an ability to interact with both E and A hormone receptors, either by binding and mimicking the biological effects of natural hormones (agonists), or by binding and blocking access of the hormones to the receptor binding sites (antagonists). Hence, the EDSP includes two *in-chemico* receptor-binding assays for the interaction of chemical compounds with the estrogen receptor (ER) and the androgen receptor (AR), with both assays using cytosolic fractions of uterus (ER-binding assay, OPPTS 890.1250 [33]) or prostate (AR-binding assay, OPPTS 890.1150 [34]). Hence, the *ex-vivo* nature of these assays means that they still require the use of animals.

The assays are not metabolically competent, as only cytosolic fractions containing the ERs or ARs are incubated with the test compounds. In addition, the information provided is limited to receptor binding, without discerning between any agonistic or antagonistic effects, or taking into account the downstream transcriptional activation cascade in physiological steroid hormone receptor function. In order to cover the latter point, the use of a transcriptional activation assay using a human HeLa transgenic cell line (OPPTS 890.1300 and OECD TG 455 [13, 35]) has been proposed in the EDSP. This test is designed to identify the transcriptional activation of the human ER α-regulated reporter gene by agonist binding, and can be used to detect ER-mediated transcriptional effects (protocols for E antagonistic effects have not yet been regulatorily accepted). However, the presented protocols/guidelines have been adopted to assess estrogenic, but not anti-estrogenic, effects. As the assay also lacks metabolic competence, it can only detect estrogenic agonistic compounds that do not require metabolic activation to exert estrogen agonism. In addition to the above-described receptor binding and ER-mediated transcriptional activation assays, similar transactivation assays for detecting effects on the AR are under consideration at the OECD Level.

As mammalian transcriptional activation assays, yeast-based reporter assays are full *in-vitro* methods, and represent a major contribution to the 3R concept. One clear drawback of cell-based systems is the potential presence of endogenous receptors that may interfere with the specific response of the transgenic cells to EDCs. In addition, some assays depend on binding to the endogenous receptors, an example being the use of increased cell proliferation in MCF-7 cell lines [36, 37]. Following exposure to E agonists, however, the presence of other receptors (apart from the ERs) precludes the system from identifying any specific ED effects. One particularly robust model is a yeast-based transgenic model that expresses human hormone receptors and a reporter gene for the detection of agonist and antagonist EDCs. For this purpose, several yeast strains have been created: one strain for the detection of EDCs that interact with the human ERα used in the yeast estrogen screen (YES) [38]; and a second strain that expresses the human AR and is used in the yeast androgen screen (YAS) [39, 40]. Both assays have undergone some degree of validation, and represent a robust alternative to the ER- and AR-binding and ER transcriptional activation assays proposed in the EDSP and OECD CF. The yeast-based assays have been described for the assessment of receptor-mediated agonism and antagonism [41], and formal validation for these end-points has been started in 2011. Yeast-based systems are cost efficient and easy to handle. Moreover, yeast does not demonstrate the endogenous expression of sex-hormone receptors that might interfere with the assay.

The *steroidogenesis assay* forms part of the EDSP Tier 1 and OECD CF Level 2. In this case, a human adenocarcinoma cell line (H295R) is used to detect substances with the potential to either inhibit or induce production of the sex steroid hormones testosterone (T) and 17β-estradiol (E2). Consequently, the assay can potentially be used to discern the target enzymes for EDCs that interfere with steroid biosynthesis (OPPTS 890.1550 and OECD TG 456 [14, 42]).

Another element of the EDSP Tier 1 *in-vitro* battery is the detection of any effects on the enzyme aromatase, which is responsible for E biosynthesis by utilizing androgens as substrates and converting them into estradiol and estrone. The proposed assay is based on the use of a recombinant human aromatase, a radioactive substrate ([^{3}H]-androstenedione), NADPH, and a reductase complex; OPPTS 890.1200 [43]). In the assay, the tritiated water that is released as androstenedione is converted to estrone provides a direct (and quantitative) measurement of aromatase activity. The main problems encountered with this assay include a lack of metabolic competence, the need to use radioactivity, and the high costs associated with the complex assay format.

In addition to the sex steroid assays that have been described above and summarized as targeted *in-vitro* assays in Table 15.1, further *in-vitro* models to address effects on the hypothalamic-pituitary-thyroid axis have also been developed. To date, certain reporter cell line assays have been created that express the thyroid hormone receptors α and β (TRα and TRβ) for the identification of TR agonists or antagonists [44], or an iodide uptake assay to identify any effects on the first step in thyroid hormone biosynthesis. New methodologies, such as *in-vitro* omics technologies, strategically combined into integrated testing batteries, are expected to cause significant improvements in the prediction of potential ED mechanisms *in vitro* [45, 46].

The inclusion of assays to assess direct and indirect ED mechanisms in testing strategies will enable the screening of newly developed compounds or compounds, and their mixtures [32, 47–50].

15.6.2 *In-Vivo* Assays

Potential mechanisms of action and/or adverse effects of EDCs can be identified *in vivo* in several standard repeated-dose toxicity (OECD TG 408 and 407 [25, 27]), reproductive toxicity (OECD TG 415, 421, 422, 416, 443 [21–23, 51, 52]), and mechanistic tests such as the Hershberger (OECD TG 441, [53]) and uterotrophic assays (OECD TG 440, [54]) and non-standard tests [9, 11, 32].

Both, the EDSP and the OECD CF consist of targeted studies in mammals that specifically assess ED effects *in vivo* (Table 15.2), namely the uterotrophic assay (OPPTS 890.1600 and OECD TG 440 [54, 55]) and the Hershberger assay (OPPTS 890.1400 and OECD TG 441 [53, 56]), as well as male (OPPTS 890.1500 [57]) and female (OPPTS 890.1450 [58]) pubertal assays and the adult male assay. For the latter assay, only a finalized standard protocol exists, with no OECD TGs yet having been adopted at the time of writing of this chapter.

Table 15.1 Targeted *in-vitro* assays.

End-point/assay	Test system	Proposed use	Test protocol(s)
Estrogen receptor binding assay			
ER binding	ER from rat uterus	EDSP Tier 1, OECD CF Level 2, other [9]	OPPTS 890.1250 [33]
ER α transcriptional activation assay			
ER-mediated transcriptional activation	HeLa cells stably transfected with hERα	EDSP Tier 1, OECD CF Level 2, other [9]	(OPPTS 890.1300 and OECD TG 455 [13, 35]
ER-mediated transcriptional activation	Yeast transformed with hERα	Other [32]	[41]
Androgen receptor binding assay			
AR binding	AR from rat prostate	EDSP Tier 1, OECD CF Level 2, other [9]	OPPTS 890.1150 [34]
AR transcriptional activation assay			
AR-mediated transcriptional activation	Yeast transformed with hAR	Other [32]	[41]
Steroidogenesis H295R assay			
Interference with steroidogenesis	H295R cells	EDSP Tier 1, OECD CF Level 2, other [9], other [32]	OPPTS 890.1550 and OECD TG 456 [14, 42]
Aromatase recombinant assay			
Aromatase inhibition	Human recombinant aromatase	EDSP Tier 1, OECD CF Level 2, other [9]	OPPTS 890.1200 [43]

AR, androgen receptor; hAR, human androgen receptor; ER, estrogen receptor; hER, human estrogen receptor.

The uterotrophic assay (US EPA Tier 1, OECD Level 3) in rats was designed to detect test substances with estrogenic potential. Two protocols exist, in which either ovariectomized adult animals or immature young females are used. In both cases, a test substance is administered to the animals at different dose levels for three consecutive days. At necropsy, the uterus weights are compared to those of untreated control animals. Test substances with estrogenic properties would cause an increased water content of the uterine cells, and consequently an increase in organ weight.

Likewise, the Hershberger assay (US EPA Tier 1, OECD Level 3) is a short-term assay in rats that allows the evaluation of both anti-androgenic and androgenic potential. In analogy to the uterotrophic assay, two Hershberger assay protocols have been developed that use groups of male rats that have been castrated shortly before puberty, or of intact but immature young males. A test substance is administered to both groups of rats at different dose levels for 10 days. To check for any possible anti-androgenic potential, the rats are treated with testosterone propionate

Table 15.2 Targeted *in-vivo* assays.

End-point/assay	Test system	Proposed use	Reference(s)
Uterotrophic assay			
Estrogenicity	Immature or ovariectomized adult female rats	EDSP Tier 1 OECD CF Level 3 other [9]	OPPTS 890.1600 and OECD TG 440 [54, 55]
Hershberger assay			
Androgenicity, anti-androgenicity, 5α-reductase inhibition	Immature or castrated male rats	EDSP Tier 1 OECD CF Level 3 other [9]	OPPTS 890.1400 and OECD TG 441 [53, 56]
Pubertal male assay			
Anti-thyroid, androgenic, or anti-androgenic activity or alterations in pubertal development via changes in gonadotropins, prolactin, or hypothalamic function	Male rats	EDSP Tier 1 OECD CF Level 4 other [9]	OPPTS 890.1500 [57]
Pubertal female assay			
Anti-thyroid, estrogenic or anti-estrogenic activity, or alterations in pubertal development via changes in gonadotropins, prolactin, or hypothalamic function	Female rats	EDSP Tier 1 OECD CF Level 4 other [9]	OPPTS 890.1450 [58]

in parallel to the test substance treatment, but no androgen is administered when the rats are screened for androgenic activity (as controls). The evaluation of test substance-mediated effects is performed by comparing the wet weights of the androgen-dependent tissues at necropsy in both treated and control animals. The tissues evaluated have included the ventral prostate, seminal vesicles, *musculi levator ani* and *bulbocavernosus*, Cowper's bulbo-urethral gland, and the glans penis.

Male and female pubertal assays (US EPA Tier 1, OECD Level 4) were also designed to detect antagonists and agonists of the estrogenic and androgenic hormone axis, but which also take into account any changes in pubertal development that may result from any interference with steroidogenesis, higher regulated feedback mechanisms, and thyroid hormone homeostasis. In both male and female pubertal assays, treatment is started shortly after weaning and continued until 7–10 days after the expected puberty onset that is typical for the rat strain used. During the in-life phase, the animals are weighed daily until the onset of puberty in order to carefully examine their growth. Following vaginal opening in females, vaginal smears are taken daily to examine the estrous cycle, while at necropsy the sex organs (of both genders), the thyroid and adrenal glands, and the liver and kidney, are each weighed. Serum levels of thyroxine (T_4) and thyroid-stimulating hormone (TSH), as well as testosterone in males, are also monitored. For both study types, all of the data are evaluated for any possible interactions with endocrine parameters that might be related to the correct growth and maturation of the developing young animals.

In the adult male assay, which is performed using rats aged about 10 weeks, the animals are treated for 15 days by gavage at three dose levels, and then examined at necropsy for any differences in both (sex) organ weights and various hormone levels, compared to controls. Histopathological examinations of the testes, epidymides and thyroid gland are also made. Although treatment-related responses might be observed, the assay is rather insensitive to weak modulators of the endocrine system.

Valuable information on ED and other toxicities can be obtained from metabolome analyses in plasma samples obtained from repeated-dose studies (e.g., OECD 407 [27]) [31, 59]. Based on more than 500 reference substances, typical metabolite changes (patterns) have been established for different toxicological modes of action, including ED. Test substance-induced metabolome changes are compared to the MoA patterns in the data base MataMap®Tox to yield information on the ED potential, as well as of other toxicities [31, 59]. In combination with *in-vitro* tests to detect EDCs, this refinement approach particularly helps to confirm or disconfirm and EDC, thereby targeting *in-vivo* testing (reduction through refinement) [32]. In addition to the more commonly studied estrogen- and androgen-mediated endocrine effects, direct and indirect thyroid toxicity can be detected using metabolome analysis [60].

The above-mentioned *in-vivo* assays in mammals mirror only certain, but sensitive, time frames of the development of young animals. However, they are less animal- and resource-intensive when compared to either one-generation (OECD TG 415 [21]) or two-generation (OECD TG 416 [22]; US EPA Tier 2; OECD Level 5) reproductive toxicity studies, which are required for the registration of new active ingredients, as well as the more recently adopted extended one-generation reproductive toxicity study (EOGRTS, OECD TG 443 [23]; OECD Level 5), a study which presents a significant reduction and refinement alternative when compared to the two-generation study [61].

As also proposed by the ECETOC approach [9], and in line with the Weybridge [3] and WHO [5] ED definitions, a substance should only be considered an ED if adverse health effects in an apical study are supported by mechanistic evidence [9]. Supporting and apical *in vivo* assays are summarized in Table 15.3.

Although no detailed discussion of non-mammalian ED effects and assays used can be elaborated here, the effects observed in studies such as are listed in Table 15.4 should be taken into consideration, if available. Likewise, any available information on the ED properties of a compound in a mammalian species should be taken into consideration during the evaluation of ED properties for non-mammalian species.

The *in-vitro* and *in-vivo* assays presented provide different qualities. While specific criteria for EDCs remain to be defined by end 2013 [6], the current scientific view considers neither mechanistic information obtained from targeted studies nor general adverse effects alone to be sufficient to determine a compound as being an EDC. If a substance does not cause adverse effects or exhibits an endocrine MoA, it is unlikely to be an EDC. On the other hand, in the cases where effects are observed

Table 15.3 Supporting and apical *in-vivo* toxicology assays.

End-point/assay	Test system	Proposed use	Reference(s)
Repeated-dose toxicity study			
Additional end-points of sex organ and accessory tissue weights and histology	28-day oral study in rat or mouse	OECD CF Level 4. other [9]	OECD TG 407 [27]
Information on putative endocrine mechanism of action	7-, 14-, or 28-day oral study in rat	Other [32]	[31]
Sub-chronic studies			
End-points of sex organ and accessory tissue weights and histology after sub-chronic treatment	90-day oral study in rodents and non-rodents	Other [9]	OECD TGs 408 and 409 [25, 26]
Prenatal development study			
General information concerning effects of prenatal exposure on pregnant test animals and on the developing organism	Continuous oral administration from implantation to 1 day prior to caesarean section	Other [9]	OECD TG 414 [24]
Chronic/oncogenicity studies			
End-points of sex organ and accessory tissue weights and histology after long-term treatment	Continuous oral administration during the majority of the lifespan of the organism	Other [9]	OECD TGs 451, 452, 453 [18–20]
Mammalian two-generation			
Integrity and performance of the male and female reproductive systems and on the growth and development of the offspring	Continuous administration of compound (orally) prior to and during mating, gestation, lactation to termination after weaning of the F_2 generation	EDSP Tier 2. OECD CF Level 5. other [9]	OECD TG 416 [22]
Mammalian one-generation			
Integrity and performance of the male and female reproductive systems and on the growth and development of the offspring	Continuous administration of compound (orally) prior to and during mating, gestation, lactation to termination after weaning of the F_1 generation	OECD CF Level 5. other [9]	OECD TG 415 [21]
Extended one-generation reproductive toxicity study			
Integrity and performance of the male and female reproductive systems and on the growth and development of the offspring	Continuous administration of compound (orally) prior to and during mating, gestation, lactation to termination. Option to extend study to a second generation	OECD CF Level 5. other [9]. other [32]	OECD TG 443 [23]

Table 15.4 Summary of targeted *in-vivo* ecotoxicology assays.

End-point/assay	Test system	Proposed use	Reference(s)
Fish screening assay			
Endocrine activity on the hypothalamic-pituitary-gonadal axis	21-day exposure under flow-through conditions	EDSP Tier 1 OECD CF Level 3 other [9]	OECD TG 230 [15]
Fish short-term reproduction assay			
Endocrine activity on the hypothalamic-pituitary-gonadal axis	21-day exposure under flow-through conditions	EDSP Tier 1 OECD CF Level 3 other [9]	OECD TG 229 [16] OPPTS 890.1350 [62]
Amphibian metamorphosis assay			
Endocrine activity on the hypothalamic-pituitary-gonadal axis	*Xenopus laevis*	EDSP Tier 1 OECD CF Level 3 other [9]	OECD TG 231 [17] OPPTS 890.1100 [63]
Fish sexual development test			
Sexual development, development, growth, and mortality	Fish embryo	OECD CF Level 4 other [9]	OECD TG 234 [64]
Fish life-cycle tests			
Effects on at least one generation (including reproduction) and development of the second generation	Fish embryo	EDSP Tier 2 OECD CF Level 5 other [9]	OPPTS 850.1500 [65]
Amphibian lifecycle tests			
Lifecycle exposure	*Xenopus laevis*	EDSP Tier 2 other [9]	–
Avian lifecycle tests			
Avian two-generation study	Birds	EDSP Tier 2 other [9]	–
Avian reproduction test	Birds	OECD CF Level 5 other [9]	OECD TG 206 [66]

the substance should be prioritized to determine the putative adverse effects, for example, in reproductive toxicity screening or other apical studies (see Tables 15.2 and 15.3).

A risk assessment procedure for mammalian ED should be based on the human (and environmental non-target species) relevance of animal study effects, taking into account potency, exposure duration, and the qualitative and quantitative nature of the adverse effect ecotoxicity. No particular consideration to ecotoxicity-related endocrine effects are given here. The ED mechanisms tend to be conservative, and ED alerts in mammalian assays should be followed up in ecotoxicity studies, and vice versa.

15.7 Proposal for Decision Criteria for EDCs: Regulatory Agencies

As discussed above, the Plant Protection Product Commission Regulation defines ED as a cut-off criterion for the approval of plant protection products [6]. Since there are currently no specific science-based legal criteria available for the assessment of substances with such properties, the German Federal Institute for Risk Assessment (BfR) has proposed a conceptual framework based on workshop recommendations [67]. The framework primarily proposed the evaluation of substances for endocrine-disrupting properties in the regulatory context [68], but also provides valuable considerations for application for screening purposes. The framework describes a stepwise approach, including: (i) the end-point-based assessment of adversity of effects; (ii) the establishment of a mode or mechanism of action in animals; (iii) the human relevance of effects; and (iv) exposure- or potency-based regulatory decisions for making an approval or disapproval of a crop protection product. Importantly, the second option for a regulatory decision is proposed to use potency as a key criterion for defining different hazard categories of endocrine disruptors, and thus distinguishes endocrine disruptors of high concern from those of a lower concern. In the decision process on ED, a substance would consequently only be approved if, on the basis of assessment of ED toxicity testing and other available information, it was categorized as ED category 1 (ED based on human evidence under low-exposure conditions), unless the exposure was negligible. Substances that would have to be categorized as ED category 2 (ED based on animal studies under moderate exposure conditions) are recommended not to fall under the cut-off legislation, but to pass to the regular risk assessment process. In the BfR framework, it is proposed to base categories, guidance values, and effects considered to support or not to support the categorization of a substance on recommendations for specific target organ toxicity by repeated exposure (STOT-RE), in accordance with the provisions of Regulation (EC) No. 1272/2008 [69] and the respective guidance documents [70]. In summary, this conceptual framework has been proposed as a possible starting point for the development of a guidance document in accordance with current regulatory procedures, as well as the OECD activities on the issue [68].

References

1. Stoker, T.E. and Kavlock, R.J. (2010) in *Hayes' Handbook of Pesticide Toxicology*, 3rd edn (ed. K. Robert), Academic Press, New York, pp. 551–569.
2. Kavlock, R.J., Daston, G.P., DeRosa, C., Fenner-Crisp, P., Gray, L.E., Kaattari, S., Lucier, G., Luster, M., Mac, M.J., Maczka, C., Miller, R., Moore, J., Rolland, R., Scott, G., Sheehan, D.M., Sinks, T., and Tilson, H.A. (1996) *Environ. Health Perspect.*, **104** (Suppl. 4), 715–740.
3. EU (1996) European Workshop on the Impact of Endocrine Disrupters on Human Health and Wildlife, No. EUR 17549, 2–4 December, Weybridge, UK.
4. EPA (1997) Special report on Environmental Endocrine Disruption: An Effects Assessment and Analysis. EPA/630/R-96/012.

5. WHO (2002) Global Assessment of the State-of-the-Science of Endocrine Disruptors, No. WHO/PCS/EDC/02.2.
6. EU (2009) *Off. J. Eur. Union*, **52**, 1–50.
7. EU (1998) *Off. J. Eur. Communities*, **L 123**, 1–63.
8. EU (2006) *Off. J. Eur. Union*, **49**, 1–849.
9. Bars, R., Broeckaert, F., Fegert, I., Gross, M., Hallmark, N., Kedwards, T., Lewis, D., O'Hagan, S., Panter, G.H., Weltje, L., Weyers, A., Wheeler, J.R., and Galay-Burgos, M. (2011) *Regul. Toxicol. Pharmacol.*, **59**, 37–46.
10. EPA (1998) *Fed. Regist.*, **63**, 248.
11. OECD (2010) Series on testing and assessment: testing for endocrine disrupters. Workshop Report No. 118 on OECD Countries Activities Regarding Testing, Assessment and Management of Endocrine Disrupters.
12. OECD (2011) Draft Guidance Document on Standardised Test Guidelines for Evaluating Chemicals for Endocrine Disruption, Version 11, May 2011.
13. OECD (2009) TG 455: The Stably Transfected Human Estrogen receptor-α Transcriptional Activation Assay for Detection of Estrogenic Agonist-Activity of Chemicals.
14. OECD (2011) TG 456: The H295R Steroidogenesis Assay, July 2011.
15. OECD (2009) TG 230: 21-day Fish Assay – A Short-Term Screening for Oestrogenic and Androgenic Activity, and Aromatase Inhibition, September 2009.
16. OECD (2009) TG 229: Fish Short Term Reproduction Assay, September 2009.
17. OECD (2009) TG 231: Amphibian Metamorphosis Assay, September 2009.
18. OECD (2009) TG 451: Carcinogenicity Studies, September 2009.
19. OECD (2009) TG 452: Chronic Toxicity Studies, September 2009.
20. OECD (2009) TG 453: Combined Chronic Toxicity/Carcinogenicity Studies, September 2009.
21. OECD (1983) TG 415: One-Generation Reproduction Toxicity Study.
22. OECD (2001) TG 416: Two-Generation Reproduction Toxicity.
23. OECD (2011) TG 443: Extended One-Generation Reproductive Toxicity Study, July 2011.
24. OECD (2001) TG 414: Prenatal Development Toxicity Study.
25. OECD (1998) TG 408: Repeated Dose 90-day Oral Toxicity Study in Rodents.
26. OECD (1998) TG 409: Repeated Dose 90-day Oral Toxicity Study in Non-Rodents, September 1998.
27. OECD (2008) TG 407: Repeated Dose 28-day Oral Toxicity Study in Rodents.
28. Borgert, C.J., Mihaich, E.M., Ortego, L.S., Bentley, K.S., Holmes, C.M., Levine, S.L., and Becker, R.A. (2011) *Regul. Toxicol. Pharmacol.*, **61** (2), 185–191.
29. ECETOC (2009) Workshop Report No. 16: Guidance on Interpreting Endocrine Disrupting Effects, 29–30 June 2009, Barcelona.
30. Klimisch, H.J., Andreae, M., and Tillmann, U. (1997) *Regul. Toxicol. Pharmacol.*, **25**, 1–5.
31. van Ravenzwaay, B., Cunha, G.C., Leibold, E., Looser, R., Mellert, W., Prokoudine, A., Walk, T., and Wiemer, J. (2007) *Toxicol. Lett.*, **172**, 21–28.
32. Kolle, S.N., Ramirez, T., Kamp, H.G., Buesen, R., Flick, B., Strauss, V., and van Ravenzwaay, B. (2012) *Regul. Toxicol. Pharmacol.*, **63**, 259–278.
33. EPA (2009) Endocrine Disruptor Screening Program Test Guideline OPPTS 890.1250: Estrogen Receptor Binding (Rat Uterine Cytosol).
34. EPA (2009) Endocrine Disruptor Screening Program Test Guideline OPPTS 890.1150: Androgen Receptor Binding (Rat Prostate Cytosol).
35. EPA (2009) Endocrine Disruptor Screening Program Test Guideline OPPTS 890.1300: Estrogen Receptor Transcriptional Activation (Human Cell Line (HeLa-9903)).
36. Soto, A.M., Sonnenschein, C., Chung, K.L., Fernandez, M.F., Olea, N., and Serrano, F.O. (1995) *Environ. Health Perspect.*, **103** (Suppl. 7), 113–122.
37. Soto, A.M., Michaelson, C.L., Prechtl, N.V., Weill, B.C., Sonnenschein, C., Olea-Serrano, F., and Olea, N. (1998) *Adv. Exp. Med. Biol.*, **444**, 9–23.
38. Routledge, E.J. and Sumpter, J.P. (1996) *Environ. Toxicol. Chem.*, **15**, 241–248.
39. Sohoni, P. and Sumpter, J.P. (1998) *J. Endocrinol.*, **158**, 327–339.

40. Bovee, T.F., Helsdingen, R.J., Hamers, A.R., van Duursen, M.B., Nielen, M.W., and Hoogenboom, R.L. (2007) *Anal. Bioanal. Chem.*, **389**, 1549–1558.
41. Kolle, S.N., Kamp, H.G., Huener, H.A., Knickel, J., Verlohner, A., Woitkowiak, C., Landsiedel, R., and van Ravenzwaay, B. (2010) *Toxicol. In Vitro*, **24**, 2030–2040.
42. EPA (2009) Endocrine Disruptor Screening Program Test Guideline OPPTS 890.1550: Steroidogenesis (Human Cell Line – H295R).
43. EPA (2009) Endocrine Disruptor Screening Program Test Guideline OPPTS 890.1200: Aromatase (Human Recombinant).
44. Freitas, J., Cano, P., Craig-Veit, C., Goodson, M.L., Furlow, J.D., and Murk, A.J. (2011) *Toxicol. In Vitro*, **25**, 257–266.
45. Naciff, J.M., Khambatta, Z.S., Reichling, T.D., Carr, G.J., Tiesman, J.P., Singleton, D.W., Khan, S.A., and Daston, G.P. (2010) *Toxicology*, **270**, 137–149.
46. West, P.R., Weir, A.M., Smith, A.M., Donley, E.L., and Cezar, G.G. (2010) *Toxicol. Appl. Pharmacol.*, **247**, 18–27.
47. Ohlsson, A., Cedergreen, N., Oskarsson, A., and Ulleras, E. (2010) *Toxicology*, **275**, 21–28.
48. Charles, G.D., Gennings, C., Zacharewski, T.R., Gollapudi, B.B., and Carney, E.W. (2002) *Toxicol. Appl. Pharmacol.*, **180**, 11–21.
49. Kolle, S.N., Melching-Kollmuss, S., Krennrich, G., Landsiedel, R., and van Ravenzwaay, R.B. (2011) *Regul. Toxicol. Pharmacol.*, **60** (3), 373–380.
50. Borgert, C.J., Quill, T.F., McCarty, L.S., and Mason, A.M. (2004) *Toxicol. Appl. Pharmacol.*, **201**, 85–96.
51. OECD (1996) TG 422: Combined Repeated Dose Toxicity Study with the Reproduction/Developmental Toxicity Screening Test.
52. OECD (1995) TG 421: Reproduction/Developmental Toxicity Screening Test.
53. OECD (2009) TG 441: Hershberger Bioassay in Rats a Short-Term Screening Assay for (Anti)Androgenic Properties 441: Hershberger Bioassay in Rats – A Short-Term Screening Assay for (Anti)Androgenic Properties.
54. OECD (2007) TG 440: Uterotrophic Bioassay in Rodents a Short-Term Screening Test for Oestrogenic Properties.
55. EPA (2009) Endocrine Disruptor Screening Program Test Guideline OPPTS 890.1600: Uterotrophic (Rat).
56. EPA (2009) Endocrine Disruptor Screening Program Test Guideline OPPTS 890.1400: Hershberger (Rat).
57. EPA (2009) Endocrine Disruptor Screening Program Test Guideline OPPTS 890.1500: Male Pubertal (Rat).
58. EPA (2009) Endocrine Disruptor Screening Program Test Guideline OPPTS 890.1450: Female Pubertal (Rat).
59. van Ravenzwaay, B., Cunha, G.C., Fabian, E., Herold, M., Kamp, H., Krennrich, G., Krotzky, A., Leibold, E., Looser, R., Mellert, W., Prokoudine, A., Strauss, V., Trethewey, R.N., Walk, T., and Wiemer, J. (2010) in *An Omics Perspective on Cancer Research* (ed. W.C.S. Cho), Springer, p. 141.
60. van Ravenzwaay, B., Herold, M., Kamp, H., Kapp, M.D., Fabian, E., Looser, R., Krennrich, G., Mellert, W., Prokoudine, A., Strauss, V., Walk, T., and Wiemer, J. (2012) *Mutat. Res.*, **746**, 144–150.
61. Moore, N., Bremer, S., Carmichael, N., Daston, G., Dent, M., Gaoua-Chapelle, W., Hallmark, N., Hartung, T., Holzum, B., Hubel, U., Meisters, M.L., Schneider, S., van, R.B., and Hennes, C. (2009) *Altern. Lab Anim.*, **37**, 219–225.
62. EPA (2009) Endocrine Disruptor Screening Program Test Guideline OPPTS 890.1350: Fish Short-Term Reproduction.
63. EPA (2009) Endocrine Disruptor Screening Program Test Guideline OPPTS 890.1100: Amphibian Metamorphosis (Frog).
64. OECD (2011) TG 234: Fish Sexual Development Test, July 2011.
65. EPA (1996) Ecological Effects Test Guidelines OPPTS 850.1500: Fish Life Cycle Toxicity, April 1996.
66. OECD (1984) TG 206: Avian Reproduction Test, April 1984.
67. BfR (2010) Establishment of Assessment and Decision Criteria in Human

Health Risk Assessment for Substances with Endocrine Disrupting Properties Under the EU Plant Protection Product Regulation, January 2010.
68. Marx-Stoelting, P., Pfeil, R., Solecki, R., Ulbrich, B., Grote, K., Ritz, V., Banasiak, U., Heinrich-Hirsch, B., Moeller, T., Chahoud, I., and Hirsch-Ernst, K.I. (2011) *Reprod. Toxicol.*, **31**, 574–584.
69. EU (2008) *Off. J. Eur. Union*, **L 353**, 1–1355.
70. DG Joint Research Centre (2009) *Off. J. Eur. Union*, **L 353**, 1–1355.

Index

Modern Methods in Crop Protection Research, First Edition.
Edited by Peter Jeschke, Wolfgang Krämer, Ulrich Schirmer, and Matthias Witschel.

e

f

g